MW01355780

IFIP Advances in Information and Communication Technology 729

IFIP Advances in Information and Communication Technology

The IFIP AICT series publishes state-of-the-art results in the sciences and technologies of information and communication. The scope of the series includes: foundations of computer science; software theory and practice; education; computer applications in technology; communication systems; systems modeling and optimization; information systems; ICT and society; computer systems technology; security and protection in information processing systems; artificial intelligence; and human-computer interaction.

Edited volumes and proceedings of refereed international conferences in computer science and interdisciplinary fields are featured. These results often precede journal publication and represent the most current research.

The principal aim of the IFIP AICT series is to encourage education and the dissemination and exchange of information about all aspects of computing.

More information about this series at https://link.springer.com/bookseries/6102

Matthias Thürer · Ralph Riedel ·
Gregor von Cieminski ·
David Romero
Editors

Advances in Production Management Systems

Production Management Systems for Volatile, Uncertain, Complex, and Ambiguous Environments

43rd IFIP WG 5.7 International Conference, APMS 2024
Chemnitz, Germany, September 8–12, 2024
Proceedings, Part II

Editors
Matthias Thürer
Chemnitz University of Technology
Chemnitz, Germany

Gregor von Cieminski
ZF Friedrichshafen AG
Friedrichshafen, Germany

Ralph Riedel
West Saxon University of Applied Sciences Zwickau
Zwickau, Germany

David Romero
Tecnológico de Monterrey
Mexico City, Mexico

ISSN 1868-4238 ISSN 1868-422X (electronic)
IFIP Advances in Information and Communication Technology
ISBN 978-3-031-65893-8 ISBN 978-3-031-65894-5 (eBook)
https://doi.org/10.1007/978-3-031-65894-5

This Springer imprint is published by the registered company Springer Nature Switzerland AG
The registered company address is: Gewerbestrasse 11, 6330 Cham, Switzerland

Preface

The year 2024 highlighted that the world is getting more and more volatile, uncertain, complex, and ambiguous. There are no simple solutions to complex problems, which require new ways to solve and manage them. New technology opens new ways as does the continuous progress of science. But to bring real benefits to people on the ground who are suffering from the consequences of conflicts and climate change, just to name two, more research is needed on how to apply and further advance this technology and knowledge. A scientific conference plays an important part in this. It is a symbol of aiming for a better future. We create new knowledge and solutions, we share all our achievements, and we meet to get to know people from all over the world, creating new ties and making new friends.

The International Conference on "Advances in Production Management Systems" (APMS) is the leading annual event of the IFIP Working Group (WG) 5.7 of the same name. At APMS 2024 in Chemnitz, Germany, hosted by Chemnitz University of Technology and West Saxon University of Applied Sciences Zwickau, more than 200 papers were presented and discussed. This is a significant step up from the first APMS Conference in 1980, which assembled just a few participants. The IFIP WG 5.7 was established in 1978 by the General Assembly of the International Federation for Information Processing (IFIP) in Oslo, Norway. Its first meeting was held in August 1979 with all its seven members present. The WG has since grown to 124 full and candidate members as well as 27 honorary members.

APMS was held for the first time in East Germany, which experienced specific challenges and disruptions, but also hopes and new futures. The Ore Mountains are a traditional industrial region, and it can be argued that the first textbook on operations management was written on the mining industry in this region. Mining always brought its own challenges in terms of sustainability, quality of work life, and process efficiency, but also major technological advancements and highly skilled labour. Many of the papers presented address similar challenges as 500 years ago. The problems did not change but are more volatile, uncertain, complex, and ambiguous today given the interconnectedness of society and production. No other technology can show a more rapid development and impact in industry and society than Information and Communication Technology (ICT).

The APMS 2024 program shows that the IFIP WG 5.7 can still make, and will continue to make, a significant contribution to production and production management disciplines. In 2024, the international review board for APMS included 95 recognized experts working in the disciplines of production and production management systems. For each paper, an average of 2.5 single-blind reviews were provided. Over two months, each submitted paper went through two rigorous rounds of reviews to allow authors to revise their work after the first round of reviews to guarantee the highest scientific quality of the papers accepted for publication. Following this process, 201 full papers were selected for inclusion in the conference proceedings from a total of 224 submissions.

APMS 2024 brought together leading international experts from academia, industry, and government in the areas of production and production management systems to discuss how to achieve responsible manufacturing, service, and logistics futures. This included topics such as innovative manufacturing, service, and logistics systems characterized by their agility, circularity, digitalization, flexibility, human-centricity, resilience, and smartification contributing to more sustainable industrial futures that ensure that products and services are manufactured, servitized, and distributed in a way that creates a positive effect on the triple bottom line.

The APMS 2024 conference proceedings are organized into 6 volumes, covering a large spectrum of research addressing the overall topic of the conference "Production Management Systems for Volatile, Uncertain, Complex, and Ambiguous Environments". We would like to thank all contributing authors for their high-quality research work and their willingness to share their findings with the APMS and IFIP WG 5.7 community. We are equally grateful for the outstanding work of all the International Reviewers, the Program Committee Members, and the Special Sessions Organizers.

September 2024

Matthias Thürer
Ralph Riedel
Gregor von Cieminski
David Romero

Organisation

Conference Co-chairs

Matthias Thürer	Chemnitz University of Technology, Germany
Ralph Riedel	West Saxon University of Applied Sciences Zwickau, Germany
Gregor von Cieminski	ZF Friedrichshafen AG, Germany

Honorary Co-chair

Egon Müller	Chemnitz University of Technology, Germany

Program Co-chairs

Philipp Wilsky	Chemnitz University of Technology, Germany
David Romero	Tecnológico de Monterrey, Mexico

Organizing Committee Co-chairs

Luise Weißflog	Chemnitz University of Technology, Germany
Felix Franke	Chemnitz University of Technology, Germany
Julia Sprigode	West Saxon University of Applied Sciences Zwickau, Germany

Doctoral Workshop Chairs

Marina Ivanova	Chemnitz University of Technology, Germany
Pierre Grzona	Chemnitz University of Technology, Germany
David Romero	Tecnológico de Monterrey, Mexico

Program Committee

Federica Acerbi
Lorenzo Agbomemewa
Natalie Cecilia Agerskans
Carla Susana Agudelo Assuad
Alireza Ahmadi
Kartika Nur Alfina
Erlend Alfnes
Farhad Ameri
Bjorn Andersen
Yevheniia Andriichenko
Oliver Antons
Veronica Arioli
Doris Aschenbrenner
Aylin Ates
Nestor Fabián Ayala
Jannicke Baalsrud Hauge
Elena Beducci
Monica Bellgran

Frederique Biennier
Alessia Bilancia
Vlad Bocanet
Jos A. C. Bokhorst
Frank Börner
Magali Bosch-Mauchand
Hafida Bouloiz
Alexandros Bousdekis
Greta Braun
Gerald Bräunig
Jessica Bruch
Adauto Bueno
Lien Bui
Guillherme Hörner Bussolo
Mario Henrique Callefi
Davide Castellano
Perawit Charoenwut
Ferdinando Chiacchio
Steve Childe
Chiara Cimini
Beatrice Colombo
Jelena Crnobrnja
Catherine da Cunha
Patrick Dallasega
Mélanie Despeisse
Mario Di Nardo
Alexandre Dolgui
Heidi Dreyer
Milos Drobnjakovic
Malin Elvin
Adrodegari Federico
Jannick Fiedler
Daniel Fischer
Martin Folz
Florian Förster
Chiara Franciosi
Jana Frank
Susanne Franke
Stefano Frecassetti
Mike Freitag
Jan Frick
Paolo Gaiardelli
Mattia Galimberti
Maryam Gallab
Mosè Gallo
Fernando José Gómez Paredes
Clarissa A. González Chávez
Jon Gosling
Pierre Grzona
Christopher Gustafsson
Harumi Haraguchi
Stefanie Hatzl
Zhiliang He
Dominik Hertel
Jan Holmström
Djerdj Horvat
Ryuichiro Ishikawa
Dmitry Ivanov
Marina Ivanova
Niloofar Jafari
Tim Maximilian Jansen
Yongkuk Jeong
HongBae Jun
Toshiya Kaihara
Matthias Kalverkamp
Kai Kang
Ibrokhimjon Khamidov
Maria Kollberg Thomassen
Juhoantti Viktor Köpman
Alexandra Lagorio
Beñat Landeta
Nicolas Leberruyer
Jan-Peter Lechner
Liu Lei
Yee Yeng Liau
Ming Lim
Ma Lin
Francesco Lolli
Damien Lovato
Egon Lüftenegger
Sivanilza Teixeira Machado
Ugljesa Marjanovic
Larisa Markov
Melissa Marques-McEwan
Antonio Masi
Sandra Mattsson
Gokan May
Nayara Cardoso Medeiros
Khaled Medini
Kunruthai Meechang
Joan Gilberto Mendes dos Reis
Nicola Mercogliano

Hajime Mizuyama
Katharina Moitzi
Elias Montini
Eiji Morinaga
Francesco Moroni
Sobhan Mostafayi Darmian
Mohamed Naim
Torbjørn Netland
Phu Nguyen
Nariaki Nishino
Sang Do Noh
Tomomi Nonaka
Maniva Oliva
Margherita Pero
Claudia Piffari
Marta Pinzone
Fabiana Pirola
Daryl Powell
Vittaldas Prabhu
Hiran Harshana Prathapage
Anna Presciuttini
Moritz Quandt
Ricardo Rabelo
Slavko Rakic
Mario Rapaccini
R. M. Chandima Ratnayake
Daniel Resanovic
Ralph Riedel
David Romero
Anita Romsdal
Christoph Roser
Martin Rudberg
Tamas Ruppert
Roberto Sala
Omkar Salunkhe
Jan Salzwedel
Adriana Saraceni
Claudio Sassanelli
Mizuho Sato
Bennet Schulz
Daniel Seifert
Marcia Terra Silva
Katrin Singer-Coudoux
Dragana Slavic
Selver Softic
Per Solibakke
Endre Sølvsberg
Dongping Song
Kenn Steger-Jensen
Oliver Stoll
Jo Wessel Strandhagen
Shota Suginouchi
Yoshitaka Tanimizu
Takashi Tanizaki
Moreira Tavares
Klaus-Dieter Thoben
Peter Thorvald
Matthias Thürer
Rodrigo Carlo Toloi
Ivan Tomašević
Mario Tucci
Ioan Turcin
Ali Turkyilmaz
Hendrik Unger
Kenneth Vidskjold
Gregor von Cieminski
Sabine Waschull
Kasuni Vimasha Weerasinghe
Shaun West
Stefan Alexander Wiesner
Joakim Wikner
Philipp Wilsky
Heiner Winkler
Thorsten Wuest
Mohammed Yaqot
Mingze Yuan
Lara Popov Zambiasi
Matteo Zanchi
Anton Zitnikov
Anne Zouggar

Contents – Part II

Evolving Workforce Skills and Competencies for Industry 5.0

Experiential Learning in Engineering Education

Smart and Sustainable Supply Chain Management in the Society 5.0 Era

Data-Driven Control System Using Machine Learning in Production Process

Takashi Tanizaki[1(✉)], Ayumu Fukuyama[1], Kunimi Uchino[2], Tetuaki Kurokawa[2], Shigemasa Nakagawa[2], and Takayuki Kataoka[1]

[1] Graduate School of Systems Engineering, Kindai University, 1 Takaya-Umenobe, Higashi-Hiroshima 739-2116, Japan
tanizaki@hiro.kindai.ac.jp

[2] Nippon Steel Texeng Co., Ltd., 8-8, Shinkawa 1-Chome, Chuo-Ku 104-0033, Japan

Abstract. Many manufacturing companies control their production machines to produce good products within quality standards by using the results of research on physical or chemical models. Those models are developed from knowledge of the physical or chemical changes that occur when the products are processed and operational knowledge. However, it is difficult for some companies to research physical or chemical models. We study data-driven control systems to enable the stable production of good products when it is difficult to study and develop physical and chemical models or to use operational knowledge. In this paper, we propose an algorithm that builds an alternative model from actual operation data using machine learning and finds the optimal operating conditions under which the product is within the quality standards range using 0–1 integer programming. The effectiveness of the proposed algorithm was verified using operation data generated using a simulator for a food manufacturing process.

Keywords: Data-driven Control System · Machine Learning · Integer Programming

1 Introduction

Many manufacturing companies control their production machines to produce good products within quality standards by using the results of research on physical or chemical models based on knowledge of the physical or chemical changes that occur when the products are processed and operational knowledge based on field experience [1]. However, high temperatures, dust, and other environmental constraints may make it difficult to collect the data needed to build physical or chemical models. In addition, human and financial constraints make it difficult for some companies to research physical or chemical models. To ensure the stable production of good products in such situations, we study a data-driven control system that builds alternative models (hereafter reaction models) to the above physical or chemical models from actual operation data using machine learning and find optimal operating conditions using 0–1 integer programming. In this paper, we propose the algorithm for the above system and describe the results of validation through computer experiments using data provided by a joint research company.

Published by Springer Nature Switzerland AG 2024
M. Thürer et al. (Eds.): APMS 2024, IFIP AICT 729, pp. 3–16, 2024.
https://doi.org/10.1007/978-3-031-65894-5_1

2 Data-Driven Control System

Physical or chemical models are built based on the results of reactions to some action on production objects. Production machines are controlled so that the characteristic values of the production objects are within the acceptable range based on these models and operational knowledge (Fig. 1). Furthermore, control is required to achieve the desired behaviour while satisfying many constraints. Model Predictive Control [2], Adaptive Control [3], and Iterative Feedback Tuning [4] have been studied as control methods for this purpose. However, as described in Sect. 1, there are companies and situations where research and development of physical and chemical models is difficult. Therefore, research is being conducted to improve control performance by utilizing actual operation data. Such a control system designed by utilizing actual operation data is called a data-driven control (DDC) system [5]. Various DDC methods have been proposed, such as simultaneous perturbation stochastic approximation (SPSA)-based DDC methods [6], Model-free adaptive control [7], and Iterative feedback tuning [8]. These methods modify the control variables sequentially when the characteristic values of the controlled object are displaced from the target values. In other words, they are not design methods that determine control variables to be in the acceptable range. On the other hand, real-world modelling methods using machine learning have been actively studied. From the above background, we are studying DDC system based on reaction models built using machine learning to determine control variable values whose characteristic values are within an acceptable range even in situations where it is difficult to build physical or chemical models (Fig. 2). As part of this research, we develop an algorithm, that consists of a reaction model built using machine learning based on actual operation data, and finding the optimal operating conditions under which the product characteristic value becomes the target value using 0–1 integer programming.

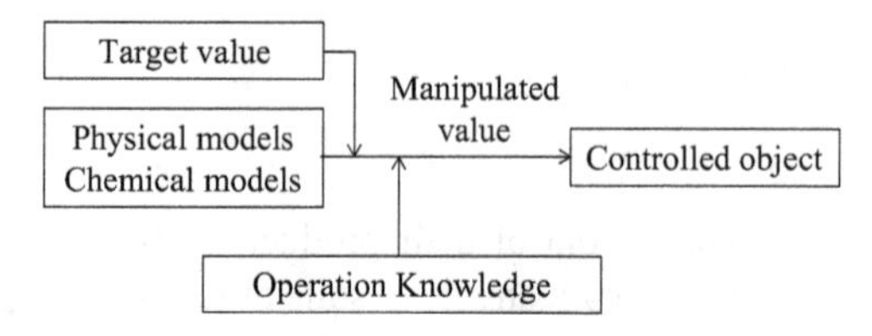

Fig. 1. Overview of control system

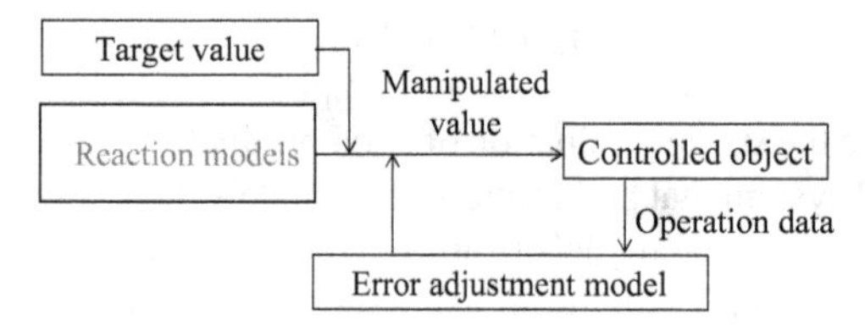

Fig. 2. Overview of data-driven control system

3 Algorithm to Find Optimal Operating Conditions

This section describes a proposed algorithm. The outline of the proposed algorithm is as follows.

S.1 Build a reaction model from operational data using machine learning.
S.2 Reconstruct the reaction model into a tree model to find the optimal solution using 0–1 integer programming.
S.3 Find the optimal operating conditions for the tree model using 0–1 integer programming.

3.1 Reaction Model

In this problem, the control parameters of the production machine are determined so that the characteristic values of the production object are within the acceptable range. Therefore, the reaction model, which models the relationship between control parameters and characteristic values, is a regression model. We used gradient boosting decision trees (hereafter GBDT) as a machine learning method to build the reaction model. GBDT build a model by sequentially generating decision trees in the direction which decreases the loss function value using gradient descent when ensemble learning is performed on multiple decision trees. The GBDT algorithm is shown below.

- Notation

M: Number of iteration
N: Number of data
T: Number of leaf nodes in the tree
$\boldsymbol{x}_i$: Explanatory variables vector for i^{th} data
y_i: Objective variable for i^{th} data
$\hat{f}^{(m)}(\boldsymbol{x})$: Predictions for m^{th} iteration
$\hat{f}_m(\boldsymbol{x})$: Predictions of decision tree for m^{th} iteration
$L(y_i, \rho)$: Loss function
ρ: Initial predictions
$\hat{g}_{im}$: First − order gradient statistics on the loss function for i^{th} data in decision tree for m^{th} iteration
$\hat{h}_{im}$: Second − order gradient statistics on the loss function for i^{th} data in decision tree for m^{th} iteration
$\hat{R}_{jm}$: Dataset for leaf node $j$of decision tree for m^{th} iteration
IG: Information gain
I_L: Instance set of the left child node
I_R Instance set of the right child node
I: Instance set of the parent node
I_{jm} Instance set of leaf node j of decision tree for m^{th} iteration
λ: Parameters to inhibit overfitting
$\hat{w}_{jm}$: weight of leaf node j of decision tree for m^{th} iteration
η: Learning rate
$\boldsymbol{I}$: Indicator function

- Algorithm

$$\hat{f}^{(0)}(\boldsymbol{x}) = \hat{f}_0(\boldsymbol{x}) = \arg\min_{\rho} \sum_{i=1}^{N} L(y_i, \rho) \quad (1)$$

for $m = 1, \cdots, M$ ***do***

$$\hat{g}_{im} = \left[\frac{\partial L(y_i, f(\boldsymbol{x}_i))}{\partial f(\boldsymbol{x}_i)}\right]_{f(\boldsymbol{x}) = \hat{f}^{(m-1)}(\boldsymbol{x})}, i = 1, \cdots, N \quad (2)$$

$$\hat{h}_{im} = \left[\frac{\partial^2 L(y_i, f(\boldsymbol{x}_i))}{\partial f(\boldsymbol{x}_i)^2}\right]_{f(\boldsymbol{x}) = \hat{f}^{(m-1)}(x)}, i = 1, \cdots, N \quad (3)$$

$$IG = \frac{1}{2}\left[\frac{\left(\sum_{i \in I_L} \hat{g}_{im}\right)^2}{\sum_{i \in I_L} \hat{h}_{im} + \lambda} + \frac{\left(\sum_{i \in I_R} \hat{g}_{im}\right)^2}{\sum_{i \in I_R} \hat{h}_{im} + \lambda} - \frac{\left(\sum_{i \in I} \hat{g}_{im}\right)^2}{\sum_{i \in I} \hat{h}_{im} + \lambda}\right] \quad (4)$$

$$\hat{w}_{jm} = \frac{\sum_{i \in I_{jm}} \hat{g}_{im}}{\sum_{i \in I_{jm}} \hat{h}_{im} + \lambda} \quad (5)$$

$$\hat{f}_m(\boldsymbol{x}) = \eta \sum_{j=1}^{T} \hat{w}_{jm} \boldsymbol{I}\left(\boldsymbol{x} \in \hat{R}_{jm}\right) \quad (6)$$

$$\hat{f}^{(m)}(\boldsymbol{x}) = \hat{f}^{(m-1)}(\boldsymbol{x}) + \hat{f}_m(\boldsymbol{x}) \quad (7)$$

end

Equation (1) calculates the initial prediction value $\hat{f}^{(0)}(\boldsymbol{x})$ of the model. Equations (2) and (3) calculate the first-order and second-order gradient statistics on the loss function, respectively. Equation (4) calculates the information gain of the candidate split points. It is split at the split point with the largest IG among the candidate points. Equation (5) calculates the weights of the leaf nodes of the decision tree. Equation (6) sums the weights of the leaf nodes containing the data and calculates the predicted value of the decision tree. Equation (7) updates the model predictions.

We used XGBoost and LightGBM, which are evaluated to have high accuracy among methods based on GBDT. XGBoost and LightGBM use different split point search algorithms. XGBoost searches for all split points to find the best split points, while LightGBM generates continuous feature values into discrete bins to narrow the search range of split points and uses these bins to find the best split points [9, 10].

3.2 Reconstruction of the Reation Model

As the first step, the reaction model was built using decision trees of depth 1. The reaction model built using the GBDT described in Sect. 3.1 consists of M depth-1 decision trees. It is necessary to reconstruct the M decision trees into a single decision tree to obtain the optimal operating conditions using 0–1 integer programming. Figure 3 shows an image of the reconstruction. By reconstructing the M decision trees into a single decision tree, the range of each explanatory variable that determines the predicted value can be obtained by checking the branches from the root node to the leaf node. As a result, optimal operating conditions can be obtained.

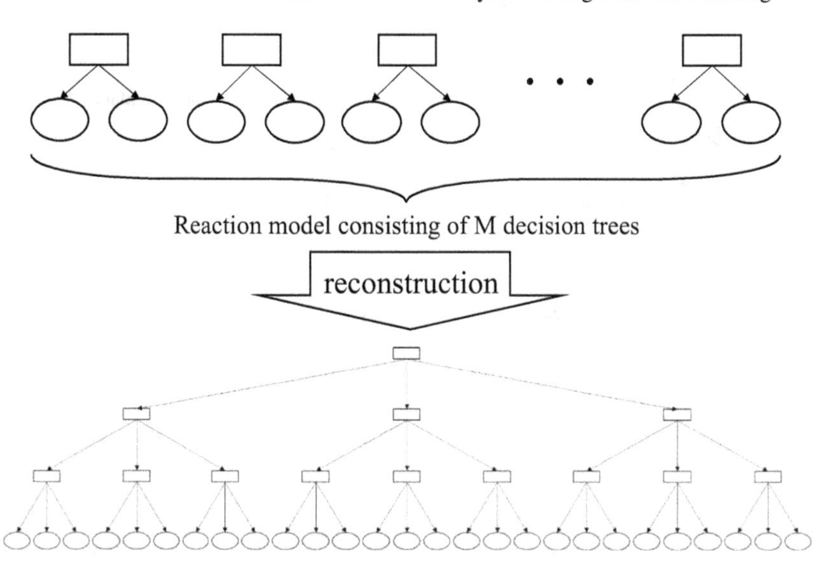

Fig. 3. Image of the reconstruction

M decision trees are reconstructed into a single decision tree by repeatedly combining two decision trees into one. The tree combining method is in the following cases, depending on the explanatory variables and thresholds that make up each of the two decision trees.

(Case1) Both explanatory variables and threshold values are the same
(Case2) The explanatory variables are the same and the threshold values are different
(Case3) Both explanatory variables and threshold values are different

Let $\mathrm{x} = \left\{x^{(1)}, x^{(2)}\right\}$ be the explanatory variables used to partition the decision tree, and α, β, γ be the threshold values (where $\alpha < \beta < \gamma$). From Eqs. (6) and (7), the predicted value of a leaf node in the decision tree after combining is the sum of the predicted values of the leaf nodes belonging to each classification.

In case 1 (Fig. 4), there are two leaf nodes because the threshold value is α only. In Case 2 (Fig. 5), there are three leaf nodes because there are two threshold values α and β. In Case 3 (Fig. 6), the root node of one decision tree is combined with the leaf node of another decision tree to reconstruct a single decision tree. Thus, there are six leaf nodes.

3.3 Finding the Optimal Operating Conditions Using 0–1 Integer Programming

To find a solution with the highest probability of producing good products (robust solution) even when there are deviations in the operation volume, we formulated this problem as a 0–1 integer programming problem whose objective function is to minimise the difference between the predicted value and the median value of the quality standards range.

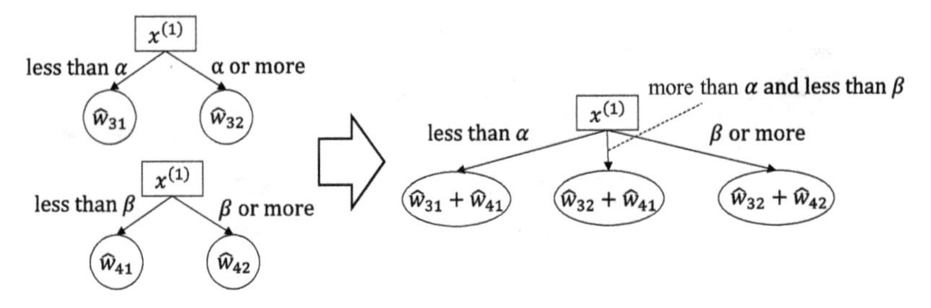

Fig. 4. Combining two decision trees (Case 1)

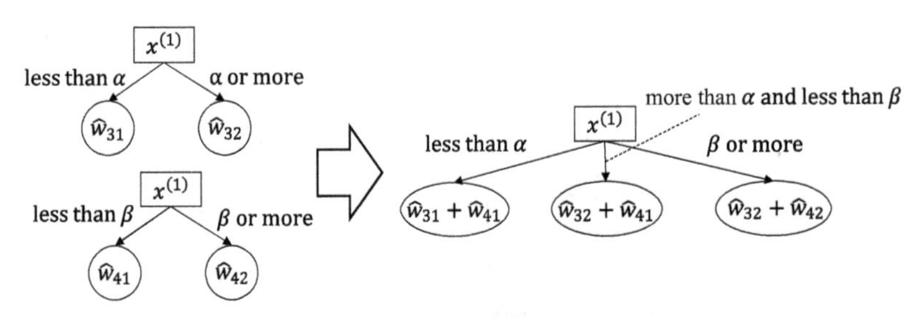

Fig. 5. Combining two decision trees (Case 2)

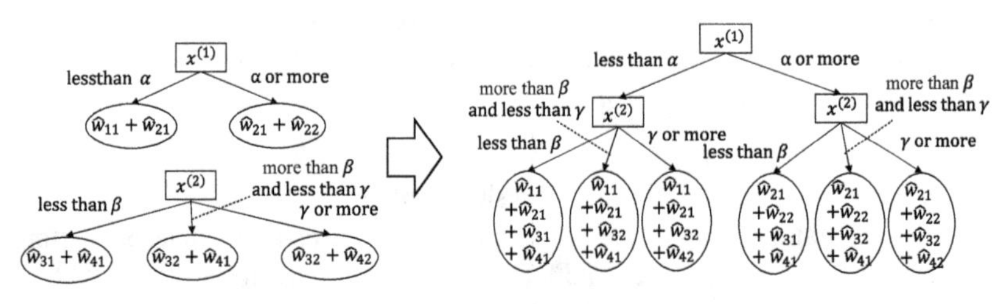

Fig. 6. Combining two decision trees (Case 3)

- Notation

a_{ij}: Contribution value (predicted value) of edge from node i to node$j$
x_{ij}: $0-1$ integer variable whether edge ij passes or not

$$x_{ij} = \begin{cases} 1\text{: pass edge } ij \\ 0\text{: not pass edge } ij \end{cases}$$

V: Node set
E : Edge set
s: Root node
T: Leaf node set
A_i: Set of child nodes of node$i$
k_{max}: Maximum value of quality standards range
k_{min}: Minimum value of quality standards range

- Objective function

$$Minimise\left|\sum_{j\in V}\sum_{i\in V}a_{ij}x_{ij} - \frac{k_{max}+k_{min}}{2}\right| \tag{8}$$

- Constraints

$$\sum_{i\in V-T}x_{ij} = \sum_{k\in V-\{s\}}x_{jk}, \quad j \in V - \{s\} - T \tag{9}$$

$$\sum_{j\in V-\{s\}}x_{ij} = 1, \quad i \in \{s\} \tag{10}$$

$$\sum_{k\in T}\sum_{j\in V-T}x_{jk} = 1 \tag{11}$$

$$\sum_{j\in A_i}x_{ij} \le 1, \quad i \in V - T \tag{12}$$

The objective function of Eq. (8) is to minimise the absolute difference between the sum of the contribution value (predicted value) of the edges on the path from the root node to the leaf node and the median of the quality standards range. Constraint (9) shows that the number of edges leaving a node is equal to the number of edges entering the node. Constraint (10) shows that there must be one edge leaving the root node. Constraint (11) shows that there must be one edge entering the leaf node. Constraint (12) shows that there must be one edge from a node to its child nodes, excluding the leaf node.

4 Computer Experiments

4.1 Overview of Numerical Experiments

The machine subject to computer experiments is the material processing machine in the food manufacturing process shown in Fig. 7. Since joint research Company A has not yet obtained actual operation data from the food manufacturing company to which it consults, the computer experiments use operation data generated using a simulator developed by Company A for its material processing machine. The input variables of the simulator are three controllable variables, such as roll speed, conveyor speed and the gap between rolls (hereafter roll gap) and two uncontrollable variables, the moisture content of the material (hereafter moisture) and air temperature (hereafter temperature). The output variable is the thickness of the product after material processing (hereafter thickness).

As a result of discussions with Company A, the thickness to be targeted was 7.3 mm, and the quality standards range to confirm the accuracy of this algorithm was set to ±1.0 mm for the wide range and ±0.3 mm for narrow with 7.3 mm as the median value. Company A expressed an opinion that the distribution of actual operation data of processes operating under normal conditions is close to a normal distribution centred on the target value. Therefore, a training data set was created by extracting 1,000 data sets so that the thickness, which is the output value of the simulator, follows a normal distribution with the target value as the mean. The following two training datasets were prepared, assuming the manufacturing capacity of the material processing machine.

- High manufacturing capacity: mean 7.3 mm, variance 0.3 mm
- Low manufacturing capacity: mean 7.3 mm, variance 1.0 mm

This data set was prepared for each combination of moisture and temperature, which are uncontrollable variables. After discussions with Company A, we selected 11 combinations of moisture (50, 52, 54, …, 70) and 3 combinations of temperature (15, 25, 35) for a total of 33 patterns.

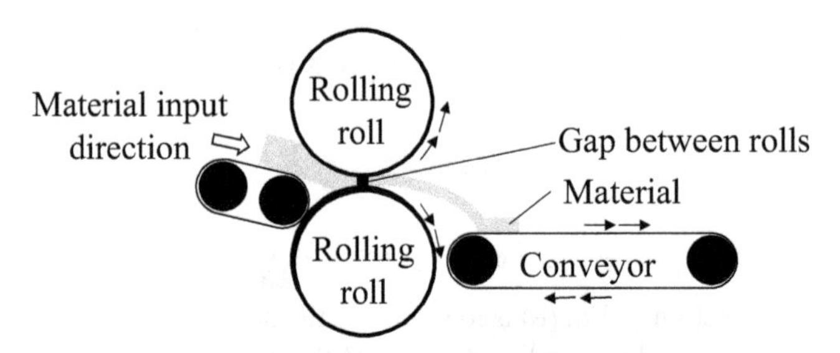

Fig. 7. Material processing machine for computer experiments

The following procedure conducted computer experiments. We used XGBoost and LightGBM as machine learning methods.

S.1 Create a training data set for each combination of moisture and temperature.
S.2 Build a reaction model using machine learning and find optimal operating conditions using 0–1 integer programming.
S.3 Create 110 verification data that satisfy the optimum operating conditions and input the data to the simulator. Calculate the ratio of the simulator's output values (thickness) that fall within the quality standards range to verification data (good product ratio).

4.2 Result of Computer Experiments 1

Table 2 shows the good product ratio resulting from the computer experiments for each combination of the variance, the machine learning method, the quality standards range, moisture and temperature. The yellow cells show the results where the good product ratio is 100%. The good product ratio for the wide quality standards range was 100% for all combinations of XGBoost and LightGBM. On the other hand, the good product ratio for the narrow quality standards was less than 100% for some combinations of both XGBoost and LightGBM.

Figure 8 shows the maximum and minimum thickness for combinations of moisture and temperature. The vertical axis is the thickness, and the horizontal axis is the combination of moisture and temperature. The black boxes in the figure show results in which the good product ratio for the narrow quality standards is less than 100%. From Fig. 8, the difference between the maximum and minimum thickness (thickness range) for the combinations in the black boxes is large. In particular, the thickness range for moisture 60 and temperature 25, plotted 17th from the left for the LightGBM variance 0.3, is large.

Table 1. Result of computer experiments 1

moisture	temperature	variance:1.0				variance:0.3			
		XGBoost		LightGBM		XGBoost		LightGBM	
		wide	narrow	wide	narrow	wide	narrow	wide	narrow
50	15	100.0	100.0	100.0	100.0	100.0	100.0	100.0	100.0
50	25	100.0	99.1	100.0	100.0	100.0	100.0	100.0	100.0
50	35	100.0	100.0	100.0	100.0	100.0	100.0	100.0	100.0
52	15	100.0	100.0	100.0	100.0	100.0	100.0	100.0	100.0
52	25	100.0	100.0	100.0	100.0	100.0	100.0	100.0	100.0
52	35	100.0	100.0	100.0	100.0	100.0	100.0	100.0	100.0
54	15	100.0	100.0	100.0	100.0	100.0	100.0	100.0	100.0
54	25	100.0	100.0	100.0	100.0	100.0	84.5	100.0	100.0
54	35	100.0	100.0	100.0	100.0	100.0	100.0	100.0	100.0
56	15	100.0	100.0	100.0	100.0	100.0	100.0	100.0	100.0
56	25	100.0	100.0	100.0	100.0	100.0	100.0	100.0	100.0
56	35	100.0	100.0	100.0	100.0	100.0	100.0	100.0	57.3
58	15	100.0	100.0	100.0	100.0	100.0	100.0	100.0	100.0
58	25	100.0	100.0	100.0	100.0	100.0	100.0	100.0	100.0
58	35	100.0	41.8	100.0	100.0	100.0	100.0	100.0	100.0
60	15	100.0	100.0	100.0	100.0	100.0	100.0	100.0	100.0
60	25	100.0	100.0	100.0	100.0	100.0	100.0	100.0	64.5
60	35	100.0	100.0	100.0	100.0	100.0	100.0	100.0	100.0
62	15	100.0	100.0	100.0	100.0	100.0	100.0	100.0	100.0
62	25	100.0	100.0	100.0	100.0	100.0	100.0	100.0	100.0
62	35	100.0	100.0	100.0	100.0	100.0	100.0	100.0	100.0
64	15	100.0	100.0	100.0	100.0	100.0	100.0	100.0	100.0
64	25	100.0	100.0	100.0	100.0	100.0	100.0	100.0	100.0
64	35	100.0	100.0	100.0	100.0	100.0	100.0	100.0	100.0
66	15	100.0	100.0	100.0	100.0	100.0	100.0	100.0	100.0
66	25	100.0	100.0	100.0	100.0	100.0	100.0	100.0	100.0
66	35	100.0	100.0	100.0	100.0	100.0	100.0	100.0	100.0
68	15	100.0	100.0	100.0	100.0	100.0	100.0	100.0	100.0
68	25	100.0	100.0	100.0	100.0	100.0	100.0	100.0	100.0
68	35	100.0	100.0	100.0	100.0	100.0	100.0	100.0	100.0
70	15	100.0	100.0	100.0	100.0	100.0	100.0	100.0	100.0
70	25	100.0	100.0	100.0	100.0	100.0	100.0	100.0	100.0
70	35	100.0	100.0	100.0	100.0	100.0	100.0	100.0	100.0
average		100.0	98.2	100.0	100.0	100.0	99.5	100.0	97.6

To examine the operating conditions' distribution of solutions whose predicted values are close to the target value, the distribution of the upper and lower bounds of the operating conditions for solutions satisfying Eq. (13) is shown in Figs. 9 and 10. Both results were predicted by the Light GBM variance of 0.3. Figure 9 shows the results under operating conditions of moisture 60 and temperature 25, where the good product ratio was less than 100%. Figure 10 shows the results under operating conditions of moisture 60 and temperature 35, where the good product ratio was 100%. Each of these figures is from two identical angles. The black dots in Fig. 9 and 10 are the maximum and minimum values of operating conditions of optimal solution that minimise the objective

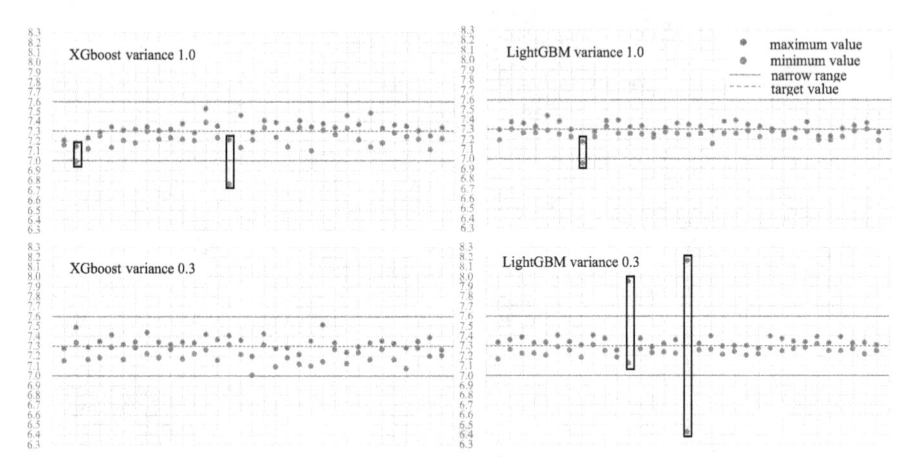

Fig. 8. Maximum and minimum thickness for combinations of moisture and temperature

function in Eq. (8). The red dots are the maximum values of operating conditions and the blue dots are the minimum values of operating conditions.

$$\left|\sum_{j\in V}\sum_{i\in V}a_{ij}x_{ij}-\frac{k_{max}+k_{min}}{2}\right|<0.001 \tag{13}$$

Figures 9 and 10 show that there exist many solutions with objective function values close to the optimal solution on almost the same curved surface. When the good product ratio is less than 100%, the optimal solution is located at the edge of the solution distribution plane (Figs. 9). On the other hand, when the good product ratio is 100%, the optimal solution is located almost in the centre of the solution distribution plane (Fig. 10). As a result of discussions with joint research Company A, it was concluded that there is a high possibility that the solution structure is multimodal due to the collinearity of the explanatory variables. As a result, if the optimal solution is located at the edge of the solution distribution plane due to a slight deviation, the good product ratio will likely be low.

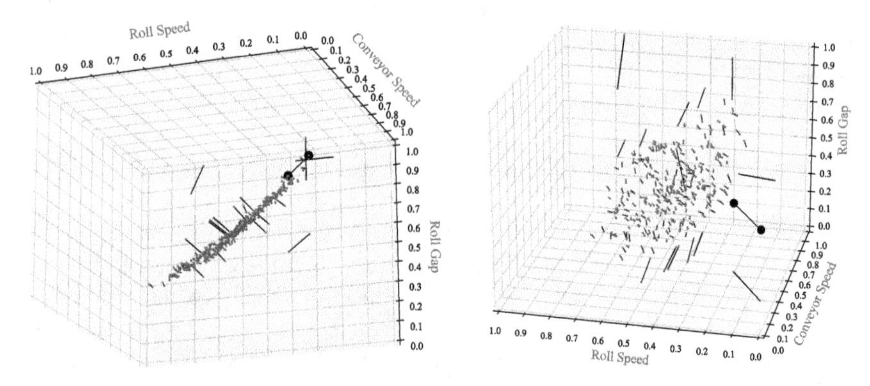

Fig. 9. Distribution of operating conditions (moisture 60, temperature 25)

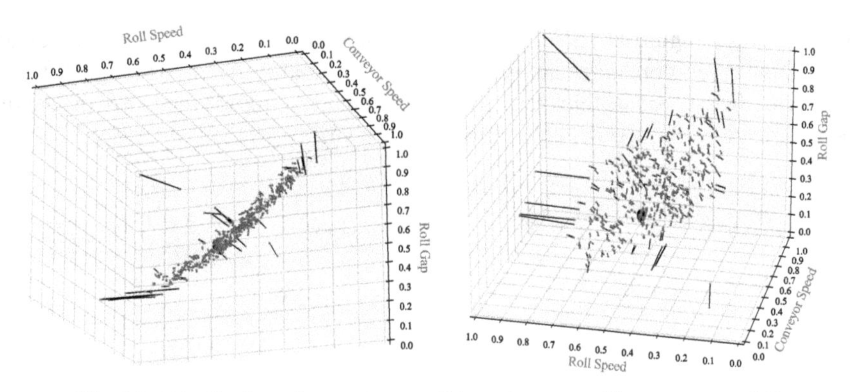

Fig. 10. Distribution of operating conditions (moisture 60, temperature 35)

4.3 Result of Computer Experiments 2

Based on the results of Sect. 4.2, we studied a solution method for operating conditions near the centre of the solution distribution plane to improve the good product ratio Table 1. The method and computer experiment results are described below.

In the P solutions satisfying Eq. (13), let $\boldsymbol{L}_i = \left(l_i^{(1)}, l_i^{(2)}, l_i^{(3)}\right)$ be the lower limit vector of explanatory variables for the ith solution and $\boldsymbol{U}_i = \left(u_i^{(1)}, u_i^{(2)}, u_i^{(3)}\right)$ be the upper limit vector. Since the controllable variables in this problem are roll speed, conveyor speed, and roll gap, the number of dimensions of the explanatory vectors is three. Mean vectors $\overline{\boldsymbol{L}}$ and $\overline{\boldsymbol{U}}$ of the lower and upper limit vectors are calculated by Eqs. (14) and (15), respectively.

$$\overline{\boldsymbol{L}} = \frac{1}{P}\sum_{i=1}^{P} \boldsymbol{L}_i \tag{14}$$

$$\overline{\boldsymbol{U}} = \frac{1}{P}\sum_{i=1}^{P} \boldsymbol{U}_i \tag{15}$$

The solution composed of the two vectors $\overline{\boldsymbol{L}}$ and $\overline{\boldsymbol{U}}$ is often not on a solution distribution plane where a solution satisfying Eq. (13) exists. The predicted value using this explanatory variable may deviate from the target value. From this, as shown in Eqs. (16) and (17), among P solutions that satisfy Eq. (13), the solution $\boldsymbol{L}^{(best)}$ and $\boldsymbol{U}^{(best)}$ that minimises the Euclidean distance to the midpoint of the solution composed of $\overline{\boldsymbol{L}}$ and $\overline{\boldsymbol{U}}$ is the optimal operating condition.

$$i_{best} = arg\min_i \left\{ \left(\frac{\boldsymbol{L}^{(i)} + \boldsymbol{U}^{(i)}}{2} - \frac{\overline{\boldsymbol{L}} + \overline{\boldsymbol{U}}}{2} \right)^2 \right\} \tag{16}$$

$$\boldsymbol{L}^{(best)}, \boldsymbol{U}^{(best)} = \boldsymbol{L}^{(i_{best})}, \boldsymbol{U}^{(i_{best})} \tag{17}$$

The computer experiments were conducted in the same manner as computer experiments 1. The good product ratio rate was 100% for all combinations of both XGBoost and LightGBM, regardless of the size of the quality standards range and the variance. Figure 11 shows the maximum and minimum thickness for combinations of moisture and temperature. Figure 11 shows that the maximum and minimum values of predicted thickness for all combinations are within the narrow quality standards range. Furthermore, the predicted product thicknesses were distributed around the median of the quality standards range, with little deviation from the median.

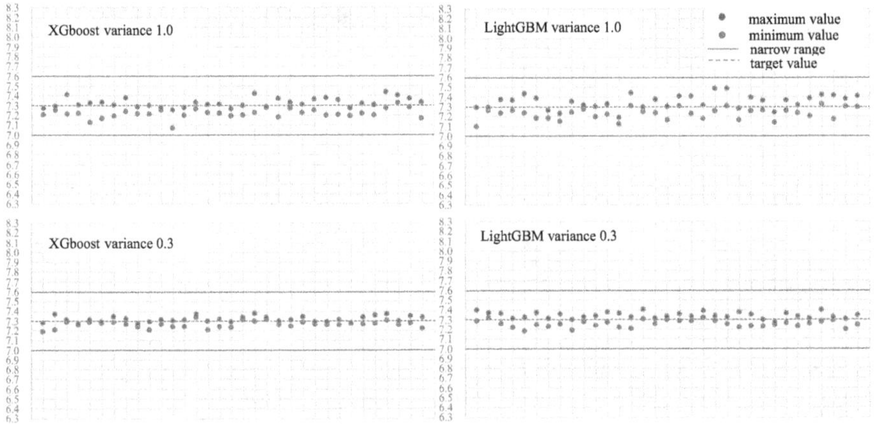

Fig. 11. Maximum and minimum thickness for combinations of moisture and temperature

The distribution of the upper and lower bounds of the operating conditions for solutions satisfying Eq. (13) is shown in Figs. 12 and 13. Both results were predicted by the Light GBM variance of 0.3. Figure 12 shows the results under operating conditions of moisture 60 and temperature 25. Figure 13 shows the results under operating conditions of moisture 60 and temperature 35. The black dots in Fig. 12 and 13 are the maximum and minimum values of operating conditions for solutions satisfying Eqs. (14), (15) and (16). The solution is located almost in the centre of the solution distribution plane in both cases.

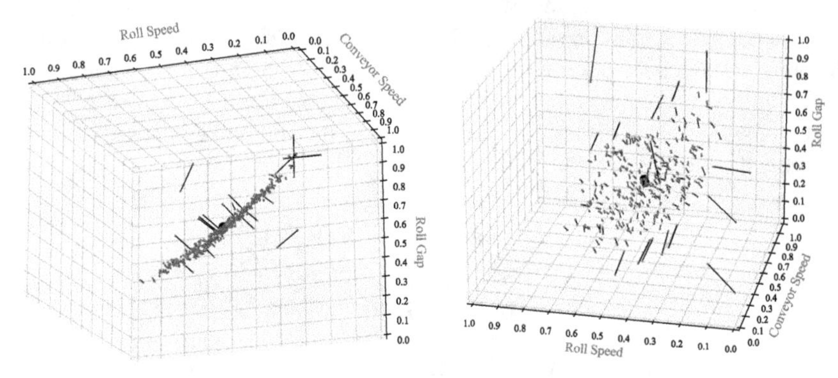

Figs. 12. Distribution of operating conditions (moisture 60, temperature 25)

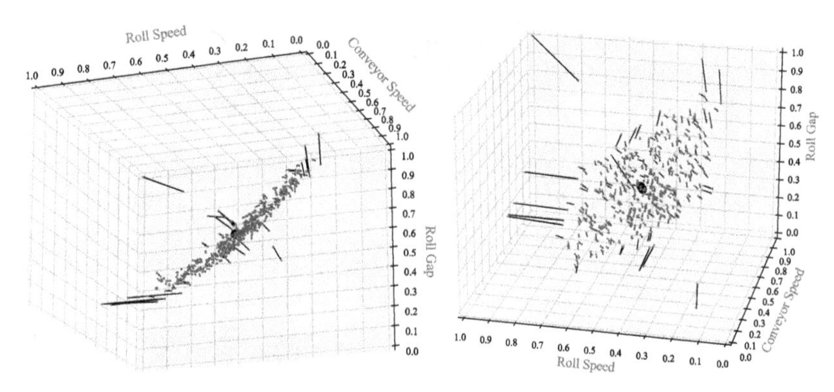

Fig. 13. Distribution of operating conditions (moisture 60, temperature 35)

Table 2 shows the least-squares errors of the target value and the predictions from computer experiments 1 and 2. The variance of the data used in the computer experiments was 0.3. Yellow-coloured cells show smaller least-squares errors in computer experiments 1 and 2. Table 2 shows that the least-squares errors are often smaller in computer experiments 2, and their mean values are also smaller. This indicates that the robustness is improved by using operating conditions for the predicted value that minimises the Euclidean distance between the midpoints of $\overline{\boldsymbol{L}}$ and $\overline{\boldsymbol{U}}$.

Table 2. Least-squares error between target and predicted values

moisture		50	50	50	52	52	52	54	54	54	56	56	56	58	58	58	60	60	60
temperature		15	25	35	15	25	35	15	25	35	15	25	35	15	25	35	15	25	35
XGBoost	Experiment 1	0.00405	0.00155	0.00061	0.00019	0.00468	0.00089	0.00205	0.06291	0.00251	0.00308	0.00152	0.00035	0.00044	0.00444	0.00015	0.00045	0.00054	0.00030
	Experiment 2	0.00439	0.00114	0.00008	0.00113	0.00004	0.00025	0.00045	0.00017	0.00184	0.00282	0.00031	0.00110	0.00119	0.00294	0.00303	0.00061	0.00137	0.00077
Light GBM	Experiment 1	0.00451	0.00043	0.00101	0.00111	0.00261	0.00224	0.00027	0.00359	0.00249	0.00079	0.00686	0.10087	0.00069	0.00221	0.00092	0.00094	0.13388	0.00056
	Experiment 2	0.00227	0.00215	0.00060	0.00113	0.00257	0.00056	0.00124	0.00052	0.00399	0.00011	0.00052	0.00115	0.00083	0.00172	0.00352	0.00032	0.00026	0.00088
moisture		62	62	62	64	64	64	66	66	66	68	68	68	70	70	70	Average		
temperature		15	25	35	15	25	35	15	25	35	15	25	35	15	25	35			
XGBoost	Experiment 1	0.00793	0.00079	0.00172	0.00045	0.00107	0.00006	0.00188	0.00209	0.00110	0.00808	0.00808	0.00182	0.00028	0.00125	0.00530	0.00402		
	Experiment 2	0.00167	0.00116	0.00039	0.00062	0.00036	0.00045	0.00051	0.00034	0.00062	0.00027	0.00053	0.00183	0.00058	0.00054	0.00131	0.00105		
Light GBM	Experiment 1	0.00148	0.00262	0.00106	0.00346	0.00342	0.00153	0.00066	0.00029	0.00017	0.00313	0.00052	0.00116	0.00022	0.00137	0.00078	0.00872		
	Experiment 2	0.00019	0.00314	0.00027	0.00026	0.00096	0.00091	0.00153	0.00151	0.00060	0.00072	0.00034	0.00218	0.00010	0.00265	0.00059	0.00122		

5 Conclusions

We study data-driven control systems to enable the stable production of good products when it is difficult to study and develop physical and chemical models or to use operational knowledge. As part of this research, this paper proposed an algorithm that builds a reaction model from actual operation data using machine learning and finds the optimal operating conditions under which the product is within the quality standards range using 0–1 integer programming. The effectiveness of the proposed algorithm was verified using operation data generated using a simulator for a food manufacturing process. Computer experiments confirmed the effectiveness of the proposed algorithm and obtained the following findings.

- In the algorithm for finding the operating conditions that minimise the difference from the median of the quality standards range (Numerical Experiments 1), the good product ratio was sometimes less than 100% when the quality standards range was narrow.
- The distribution of solutions that were close to the optimal solution and objective function value in Numerical Experiments 1 revealed a multimodal solution structure with many solutions on almost the same curved surface. The operating conditions of the solutions with a good product ratio of less than 100% were located at the edges of the distribution surface. The reason for this is considered to be the high possibility of collinearity in the explanatory variables. As a result, there were cases in which a slight deviation did not result in a good product ratio of 100%.
- The algorithm (Computer Experiments 2), which finds the point on the solution surface with the nearest Euclidean distance to the midpoint of the mean of the upper and lower limits of the prediction variables whose objective function values were close to the optimal solution of Computer Experiments 1, resulted in 100% of the good product ratio even when the quality standards range was narrow. Furthermore, the predicted values using this algorithm were near the median value of the quality standards range.

In the future, we will study data-driven control systems that reflect state variables that cannot be observed (e.g., state changes during continuous changes in machines) to make the model more applicable to actual operating machines.

References

1. Kano, M., Nakagawa, Y.: Data-based process monitoring, process control, and quality improvement: recent developments and applications in steel industry. Comput. Chem. Eng. **32**, 12–24 (2008)
2. Camacho, E.F., Bordons, C.: Model Predictive Control. Springer, London (2013). https://doi.org/10.1007/978-0-85729-398-5
3. Åströom, K.J., Wittenmark, B.: Adaptive Control, 2nd edn. Dover Publications (2008)
4. Hjalmarsson, H., Gevers, M., Gunnarsson, S., Lequin, O.: Iterative feedback tuning: theory and applications. IEEE Control. Syst. Mag. **18**(4), 26–41 (1998)
5. Hou, Z.-S., Wang, Z.: From model-based control to data-driven control: survey, classification and perspective. Inf. Sci. **235**, 3–35 (2013)
6. Spall, J.C.: Multivariate stochastic approximation using a simultaneous perturbation gradient approximation. IEEE Trans. Autom. Control **37**(3), 332–341 (1992)
7. Hou, Z., Jin, S.: Model-free adaptive control for a class of nonlinear discrete-time systems based on the partial form linearization. IFAC Proc. Vol. **41**(2), 3509–3514 (2008)
8. Hjalmarsson, H., Gunnarsson, S., Gevers, M.: A convergent iterative restricted complexity control design scheme. In: Proceedings of the 33rd IEEE Conference on Decision and Control, pp. 1735–1740 (1994)
9. Chen, T., Guestrin, C.: XGBoost: a scalable tree boosting system. In: Proceedings of the 22nd ACM SIGKDD International Conference on Knowledge Discovery and Data Mining (2016)
10. Ke, G., et al.: LightGBM: a highly efficient gradient boosting decision tree. In: Proceedings of the 31st International Conference on Neural Information Processing Systems (2017)

Data-Driven Scheduling of Cellular Manufacturing Systems Using Process Mining with Petri Nets

Hidefumi Kurakado, Tatsushi Nishi(✉), and Ziang Liu

Okayama University, 3-1-1 Tsushima-Naka, Kita-ku, Okayama 700-8530, Japan
nishi.tatsushi@okayama-u.ac.jp

Abstract. Petri net is a mathematical model for representing parallel, asynchronous, and distributed systems. Petri nets can model parallel and synchronous activities in manufacturing systems at various levels of abstraction. In this study, we propose data-driven modeling and scheduling for cellular manufacturing systems using process mining with Petri nets. In the proposed method, the event log data is extracted from a virtual plant and then the Petri net model considering the movement of products and operators is developed by using the process mining technique with the Petri net model. We also derived an approximate solution for the derived Petri net model from the event log using a local search method using a Petri net simulator. The analysis and modification of the model are conducted in the proposed method. Near-optimal schedules are derived using Petri net simulations. The validity of the proposed model is evaluated.

Keywords: cellular manufacturing · data-driven scheduling · Petri nets · process mining

1 Introduction

In recent years, many manufacturing companies have been conducting high-mix, low-volume production in order to quickly respond to a wide range of needs and the latest trends. For cellular manufacturing in high-mix, low-volume production, each process is different for each product, making it difficult to effectively manage the manufacturing processes. Therefore, the need to optimally and flexibly manage cellular manufacturing systems for high-mix, low-volume production is increasing.

The characteristics of cell production include immediacy to production fluctuations, rapid reduction of losses, strict efficiency control, and independence and completeness of work, which makes it possible to respond to various changes such as layout changes and production volume fluctuations [1]. However, in actual cell production systems, it is difficult to accurately model the target system. For example, in cell production systems that have been studied previously, the movement of operators between processes is often not considered. However, in the cell production line to produce large products, the movement of products and operators is highly correlated. It is significant to schedule

Published by Springer Nature Switzerland AG 2024
M. Thürer et al. (Eds.): APMS 2024, IFIP AICT 729, pp. 17–28, 2024.
https://doi.org/10.1007/978-3-031-65894-5_2

managing product flow and operators simultaneously for complex cellular production lines.

Conventional production simulation and optimization methods require a lot of expert knowledge to develop an appropriate simulation model. Especially, the development of simulation models for industrial use requires several steps with investigation of production systems, interviewing experts, extracting data, finding key components and a lot of validation experiments to fit the simulation data. If an appropriate simulation and optimization model is easily constructed from event log data, it is useful to reduce a lot of efforts to construct a simulation and optimization model for actual production environments.

Therefore, in this study, we propose a data-driven scheduling method using process mining with Petri nets. The proposed method generates a Petri net model from event logs from cell production simulation data and automatically extracts Petri net models using process mining. To find the feasibility of our data-driven scheduling method, we regard Simens Plant Simulation as a virtual plant.

Generally, it is extremely hard to develop an appropriate simulation model for complex cellular production systems considering the product flow and workforce management simultaneously. Also, the optimization with the simulation model requires a huge computational effort. Therefore, in this study, we develop a production simulation model using event log data using process mining with Petri nets.

Petri net is a mathematical model for representing parallel, asynchronous, and distributed systems. Therefore, Petri nets are one of the ways to represent complex production systems, and various studies have been conducted [2–5]. In this study, we model cell production using Petri nets, taking into account the movement of products and operators, and also examine the methodology of Petri net modeling using data from a production simulator and the solution of the optimization of optimal firing sequence problem for Petri nets.

We apply our method for a cellular production system with a multi-stage production system considering workforce scheduling and production input sequence simultaneously under several practical constraints. To investigate the feasibility of the proposed data-driven scheduling method, we first develop a Petri net model with a small production line with five or six products. The model assumes a cyclic schedule. Therefore, the developed model under the environment is useful for the optimization of actual production lines to investigate the feasibility of the proposed data-driven scheduling method.

The rest of the paper consists of the following sections. Section 2 states the problem description and scheduling problem for the cellular manufacturing system. Section 3 introduces data-driven scheduling method using process mining with Petri nets. Section 4 presents model validation and the comparison of scheduling results with conventional mathematical programming modeling. Section 5 concludes this research.

2 Problem Description

2.1 Problem Setting

The cellular manufacturing system treated in this research is explained below. The cell production system consists of four consecutive machines. Each machine has one or two work stations where some operators are located. Each product is put into a machine specified in the process order, and passes through the production line according to the technical order of each operation. Each process has one machine, and each operation requires one or two operators at the same time. The time required to move between machines or complete a task is expressed in the unit of work time (wt). The working time can be set to any time depending on the production system, such as by setting [wt] = second. The detailed conditions of the problem are explained below.

Definition of Machine

The production line has four consecutive processes: 1, 2, 3, and 4. Each machine can perform a specific process. Each process P_i is defined as a process on the machine i. Process P_1 on machine 1 and process P_3 on machine 3 must be performed by one operator, and the process P_2 on machine 2 and process P_4 on machine 4 must be performed by two operators simultaneously.

Definition of Station

The station is the place where all operators are assigned at time 0, and each operator moves from the station to each work station. There are as many stations as there are operators. The travel time from the station to each machine is arranged at intervals of 1 [wt] from machine 1.

Defining the Workplace

Each machine has one or two workplaces. Each workplace is a place where operators perform their work. Operators must not leave the work area until the work is completed. After operators finish their work, if they need to do other work, they move to another work area.

In a workplace, it can be expressed that the movement of operators affects the start time of work. To perform process 1, the operator must be in workstation 1 of machine 1. Also, operators must stay at work area 1 until process 1 is completed. To perform process 2 after process 1 is completed, the operator must move to either work station 2 or 3 of machine 2. Additionally, since process 2 requires two operators, it cannot start unless there are operators in both workplaces 2 and 3.

Definition of Operator

There are two operators for each cell. The time it takes for an operator to move between adjacent machines is 1 [wt]. All operators are multi-skilled and can initiate all tasks. After the product arrives, the operator begins moving to the product.

Definition of Product

Each product must follow a determined work process. When the process is completed, the product is moved to the machine that performs the next process. The unit of time for a product to move between machines is [wt].

Definition of Input Field

All products are fed into each machine from the feeding area. The travel time from the loading area to each machine is arranged at intervals of 1 [wt] from machine 1.

Definition of Process Stage

The process stage is determined for each product. An example of the process for each product is as follows: Product A ($P_1 \rightarrow P_2 \rightarrow P_4$), Product B ($P_2 \rightarrow P_3 \rightarrow P_1$), Product C ($P_1 \rightarrow P_3$), Product D ($P_1 \rightarrow P_3 \rightarrow P_4$), Product E ($P_1 \rightarrow P_2 \rightarrow P_3$) (Fig. 1).

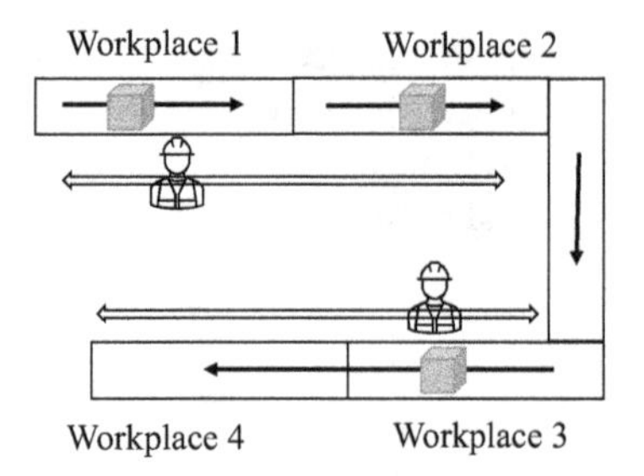

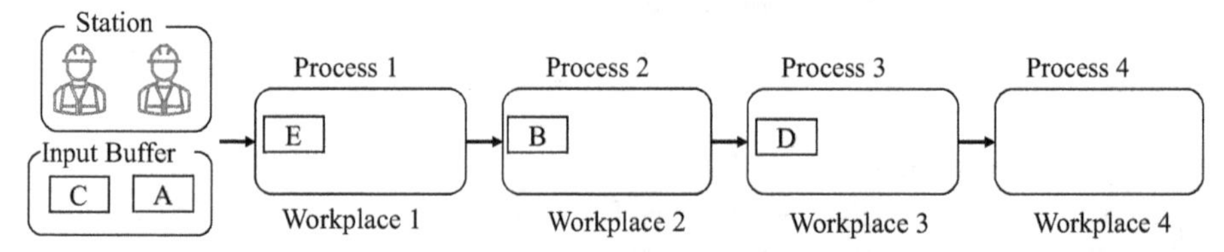

Fig. 1. Illustration of targeted cellular production system

An example of the process flow of product A is explained as follows. Even if the product has arrived at the machine, work cannot begin unless an operator is present at the machine. However, in this example, the movement of operators is not considered. Product A, like other products, travels from the input area to the machine corresponding to the process specified in the work process. Product A first arrives at machine 1 from the input area at time 1 and executes process P_1 of product A. After P_1 completes, it arrives at machine 2 at time 3 to execute P_2. After arriving at machine 2, execute P_2. After P_2 completes, it arrives at machine 4 at time 7 to execute P_4. After arriving at machine 4, P_4 is executed and product A completes all work steps.

The definition of scheduling problem for cellular manufacturing systems is stated as follows.

Information Given: number of products, working steps for each product, distance between each machine, number of operators required for the process, processing time for each task, number of operators in the cell.

Decision Variables: Product introduction order and operator assignment to each work area.

Objective Function: Makespan (maximum completion time).

Constraints: There are some constraints that must be met. This makes the problem difficult to solve. In this study, we consider the following constraints (i) to (ix).

(i) Restrictions on joint work
There are two types of work: work performed by one operator alone and work performed jointly by two operators.

(ii) Restrictions on product movement
After each product is finished working, it is moved to the next designated machine. The next work cannot be done until the next machine is reached.

(iii) Work restrictions in the workplace
Operators must remain at the workplace until the work is completed.

(iv) Restrictions on movement of operators
After finishing their work, operators can move to the next work area. Operators cannot work while moving.

(v) Restrictions on the number of products handled by the machine
Each process can only handle one product.

(vi) Work execution constraints
Each process specified for each product is executed only once.

(vii) Input site constraints
Each product is introduced only once. One product is introduced every hour.

(viii) Restrictions on setup of loading area
Each product cannot start its designated process until it reaches the designated (first executed) process from the input area.

(ix) Workplace restrictions
Each operator is placed at a worksite (set for each operator) at time 0, and each operator moves from the station to the workplace.

2.2 Simens Plant Simulation

Siemens' Plant Simulation was used to create the simulation model. Plant Simulation is a commercial simulation used in research on many production models [6–8]. Figure 2 shows the created simulation model. In the simulation, the number of products, operators, machines (stations), travel time between products and operators, etc., can be defined. The blue lines in Fig. 2 represent connections between objects called "connectors". The connector connecting Foot Path to Workplace represents the movement flow of operators passing through Foot Path and arriving at Workplace. The connector connecting the Conveyor to the Station represents the movement flow of products passing through the Conveyor and arriving at the Station.

When running a simulation, only one product is input from the Source at each time, and Flow Control selects the Conveyor connected to the Station that corresponds to the product's work process. After the product arrives at the Station, the operator starts moving by passing through the Foot Path that connects the Workplace from the Operator Pool to the Workplace installed at the Station where the product arrived. Operators begin work in the order in which the products arrive at the station. After all work processes are completed, the product is put into the sink.

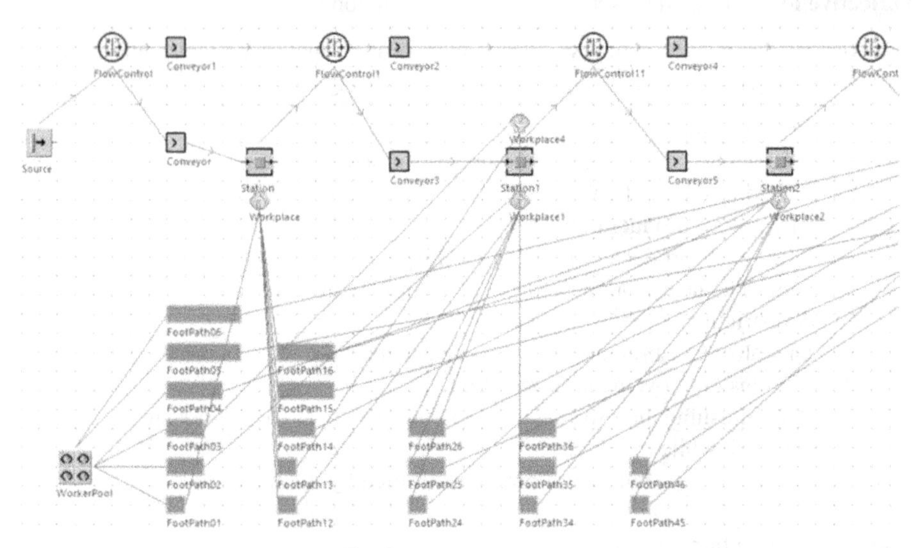

Fig. 2. Simulation model

3 Data-Driven Modeling of Cellular Manufacturing Systems

3.1 Definition of Petri Nets

A Petri net consists of a set of places P and a set of transitions T, which are connected by a set of arcs A. Places are represented by circles, transitions are represented by squares, and arcs are represented by arrows. A Petri net can be described as a directed bipartite graph consisting of a set of places and a set of transitions. Also, tokens are placed inside the places. A place represents the local state of the system by the number of tokens that enter it. A token corresponds to the existence of a resource. If a token is placed in a certain place, it means that a resource exists there.

The arrangement of tokens throughout the Petri net is called marking. The marking is a non-negative integer vector of dimension number of places. It is expressed by (1) where n is the total number of places and $m(p_i)$ is the number of tokens for p_i.

$$M = [m(p_1), m(p_2), \ldots, m(p_i), \ldots, m(p_n)] \quad (1)$$

$M(p)$ is the mapping from place p to non-negative integer. For place p and transition t, if $A(p, t) \geq 1$, then p is said to be the input place of t, and t is said to be the output transition of p.

To represent the behavior of the system, the markings of the Petri net change according to the following transition firing law:

1. A transition occurs when there are more tokens than the number of arcs in each input place. A transition t is said to be fireable when

$$M(p) \geq A(p,t), \forall p \in P \tag{2}$$

2. Transitions that can fire may or may not fire.
3. When the transition fires, tokens equal to the number of arcs are lost from each input place, and tokens equal to the number of arcs are given to the output place. Therefore, the firing of transition t changes to the next marking M'. We denote that the marking M is changed into the marking M'. by the firing of transition t as

$$M[t > M' \tag{3}$$

Then, the marking is changed according to the following equation.

$$M'(p) = M(p) + A(t,p) - A(p,t) \tag{4}$$

For a marking M with a Petri net and a sequence of transitions $\sigma = t(i_1), t(i_2), \ldots, t(i_n)$, if each transition from M to σ can be fired sequentially, that is,

$$M[t_{i_1} > M_1, M_1[t_{i_2} > M_2, \ldots, M_{i-1}[t_{i_n} > M_n \tag{5}$$

If it holds, then σ is said to be a possible firing sequence from M.

Two transitions are said to be non-conflicting if they do not have a common input place, or if they do have a sufficient number of tokens to fire both transitions.

For $PN = (P, T, A, M_0)$, two non-negative alignment matrices B^+_{PN} and B^-_{PN} are defined as follows.

$$B^+_{PN}(p,t) = A(t,p), B^-_{PN}(p,t) = A(p,t)$$

Also,

$$B_{PN} = B^+_{PN} - B^-_{PN}$$

Suppose that B^+_{PN}, B^-_{PN}, and B_{PN} are called the input connection matrix, output connection matrix, and connection matrix of *PN*, respectively. Each column of the connection matrix represents the change in marking associated with the firing of the corresponding transition. The optimal firing sequence problem is the problem of setting the final marking and finding a possible firing sequence from the initial marking to the final marking. If the initial marking is M_0 and the final marking is M, the problem of finding the solution with the minimum number of firings can be formulated as follows.

$$\text{min.} \sum_{i=1}^{n} |\sigma_i| \tag{6}$$

$$\text{s.t.} M - M_0 = B_{PN} \cdot \sigma \tag{7}$$

3.2 Process Mining

Process mining plays an important role in integrating data mining, business process modeling, and analysis. The process discovery algorithm, which is a field in process mining, is a function that maps event logs to process models, so the (Petri net) model represents the behavior seen in the event logs [9]. Recently, the number of researchers working on process mining has increased, and many software vendors have begun to incorporate process mining functionality into their own process mining tools. PM4Py is a Python library that provides a comprehensive set of tools for process mining [10]. In this research, we use PM4Py to generate Petri nets from event logs output from SIEMENS Plant Simulation. The event log includes events representing product movement, product work, and operator movement. We use the α algorithm, which is a basic algorithm for process discovery, to generate a Petri net from event logs.

Figure 3 shows the Petri net generated from the extracted event log. Although the arc in Fig. 3 is cut out to improve visibility, the generated Petri net is a straight line horizontally. The first transition represents the movement of the product, the next transition represents the movement of the operator, and the next place represents the work on product A. In the firing sequence, the transition representing the movement of the operator can be fired after the transition representing the movement of the product is fired, so it expresses that the operator starts moving after the product arrives. Therefore, it is a Petri net that represents cell production that takes into account the movement of products and operators.

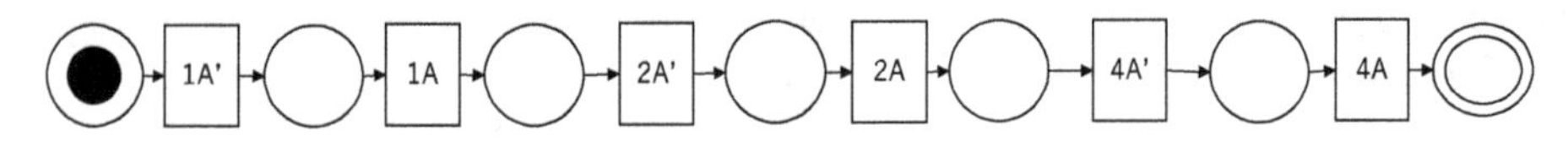

Fig. 3. Petri Net model of Product A derived from event logs

Figure 3 is a Petri net representing cell production when only product A is handled. However, since the cell production we are dealing with in this article deals with multiple products, it is necessary to express the movement of multiple products using the generated Petri net as a reference.

3.3 Proposed Method for Cellular Manufacturing Systems

The procedure for creating a Petri net that represents cell production that handles multiple products is described below, using cell production that handles products A and B as an example.

<Step 1> Generate Petri net from the event log of each product

Generate Petri nets for product A and product B from the simulation event log.

<Step 2> Divide the generated Petri net into product movement, work and operator movement. Each transition represents the movement of products, movement of operators, and work. Here, we will consider the movement of products, work, and movement of operators separately.

<Step 3> Create tokens representing products and operators

Create tokens and places that represent products and operators. Since there are two operators, the number of tokens representing the operators is two.

<Step 4> Express the movement of operators and the workplace

A place is defined to represent the workplace where operators move, and connect arcs to each transition. When the firing sequence is confirmed, the transition representing the movement of the product is fired, and then the transition representing the movement of the operator can be fired. Therefore, although the places and tokens have changed, the firing sequence remains the same as the original Petri net.

<Step 5> Correspond to the workplace from the Petri net of each product
<Step 6> Add places and tokens considering the order of product introduction

There is a restriction that one product is introduced at each time. Therefore, add a place and token that correspond to that constraint. The transitions representing the movement of product A and product B from the input area to the machine conflict with the added tokens, and the two transitions cannot fire at the same time. This allows it to be expressed that one product is introduced at each time.

<Step 7> Add places and tokens considering the constraints on the number of products handled by the machine

Each machine has a restriction that it cannot handle two or more products at the same time. Therefore, add a place and token that correspond to that constraint. Fig. 4 represents a Petri net with additional places and tokens considering the number of products that the machine can handle. The transitions representing work 2 and work 4 of each product conflict with the added token, and the two transitions cannot fire at the same time. This allows us to express that the machine cannot handle two or more products at the same time.

4 Model Validation Experiments

Before starting the Petri net simulator, set the number of iterations, initial marking, and final marking. The simulator solves the optimal firing sequence problem up to the set maximum number of iterations (Max itr) and finds the optimal firing sequence that minimizes the time at which the final marking is reached. The firing rule randomly determines which transitions to fire until there are no more ignitable transitions (Ignitable_t). If the final marking is not reached even after period N, the firing sequence and marking are initialized and the next iteration is executed. Table 1 shows the output results of (Resource Constrained Project Scheduling Problem: RCPSP formulation and Petri net for each case. The simulator was run with the maximum number of iterations set to 10 and 100. Since the simulator randomly determines which transitions to fire, as the number of iterations increases, it is possible to reduce the makespan.

As a result of comparing with the output results of RCPSP, when the number of iterations is 100, the value is similar to the exact solution derived by RCPSP. It was

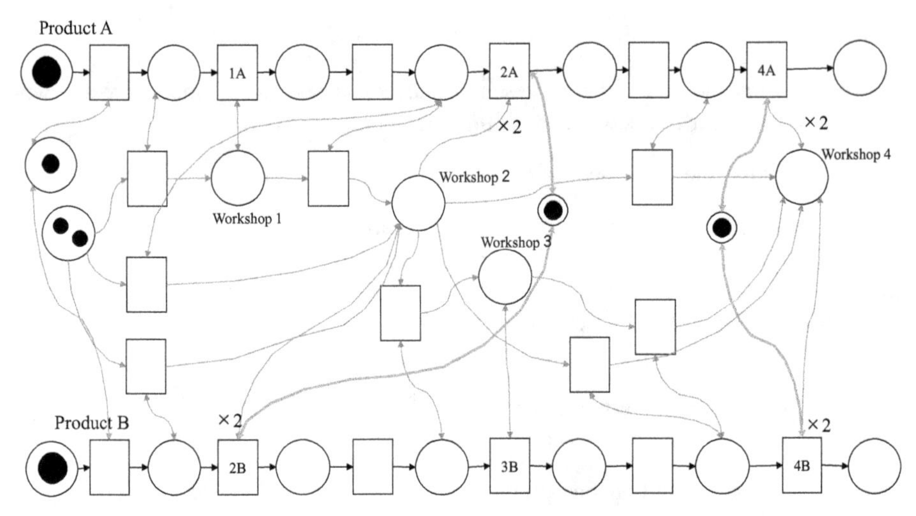

Fig. 4. Petri Net model with the number of product constraints handled by the machine

found that the calculation time was less than RCPSP. It could be seen from the results that the work order obtained with Petri net is similar to that of RCPSP. However, in case 5, a feasible solution could not be obtained using Petri nets.

From the result of the infeasible case, it can be seen that the work for the majority of the products has not been done, and the work for process 2 has not started. There are two possible reasons why a feasible solution cannot be obtained. One is for the simulator to fire all possible transitions. Therefore, a transition representing a movement of an operator that is not necessary to start work at the next time may be fired.

Table 1. Computational results from Petri Net simulator and RCPSP

Cases	Products	Petri Net				RCPSP	
		Itr = 10		Itr = 100			
		Value	CPU-time	Value	CPU-time	Value	CPU-time
Case 1	A, B, C	19	0.5	17	2.4	17	0.3
Case 2	A, B, C, D	22	1.1	21	3.0	21	1
Case 3	A, B, C, D, E	26	1.3	25	3.2	25	30
Case 4	A, B, C, D, E, F	28	1.6	26	3.0	26	15
Case 5	A, B, C, D, E, F, G	–	1.6	–	3.0	28	176

From Fig. 2, it can be seen that the operator moved to the work area for process 3 even though the work for process 2 was the next time. The second thing is. The created Petri net cannot represent the movement of operators from the workplace in the subsequent process to the workplace in the previous process. Even if a transition that represents an unnecessary movement of a transition fires, there is no problem as long as the transition that represents the movement of the operator to the original work area fires at the next time. In Fig. 3, even if the operator moves to process 3, there is a possibility that a feasible solution can be obtained by moving the operator to process 2 at the next time. However, the movement of operators in the Petri net generated from process mining only represents the movement from the workplace of the previous process to the workplace of the subsequent process, and does not represent the movement of operators to the original workplace. These are considered to be the reasons why a feasible solution cannot be obtained in case 5.

The Petri net model used in this method can represent up to six products in cell production, taking into account the movement of products and operators, and can obtain a feasible solution to the optimal firing sequence problem. However, when the number of products exceeds seven, it is not possible to obtain a feasible solution to the optimal firing sequence problem. Future challenges include improving the Petri net simulator and comprehensively representing the movement of operators in the generated Petri net.

5 Conclusion and Future Works

In this paper, we have developed a data-driven scheduling method using process mining with Petri nets. Event log data is taken from the simulator. Then the data is used to build the scheduling model using process mining with Petri nets. The model validation and the optimization of the derived model are conducted using the Petri net simulator. The results show that the performance of the derived schedule is close to the solution derived by the mathematical formulation of the problem. The proposed method can derive a near-optimal solution for small-scale problem instances. However, in some cases, a feasible solution cannot be derived. Our future work is to obtain a feasible solution by modifying the Petri net model for large-scale instances.

Acknowledgments. The authors would like to thank the funding provided by JSPS KAKENHI KIBAN (B) 23K22983.

References

1. Liu, C., Lian, J., Yin, Y., Li, W.: Seru seisan - an innovation of the production management mode in Japan. Asian J. Technol. Innov. **18**(2), 89–113 (2010)
2. Cecil, J.A., Srihari, K., Emerson, C.R.: A review of Petri-net applications in manufacturing. Int. J. Adv. Manuf. Technol. **7**(3), 168–177 (1992)
3. Grobelna, I., Karatkevich, A.: Challenges in application of Petri nets in manufacturing systems. Electronic **10**(18), 2305 (2021)
4. Alhourani, F.: Cellular manufacturing system design considering machines reliability and parts alternative process routings. Int. J. Prod. Res. **54**(3), 846–863 (2016)
5. Süer, G.A., Ates, O.K., Mese, E.M.: Cell loading and family scheduling for jobs with individual due dates to minimise maximum tardiness. Int. J. Prod. Res. **52**(19), 5656–5674 (2014)
6. Ferro, R., Cordeiro, G.A., Ordóñez, R.E.C., Beydoun, G., Shukla, N.: An optimization tool for production planning: a case study in a textile industry. Appl. Sci. **11**(18), 8312 (2021)

7. Kloud, T., Koblasa, F.: Solving job shop scheduling with the computer simulation. Int. J. Transp. Logist. **20**(11), 7–17 (2018)
8. Haraszkó, C., Németh, I.: DES configurators for rapid virtual prototyping and optimization of manufacturing systems. Periodica Polytechnica Mech. Eng. **59**(3), 143–152 (2015)
9. Van Der Aalst, W.M.P.: Process Mining: Data Science in Action, Springer, Heidelberg (2018). https://doi.org/10.1007/978-3-662-49851-4
10. Berti, A., Zelst, S.V., Schuster, D.: PM4Py: a process mining library for Python. Softw. Impacts **17**, 100556 (2023)

A Study on Sophisticated Production Management for Engineer-to-Order Production: A Mixed Integer Programming Formulation for Production Scheduling

Eiji Morinaga[1(✉)], Koji Iwamura[1], Yoshiyuki Hirahara[2], Masamitsu Fukuda[2], Ayumu Niinuma[2], Hirotomo Oshima[2], and Yasuo Namioka[3]

[1] Osaka Metropolitan University, Sakai, Osaka 599-8531, Japan
morinaga.e@omu.ac.jp
[2] Toshiba Corporation, Yokohama, Kanagawa 235-0017, Japan
[3] Advanced Institute of Industrial Technology, Shinagawa-ku, Tokyo 140-0011, Japan

https://www.omu.ac.jp/i/en/dii/

Abstract. Engineer-to-order (ETO) production in which products are designed and manufactured in response to customer orders is required to respond flexibly to customer requests at various stages from design to maintenance. This characteristic makes it difficult to apply a standard production planning strategy which divides the planning into three phases, i.e., long-term, medium-term and short-term planning (production scheduling), because there are large discrepancies among the phases and rescheduling requires a lot of man-hours. We proposed a production planning framework that unifies the granularity of resources and unit time in all of the planning phases aiming to reduce the discrepancies, and a model that is commonly used in the three phases of planning was organized as flexible job-shops. This paper provides a mixed integer programming formulation of the production scheduling problem based on the model considering the following characteristics of the target ETO production site: (1) The planner has discretion in shortening required processing time; (2) Operation time is limited to day time of weekdays; (3) Overtime works can be accepted if necessary; (4) Some operations of multiple parts must be processed at the same time on the same machine. A numerical experiment showed validity of the model.

Keywords: Flexible job-shops · Production scheduling · Field-oriented method · Mixed integer programming · Engineer-to-order production

M. Thürer et al. (Eds.): APMS 2024, IFIP AICT 729, pp. 29–43, 2024.
https://doi.org/10.1007/978-3-031-65894-5_3

1 Introduction

A variety of methodologies for appropriate planning, efficient operation and proper management of production systems have been discussed for a long time. Many methods have been proposed and sophistication of them has been attempted. However, many of them have not been put into practical use at production sites. Generally, in method development, various requirements and constraints that exist in production sites are extracted and selected, the problem is idealized, and a solution is then derived. But, the viewpoint of the sites is not sufficiently reflected in the process of extraction and selection. This is thought to be one of the reasons why the methods are not sufficiently applied to actual production. In order to achieve highly-competitive production activities, it is important to develop field-oriented methods that place greater emphasis on reflecting the requirements, constraints, and circumstances at the production site.

At sites of engineer-to-order (ETO) production such as power plants and shipbuilding, where large-scale infrastructure products and equipment are designed and manufactured in response to customer orders, the following problems exist:

- Detailed design of the product is carried out after specifications have been determined in accordance with the customer's requirements. Thus, it is not possible to list and prepare necessary parts before the designing.
- Accurate estimation of production lead time is difficult because additional processes may be required for changes in specifications just before or during manufacturing or assembly adjustments.
- Even after delivery to the customer, various maintenance work must be carried out, such as periodic inspections and parts replacement in case of unexpected malfunctions.

Thus, ETO production is characterized by the need to respond flexibly to customer requests at various stages from design to maintenance.

For ETO production, which requires flexible response, three kinds of production planning are performed currently: long-term planning which covers from six months to a few years ahead, medium-term planning covering a period of several months to six months, and short-term planning (production scheduling) with a horizon of a few weeks to a few months. However, there are large discrepancies among each of the plans. In addition, rescheduling requires a lot of man-hours. Considering these issues, we aimed to build a model that can apply common resource granularity and unit time to each planning, improved the production planning framework, and organized the motif problem as a flexible job-shop [1]. In order to organize the framework by a mathematical model, this paper provides a mathematical programming formulation of the production scheduling problem considering characteristics of the target ETO production site.

2 A Production Management Framework for Engineer-to-Order Production

Generally, the following strategy for production planning is adopted: The planning problem is divided into three phases of long-term planning, medium-term planning and short-term planning depending on the planning horizon, and each of the planning is then performed in order from long-term/rough planning to short-term/detailed planning. However, due to the difference in the granularity of target resources, for example, it may turn out that a plan which is supposed to be able to meet the due date in the long-term planning is unreasonable at the phase of medium-term and short-term planning. As a result, a lot of overtime may be required to achieve the plan, or conversely, the operating rate of machines and workers may be lower than expected. In ETO production, product specifications changes, additional production, parts delivery delays and additional processing occur frequently. Hence, it is difficult to generate a plan and revise it each time in accordance with the planning strategy. For this reason, we aimed to construct a framework that unifies the granularity of resources and unit time in all of the planning phases and organized it as shown in Fig. 1 [1]. A model that is commonly used in the long-term, medium-term and short-term planning was organized as flexible job-shops as follows:

- Product number: a $(1, \cdots, A)$
- Parts for product a: $Prod_j^a$ $(j = 1, \cdots, J_a)$
- Release date of $Prod_j^a$: Rd_j^a
- Due date of $Prod_j^a$: Dd_j^a
- The total number of machines: M
- Information about operations of $Prod_j^a$:
  - Required operations of $Prod_j^a$: $\{1, \cdots, O_j^a\}$
  - Machines which can process the o-th operation of $Prod_j^a$: $\overline{M}_{jo}^a$
  - Time required for processing the o-th operation of $Prod_j^a$ with machine μ: $P_{jo}^{a\mu}$

3 Literature Review

Production scheduling for flexible job-shops has been discussed lively in recent three decades [2,3]. As well as other types of scheduling problems, many methods have been developed based on various approaches such as mathematical programming [4,5], meta-heuristics [6–10], artificial intelligence [11]. In the development, enhancement considering various needs in production sites has been discussed. A direction of this enhancement is to cope with uncertainty. Pan, et al. [12] took uncertainty of operation time and delivery time into account and provided a solving method based on quantum particle swarm optimization algorithm. Li, et al. [13] were focused on uncertainty of processing times and proposed a method using hybridized genetic algorithm and binary particle swarm optimization. Chen, et al. [14] considered uncertainty of processing times also. They

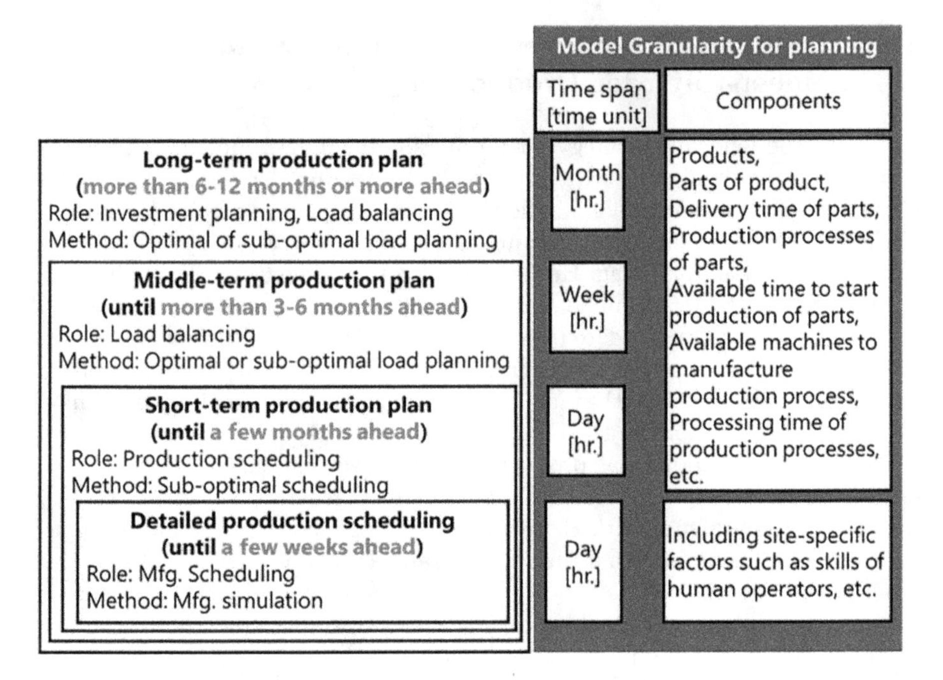

Fig. 1. Framework of production planning for ETO production [1]. (The detailed production scheduling in this figure is not considered in this research.)

characterized the uncertainty by a generalized grey number and developed an elite genetic algorithm for finding good solutions of the scheduling problem. He and Sun [15] were focused on machines breakdown and proposed an algorithm based on an approach with multi-strategies to make the scheduling more robust and stable. Sun, et al. [16] applied non-cooperative game theory with complete information to the scheduling problem and build a new model for obtaining better robust and stable performance. Ayyoubzadeh, et al. [17] considered machines breakdown also and proposed a hybrid metaheuristic algorithm based on genetic algorithm and particle swarm optimization. Zhang, et al. [18] proposed a two-stage algorithm based on convolutional neural network for solving flexible job-shop scheduling problems (FJSPs) with machine breakdown.

Another direction is to deal with various and multiple criteria. In addition to makespan, *regular criteria*, which are non-decreasing functions of the completion times of the jobs [3], have been taken into account, and minimization of total flow time [19], lateness [20], tardiness [21] and so on have been discussed. Furthermore, minimization of *non-regular criteria* such as workload of machines [22,23], earliness [24,25] and in-process inventory [26], which are for just-in-time production, and energy consumption [27] have been researched.

Consideration of additional constraints which are relevant in practice has been discussed also. A major kind of them is time lag constraints, and problems with no-wait constraint has been dealt with for pharmaceutical industry [28], glass factory [29] and hot rolling mill process [30]. Minimum time lag has been considered also for coping with delays imposed for process considerations in semiconductor fabrication [31]. Another major sort of the additional constraints is setup time constraint, and methods considering sequence-dependent setup time [32–34], separable setup time [35] and inseparable setup time [36] have been proposed.

Integration of FJSP with another decision-making problem has been researched also [37]. Baykasoğlu widened FJSP by considering alternative process plans for each part, and proposed methods for solving the problem [38,39]. He and Mandeğlu provided a procedure for simultaneous dynamic scheduling of operations and preventive maintenance activities also [40]. Saber, et al. focused on integration with operator scheduling and provided a mixed integer programming model [41]. Morinaga, et al. [42] dealt with integration with operator scheduling also and proposed a method for obtaining a production schedule with teams of workers.

As stated above, various researchers have challenged development of practical methods for flexible job-shop scheduling. However, there are other elements to be considered for development of practical methods depending on actual production sites. The ETO production considered in this research have the following characteristics:

C1 The planner has a degree of freedom on shortening processing time, that is, can shorten given processing time within a given range in scheduling.
C2 Production must be performed only in daytime (from 8:00 to 17:00) of weekdays.
C3 The planner may set a given range of overtime after 17:00 of weekdays.
C4 Some operations of parts of a product need *coupling*, that is, the parts are assembled temporarily just before the operations and then disassembled just after them.

It is necessary to develop a scheduling method for FJSP with these characteristics.

4 Mixed Integer Programming Formulation of the Scheduling Problem

In this section, a mathematical model of the FJSP with the characteristics C1–C4 is provided as a first step of development of a scheduling method for the ETO production. By introducing continuous, binary and integer variables, the problem can be formulated as a mixed integer linear programming (MILP) as follows:

Constants and Symbols:

- $\mathbb{R}^+$: the set of non-negative real numbers
- $\mathbb{Z}^+$: the set of non-negative integers

- $\alpha_1, \alpha_2, \cdots, \alpha_5$: weights for terms of the objective function
- $W(\leq 7)$: the upper limit of overtime on a day ([h])
- $\overline{V}^a_{jo}$: the upper limit of shortening processing time of an operation ([h])
- C^v: the unit cost for shortening processing time
- C^w: the unit cost for overtime work
- C_a: the total number of combinations of parts and operations of product a that requires a coupling process
- $\overline{C}^a_c (c = 1, \ldots, C_a)$: a combination of parts and operations of product a that requires a coupling process
- B: a big positive number

Decision variables:

- $S_{jo} \in \mathbb{R}^+$: starting time of operation o of $Prod^a_j$
- $C_{max} \in \mathbb{R}^+$: completion time of the part which finishes last among all of the parts
- $S_{min} \in \mathbb{R}^+$: starting time of the part which starts first among all of the parts
- $X_{a_1 j_1 o_1 a_2 j_2 o_2} \in \{0, 1\}$: precedence relationship between operation o_1 of $Prod^{a_1}_{j_1}$ and operation o_2 of $Prod^{a_2}_{j_2}$ which are processed by the same machine (This variable takes 1 when the operation o_1 is processed before the operation o_2, and 0 otherwise.)
- $Y^{a\mu}_{jo} \in \{0, 1\}$: a variable for representing whether operation o of $Prod^a_j$ is processed by machine μ (This value takes 1 when the operation o is processed by the machine, and 0 otherwise.)
- $V^a_{jo} \in \mathbb{R}^+$: the amount of shortening processing time of operation o of $Prod^a_j$ ([h])
- $Ds^a_{jo} \in \mathbb{Z}^+$: a variable for representing what day of the planning period operation o of $Prod^a_j$ is started
- $Df^a_{jo} \in \mathbb{Z}^+$: a variable for representing what day of the planning period operation o of $Prod^a_j$ is finished
- $Wf^a_{jo} \in \mathbb{R}^+$: the overtime work for operation o of $Prod^a_j$ that is necessary to be performed in the day on which operation o of $Prod^a_j$ is finished ([h])
- $Wp^a_{jo} \in \mathbb{R}^+$: the overtime work for operation o of $Prod^a_j$ that is necessary to be performed in the days before Df^a_{jo} ([h])
- $Ws^a_{jo} \in \mathbb{R}^+$: the amount of difference between the end of the daytime and the starting time of operation o of $Prod^a_j$ in the case that the operation does not start before 17:00 on Ds^a_{jo} ([h])

Objective function:

$$\begin{aligned}
&\text{Minimize} \\
&\alpha_1 \left(C_{max} - S_{min}\right) \\
&+ \alpha_2 \sum_{a=1}^{A} \sum_{j=1}^{J_a} \left[\left\{ S^a_{jO_j} + \sum_{\mu=1}^{M} Y^{a\mu}_{jO_j} P^{a\mu}_{jO_j} - V^a_{jO_j} \right.\right. \\
&\qquad\qquad \left.\left. + 15 \left(Df^a_{jO_j} - Ds^a_{jO_j} \right) - Wp^a_{jO_j} \right\} - S^a_{j1} \right]
\end{aligned}$$

$$+\alpha_3 \sum_{a=1}^{A} \sum_{j=1}^{J_a} \left[Dd_j^a - \left\{ S_{jO_j}^a + \sum_{\mu=1}^{M} Y_{jO_j}^{a\mu} P_{jO_j}^{a\mu} - V_{jO_j}^a + 15\left(Df_{jO_j}^a - Ds_{jO_j}^a \right) - Wp_{jO_j}^a \right\} \right]$$

$$+\alpha_4 \sum_{a=1}^{A} \sum_{j=1}^{J_a} \sum_{o=1}^{O_j} C^v V_{jo}^a + \alpha_5 \sum_{a=1}^{A} \sum_{j=1}^{J_a} \sum_{o=1}^{O_j} C^w \left(Wp_{jo}^a + Wf_{jo}^a \right) \tag{1}$$

Constraints:

$$S_{j1}^a \geq Rd_j^a, \ \forall j, \ \forall a \tag{2}$$

$$S_{jO_j}^a + \sum_{\mu=1}^{M} Y_{jO_j}^{a\mu} P_{jO_j}^{a\mu} - V_{jO_j}^a + 15\left(Df_{jO_j}^a - Ds_{jO_j}^a \right) - Wp_{jO_j}^a \leq Dd_j^a, \ \forall j, \ \forall a \tag{3}$$

$$S_{jo}^a + \sum_{\mu=1}^{M} Y_{jo}^{a\mu} P_{jo}^{a\mu} - V_{jo}^a + 15\left(Df_{jo}^a - Ds_{jo}^a \right) - Wp_{jo}^a \leq S_{j(o+1)}^a, \forall o \in \{1, \cdots, O_j - 1\}, \ \forall j, \ \forall a \tag{4}$$

$$\sum_{\mu=1}^{M} Y_{jo}^{a\mu} = 1, \forall o, \ \forall j, \ \forall a \tag{5}$$

$$Y_{jo}^{a\mu} = 0, \forall a, \ \forall j, \ \forall o, \ \forall \mu \notin \overline{M}_{jo}^a \tag{6}$$

$$S_{j_1 o_1}^{a_1} + \sum_{\mu=1}^{M} Y_{j_1 o_1}^{a_1 \mu} P_{j_1 o_1}^{a_1 \mu} - V_{j_1 o_1}^{a_1} + 15\left(Df_{j_1 o_1}^{a_1} - Ds_{j_1 o_1}^{a_1} \right) - Wp_{j_1 o_1}^{a_1} \leq S_{j_2 o_2}^{a_2} + B\left(1 - X_{a_1 j_1 o_1 a_2 j_2 o_2}\right) + B\left(1 - Y_{j_1 o_1}^{a_1 \mu'}\right) + B\left(1 - Y_{j_2 o_2}^{a_2 \mu'}\right), \ \forall a_1, \ \forall j_1, \ \forall o_1, \ \forall a_2, \ \forall j_2 \neq j_1, \ \forall o_2, \ \forall \mu' \tag{7}$$

$$X_{a_1 j_1 o_1 a_2 j_2 o_2} + X_{a_2 j_2 o_2 a_1 j_1 o_1} \geq 1 - B\left(2 - Y_{j_1 o_1}^{a_1 \mu} - Y_{j_2 o_2}^{a_2 \mu}\right), \ \forall a_1, \ \forall j_1, \ \forall o_1, \ \forall a_2, \ \forall j_2, \ \forall o_2 \tag{8}$$

$$X_{a_1 j_1 o_1 a_2 j_2 o_2} + X_{a_2 j_2 o_2 a_1 j_1 o_1} \leq 1 + B\left(2 - Y_{j_1 o_1}^{a_1 \mu} - Y_{j_2 o_2}^{a_2 \mu}\right), \ \forall a_1, \ \forall j_1, \ \forall o_1, \ \forall a_2, \ \forall j_2, \ \forall o_2 \tag{9}$$

$$S_{min} \leq S_{j1}^a, \ \forall a, \forall j \tag{10}$$

$$S_{min} \geq \min_{a,j} S_{j1}^a, \ \forall a, \forall j \tag{11}$$

$$C_{max} \geq S_{jO_j}^a + \sum_{\mu=1}^{M} Y_{jO_j}^{a\mu} P_{jO_j}^{a\mu} - V_{jO_j}^a + 15\left(Df_{jO_j}^a - Ds_{jO_j}^a \right) - Wp_{jO_j}^a, \ \forall j, \ \forall a \tag{12}$$

$$C_{max} \leq \max_{a,j} \left\{ S^a_{jO_j} + \sum_{\mu=1}^{M} Y^{a\mu}_{jO_j} P^{a\mu}_{jO_j} - V^a_{jO_j} \right.$$
$$\left. + 15 \left(Df^a_{jO_j} - Ds^a_{jO_j} \right) - Wp^a_{jO_j} \right\} \quad (13)$$

$$Ws^a_{jo} = \max \left\{ S^a_{jo} - \left(17 + 24Ds^a_{jo} \right), 0 \right\}, \ \forall a, \ \forall j, \ \forall o \quad (14)$$

$$Wf^a_{jo} = \max \left\{ S^a_{jo} + \sum_{\mu=1}^{M} Y^{a\mu}_{jo} P^{a\mu}_{jo} - V^a_{jo} + 15 \left(Df^a_{jo} - Ds^a_{jo} \right) \right.$$
$$\left. - Wp^a_{jo} - \left(17 + 24Df^a_{jo} \right), 0 \right\}, \ \forall a, \ \forall j, \ \forall o \quad (15)$$

$$Ds^a_{jo} \leq Df^a_{jo}, \ \forall o, \ \forall j, \ \forall a \quad (16)$$

$$V^a_{jo} \leq \overline{V}^a_{jo} \ \forall a, \ \forall j, \ \forall o \quad (17)$$

$$8 + 24Ds^a_{jo} \leq S^a_{jo} < 17 + 24Ds^a_{jo} + W, \ \forall a, \ \forall j, \ \forall o \quad (18)$$

$$8 + 24Df^a_{jo} < S^a_{jo} + \sum_{\mu=1}^{M} Y^{a\mu}_{jo} P^{a\mu}_{jo} - V^a_{jo} + 15 \left(Df^a_{jo} - Ds^a_{jo} \right) - Wp^a_{jo}$$
$$\leq 17 + 24Df^a_{jo} + W, \ \forall a, \ \forall j, \ \forall o \quad (19)$$

$$Ws^a_{jo} + Wp^a_{jo} \leq W \left(Df^a_{jo} - Ds^a_{jo} \right), \ \forall a, \ \forall j, \ \forall o \quad (20)$$

$$Wf^a_{jo} \leq W, \ \forall a, \ \forall j, \ \forall o \quad (21)$$

$$S^a_{j_1 o_1} = S^a_{j_2 o_2}, \ \forall j_1, \ \forall o_1, \ \forall j_2, \ \forall o_2 \in \overline{C}^a_c, \ \forall c, \ \forall a \quad (22)$$

$$Y^{a\mu}_{j_1 o_1} = Y^{a\mu}_{j_2 o_2}, \ \forall \mu, \ \forall j_1, \ \forall o_1, \ \forall j_2, \ \forall o_2 \in \overline{C}^a_c, \ \forall c, \ \forall a \quad (23)$$

$$V^a_{j_1 o_1} = V^a_{j_2 o_2}, \ \forall j_1, \ \forall o_1, \ \forall j_2, \ \forall o_2 \in \overline{C}^a_c, \ \forall c, \ \forall a \quad (24)$$

$$Ds^a_{j_1 o_1} = Ds^a_{j_2 o_2}, \ \forall j_1, \ \forall o_1, \ \forall j_2, \ \forall o_2 \in \overline{C}^a_c, \ \forall c, \ \forall a \quad (25)$$

$$Df^a_{j_1 o_1} = Df^a_{j_2 o_2}, \ \forall j_1, \ \forall o_1, \ \forall j_2, \ \forall o_2 \in \overline{C}^a_c, \ \forall c, \ \forall a \quad (26)$$

$$Wp^a_{j_1 o_1} = Wp^a_{j_2 o_2}, \ \forall j_1, \ \forall o_1, \ \forall j_2, \ \forall o_2 \in \overline{C}^a_c, \ \forall c, \ \forall a \quad (27)$$

$$Ws^a_{j_1 o_1} = Ws^a_{j_2 o_2}, \ \forall j_1, \ \forall o_1, \ \forall j_2, \ \forall o_2 \in \overline{C}^a_c, \ \forall c, \ \forall a \quad (28)$$

$$Wf^a_{j_1 o_1} = Wf^a_{j_2 o_2}, \ \forall j_1, \ \forall o_1, \ \forall j_2, \ \forall o_2 \in \overline{C}^a_c, \ \forall c, \ \forall a \quad (29)$$

Because this research is targeted at production of large products, the objective needs to be defined considering cost and area limitation of storage of work-in-processes and completed products. Hence, in the objective function (1) given as a weighted sum of five terms, the total flow time and the total earliness are included by the second and third terms. In these terms, $\sum_{\mu=1}^{M} Y^{a\mu}_{jO_j} P^{a\mu}_{jO_j} - V^a_{jO_j}$ stands for the shortened processing time of the final operation of each part, and the total cost for the shortening is represented by the fourth term. $15(Df^a_{jO_j} - Ds^a_{jO_j})$ calculates the non-working hours of the plant (17:00–8:00) in the days required for the final operation of each part. Because this value is shortened by overtime in the days before the day on which the operation is completed, $-Wp^a_{jO_j}$ is included in the second term. Similar representation is used in the third term. The fifth term of the function provides the total cost for overtime.

The constraint (2) imposes that the first operation of each part does not start earlier than its release date, and (3) that each part is completed by its due date, respectively. Inequality (4) assures required operations of a part are processed in the given sequence. Equations (5) and (6) are for ensuring that each operation is assigned to only one of the machines which can process it. Inequalities (7)–(9) impose that the precedence relationship between operations processed by the same machine coincide with the magnitude relationship between the starting and finishing times of those operations. The big-M method is used, and thus these inequalities are always satisfied by a sufficiently-large positive number B for operations which are not processed by the same machine. The pair of constraints (10) and (11) is to make S_{min} represents the value as defined, and the pair (12) and (13) plays the same role for C_{max}. Equations (14) and (15) are for ensuring the definition of Ws^a_{jo} and Wf^a_{jo}, respectively[1]. The inequality (16) assures the magnitude relationship between the starting date and the finishing date of each operation, and (17) imposes that the amount of shortening processing time of each operation does not exceed the maximum value. The constraints (18)–(21) are related to make the overtime acceptable and reasonable. Inequality (18) assures that each operation starts in the working hours including the upper limit of overtime (8:00–(17+W):00) of the starting date, (19) ensures the similar relationship about the finish of each operation, and (20) and (21) impose that the total overtime on each operation does not exceed the maximum amount of overtime accepted for it. Inequalities (22)–(29) ensure that the operations which need coupling are processed on the same condition.

In this formulation, weekends are not taken into account. Thus, a solution of this MILP represents a schedule as if a week is composed of five days from Monday to Friday. The schedule can be shaped into a schedule including weekends by inserting forty-eight hours at 24:00 of every Friday. Note that the value of release date Rd^a_j and due date Dd^a_j need to be given by reducing forty-eight hours for every weekend in advance. Furthermore, for simplicity, lunch time (12:00–13:00) and short breaks in the morning and the afternoon are ignored also. But, it is possible to consider those by including the time required for them into processing time of each operation in advance. It is also possible to take lunch break into consideration by assuming that a day is not composed of 24 h but 23 h and replacing "17" and "24" in the constraints with "16" and "23", respectively. In addition, since the obtained value of Wp^a_{jo} stands for the total value of overtime for each operation in the days before the day on which the operation is completed, it needs to be split and distributed to those days so that overtime hours on each day does not exceed the upper limit W.

[1] The inequalities (11) and (13) and the equations (14) and (15) are not linear, since they include the max and min functions. However, they can be equivalently transformed into linear forms by introducing additional binary variables.

5 Numerical Evaluation

The MILP model was applied to a very small-sized example of ETO production for large products as follows:

- the total number of production numbers: 1 ($A = 1$)
- the total number of parts for the production number: 2 ($J_a = 2$)
- the total number of machines: 38 ($M = 38$)
- the total number of required operations of the parts: 26 or 36 ($O_j^a = \{26, 36\}$)
- the number of machines that can process an operation: 1 or 2 (card $\overline{M}_{jo}^a = \{1, 2\}$)
- processing time of an operation: a multiple of 8 h from 8 to 280; the same value was given to the machines which can process the operation ($P_{jo}^{a\mu} = \{8, 16, 24, \ldots, 280\}$).
- the sum of the processing times for each part: approximately from 1200 to 1500 h (50–63 days)
- release date of a part: 0 ($Rd_j^a = 0$)
- due date of a part: 2500 ($Dd_j^a = 2500$)
- the upper limit of overtime on a day: 4 h ($W = 4$)
- the upper limit of shortening processing time of an operation ($\overline{V}_{jo}^a$): 20% of $P_{jo}^{a\mu}$ rounded off to the nearest whole number for the operations with $P_{jo}^{a\mu} > 8$; 0 for the operations with $P_{jo}^{a\mu} = 8$

All the weights in the objective functions were set to 1, and the costs C^v and C^w were set to 100.

This MILP was solved by using a commercial solver (IBM ILOG CPLEX Optimization Studio 22.1.0.0) and a generic workstation (Intel Core i7-9700K 3.60GHz, 64GB RAM), and the optimal solution was obtained in 2 sec. Due to the space limitation, a summary of the obtained schedule from the point of view of checking the characteristics C1-C4 is provided in this section. The optimal value of C_{max} was 2489. Table 1 shows the optimal values of V_{jo}^a and $Wp_{jo}^a + Wf_{jo}^a$ and the given values of $P_{jo}^{a\mu}$ for the machine μ which satisfies $Y_{jo}^{a\mu} = 1$. It can be confirmed that the degree of freedom on shortening processing time and overtime was actively taken advantage of for keeping the due dates (2500), especially in the operations which take long processing time. Table 2 shows the optimal values of Ds_{jo}^a, Df_{jo}^a and $S_{jo}^a - 24Ds_{jo}^a$ which is starting time of each operation in the Ds_{jo}^a-th day. The first operation of both of the parts starts at 8:00 in the first date of the planning horizon, the starting times of other operations were between 8:00 and $(17{+}W)$:00 of the day in which the operation starts. It is confirmed that a schedule considering the working hours can be generated. In this example, the 17, 18 and 19th operations of part 1 and the 27, 28 and 29th operations of part 2 require coupling. It can be confirmed also that the same values were given to those variables of these operations.

Table 1. The optimal values of V^a_{jo} and $Wp^a_{jo} + Wf^a_{jo}$ and the given values of $P^{a\mu}_{jo}$.

j	1																			
o	1	2	3	4	5	6	7	8	9	10	11	12	13	14	15	16	17	18	19	20
V^a_{jo}	5	24	19	45	0	7	0	6	5	1	4	0	0	4	0	19	3	6	0	0
$Wp^a_{jo} + Wf^a_{jo}$	1	24	23	0	0	0	0	8	0	4	0	0	0	0	0	12	4	5	0	0
$P^{a\mu}_{jo}$	24	120	96	280	8	80	8	32	32	32	24	16	56	56	16	96	16	32	8	16
o	21	22	23	24	25	26	–	–	–	–	–	–	–	–	–	–	–	–	–	–
V^a_{jo}	3	3	5	0	3	0	–	–	–	–	–	–	–	–	–	–	–	–	–	–
$Wp^a_{jo} + Wf^a_{jo}$	4	4	1	4	0	0	–	–	–	–	–	–	–	–	–	–	–	–	–	–
$P^{a\mu}_{jo}$	16	16	24	40	16	32	–	–	–	–	–	–	–	–	–	–	–	–	–	–
j	2																			
o	1	2	3	4	5	6	7	8	9	10	11	12	13	14	15	16	17	18	19	20
V^a_{jo}	10	19	16	26	18	0	3	8	8	8	13	14	5	3	5	10	10	5	3	0
$Wp^a_{jo} + Wf^a_{jo}$	2	23	19	30	20	4	3	8	12	5	14	16	8	4	0	12	2	4	1	0
$P^{a\mu}_{jo}$	48	96	80	128	88	8	16	40	40	40	64	72	24	16	24	48	48	24	16	16
o	21	22	23	24	25	26	27	28	29	30	31	32	33	34	35	36	–	–	–	–
V^a_{jo}	3	16	5	22	0	10	3	6	0	5	6	5	6	8	5	6	–	–	–	–
$Wp^a_{jo} + Wf^a_{jo}$	2	19	0	18	0	9	4	5	0	3	8	1	0	12	2	8	–	–	–	–
$P^{a\mu}_{jo}$	16	80	32	112	16	48	16	32	8	24	32	24	32	40	24	32	–	–	–	–

Table 2. The optimal values of Ds^a_{jo}, Df^a_{jo} and $S^a_{jo} - 24Ds^a_{jo}$.

j	1																			
o	1	2	3	4	5	6	7	8	9	10	11	12	13	14	15	16	17	18	19	20
$S^a_{jo} - 24Ds^a_{jo}$	8	17	8	8	9	8	9	17	8	8	8	10	8	10	8	15	8	8	11	10
Ds^a_{jo}	0	1	10	16	42	42	51	51	53	56	60	62	64	70	76	77	85	86	88	89
Df^a_{jo}	1	10	15	42	42	51	51	53	56	59	62	63	70	75	77	84	85	88	89	90
o	21	22	23	24	25	26	–	–	–	–	–	–	–	–	–	–	–	–	–	–
$S^a_{jo} - 24Ds^a_{jo}$	8	8	8	8	8	12	–	–	–	–	–	–	–	–	–	–	–	–	–	–
Ds^a_{jo}	91	92	93	95	99	100	–	–	–	–	–	–	–	–	–	–	–	–	–	–
Df^a_{jo}	91	92	95	98	100	103	–	–	–	–	–	–	–	–	–	–	–	–	–	–
j	2																			
o	1	2	3	4	5	6	7	8	9	10	11	12	13	14	15	16	17	18	19	20
$S^a_{jo} - 24Ds^a_{jo}$	8	8	8	8	8	13	8	9	15	8	8	9	15	8	8	9	8	8	14	8
Ds^a_{jo}	0	4	10	15	23	28	29	30	32	35	38	42	46	48	49	51	54	58	59	61
Df^a_{jo}	3	9	14	22	28	28	30	32	35	37	42	46	47	49	51	54	57	59	60	62
o	21	22	23	24	25	26	27	28	29	30	31	32	33	34	35	36	–	–	–	–
$S^a_{jo} - 24Ds^a_{jo}$	15	8	8	8	8	15	8	8	11	10	8	8	8	16	9	17	–	–	–	–
Ds^a_{jo}	62	64	69	72	80	81	85	86	88	89	91	93	95	97	100	101	–	–	–	–
Df^a_{jo}	63	68	71	79	81	84	85	88	89	91	93	94	97	100	101	103	–	–	–	–

6 Conclusion

ETO production of large-scale products takes a long lead time which cannot be estimated accurately in advance. It is hence difficult to apply the standard production planning strategy that divides the planning into three phases of different planning horizons to this type of production, because of large discrepancies among each of the phases. In order to reduce the discrepancies, a production planning framework that unifies the granularity of resources and unit time in all of the phases has been considered, and a model that is commonly used in the three phases was organized as flexible job-shops. This paper provided a MILP formulation of the production scheduling problem based on the model considering special characteristics of the target ETO production site, and the validity of the formulation was confirmed by a numerical experiment.

In this paper, the proposed mathematical model was applied simply to an example of the whole production for the sake of just evaluating its validity about a mathematical description. We have been developing a method for the long-term and medium-term planning [43,44], and it is necessary to consider how to incorporate the proposed model for production scheduling to those methods. In addition, application of the model to an industrial-size problem involves a huge computational load even if it is used for the short-term planning. It is required to develop a reasonable solution method for the model also. These issues will be discussed in future works.

References

1. Hirahara, Y., et al.: A study on sophisticated production management for engineer-to-order production. In: Proceedings of the Japan Joint Automatic Control Conference (2022). 2E2-1 (In Japanese). https://doi.org/10.11511/jacc.65.0_1156
2. Chaudhry, I.A., Khan, A.A.: A research survey: review of flexible job shop scheduling techniques. Int. Trans. Oper. Res. **23**(3), 551–591 (2016). https://doi.org/10.1111/itor.12199
3. Dauzère-Pérès, S., Ding, J., Shen, L., Tamssaouet, K.: The flexible job shop scheduling problem: a review. Eur. J. Oper. Res. **314**(2), 409–432 (2024). https://doi.org/10.1016/j.ejor.2023.05.017
4. Özgüven, C., Özbakir, L., Yavuz, Y.: Mathematical models for job-shop scheduling problems with routing and process plan flexibility. Appl. Math. Model. **34**(6), 1539–1548 (2010). https://doi.org/10.1016/j.apm.2009.09.002
5. Demir, Y., İşleyen, S.K.: Evaluation of mathematical models for flexible job-shop scheduling problems. Appl. Math. Model. **37**(3), 977–988 (2013). https://doi.org/10.1016/j.apm.2012.03.020
6. Pezzella, F., Morganti, G., Ciaschette, G.: A genetic algorithm for the flexible job-shop scheduling problem. Comput. Oper. Res. **35**(10), 3202–3212 (2008). https://doi.org/10.1016/j.cor.2007.02.014
7. Roshanaei, V., Azab, A., ElMaraghy, H.: Mathematical modelling and a meta-heuristic for flexible job shop scheduling. Int. J. Prod. Res.**51**(20), 6247–6274 (2013). https://doi.org/10.1080/00207543.2013.827806

8. Gaham, M., Bouzouia, B., Achour, N., Tebani, K.: Meta-heuristics approaches for the flexible job shop scheduling problem. In: Talbi, E.-G., Yalaoui, F., Amodeo, L. (eds.) Metaheuristics for Production Systems. ORSIS, vol. 60, pp. 285–314. Springer, Cham (2016). https://doi.org/10.1007/978-3-319-23350-5_13
9. Kaweegitbundit, P., Eguchi, T.: Flexible job shop scheduling using genetic algorithm and heuristic rules. J. Adv. Mech. Des. Syst. Manuf. **10**(1), JAMDSM0010 (2016). https://doi.org/10.1299/jamdsm.2016jamdsm0010
10. Zaidi, M., Amirat, A., Jarboui, B., Yahyaoui, A.: A hybrid meta-heuristic to solve flexible job shop scheduling problem. In: Alharbi, I., Ben Ncir, CE., Alyoubi, B., Ben-Romdhane, H. (eds.) Advances in Computational Logistics and Supply Chain Analytics. Unsupervised and Semi-Supervised Learning. Springer, Cham (2024). https://doi.org/10.1007/978-3-031-50036-7_4
11. Song, W., Chen, X., Li, Q., Cao, Z.: Flexible job-shop scheduling via graph neural network and deep reinforcement learning. IEEE Trans. Industr. Inf. **19**(2), 1600–1610 (2023). https://doi.org/10.1109/TII.2022.3189725
12. Pan, F.S., Ye, C.M., Yang, J.: Flexible job-shop scheduling problem under uncertainty based on QPSO algorithm. Adv. Mater. Res. **605**, 487–492 (2012). https://doi.org/10.4028/www.scientific.net/amr.605-607.487
13. Li, W.-L., Murata, T., Bin Md Fauadi, M.H.F.: Robust scheduling for flexible job-shop problems with uncertain processing times. IEEJ Trans. Electron. Inf. Syst. **135**(6), 713–720 (2015). https://doi.org/10.1541/ieejeiss.135.713
14. Chen, N., Xie, N., Wang, Y.: An elite genetic algorithm for flexible job shop scheduling problem with extracted grey processing time. Appl. Soft Comput. **131**, 109783 (2022). https://doi.org/10.1016/j.asoc.2022.109783
15. He, W., Sun, D.: Scheduling flexible job shop problem subject to machine breakdown with route changing and right-shift strategies. Int. J. Adv. Manuf. Technol. **66**(1–4), 501–514 (2013). https://doi.org/10.1007/s00170-012-4344-4
16. Sun, D.-H., He, W., Zheng, L.-J., Liao, X.-Y.: Scheduling flexible job shop problem subject to machine breakdown with game theory. Int. J. Prod. Res. **52**(13), 3858–3876 (2013). https://doi.org/10.1080/00207543.2013.784408
17. Ayyoubzadeh, B., Ebrahimnejad, S., Bashiri, M., Baradaran, V., Hosseini, S.M.H.: A reactive approach for flexible job shop scheduling problem with tardiness penalty under uncertainty. Sci. Iranica **29** (2022). https://doi.org/10.24200/sci.2022.58491.5754
18. Zhang, G., Lu, X., Liu, X., Zhang, L., Wei, S., Zhang, W.: An effective two-stage algorithm based on convolutional neural network for the bi-objective flexible job shop scheduling problem with machine breakdown. Expert Syst. Appl. **203**, 117460 (2022). https://doi.org/10.1016/j.eswa.2022.117460
19. Deng, G., Zhang, Z., Jiang, T., Zhang, S.: Total flow time minimization in no-wait job shop using a hybrid discrete group search optimizer. Appl. Soft Comput. **81**, 105480 (2019). https://doi.org/10.1016/j.asoc.2019.05.007
20. Lee, S., Moon, I., Bae, H., Kim, J.: Flexible job-shop scheduling problems with 'AND'/'OR' precedence constraints. Int. J. Prod. Res. **50**(7), 1979–2001 (2012). https://doi.org/10.1080/00207543.2011.561375
21. Mason, S.J., Fowler, J.W., Matthew, C.W.: A modified shifting bottleneck heuristic for minimizing total weighted tardiness in complex job shops. J. Sched. **5**(3), 247–262 (2002). https://doi.org/10.1002/jos.102
22. Xing, L.N., Chen, Y.W., Yang, K.W.: Double layer ACO algorithm for the multi-objective FJSSP. N. Gener. Comput. **26**, 313–327 (2008). https://doi.org/10.1007/s00354-008-0048-6

23. Davarzani, Z., Akbarzadeh-T, M.-R, Khairdoost, N.: Multiobjective artificial immune algorithm for flexible job shop scheduling problem. Int. J. Hybrid Inf. Technol. **5**(3), 75–88 (2012). https://doi.org/10.14257/ijhit.2012.5.3.08
24. Gao, K.Z., Suganthan, P.N., Pan, Q.K., Chua, T.J., Cai, T.X., Chong, C.S.: Pareto-based grouping discrete harmony search algorithm for multi-objective flexible job shop scheduling. Inf. Sci. **289**, 76–90 (2014). https://doi.org/10.1016/j.ins.2014.07.039
25. Zambrano Rey, G., Bekrar, A., Trentesaux, D., Zhou, B.-H.: Solving the flexible job-shop just-in-time scheduling problem with quadratic earliness and tardiness costs. Int. J. Adv. Manuf. Technol. **81**(9), 1871–1891 (2015). https://doi.org/10.1007/s00170-015-7347-0
26. Gomes, M.C., Barbosa-Póvoa, A.P., Novais, A.Q.: Reactive scheduling in a make-to-order flexible job shop with re-entrant process and assembly: a mathematical programming approach. Int. J. Prod. Res. **51**(17), 5120–5141 (2013). https://doi.org/10.1080/00207543.2013.793428
27. Shen, L., Dauzère-Pérès, S., Maecker, S.: Energy cost efficient scheduling in flexible job-shop manufacturing systems. Eur. J. Oper. Res. **310**(3), 992–1016 (2023). https://doi.org/10.1016/j.ejor.2023.03.041
28. Raaymakers, W.H.M., Hoogeveen, J.A.: Scheduling multipurpose batch process industries with no-wait restrictions by simulated annealing. Eur. J. Oper. Res. **126**(1), 131–151 (2000). https://doi.org/10.1016/S0377-2217(99)00285-4
29. Alvarez-Valdes, R., Fuertes, A., Tamarit, J.M., Giménez, G., Ramos, R.: A heuristic to schedule flexible job-shop in a glass factory. Eur. J. Oper. Res. **165**(2), 525–534 (2005). https://doi.org/10.1016/j.ejor.2004.04.020
30. Aschauer, A., Roetzer, F., Steinboeck, A., Kugi, A.: Scheduling of a flexible job shop with multiple constraints. IFAC-PapersOnLine **51**(11), 1293–1298 (2018). https://doi.org/10.1016/j.ifacol.2018.08.354
31. Tamssaouet, K., Dauzère-Pérès, S., Knopp, S., Bitar, A., Yugma, C.: Multiobjective optimization for complex flexible job-shop scheduling problems. Eur. J. Oper. Res. **296**(1), 87–100 (2022). https://doi.org/10.1016/j.ejor.2021.03.069
32. Mousakhani, M.: Sequence-dependent setup time flexible job shop scheduling problem to minimise total tardiness. Int. J. Prod. Res. **51**(12), 3476–3487 (2013). https://doi.org/10.1080/00207543.2012.746480
33. Allahverdi, A.: The third comprehensive survey on scheduling problems with setup times/costs. Eur. J. Oper. Res. **246**(2), 345–378 (2015). https://doi.org/10.1016/j.ejor.2015.04.004
34. Winklehner, P., Hauder, V.A.: Flexible job-shop scheduling with release dates, deadlines and sequence dependent setup times: a real-world case. Procedia Comput. Sci. **200**, 1654–1663 (2022). https://doi.org/10.1016/j.procs.2022.01.366
35. Rossi, A., Dini, G.: Flexible job-shop scheduling with routing flexibility and separable setup times using ant colony optimisation method. Robot. Comput. Integr. Manuf. **23**(5), 503–516 (2007). https://doi.org/10.1016/j.rcim.2006.06.004
36. Özgüven, C., Yavuz, Y., Özbakir, L.: Mixed integer goal programming models for the flexible job-shop scheduling problems with separable and non-separable sequence dependent setup times. Appl. Math. Model. **36**(2), 846–858 (2012). https://doi.org/10.1016/j.apm.2011.07.037
37. Li, X., et al.: Survey of integrated flexible job shop scheduling problems. Comput. Ind. Eng. **174**, 108786 (2022). https://doi.org/10.1016/j.cie.2022.108786
38. Baykasoğlu, A.: Linguistic-based meta-heuristic optimization model for flexible job shop scheduling. Int. J. Prod. Res. **40**(17), 4523–4543 (2002). https://doi.org/10.1080/00207540210147043

39. Baykasoğlu, A., Özbakir, L., Sönmez, A.İ: Using multiple objective tabu search and grammars to model and solve multi-objective flexible job shop scheduling problems. J. Intell. Manuf. **15**(6), 777–785 (2004). https://doi.org/10.1023/B:JIMS.0000042663.16199.84
40. Baykasoğlu, A., Madenoğlu, F.S.: Greedy randomized adaptive search procedure for simultaneous scheduling of production and preventive maintenance activities in dynamic flexible job shops. Soft. Comput. **25**(23), 14893–14932 (2021). https://doi.org/10.1007/s00500-021-06053-0
41. Ghorbani Saber, R., Leyman, P., Aghezzaf, E.-H.: Mixed integer programming for integrated flexible job-shop and operator scheduling in flexible manufacturing systems. In: Alfnes, E., Romsdal, A., Strandhagen, J.O., von Cieminski, G., Romero, D. (eds.) Advances in Production Management Systems. Production Management Systems for Responsible Manufacturing, Service, and Logistics Futures: IFIP WG 5.7 International Conference, APMS 2023, Trondheim, Norway, September 17–21, 2023, Proceedings, Part IV, pp. 460–470. Springer Nature Switzerland, Cham (2023). https://doi.org/10.1007/978-3-031-43688-8_32
42. Morinaga, E., Sakaguchi, Y., Wakamatsu, H., Arai, E.: A method for flexible job-shop scheduling considering workers and teams. In: Proceedings of International Conference on Leading Edge Manufacturing in 21st century : LEM21, 126 (2017). https://doi.org/10.1299/jsmelem.2017.9.126
43. Hirahara, Y., et al.: A study on sophisticated production management for engineer-to-order production—A load planning method based on integer programming problem for long-term production plan. In: Proceedings of the Scheduling Symposium (2023). GS6-1. (In Japanese)
44. Iwamura, K., et al.: A study on hierarchical production planning framework for engineer-to-order production of large products. In: Proceedings of International Symposium on Flexible Automation (2024). Accepted

Three-Dimensional Bin Packing Problems with the Operating Time of a Robot Manipulator

Naoya Mikyu, Tatsushi Nishi(✉), Ziang Liu, and Tomofumi Fujiwara

Faculty of Environmental, Life, Natural Science and Technology, Okayama University, Okayama, Japan
nishi.tatsushi@okayama-u.ac.jp

Abstract. Many optimization algorithms for solving three-dimensional bin-packing problems have been addressed. The problem is to minimize the cost of the bins in which packages are placed with the aim of reducing transportation costs and improving productivity. From the viewpoint of actual optimization of the entire factory, it is necessary to consider the motion planning of a robot manipulator to handle the items in the three-dimensional bin packing problem. This study considers a three-dimensional robotic bin packing problem, which minimizes the weighted sum of bin cost and the robot's operating time. A genetic algorithm is developed using Sequence-Triple representation, a layout representation of a three-dimensional arrangement. For robot motion planning, the Rapidly Exploring Random Tree Star is used for trajectory generation to obtain near-optimal solutions. Computational experiments show the superiority of the proposed optimization method by comparing it with a conventional sequential optimization method.

Keywords: Packing problem · Sequence-Triple · motion planning

1 Introduction

Japan has experienced a rapidly aging society with a declining birthrate, and there is concern about a decrease in the productive workforce. In addition, many production sites in the manufacturing industry are changing their production methods from high-mix low-volume production to variable-mix, variable-volume production that can meet the various needs of customers. To solve these problems, it is necessary to promote manpower and labor-saving by replacing tasks previously performed by humans with robots. Furthermore, introducing robots will make it easier to manage the entire factory, leading to higher productivity.

Recently, robot arms have been introduced to sort goods to improve the efficiency of factory operations. The robot arm picks goods flowing on a conveyor belt according to the region and customer to which they are to be transported, and packs them into containers. In this way, goods can be transported in batches reducing transportation costs and improving productivity. Furthermore, further efficiency can be expected by

Published by Springer Nature Switzerland AG 2024
M. Thürer et al. (Eds.): APMS 2024, IFIP AICT 729, pp. 44–60, 2024.
https://doi.org/10.1007/978-3-031-65894-5_4

maximizing the use of the space inside the container in the packing process. The problem of improving the filling rate of containers and minimizing the total cost of containers is called the bin packing problem. This problem is known to be NP-hard, and there is currently no polynomial-time algorithm to obtain an exact solution [1].

Many previous studies have proposed new methods to solve the three-dimensional bin packing problem; Hong et al. proposed an online packing system designed to recognize packages in real time and accurately place them inside containers [2]. Seda et al. proposed a heuristic based on a novel genetic algorithm that focuses on three factors: family unity, package balance, and minimizing the number of bins [3]. Furthermore, Zhao et al. formulated the online three-dimensional bin packing problem as a constrained Markov decision process and used deep reinforcement learning as a solution method [4]. These three-dimensional bin-packing problems have the common goal of minimizing the number of bins and do not consider the motion time of the picking robot. However, in the current logistics system, many boxes need to be packed, and thus the robot's operating time needs to be reduced. The following are some studies that consider the robot's operating time.

Weng et al. proposed a framework for the three-dimensional robot bin-packing problem using dual arms to reduce the time required for packing packages [5]. Kawabe et al. proposed a layout planning method using Sequence-Pair, which determines the relative positions of packages [6]. Since Sequence-Pair is a two-dimensional layout representation, their method cannot be applied to three-dimensional problems.

In this study, a novel genetic algorithm (GA) is proposed to optimize both the bin cost and the robot's operating time in the three-dimensional bin-packing problem. Each candidate solution in GA is encoded as three permutations of the Sequence-Triple and a variable for bin selection. Sequence-Triple is a three-dimensional layout representation method. We also use random forests and RRT* (Rapidly Exploring Random Tree Star) to construct a surrogate model that can estimate the robot's operating time in less computation time required to obtain the solution. Then, we examine the efficiency of the proposed method by comparing it with a bi-level optimization method using Gurobi.

2 Three-Dimensional Bin Packing Problems with Robot Operating Time

2.1 Problem Description

We consider a three-dimensional space with a robot, a conveyor belt, rectangular packages, and bins (see Fig. 1) in the optimization problem. In the virtual space, the packages are flowing on a conveyor belt, and the robot has the task of picking the packages at a certain point and distributing them into a bin located next to the single robot manipulator. The packages are arrived in a given order. In this experiment, the bottle is a cube-shaped container. There are multiple bins, and multiple bins may be used to pack all the packages into multiple bins. The objective is to minimize the weighted sum of two objective functions: the cost of the bins used, and the robot operating time required for the robot to pack all the packages. The coordinates of containers and the trajectory of the robot manipulator should be determined.

In this study, the following constraints are considered for the optimization problem.

Constraints

- The packages must be placed so that they do not overlap each other in three-dimensional space.
- Each bin has capacity. All packages must be assigned to bins.
- The package must fit in the bin.
- The packages must be within the robot's range of motion.

In addition, the following assumptions are made in the problem set-up of this study.

Assumptions

- The size of the package to be placed is known.
- The size of the packages and bins to be placed are rectangular.
- The order of placement is given from the beginning.
- The packages to be placed in the same bin flow continuously on the conveyor belt. In this case, the packages placed at the bottom of the bin are first placed on the belt conveyor.
- The packages are placed on the conveyor belt first, starting with the package placed at the bottom.
- The robot picks the bins at a fixed point on the conveyor.
- The robot should not take a long time to change bins.

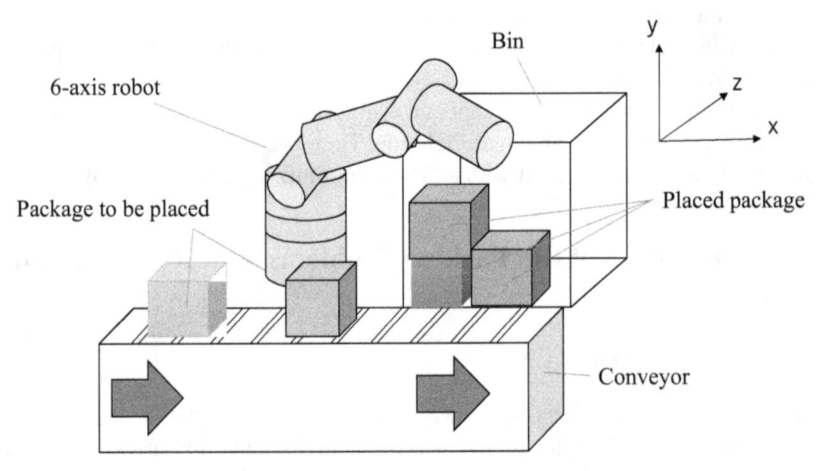

Fig. 1. Example of a three-dimensional Robot Bin Packing Problem

2.2 Problem Formulation

Let the width w_i, height h_i, depth d_i of each rectangular package $i \in I$ contained in the packages set $I = \{1,2,\ldots,n\}$ and the width W_j, height H_j, depth D_j, and cost C_j of each bin $j \in B$ contained in bin set $B = \{1,2,\ldots,u\}$. In addition, each package and each bin are a rectangle and determine $k \in I(k \neq i)$. Next, we define the coordinates $\boldsymbol{x} = [x_1,\ldots,x_n], \boldsymbol{y} = [y_1,\ldots,y_n], \boldsymbol{z} = [z_1,\ldots,z_n]$ of the vertex at which the sum of the x, y, z coordinate values of each package is the smallest. Also, determine the 0–1 variable s_{ij}, which is 1 when package i is packed in bin j and 0 otherwise, the 0–1 variable t_j, which indicates whether bin j can be used, and the 0–1 variable v_{ik}, which is 1 when package i and package k are packed in the same bin and 0 otherwise. The constants p and q and the large constant M are given in advance in the problem. Furthermore, we determine the robot's motion time O, the simulation part θ, and a 0–1 variable p_{ik}^r that is 1 when package i is in the r-direction ($r \in R = \{left, right, below, above, front, back\}$) of package k. The objective function and constraints of the three-dimensional robot bin-packing problem are then defined as follows.

$$\text{Minimize} \sum_{j=1}^{u} pC_j t_j + qE\left[O(\boldsymbol{x},\boldsymbol{y},\boldsymbol{z},\theta)\right] \tag{1}$$

s.t.

$$\sum_{j=1}^{u} s_{ij} = 1 \tag{2}$$

$$x_i - M\left(1 - s_{ij}\right) \leq W_j - w_i \tag{3}$$

$$y_i - M\left(1 - s_{ij}\right) \leq H_j - h_i \tag{4}$$

$$z_i - M\left(1 - s_{ij}\right) \leq D_j - d_i \tag{5}$$

$$\sum_{r \in R} p_{ik}^r \geq 1 \tag{6}$$

$$p_{ik}^{left} = 1 \Rightarrow x_i + w_i \leq x_k + M(1 - v_{ik}) \tag{7}$$

$$p_{ik}^{right} = 1 \Rightarrow x_k + w_k \leq x_i + M(1 - v_{ik}) \tag{8}$$

$$p_{ik}^{below} = 1 \Rightarrow y_i + h_i \leq y_k + M(1 - v_{ik}) \tag{9}$$

$$p_{ik}^{above} = 1 \Rightarrow y_k + h_k \leq y_i + M(1 - v_{ik}) \tag{10}$$

$$p_{ik}^{front} = 1 \Rightarrow z_i + d_i \leq z_k + M(1 - v_{ik}) \tag{11}$$

$$p_{ik}^{back} = 1 \Rightarrow z_k + d_k \leq z_i + M(1 - v_{ik}) \tag{12}$$

$$v_{ik} = v_{ki} \tag{13}$$

$$p_{ik}^{left} = p_{ki}^{right} \quad (14)$$

$$p_{ik}^{below} = p_{ki}^{above} \quad (15)$$

$$p_{ik}^{front} = p_{ki}^{back} \quad (16)$$

$$\frac{1}{n}\sum_{i=1}^{n} s_{ij} \leq t_j < \frac{1}{n}\sum_{i=1}^{n} s_{ij} + 1 \quad (17)$$

$$s_{ij}, t_j, v_{ik}, p_{ik}^{r} \in \{0,1\} \quad (18)$$

$$0 \leq x_i, y_i, z_i \quad (19)$$

Equation (1) is the objective function to be minimized. Equation (2) ensures that each package is packed in only one of the bins. Equations (3) to (5) ensure that each package does not exceed the maximum capacity of a bin. Equations (6) through (12) are the constraints to prevent two pieces of package from overlapping. Equations (7) through (12) contain logical relations, so when solving this problem in the solver, it is necessary to convert them into simultaneous equations and inequalities. Equations (13) through (16) are constraints on baggage positioning. Equations (17) to (19) are the variable domain constraints. To satisfy the above constraints guarantees that all packages are packed in the bins.

3 Framework of Proposed Method

3.1 Overall Algorithm

Figure 2 illustrates the flow of the proposed method addressed in this study. The proposed method consists of two main components: creation of a surrogate model and execution of a GA. The surrogate model to estimate the robot operating time from the given 3D layout. It is used to reduce the computational cost of the robot's operating time, and the robot's operating time can be estimated from the angle difference between the robot's initial and target posture at each joint. There are two main reasons for using GA in this study. The first is to obtain a globally optimal solution. The second reason is that GA does not require the introduction of complex constraint formulas, as in mathematical programming, when determining the placement relationship of package, and thus Sequence-Triple can be easily used. The proposed algorithm consists of the following steps

Step 1: A surrogate model is constructed by motion planning using RRT* and learning from the obtained data using random forests.
Step 2: Generate a set of solutions in the form of Sequence-Triple-Bin in the initialization. Sequence-Triple is a representation of three-dimensional layout of packages. The detail of Sequence-Triple-Bin is explained in Sect. 3.2.
Step 3: The critical path length is determined by the Sequence-Triple placement rule, which provides a layout representation of the components in each bin.

Step 4: Based on the critical path length, the cost of bins is determined if the components fit into bins.
Step 5: A surrogate model is applied to the obtained layout to derive the robot's operating time and determine its feasibility.
Step 6: Crossover and mutation are performed on the solutions, and the more adaptive solution is selected for the next generation.
Step 7: If the termination condition is not met, steps 3 through 7 are performed again. Once the termination conditions are met, motion planning and simulation are performed to determine the exact robot motion time.

The flow of the proposed GA method is shown in Fig. 2.

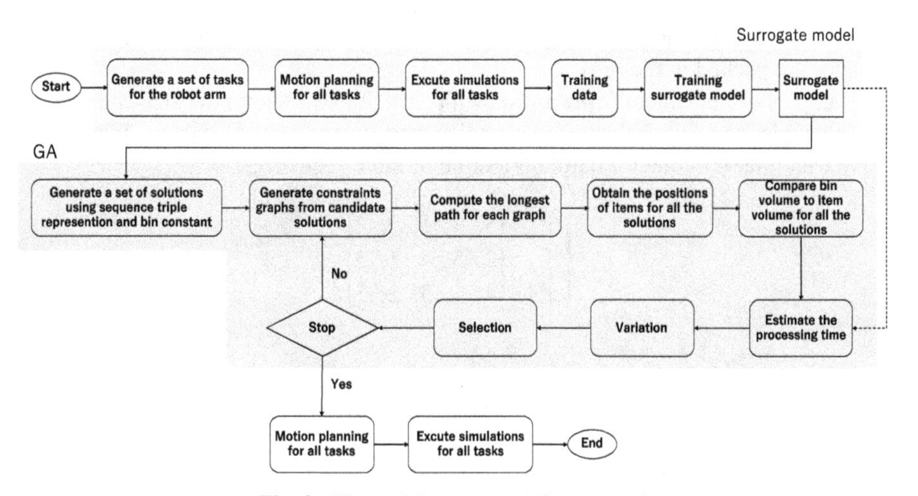

Fig. 2. Flow of the proposed GA method

3.2 Sequence-Triple

Sequence-Triple is an extension of Sequence-Pair, a method for representing two-dimensional arrangements of rectangles to three dimensions [7]. As an extension of this method, Yamazaki et al. proposed the Sequence-Triple representation method, which uses three permutations to represent the three-dimensional arrangement of rectangles [1].

Sequence-Triple is a method to represent a rectangular arrangement by assigning top-bottom, left-right, and front-back positions to rectangular objects in the three permutations Γ_1, Γ_2 and Γ_3. Table 1 shows the decoding rules for Sequence-Triple.

Since this study assumes that the bins are filled with rectangular packages, it is necessary to minimize the layout volume while satisfying the relative positions of the packages using Sequence-Triple. Therefore, the bins are placed as close as possible to one of the rectangles.

Table 1. Sequence-Triple decoding rules

	x-direction	y-direction	z-direction
Γ_1	....	...$\boldsymbol{i}$...j...	...j...$\boldsymbol{i}$...
Γ_2	...$\boldsymbol{i}$...j...	...j...$\boldsymbol{i}$...	...j...$\boldsymbol{i}$...
Γ_3	...j...$\boldsymbol{i}$...	...j...$\boldsymbol{i}$...	...j...$\boldsymbol{i}$...
arrangement	i is right to j	i is above j	i is in the rear of j

One of the characteristics of the Sequence-Triple arrangement is the critical path. A critical path is the longest path in each direction when the path length is the sum of the lengths of the rectangles in the path. This makes it possible to determine if all the packages are packed in the bins.

Examples of Sequence-Triple are shown below. Table 2 shows the size and color of each rectangle, and Fig. 3 shows an example arrangement using Sequence-Triple.

An example of Sequence-Triple is a tuple of three sequences:

$$\begin{cases} \Gamma_1 : [5,\ 4,\ 1,\ 3,\ 2] \\ \Gamma_2 : [2,\ 5,\ 4,\ 1,\ 3] \\ \Gamma_3 : [4,\ 1,\ 3,\ 2,\ 5] \end{cases}$$

Table 2. The size and color of each rectangle,

number	x-axis:width (mm)	y-axis:hight (mm)	z-axis:depth(mm)	color
1	20	30	30	blue
2	30	20	30	pink
3	30	30	20	purple
4	30	30	30	black
5	20	20	20	yellow

From the decoding rules in Table 2, rectangles 2 and 5 are placed to the right of rectangles 1, 3, and 4. Similarly, rectangle 5 is placed above rectangle 2, and rectangle 3 is placed further behind rectangle 1 behind rectangle 4. The final arrangement is the one in which all the rectangles are brought together into a black rectangle, keeping this positional relationship satisfied.

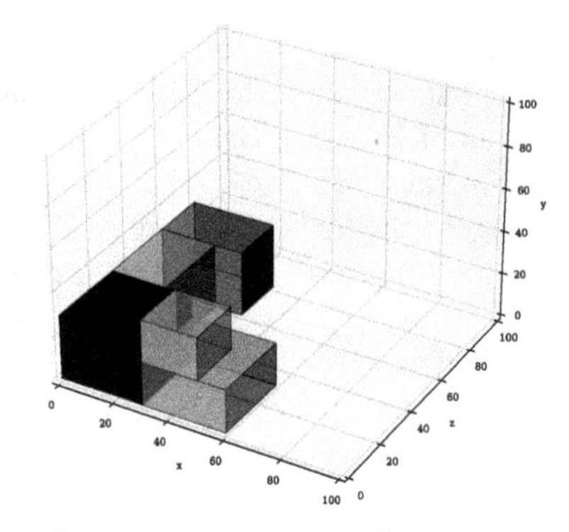

Fig. 3. Example arrangement using Sequence-Triple

To handle the three-dimensional bin packing problem, we modified Sequence-Triple-Bin by incorporating a bin selection element into the conventional Sequence-Triple. When the number of bins is u, the relationship between each rectangle and bin is represented by introducing $u - 1$ (individual) '-1' at random positions for the permutations in Sequence-Triple.

Figure 4 shows an example of a modified Sequence-Triple. This figure shows an example of Sequence-Triple-Bin with 6 rectangles and 5 bins, showing the packing of rectangles 5, 6 and 2 into bin 2, rectangle 1 into bin 3, and rectangles 4 and 3 into bin 5.

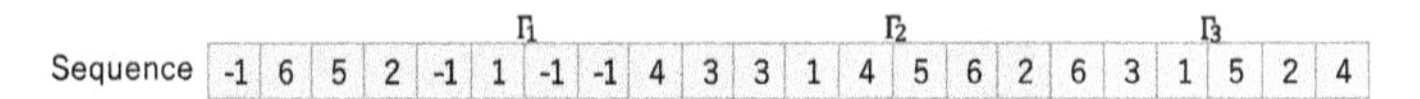

Fig. 4. Example of Sequence-Triple-Bin

3.3 Motion Planning

This section describes Rapidly Exploring Random Trees Star (RRT*) used for robot motion planning. RRT* guarantees a certain degree of optimality because the shortest path to an existing node is recomputed after each search. RRT* also has the advantages of considering restrictions and obstacles, and of obtaining a path between initial and target postures by simply specifying the initial posture and the target posture. However, RRT* has the disadvantage of requiring about twice as much computation as RRT.

Figure 5 shows an overview of the RRT* algorithm. The flow of the RRT* algorithm is shown in Algorithm 1 [6]. First, a tree $\mathcal{T} = (V, E)$ is constructed as in the RRT algorithm, where V are nodes and E are arcs. Next, a point is randomly selected in the space and designated z_{rand}. Then, the node closest to z_{rand} is $z_{nearest}$, and z_{new} is created at a fixed distance ε from $z_{nearest}$ on the line connecting $z_{nearest}$ and z_{rand}. Furthermore, the interior of a sphere of radius r centered at z_{new} is searched, and the set of nodes contained in the sphere is denoted by Z_{near}. Finally, the node with the lowest cost in the set Z_{near} is selected and connected to z_{new}.

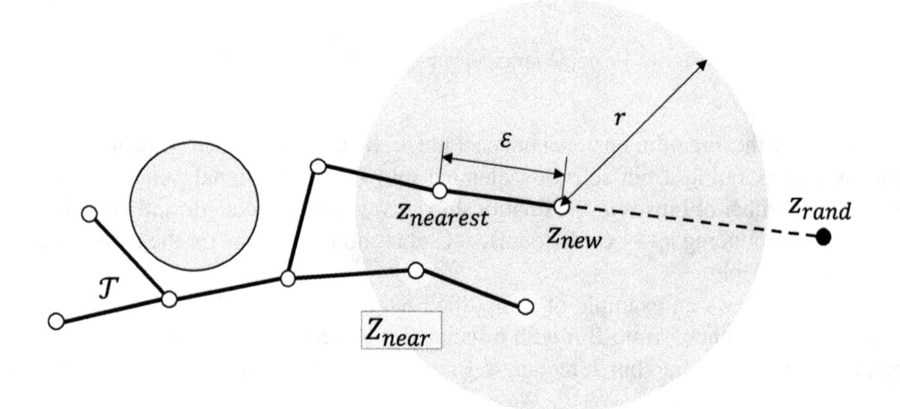

Fig. 5. Overview of RRT* Algorithm

The overall algorithm of the RRT* algorithm is written by Algorithm 1 where Algorithm 2 and Algorithm 3 are the functions of Algorithm 1. The pseudo-codes of those algorithms are shown as follows.

Algorithm 1 $\mathcal{T} = (V, E) \leftarrow$ **RRT*(** z_{init}**)**

1: $\mathcal{T} \leftarrow$ InitializeTree();
2: $\mathcal{T} \leftarrow$ InsertNode($\emptyset$, z_{init}, $\mathcal{T}$);
3: **for** $i = 1$ *to* $i = N$ **do**
4: $z_{rand} \leftarrow$ Sample(i);
5: $z_{nearest} \leftarrow$ Nearest($\mathcal{T}$, z_{rand});
6: $(x_{new}, u_{new}, T_{new}) \leftarrow$ Steer($z_{nearest}$, z_{rand});
7: **if** Obstaclefree(x_{new}) **then**
8: $Z_{near} \leftarrow$ Near($\mathcal{T}$, z_{rand}, $|V|$);
9: $z_{min} \leftarrow$ Chooseparent(z_{near}, $z_{nearest}$, z_{new}, x_{new});
10: $\mathcal{T} \leftarrow$ InsertNode(z_{min}, x_{new}, $\mathcal{T}$);
11: $\mathcal{T} \leftarrow$ Rewire($\mathcal{T}$, Z_{near}, z_{min}, z_{new});
12: **end for**
13: **end for**
14: **return** $\mathcal{T}$

Algorithm 2 $z_{min} \leftarrow$ **ChooseParent(** Z_{near}, $z_{nearest}$, z_{new}, x_{new}**)**

1: $z_{min} \leftarrow z_{nearest}$;
2: $c_{min} \leftarrow$ Cost $(z_{nearest}) + c(x_{new})$
3: **for** $z_{near} \in Z_{near}$ **do**
4: $(x', u', T') \leftarrow$ Steer(z_{near}, z_{new});
5: **if** ObstacleFree (x') *and* $x'(T') = z_{new}$ **then**
6: $c' =$ Cost $(z_{near}) + c(x')$;
7: **if** $c' <$ Cost (z_{new}) *and* $c' < c_{min}$ **then**
8: $z_{min} \leftarrow z_{near}$;
9: $c_{min} \leftarrow c'$;
10: **end if**
11: **end if**
12: **end for**
13: **return** z_{min}

Algorithm 3 $\mathcal{T} \leftarrow$ **ReWire(**$\mathcal{T}$, Z_{near}, z_{min}, z_{new}**)**

1: **for** $z_{near} \in Z_{near} \setminus \{z_{min}\}$ **do**
2: $(x', u', T') \leftarrow$ Steer (z_{new}, z_{near});
3: **if** ObstacleFree (x') *and* $x'(T') = z_{near}$ *and* Cost$(z_{new}) + c(x') <$ Cost(z_{near}) **then**
4: $\mathcal{T} \leftarrow$ ReConnect(z_{min}, z_{near}, $\mathcal{T}$);
5: **end if**
6: **end for**
7: **return** $\mathcal{T}$

3.4 Surrogate Model

Surrogate model is a model for estimating the robot's operating time. When estimating the robot's operating time, the RRT* algorithm for each solution is computationally expensive. Therefore, in this study, the surrogate model is constructed in the following steps to estimate the robot's operating time based on the angle difference between the robot's initial and target postures at each joint. A surrogate model is constructed by the following steps.

- Step 1: Generate random initial and target postures.
- Step 2: Motion planning is performed using RRT*, and the robot's motion time is calculated.
- Step 3: The angle difference between the initial and target postures and the robot's motion time are stored as sample data.
- Step 4: Repeat Steps 1 to 3 until the end condition is reached.
- Step 5: Construct a surrogate model from the sample data using a random forest.

In this study, 10,000 sample data were collected, 8,000 were used as training data and 2,000 as test data. The model trained using random forests has an accuracy of $R^2 = 0.98$, which is expected to be accurate enough to estimate the robot's motion time from joint angles.

3.5 Proposed Genetic Algorithm Using Sequence-Triple-Bin

GA is a meta-heuristic algorithm proposed by John Henry Holland in 1975 to search for approximate solutions. In GA, multiple individuals are prepared with genetic representations of the data, and individuals with high adaptability are preferentially selected for crossover and mutation operations to search for solutions. Figure 6 shows the flow of a GA.

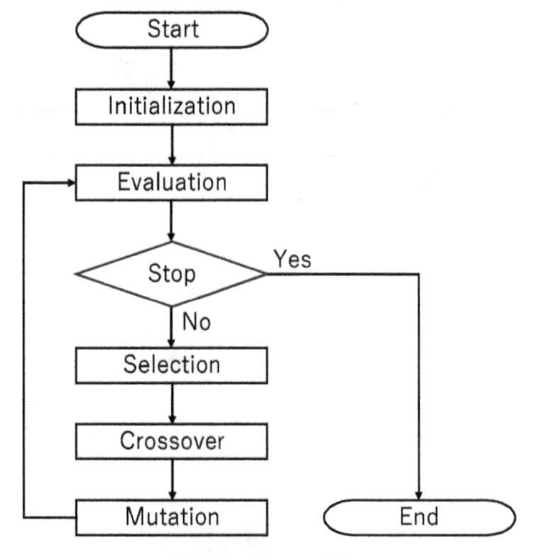

Fig. 6. Flow of GA

The detailed procedure of the GA method is described by the following steps.

Step 1: Initial population generation

The parent population of the initial set of individuals of size N is randomly generated. In this study, Sequence-Triple-Bin is used as the data, and the number of initial individuals is set to 200.

Step 2: Valuation of adaptation

Adaptation is evaluated for all individuals of the generation. The objective function in this study is a weighted sum of "bin cost" and "robot's operating time". "bin cost" is predetermined by the size of the bin, and the larger the bin, the higher the bin cost. In other words, the higher the bin cost, the more freely the rectangles can be placed and the shorter the robot's operating time. From this, it is expected that there is a trade-off between the bin cost and the robot's operating time.

Step 3: Selection

Individuals to form the next generation population are selected based on the degree of adaptation. Selection algorithms include roulette selection, ranking selection, and tournament selection. In this study, we use tournament selection. In tournament selection, a predetermined number of individuals are randomly selected from the population, and the individual with the highest fitness is selected.

Step 4: Crossover

Crossover is a method to obtain individuals that are closer to the optimal solution by replacing some of the genes of the individuals. There are four types of crossovers: one-point crossover, two-point crossover, uniform crossover, and order crossover. In this study, order crossover is used. Order crossover is a mating method proposed by Davis, in which a portion of the mutant is inherited directly from one parent, while the relative order of the remaining parts is preserved [8].

Step 5: Mutation

Mutation is an operation that changes a part of the genes of an individual, with the effect of preventing it from falling into a locally optimal solution. There are also two types of mutation

Interchange: Two randomly selected values are interchanged with each other.

Insertion: A randomly selected value is taken out, its position is filled, and inserted into the position of another randomly selected value, delaying the subsequent order by one.

The termination condition in this study is that the number of generations reaches a specified number.

Sequence-Triple crossovers and mutations used in previous studies are listed below.

- Crossover

Figure 7 shows an ordered crossover in Sequence-Triple: In Sequence-Triple, two points (start and end) are determined randomly, and the string between start and end is transferred from one parent to the child. Next, the other parent inserts the string from the top while maintaining the order of the parts other than the selected string. Then, the next parent to be selected is switched, and the same procedure is performed.

- Mutation

Figure 8 shows mutation in Sequence-Triple. It employs the method of exchange, randomly replacing two values in the Sequence-Triple permutation.

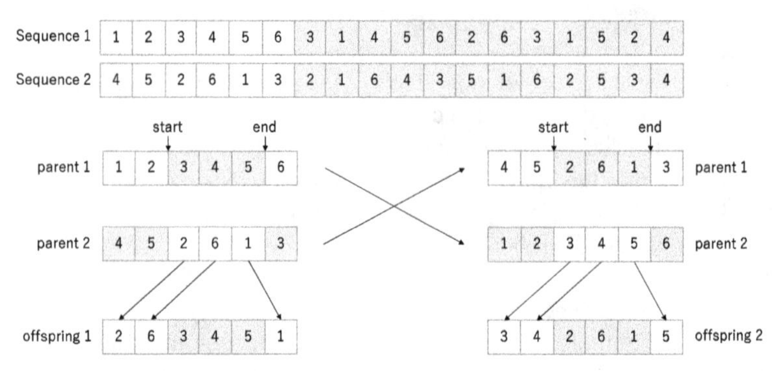

Fig. 7. Order crossover in Sequence-Triple

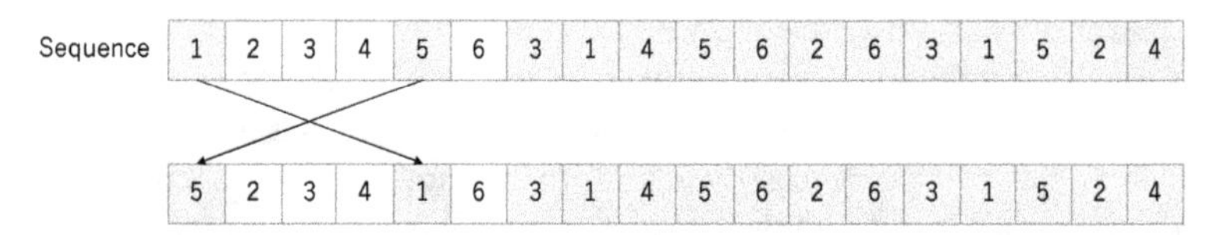

Fig. 8. Mutation in Sequence-Triple

In this study, in addition to the conventional crossover and mutation on Sequence-Triple, we introduce a mutation on bin selection. Since the genetic algorithm data used in this study are Sequence-Triple-Bin, we perform the conventional crossover and mutation on the Sequence-Triple after removing the constant for bin selection from Sequence-Triple-Bin. In addition, the mutation on bin selection does not change the position of -1 in Sequence-Triple-Bin's Γ_1 for half of the bins. The other half is changed randomly. This produces a child that inherits the characteristics of the parent. Figure 9 shows the mutation on bin selection. Since the number of -1 s in Γ_1 of Sequence-Triple-Bin in Fig. 9 is four. Therefore, the position of -1 is changed in two places.

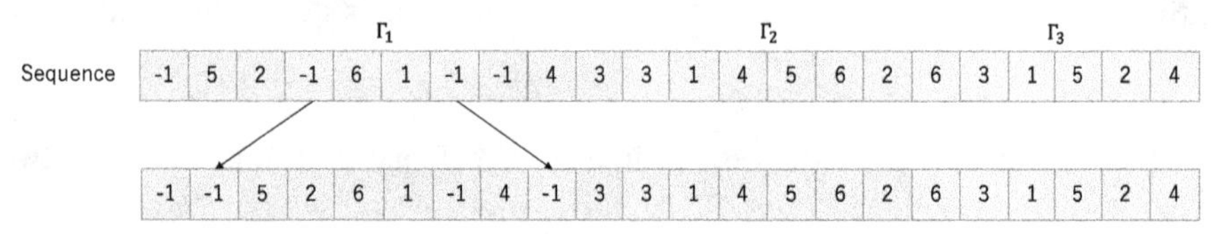

Fig. 9. Mutation on bin selection

3.6 Conventional Approach

In this study, two comparison methods are used to demonstrate the effectiveness of the proposed method. This method performs optimization using a GA as in the proposed method, but instead of using a surrogate model, it uses RRT* to generate trajectories and simulate them to obtain the robot's operating time. The second comparison method is used to verify the validity of the solution and performs the optimization using Gurobi, a mathematical optimization solver. Gurobi is used to solve the three-dimensional bin packing problem to minimize the bin cost. RRT* is applied to the obtained solution to obtain the robot's operating time. Unlike the above method, this method does not perform simultaneous optimization of bin cost and robot operating time but is expected to yield a solution that minimizes bin cost.

4 Computational Experiments

There are 10 rectangles and 8 bins used in this study. The objective function in this study is set to $p = 1$ and $q = 15$. Table 3 shows a comparison of the GA with and without the surrogate model when the number of individuals is 200 and the number of generations is 100. The results show that the implementation of a surrogate model reduces the computation time by about one-eighth. There is almost no difference in the objective function value with and without the surrogate model, suggesting that it is possible to accurately estimate the robot's operating time in a short time.

Table 3. Comparison with and without surrogate model

	surrogate model	No surrogate model
computation time [s]	2811.12	21904.85
objective function value	611.99	611.41

Table 4 compares the GA and Gurobi methods when the GA is run with 200 individuals and 2500 generations. It is confirmed that the solution using GA has a smaller value of objective function than the solution using Gurobi. Therefore, the GA method is more effective for problems where the robot's operating time must be considered. An exact solution method such as Gurobi is not suitable when dealing with complex decision variables such as the robot's movement time. However, the computation time of the solution using Gurobi was found to be about 15 [s], whereas the computation time of the GA was 19.06 [h]. This may be because the solution using GA simultaneously optimizes the bin cost and the robot's operating time, while the solution using Gurobi only minimizes the bin cost. Other methods of crossover and mutation in Sequence-Triple should be explored to reduce the robot's operating time.

Table 4. Comparison of GA methods and Gurobi methods

	GA	Gurobi
bin cost	84	50
robot's operating time [s]	32.31	36.31
objective function value	568.69	594.71

Figure 10 shows an example of layout diagrams of Bin 1, Bin 3, and Bin 5 obtained by using GA. Figure 11 shows layout diagrams of Bin 1, Bin 2, and Bin 3 obtained by using Gurobi. In Fig. 11, the items are stored in bin 1 and bin 2, respectively, because the sizes of bin 1 and bin 2 are (40, 40, 40), while the sizes of the items are (30, 30, 30). Also, the two items could be stored in one bin of larger size, but this would result in higher costs. From Figs. 10 and 11, the method using Gurobi resulted in a layout diagram with a higher weighting factor for robot's operating time and lower bin cost than the method using GA. Figure 12 also shows an example of a motion plan.

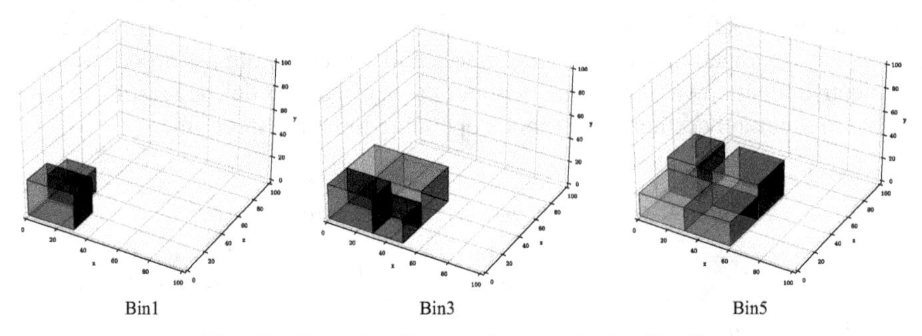

Fig. 10. Example of layout diagram obtained by GA

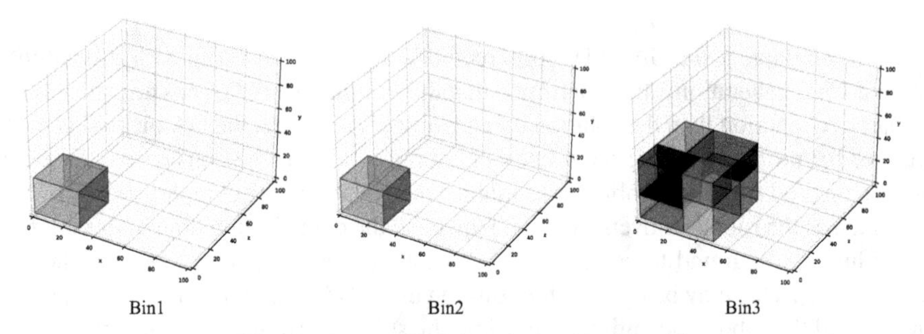

Fig. 11. Layout diagram obtained by Gurobi

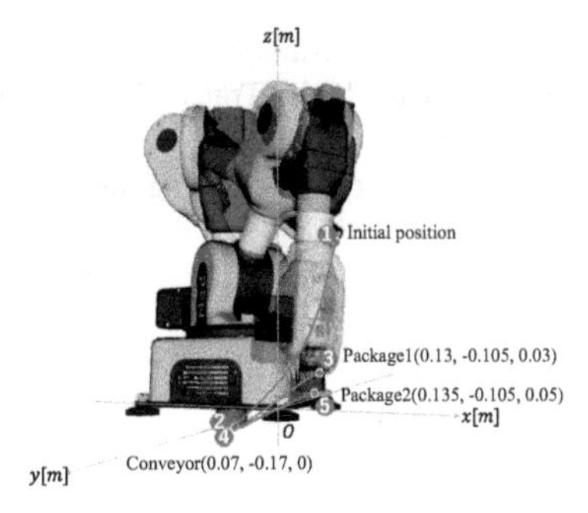

Fig. 12. Example of motion plan

5 Conclusion

In this study, we have proposed a method that combines Sequence-Triple, a three-dimensional layout representation, and RRT* for trajectory generation to optimize the three-dimensional robot bin-packing problem considering the robot's operating time. The implementation of the surrogate model enabled us to reduce the computation time to about one-eighth without compromising the accuracy of the computation. Furthermore, by comparing the proposed method with a method that derives the robot's operating time after solving the bin cost minimization problem, we showed that the proposed method is effective for problems where the robot's operating time must be considered. By adding new constraints such as rotating packages to the three-dimensional robot bin-packing problem, the proposed method will be utilized in actual factory settings.

References

1. Yamazaki, H., Sakanushi, K., Nakatake, S., Kajitani, Y.: The 3D-packing by meta data structure and packing heuristics. IEICE Trans. Fundam. Electron. Commun. Comput. Sci. **83**(4), 639–645 (2000)
2. Hong, Y.-D., Kim, Y.-J., Lee, K.-B.: Smart pack: online autonomous object-packing system using RGB-D sensor data. Sensor **20**(16), 4448 (2020)
3. Erbayrak, S., Özkır V., Yıldırım, U.M.: A hybrid genetic algorithm-based heuristics for bin packing problems with load balance and product unity constraints (2023). https://doi.org/10.2139/ssrn.4544042
4. Zhao, H., She, Q., Zhu, C., Zhu, Y., Xu, K.: Online 3D bin packing with constrained deep reinforcement learning. In: Proceedings of the AAAI Conference on Artificial Intelligence, vol. 35, pp. 741–749 (2021)
5. Weng, C.-Y., Yin, W., Zhong, J.L., Chen, I.-M.: A framework for robotic bin packing with a dual-arm configuration. In: Advances in Mechanism and Machine Science: Proceedings of the 15th IFToMM World Congress on Mechanism and Machine Science, vol. 15, pp. 2799–2808 (2019)
6. Kawabe, T., Liu, Z., Nishi, T., Alam, M.M., Fujiwara, T.: Optimal motion planning and layout design in robotic cellular manufacturing systems. In: Proceedings of 2022 IEEE International Conference on Industrial Engineering and Engineering Management (IEEM), pp. 1541–1545 (2022)

7. Murata, H., Fujikoshi, K., Nakatake, S., Kajitani, Y.: VLSI module placement based on rectangle-packing by the sequence-pair. IEEE Trans. Comput. Aided Des. Integr. Circuits Syst. **15**(12), 1518–1524 (1996)
8. Yagiura, M., Ibaraki, T.: A consideration on the crossing method in genetic Algorithms. Lect. Inst. Math. Anal. **871**, 190–196 (1994)

AI Applications in the Healthcare Logistics and Supply Chain Sectors

Claudia Piffari[1], Alexandra Lagorio[1], and Anna Corinna Cagliano[2](✉)

[1] Department of Management, Information and Production Engineering, University of Bergamo, Viale Marconi 5, 24044 Dalmine, BG, Italy
{claudia.piffari,alexandra.lagorio}@unibg.it

[2] Department of Management and Production Engineering, Politecnico di Torino, Corso Duca degli Abruzzi 24, 10129 Torino, Italy
anna.cagliano@polito.it

Abstract. Artificial Intelligence (AI) has recently been established in healthcare management to support clinical activities and pharmaceutical research and development. Moreover, AI can potentially improve decision-making in healthcare supply chains (HSCs) by leveraging the information provided by various sources. However, research on the application of AI to HSCs is still in its infancy. This work presents a Systematic Literature Review to identify the main trends and future research directions. The analysis of the 23 pertinent papers suggests that more quantitative case studies on AI implementation in HSC are necessary. Additionally, the role of AI in facilitating logistics and supply chain management activities, promoting supply chain resilience, and ultimately creating integrated and agile HSCs should be investigated. Further literature reviews on AI-driven HSC management will help to keep the focus on this research field and its relevant developments.

Keywords: Artificial Intelligence · Healthcare Supply Chain · Systematic Literature Review

1 Introduction

In recent years, we have witnessed an increase in applications based on the use of Artificial Intelligence (AI) in the most diverse fields: from cybersecurity to demand forecasting, from predictive maintenance to agriculture.

One of the sectors where AI has undoubtedly proven most promising is healthcare. The advancement of research in the medical field has always been tied to a great need for economic investments, human resources, and time to test solutions and scenarios and analyse their effectiveness. AI applications allow for benefits at a lower cost, actively supporting researchers in choosing scenarios and alternatives to test, significantly reducing the time required for data analysis [1]. In particular, AI is increasingly being integrated into the healthcare sector, with applications in diagnostic assistance, drug discovery,

Published by Springer Nature Switzerland AG 2024
M. Thürer et al. (Eds.): APMS 2024, IFIP AICT 729, pp. 61–75, 2024.
https://doi.org/10.1007/978-3-031-65894-5_5

clinical trials, patient care, and medical imaging [2]. These applications can potentially improve health outcomes and patient experiences, particularly in the context of pandemics like COVID-19 [3]. In recent years, we have also witnessed the birth and spread of web and smartphone applications based on AI designed for telemedicine or remote assistance. Through these applications, it is possible to assist and help patients being treated for various diseases, thanks to the ability to collect data through devices and machinery located at the patient's home when medical staff are unavailable [4].

Despite the many benefits and the possibilities of application in the healthcare sector, several authors [2, 5, 6] agree that there are many challenges to consider to fully exploit the potential offered by AI without running the risk of failure or not achieving the expected results. Among these challenges is the issue of system use accountability: it is necessary to clarify who is responsible for errors or accidents with AI technologies. This aspect is complex, raising ethical and managerial concerns. We must also consider the so-called AI Divide: patients' trust in medical professionals poses a challenge when AI systems replace human doctors. Building trust between patients and AI systems is crucial. There are also aspects related to Cybersecurity for Privacy and Security. Privacy concerns arise due to extensive data use by AI systems. Balancing confidentiality rules with AI development requires ethical considerations. Even the Loss of Managerial Control can be a significant challenge. Integrating AI blurs traditional healthcare boundaries, requiring a new governance approach. Hospital administrators may struggle with this dynamic care provision system. Finally, among the challenges, we must also remember the aspects that emerge when introducing any new technology that assists or makes some tasks autonomous for operators and managers, namely the risk of job loss and the ability to understand the new job opportunities. Other challenges include the need for investments in training and education to fill the gaps related to skills, difficulties in having staff adequately prepared for interaction with the new technology and the pain of transformation that can lead to resistance from workers and patients to AI applications.

1.1 Research Study Motivations

Although most applications that have involved AI in healthcare so far have focused on diagnostics and patient care in general, the benefits of AI in improving the managerial aspects of healthcare have become increasingly evident, thanks to technological advancements in the industrial sector. AI not only allows for automatic checks on the correctness of data entry processes from medical records to administrative documents, reducing the chances of human errors that could have significant medical and economic repercussions [5] but also allows for the creation of new potential for resource management optimisation and cost reduction [6]. However, despite many authors highlighting the potential of AI in resource management in healthcare and, more generally, in the management of logistics and healthcare supply chains (HSCs), the scientific literature investigating these aspects is still limited and fragmented, probably due to the still scarce real applications of AI in these areas and the many challenges identified in the previous section that this type of implementation entails. These are the main reasons that led us to formulate the research questions at the base of this research:

RQ1: What are AI's main applications in logistics and HSC management? For what purposes are they implemented?

RQ2: What are the main enabling factors and barriers to this type of application?
RQ3: What are the most promising future research directions regarding AI's application in logistics and HSC management?

To answer these research questions, after introducing the topic and analysing the research study motivations (Sect. 1), a systematic analysis of the scientific literature was developed, the methodology of which is illustrated in Sect. 2. The results of the literature analysis that allow us to answer RQ1 and RQ2 are instead reported in Sect. 3, while the discussion in Sect. 4 allows us to answer RQ3. Finally, we have provided the conclusions of our research work (Sect. 5).

2 Methodology

The knowledge and information presented in this paper are based on a Systematic Literature Review (SLR) approach, aiming at isolating and understanding the significant trends and the existing gaps in the scientific literature related to AI in healthcare logistics and supply chains. From the methodological perspective, an SLR provides a replicable research protocol [7], and the detailed documentation of the performed steps within the SLR enables an in-depth evaluation of the conducted study. A review is deemed "systematic" when it is based on clearly formulated questions, identifies relevant studies, appraises their quality, and summarises the evidence using a specific methodology [8]. This study followed the three-step protocol developed to allow for replication [9]. In the first step of the methodology, a preliminary list of keywords and inclusion criteria was identified, making the research as comprehensive as possible [10]. Scopus, the scientific literature database, was selected for analysis. Based on the first set of criteria reported in Table 1, the query was launched on the database, resulting in a corpus of 338 papers.

Table 1. SLR inclusion criteria.

Inclusion Criteria	Description				
Keywords	Artificial Intelligence	*AND*	Healthcare	*AND*	Logistic*
	OR		*OR*		*OR*
	AI		Health Care		Supply chain*
Language	English				
Document types	Articles				
Source types	Peer-reviewed journals and conferences, book chapters				
Time interval	2014 – February 2024				

In the second step of the research, each of the three authors of this paper reviewed the titles and abstracts of the papers that resulted from Step 1. Following a discussion among the authors, papers out of the research scope were removed from the corpus. In particular, 263 papers that did not focus strictly on AI application in the logistics or supply chain operations were excluded (i.e., papers focused on specific issues related to diagnosis, cure assistance, blockchain technologies and papers too general concerning the AI applications in the healthcare sector). The last step of the protocol involved refining the list of selected papers. After reading the full versions of papers emerging from Step 2, 52 papers resulted out of the scope of our research and were thus excluded. These papers were excluded because they mainly focused on applications of blockchain technologies or did not refer to logistics or supply chains. The remaining corpus of 23 papers has been then analysed. The details of these analyses and how they were performed are reported in the following sections.

3 Results

This section presents the results of the literature review concerning AI applications within the HSC. Section 3.1 encompasses a descriptive analysis of the paper corpus and research methodologies. Section 3.2 offers an overview of the AI application fields, the products or services analysed within the HSC, and the echelons and processes of the considered supply chains. Section 3.3 delves into the AI characteristics, focusing on types of AI, the drivers motivating their implementation, and the goals of AI applications, answering RQ1. Furthermore, this section provides an in-depth assessment of the challenges hindering the AI application within the HSC, answering RQ2. For these classifications, each author analysed the papers and classified them inductively, identifying recurring patterns or common themes. Subsequently, the authors compared and discussed their classifications to ensure agreement and consistency.

3.1 Corpus Descriptive Analysis

The analysed topic is relatively recent; Sect. 2 demonstrates that the inclusion criteria used to select the papers cover the period from 2014 to 2024. However, the least recent papers in the selection were published in 2019. Therefore, AI implementation within HSCs is an emerging topic. This conclusion is further supported by the distribution of the selected papers by years of publication. The analysis reveals a growing interest in AI applications within HSC in 2023, constituting 65% of the paper corpus. The number of publications significantly increased compared to previous years.

The corpus primarily comprises journal papers (65%), followed by 30% of conference papers, with only one contribution being a book chapter.

Concerning the employed research methodology, the analysis reveals that most of the research is qualitative, constituting 54% of the paper corpus, while quantitative methods comprise 36%. Additionally, 11% of papers applied mixed-method research, combining qualitative and quantitative approaches. Delving deeper into this classification, it was identified that 22% of papers are based on a literature review. The second most frequently applied research methodology is based on surveys. Indeed, this research method accounts for 19% of the corpus. Conceptual frameworks are employed in 13% of the papers, such as using the Technology-Organisations-Environment framework. Finally, additional methodologies commonly employed encompass the development of algorithms tailored for AI implementation and utilising optimisation algorithms. The composition of the paper corpus based on applied research methodologies is illustrated in Fig. 1. It is essential to note that each paper may employ multiple research methodologies.

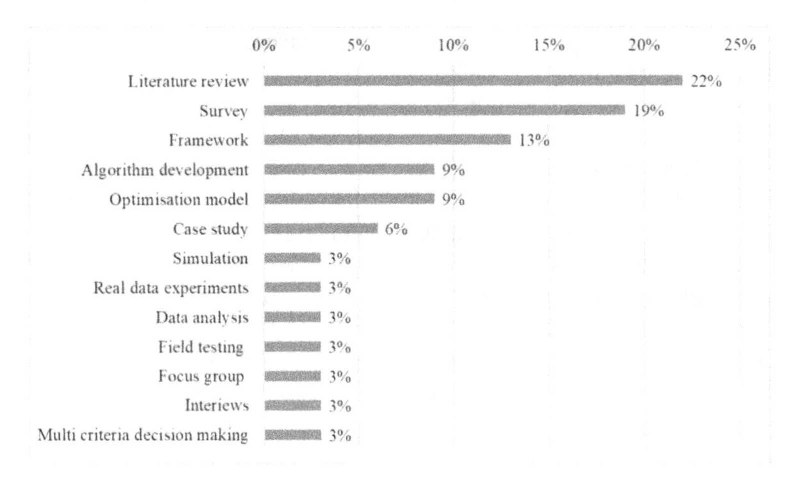

Fig. 1. Research methodologies applied in the paper corpus.

3.2 AI Application in the HSC: Products, Echelons, and Processes

HSCs encompass an intricate network of processes and components to ensure timely manufacturing, distribution, and provision of medicines and healthcare supplies to patients. HSCs differ from other supply chains in terms of complexity, diversity, types of services, uncertainty, and objectives [11]. The current section will examine the HSCs covered in the paper corpus from the perspective of products and services offered, echelons, and processes analysed. These three aspects have been chosen because they are among the most commonly characteristics used to described and categorized a supply chain in the scientific literature [12]. Each author has classified all the papers and selected information regarding the different aspects to take into consideration. Then, a discussion was held for the papers on which there was a disagreement between the authors until an agreement was reached.

HSC Products and Services. Most analysed papers focus on supply chains for medical supplies and equipment, including consumables, surgical instruments, sterile materials, and patient bedding. Indeed, supply chains for medical supplies and equipment constitute 49% of the paper corpus, followed by supply chains for drugs and medicines. The mentioned supply chains are inherently complex and frequently susceptible to shortages, underscoring the criticality of their optimisation in healthcare. 38% of the corpus focuses on the application of AI for managing drug supply chains. Additionally, some papers include AI applications within the blood supply chain (5%) and for managing bed occupancy in hospitals (5%). Furthermore, one paper delves into analysing how AI can support surgical operations, reduce surgery duration, and alleviate overcrowding in the operating room.

HSC Echelons and Processes. In 40% of the selected papers, the analysis pertains to a generic supply chain without specific reference to echelons within that supply chain. This result could be attributed to the preliminary nature of these explorations, wherein the potential benefits of integrating AI are posited but not yet implemented. Excluding papers that do not refer to specific echelons, the hospital emerges as the most frequently mentioned echelon, encompassing the general hospital (20%), the central warehouse of the entire hospital (16%), and the warehouses of individual departments (8%).

From the perspective of the processes analysed, 40% of the papers address the application of AI in supply management, i.e., how to optimise order quantities, support decision-making regarding orders, and automate the ordering process [13]. Additionally, 23% of the papers examine the application of AI to inventory management, wherein AI supports stock monitoring, storage decisions, and stockout management [14]. Demand management, covered in 17% of the papers, uses AI to optimise patient management and bed occupancy, forecast demand for products or services, and facilitate delivery [15]. Finally, in 7% of the papers, AI supports product distribution operations, optimising scheduling and routes [16]. The analysis of these processes remains consistent across product types, with no clear indication of AI application prevalence in specific processes for different products.

3.3 AI Application in the HSC: Drivers, Goals, and Challenges

This section analyses the types of AI applied in different areas of the supply chain, the drivers behind their implementation, the objectives of their application, and the challenges hindering their successful implementation. These aspects emerged inductively by reading the papers of the corpus and then have been discussed among the authors in order to perform a correct and unbiased classification and analysis of the aspects emerged by the analysed papers. Also in this case, each author has classified all the papers and selected information regarding the different aspects to take into consideration. Then, a discussion was held for the papers on which there was a disagreement between the authors until an agreement was reached.

AI Types. Although the literature concerning the application of AI in HSCs is somewhat limited, several instances of AI integration within HSCs can be identified. Approximately 40% of the papers reviewed indicate the AI implementation within Decision Support Systems (DSSs). DSSs, equipped with predictive models, aid healthcare professionals, managers, and decision-makers to apply timely interventions and make informed decisions [15]. Specifically, these DSSs are used for predicting patients' length of stay [15], forecasting demands [17], supporting purchasing decisions [18], and enhancing disruption management [11]. The adoption of AI within DSSs is attributable to its capability to observe, analyse, learn, and make decisions regarding complex situations, thereby supporting management in decision-making processes. Other applications include systems used for intra-surgical assistance [19], medical drones [20], e-commerce, collaborative, and e-healthcare platforms supported by AI [6, 21, 22]. These represent key instances where AI seamlessly integrates into HSC and logistics management [22]. The most commonly implemented AI is machine learning, covered in 41% of the papers. Machine learning includes algorithms such as deep learning and neural networks [11, 15]. Machine learning involves creating a system capable of learning from examples and solving new problems without explicitly programming each behaviour into the software. These algorithms can learn and develop models using database data [16]. Expert systems, discussed in 10% of the papers, represent another AI implementation within HSCs.

These systems, designed to mimic the decision-making abilities of experts in a specific field of knowledge, are employed to support decision-making processes. An example includes the use of expert systems in order to optimise medicine dispatching flow within hospitals [16]. Some applications also leverage natural language processing, enabling machines to understand, interpret, and generate human language and speech recognition for voice-controlled virtual assistants [19]. The last innovations based on artificial intelligence applied within HSCs include drones and robotics [20]. An overview of the AI applied within HSC is reported in Fig. 2.

AI adoption within the healthcare sector, particularly in supply chain management, is still evolving, resulting in a considerable portion of the selected literature (28%) comprising generic analyses of AI. These publications provide an overview of the relationship between AI adoption and organisational resources and AI adoption in inter-organisational networks [6]. Consequently, rather than focusing on the study of specific AI technologies, they more often provide assessments of the variables that can support or hinder the adoption of AI in HSCs.

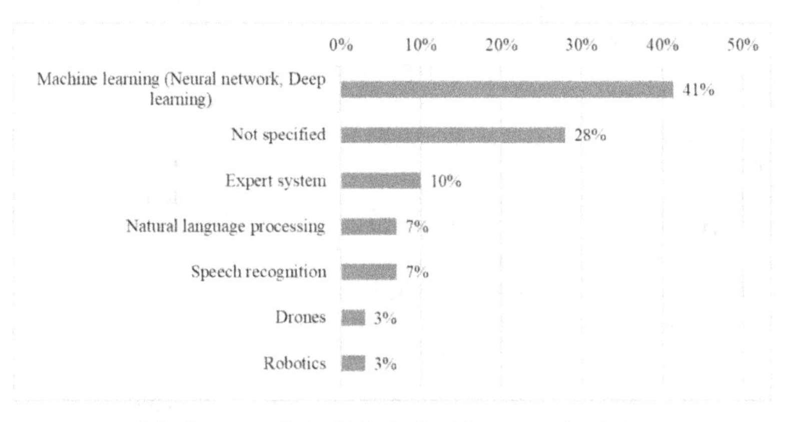

Fig. 2. AI applied within the healthcare supply chain.

Drivers of AI Implementation. Several drivers pushing AI adoption within HSC management are reported in the literature. They can be distinguished between external and internal drivers. External drivers influence the organisation's environment and stem from external elements such as competitors and regulatory pressures. Among the most influential external drivers are cost pressure, market dynamics, and competitive pressure. The healthcare sector is often driven towards maximising efficiency and reducing costs. This aspect makes the competitive pressure in the sector very high, which leads companies to look for sources of competitive advantage. The adoption of AI-based innovations is one of them [21]. Additionally, demographic shifts such as an ageing population and population growth influence AI adoption intentions within HSCs [23]. Furthermore, the inherent characteristics of the HSC, based on numerous actors and interactions, influence adoption intentions. In this regard, the HSC is very complex and dynamic, and the implementation of AI simplifies its management [18]. Conversely, internal drivers are internal to the organisation itself, and the organisation can influence them through strategic choices and decisions. Examples of internal drivers promoting the adoption of AI technologies are the organisational culture and management commitment. A company characterised by managers who are hostile to change is less likely to adopt AI [24]. In particular, absorptive capacity, which refers to the ability of an enterprise to recognise the value of new external information, absorb information, and apply it to business purposes, positively influences AI adoption intention [21]. Culture positively influences AI application; however, the organisation and the people who make up it must have adequate tools and skills to implement AI in HSC operations [24] effectively.

Goals of AI Implementation. The objectives pursued through the implementation of AI in HSCs are summarised in Table 2, along with their description and frequency within the corpus of papers. The latter takes into account that each document could address multiple objectives. Overall, it can be concluded that in most cases, AI is introduced to optimise a process, such as increasing efficiency and effectiveness, reducing waste and costs, and improving quality. AI is often introduced to support forecasting demand for products or services and predict the length of hospital stays. Here, process optimization and process forecasting are considered separately because of their different meanings. On the one hand, process optimization in supply chain management involves maximizing efficiency and minimizing costs by refining and improving the processes involved in inventory management, production control, purchasing decisions, and delivery management [25]. On the other hand, forecasting in supply chain management refers to predicting future events or trends related to the movement of products, services, and information within interrelated organizations [26]. Furthermore, AI enhances data exploitation, as healthcare organisations frequently collect vast amounts of underutilised data. AI technologies make healthcare data available and provide rapid and accurate analysis. AI also compensates for the lack of experts and thus supports practitioners and operators in decision-making processes. Indeed, recent advances in AI have brought forth novel analytical tools that effectively support healthcare professionals in their daily operations [1].

Challenges of AI Implementation. Despite the advantages AI could introduce within the HSC, its application remains limited [24]. Such a lag compared to other industries can be attributed to several barriers hindering the adoption of AI. In the first place, there are technological and infrastructural aspects. The lack of data, difficulties in data retrieval, and inadequate technological infrastructure are the main challenges to ensuring the successful implementation of these technologies [1, 11, 15]. The interaction between the technology and healthcare or non-healthcare operators determines another challenge that could hinder the implementation of AI in HSCs. The lack of AI skills and the consequent need for training could limit the adoption of AI and increase its implementation costs [19, 20]. Moreover, even once adopted, there could be problems with accepting these new technologies. Operators might be reluctant to adopt the new procedures and refuse to integrate them into their activities. These human aspects represent significant challenges in AI adoption across all sectors, particularly in the health sector [1, 27].

Finally, there are challenges caused by the peculiarities of the health sector. While achieving an efficient supply chain poses a persistent challenge across various industries and businesses, the complexity increases within the HSC due to its involvement with patient safety and related health outcomes. This complexity is further exacerbated by the

Table 2. Goals of AI implementation within HSC.

Goal of AI application within HSC	Description	Frequency	Reference
Process optimisation	Optimise performance, such as waste reduction and cost reduction, improve safety, quality, efficiency, dispatch optimisation, and reduce overcrowding	25%	[1, 13, 17, 28]
Forecasting	Predict length-of-stay, HSC sustainability, blood demand, and medical equipment demand	13%	[3, 11, 15, 17]
Improve data exploitation	Make healthcare data available and provide rapid, accurate analysis	13%	[1, 24, 28]
Support decision making	Support HSC management decisions, such as purchasing decisions and dispatching decisions	13%	[13, 14, 29]
Support healthcare operators	Compensate for human resource shortages and support operators in their daily activities	11%	[18, 29]
Resilience improvement	Improve risk management, disruption management, and responsiveness	7%	[3, 29]
Sustainability improvement	Improve environmental and financial sustainability	5%	[11, 13]
Process monitoring	Observation and tracking of activities, procedures, or systems	6%	[13]
Collaboration improvement	Improve buyer-supplier relationships and data transfer	4%	[28]
Improve patients' safety	Enhance the well-being of patients within healthcare settings through the implementation of technologies aimed at reducing risks	4%	[19]
Medical product delivery	Distributing pharmaceuticals, medical devices, or other healthcare-related goods	2%	[20]

involvement of multiple stakeholders operating under different regulatory jurisdictions [24]. Therefore, issues of protecting data privacy and ensuring security, along with better integration, pose significant obstacles to implementing AI in this field [19]. Finally, the last characteristic element of the health sector that makes the adoption of AI critical is determined by the stringent legal and regulatory constraints typical of this sector [21]. The implementation of AI in e-healthcare is still hindered by issues such as the requirement to abide by stringent regulations [22].

4 Discussion

This section discusses the key research trends and future research directions emerging from the SLR conducted, thus answering RQ3. Moreover, the implications of the study for both academics and practitioners are presented.

4.1 Research Insights

The attention of the literature to the application of AI to HSC is very recent. As with many management innovations introduced in HSCs over the past decades, such as warehouse automation or Lean Management, AI has started to be applied in the healthcare sector following its success in manufacturing. In particular, a two-step approach can be identified. AI was first employed to support clinical activities and research and development in the pharmaceutical industry [22], and then it was extended to material management. The huge pressure of the COVID-19 pandemic on healthcare systems worldwide and the associated negative impacts on HSCs were the main drivers for developing studies on the topic. This aspect could explain the surge in the literature published in 2023 when the pandemic crisis could be considered definitively over.

Today, the benefits of AI in managing highly complex HSCs are beginning to be realised. Thus, on the one hand, researchers should continue to deepen AI applications in areas already explored and investigate new application domains. On the other hand, healthcare supply chain managers should participate in pilot implementations of AI algorithms to support and improve their decision-making and operational tasks. Table 3 shows the future research directions that can be derived from the SLR performed.

The research methodologies adopted by the reviewed papers reflect the topic's novelty. Since the number of field applications is still limited, few contributions discuss case studies on the implementation of AI in HSC. Additionally, accepting advanced technologies is a crucial barrier to overcome in promoting the widespread use of AI to support HSC management. Therefore, it is important to conduct surveys to understand healthcare managers' and professionals' perceptions and readiness to adopt AI, identify effective strategies to increase their confidence in the technology and establish organisational culture and managerial commitment to support AI applications. Close cooperation between researchers and healthcare professionals will be needed in future to build case histories of more or less successful AI implementations in HSCs. In such a way, the advantages and disadvantages of supply chain management supported by AI will be quantitatively proven. Most current works focus on drugs, medical supplies, and equipment. Such an outcome is likely to be influenced by the recent COVID-19 pandemic, which highlighted the importance of timely availability of essential healthcare products and the use of digital technologies to ensure resilient supply chain management. Surprisingly, there are few

Table 3. Future research directions.

Suggested research topics
Quantitative case studies of AI implementations in HSCs
Surveys about the perceptions and readiness of healthcare professionals to adopt AI
How AI can improve the management of drugs, blood, and implants
Explore the role of AI in those decision-making processes that are closer to the patient, including product distribution management and product tracking
How AI can compensate for the lack of expertise and alleviate pharmacists and other clinical professionals from supply chain activities
How AI promotes resilience and supports disruption management in HSCs
How AI can support data collection and analysis in HSCs by being compliant with data protection and privacy regulations
Exploring the potential of AI to create integrated, efficient, and agile HSCs

literature contributions on the blood supply chain, as blood is another important product in the healthcare sector. This evidence calls for more studies on how AI can improve the management of healthcare materials, which play a fundamental role in saving lives. The attention to some of the key healthcare products is not accompanied by a strong focus on the specific supply chain echelons at which these products are handled, such as central hospital warehouses or individual department storerooms. Additionally, research does not address upstream supply chain echelons (e.g., drug suppliers and distributors). However, when looking at the individual supply chain processes, it is possible to see a greater focus on procurement and inventory management. On the contrary, distribution logistics is rarely addressed, even though it is responsible for making healthcare products available to users at the right time. In this context, research efforts should focus on the application of AI to all echelons of the HSC to understand how it can facilitate and make more reliable the relationships between them.

According to the performed SLR, process optimisation and demand forecasting are the main goals of the current applications of AI to HSCs. However, there are still understudied objectives that deserve further research attention. First, the role of AI in compensating for the lack of expertise that may exist in decision-making, for example, concerning purchasing and delivery, is a little-discussed topic. This aspect brings up another important issue. In some healthcare warehouses, purchasing, stock management, and distribution decisions are made by pharmacists or other professionals whose skills are more focused on patient care. They may have developed good logistics skills, but that is not their job. Introducing AI in their decision-making processes can potentially alleviate these professionals from logistics tasks, allowing them more time to spend on clinical activities. Second, some authors recognise that AI increases HSC resilience [1], but the literature rarely addresses practical applications. This result is obtained in the context of sparse literature on approaches to increase resilience in HSC. Therefore, as suggested by [30], further studies on the role of digital technologies and AI in promoting resilience are encouraged. In addition, studying AI-based product tracking systems would help limit phenomena such as expired products or misappropriation of materials that still affect many healthcare warehouses. This will in turn enhance both environmental and financial sustainability of HSC operations.

Finally, healthcare organisations often collect data about their supply chains without a rigorous approach and a clear goal, wasting time and effort. Studies on how AI can support a structured collection and analysis of these data to extract meaningful information in a reliable and timely manner should be carried out. However, they should not neglect the need to comply with data protection and privacy regulations.

The literature review brought about drivers and challenges that need further investigation. Research is required to address the challenges associated with the adoption of AI technologies in healthcare organisations. Future research should identify strategies for the effective management and reduction of these challenges, thereby facilitating the wider deployment of AI in healthcare. Among the challenges, the concept of supply chain complexity is exacerbated in healthcare as it involves delivering high-quality services while maintaining cost-effectiveness. The medical supply chain is also fragmented as it includes multiple stakeholders, e.g., manufacturers, hospitals, providers, insurance companies, and regulatory agencies. Exploring the potential of AI to create an integrated, efficient, and agile HSC will generate knowledge on how to reduce complexity by leveraging large amounts of diverse supply chain data. The need to effectively manage the growing complexity of the supply chain is also imposed by the expectation that an increasingly ageing population will drive demand for healthcare services in the coming decades.

4.2 Implications for Academics and Practitioners

The application of AI to HSC is still in its infancy. However, the existing body of literature makes it interesting to trace a state of the art to direct future research. As such, the present work provides guidelines on the most pressing topics that academics should focus on to facilitate the diffusion of AI to support decision-making in HSCs. In particular, it could encourage studies that present not only conceptual frameworks for the application of AI but also, most importantly, algorithms and quantitative case studies that enable real-world evaluation of the advantages and disadvantages of AI-based supply chain management. In addition, researchers should not neglect the investigation of the technological, organisational, and contextual factors that influence the adoption of AI in healthcare environments, as well as the relationships between AI and supply chain viability and resilience. From a technological standpoint, the research remains at an early stage of development and, consequently, the specifics of the technologies employed are not yet fully understood. Future research may delve more deeply into the characteristics of the proposed AI in order to converge toward higher performance. Finally, this research work might foster the development of future literature reviews, hopefully including a larger number of papers that will be published in the coming years. Such efforts will allow for the systematisation of the various ramifications of the topic that will emerge over time. From a practitioners' point of view, this paper introduces them to the wide range of opportunities AI can bring to HSC. Moreover, it suggests the most important echelons, processes, and products in the supply chain to which AI could be applied. Also, supply chain professionals might find this literature review beneficial in identifying the technological transformations that can support the development of resilience strategies to enhance responsiveness under the events causing supply chain disruptions, including healthcare emergencies, poor availability of active drug ingredients, and constraints to international commercial channels. Furthermore, for seamless AI adoption within HSC, it is of paramount importance for organisations to verify the possession of the requisite infrastructure and technology to facilitate the effective implementation of AI. In addition, organisations must ensure the compliance with the relevant regulations. It is also necessary to consider the human aspect of AI implementation, including training of the workforce, to ensure that the introduction of AI is supportive.

5 Conclusion

The present SLR brings to the attention of the supply chain management community HSCs and how they can leverage the technologies now being consolidated in the manufacturing sector, particularly AI. The small amount of literature on the topic is the main limitation to generalising the outcomes obtained. Moreover, it prevents performing a meaningful geographical analysis of the contributions, which could uncover whether the healthcare systems in some regions of the world are more prone to integrate AI in their supply chains. Future research will be directed towards extending the present literature review to include the latest developments in AI applications to HSCs. Moreover, the authors would like to address the highlighted gap in supply chain resilience by proposing strategies to improve hospital supply chain resilience, along with KPIs to measure it. In this context, AI will support strategy development and assist in data collection and analysis to calculate KPIs.

References

1. Deveci, M.: Effective use of artificial intelligence in healthcare supply chain resilience using fuzzy decision-making model. Soft. Comput. (2023). https://doi.org/10.1007/s00500-023-08906-2
2. Lee, D., Yoon, S.N.: Application of artificial intelligence-based technologies in the healthcare industry: opportunities and challenges. Int. J. Environ. Res. Public Health **18**, 271 (2021). https://doi.org/10.3390/ijerph18010271
3. Koç, E., Türkoğlu, M.: Forecasting of medical equipment demand and outbreak spreading based on deep long short-term memory network: the COVID-19 pandemic in Turkey. SIViP **16**, 613–621 (2022). https://doi.org/10.1007/s11760-020-01847-5
4. Lee, D.: Strategies for technology-driven service encounters for patient experience satisfaction in hospitals. Technol. Forecast. Soc. Change **137**, 118–127 (2018)
5. Sharma, P., Namasudra, S., Gonzalez Crespo, R., Parra-Fuente, J., Chandra Trivedi, M.: EHDHE: Enhancing security of healthcare documents in IoT-enabled digital healthcare ecosystems using blockchain. Inf. Sci. **629**, 703–718 (2023). https://doi.org/10.1016/j.ins.2023.01.148
6. Cannavale, C., Esempio Tammaro, A., Leone, D., Schiavone, F.: Innovation adoption in inter-organizational healthcare networks – the role of artificial intelligence. Eur. J. Innov. Manag. **25**, 758–774 (2022). https://doi.org/10.1108/EJIM-08-2021-0378

7. Tranfield, D., Denyer, D., Smart, P.: Towards a methodology for developing evidence-informed management knowledge by means of systematic review. Br. J. Manag. **14**, 207–222 (2003). https://doi.org/10.1111/1467-8551.00375
8. Khan, K.S., Kunz, R., Kleijnen, J., Antes, G.: Five steps to conducting a systematic review. J. R. Soc. Med. **96**, 118–121 (2003)
9. Lagorio, A., Pinto, R., Golini, R.: Research in urban logistics: a systematic literature review. Int. J. Phys. Distrib. Logist. Manag. **46**, 908–931 (2016). https://doi.org/10.1108/IJPDLM-01-2016-0008
10. Wong, C.Y., Wong, C.W., Boon-itt, S.: Integrating environmental management into supply chains: a systematic literature review and theoretical framework. Int. J. Phys. Distrib. Logist. Manag. **45**, 43–68 (2015). https://doi.org/10.1108/IJPDLM-05-2013-0110
11. Azadi, M., Yousefi, S., Farzipoor Saen, R., Shabanpour, H., Jabeen, F.: Forecasting sustainability of healthcare supply chains using deep learning and network data envelopment analysis. J. Bus. Res. **154**, 113357 (2023). https://doi.org/10.1016/j.jbusres.2022.113357
12. Fisher, M.: What Is the Right Supply Chain for Your Product? (1997). https://hbr.org/1997/03/what-is-the-right-supply-chain-for-your-product
13. Jordon, K., Dossou, P.-E., Junior, J.C.: Using lean manufacturing and machine learning for improving medicines procurement and dispatching in a hospital. Procedia Manuf. **38**, 1034–1041 (2019). https://doi.org/10.1016/j.promfg.2020.01.189
14. Taertulakarn, S., Sritart, H., Tosranon, P., Pongpaiboon, K., Subenja, K.: The design and development of an AI-based medical laboratory inventory monitoring system. In: 2023 15th Biomedical Engineering International Conference (BMEiCON), pp. 1–5 (2023). https://doi.org/10.1109/BMEiCON60347.2023.10321987
15. Alnsour, Y., Johnson, M., Albizri, A., Harfouch, A.: Predicting patient length of stay using artificial intelligence to assist healthcare professionals in resource planning and scheduling decisions. J. Glob. Inf. Manag. **31** (2023). https://doi.org/10.4018/JGIM.323059
16. Pinheiro, J.C., Dossou, P.-E., Junior, J.C.: Methods and concepts for elaborating a decision aided tool for optimizing healthcare medicines dispatching flows. Procedia Manuf. **38**, 209–216 (2019). https://doi.org/10.1016/j.promfg.2020.01.028
17. Benelmir, W., Hemmak, A., Senouci, B.: Smart platform for blood management in healthcare using AI/ML approach. In: 2023 International Conference on Artificial Intelligence in Information and Communication (ICAIIC), pp. 007–011 (2023). https://doi.org/10.1109/ICAIIC57133.2023.10067054
18. Ghaderi, F., Ghatari, A.R., Radfar, R.: An intelligent decision support system based on fuzzy techniques and neural networks for purchasing medical supplies **25** (2023)
19. Kim, J.H., et al.: Development of a smart hospital assistant: integrating artificial intelligence and a voice-user interface for improved surgical outcomes. Proc. SPIE Int. Soc. Opt. Eng. **11601**, 116010U (2021). https://doi.org/10.1117/12.2580995
20. Damoah, I.S., Ayakwah, A., Tingbani, I.: Artificial intelligence (AI)-enhanced medical drones in the healthcare supply chain (HSC) for sustainability development: a case study. J. Clean. Prod. **328**, 129598 (2021). https://doi.org/10.1016/j.jclepro.2021.129598
21. Kong, Y., Hou, Y., Sun, S.: The adoption of artificial intelligence in the e-commerce trade of healthcare industry. In: Wang, Y., Wang, W.Y.C., Yan, Z., Zhang, D. (eds.) DHA 2020. CCIS, vol. 1412, pp. 75–88. Springer, Singapore (2021). https://doi.org/10.1007/978-981-16-3631-8_8
22. Painuly, S., Sharma, S., Matta, P.: Artificial intelligence in e-healthcare supply chain management system: challenges and future trends. In: 2023 International Conference on Sustainable Computing and Data Communication Systems (ICSCDS), pp. 569–574 (2023). https://doi.org/10.1109/ICSCDS56580.2023.10104746
23. Maheshwari, S., Kaur, G., Kotecha, K., Jain, P.K.: Bibliometric survey on supply chain in healthcare using artificial intelligence. Libr. Philos. Pract. **2020**, 1–18 (2020)

24. Kumar, A., Mani, V., Jain, V., Gupta, H., Venkatesh, V.G.: Managing healthcare supply chain through artificial intelligence (AI): a study of critical success factors. Comput. Ind. Eng. **175**, 108815 (2023). https://doi.org/10.1016/j.cie.2022.108815
25. Alkahtani, M.: Mathematical modelling of inventory and process outsourcing for optimization of supply chain management. Mathematics **10**, 1142 (2022). https://doi.org/10.3390/math10071142
26. Shibu, N., Agarwal, R.: Analysing and visualising trends for supply chain demand forecasting. In: 2023 International Conference on Computational Intelligence and Sustainable Engineering Solutions (CISES), pp. 901–906 (2023). https://doi.org/10.1109/CISES58720.2023.10183439
27. Piffari, C., Lagorio, A., Cimini, C., Pinto, R.: The role of human factors in the human-centred design of service processes: a focus on the healthcare sector. Presented at the Proceedings of the Summer School Francesco Turco (2022)
28. Bag, S., Dhamija, P., Singh, R.K., Rahman, M.S., Sreedharan, V.R.: Big data analytics and artificial intelligence technologies based collaborative platform empowering absorptive capacity in health care supply chain: an empirical study. J. Bus. Res. **154** (2023). https://doi.org/10.1016/j.jbusres.2022.113315
29. Painuly, S., Sharma, S., Matta, P.: Deep learning tools and techniques in e-healthcare supply chain management system. Presented at the 2023 IEEE 8th International Conference for Convergence in Technology, I2CT 2023 (2023). https://doi.org/10.1109/I2CT57861.2023.10126339
30. Arji, G., Ahmadi, H., Avazpoor, P., Hemmat, M.: Identifying resilience strategies for disruption management in the healthcare supply chain during COVID-19 by digital innovations: a systematic literature review. Inform. Med. Unlocked. **38**, 101199 (2023). https://doi.org/10.1016/j.imu.2023.101199

Analysis of Critical Success Factors of Sustainable and Resilient Aioe-based Supply Chain in Industry 5.0

Hamed Nozari[1], Reza Tavakkoli-Moghaddam[2](✉), and Alexandre Dolgui[3]

[1] Department of Management, Azad University of the Emirates, Dubai, United Arab Emirates
[2] School of Industrial Engineering, College of Engineering, University of Tehran, Tehran, Iran
tavakoli@ut.ac.ir
[3] IMT Atlantique, LS2N – CNRS, La Chantrerie, Nantes, France
alexandre.dolgui@imt-atlantique.fr

Abstract. Critical success factors (CSFs) are limited components that are critical to the success of the organization. If the organization needs its presence, it should give them. Therefore, organizations should consider these factors in their operational processes. In today's world, which is mixed with transformative technologies and business and supply chain processes have different conditions than in the past, understanding the CSFs for smart processes in the supply chain is very necessary because it can guarantee the success of smart organizations. Understanding these factors and their position can greatly help the resilience and sustainability of the supply chain, which are key indicators of Industry 5.0. Therefore, this research aims to extract and analyze the critical success factors in resilient, sustainable, and intelligent supply chains based on Artificial Intelligence is Everything (AIoE) in Industry 5.0. To evaluate the critical factors of success, this research has prioritized these factors using a fuzzy nonlinear approach based on the hierarchical analysis method and pairwise comparison tables based on linguistic variables. The results show that the existence of technical infrastructure and hardware have the highest priority in the implementation of these systems. Understanding the framework presented in this research can provide a deep insight into the effective implementation of these smart supply chain systems in the age of transformative technologies.

Keywords: Critical Success Factors · Intelligent Supply Chain · Artificial Intelligence of Everything (AIoE) · Industry 5.0

1 Introduction

Organizations attempt to establish their maintainability through distinctive methodologies such as environmental management frameworks, lean, agile, adaptable, green, compatibility with the environment, and competence to maintain their position in complex and changing environments. For a long time, money-related clashes, administrative

Published by Springer Nature Switzerland AG 2024
M. Thürer et al. (Eds.): APMS 2024, IFIP AICT 729, pp. 76–90, 2024.
https://doi.org/10.1007/978-3-031-65894-5_6

and competitive pressures, increased client demand, and complex natural controls have expanded attention to sustainable supply chains. Leading supply chain businesses always seek a new advantage [1]. They know that even a small or medium-sized technical or process improvement can be multiplied at the scale of sizeable modern supplier networks, providing a competitive advantage or significant cost savings. The issue that supply chain managers have problems with is the quantification of the results of these optimizations and the ability to compare them [2].

Considering the importance of supply chain processes, the resilience and stability of the supply chain are among the features that should be given special attention due to the development of transformative technologies. A resilient supply chain may not be the slightest costly supply chain, but a resilient supply chain can overcome vulnerabilities and disturbances within the commerce environment. The competitive advantage of the supply chain does not depend as it were on moo costs, tall quality, decreased delay time, and high level of benefit. Rather, it has the capacity of the chain to maintain a strategic distance from catastrophes and overcome basic conditions, which is the supply chain's resilience [3].

Along with the globalization of the economy, supply chain management has become important in terms of responsibility and sustainability. As supply chains grow and become complex networks, they become more difficult to manage. This challenge is due to problems such as lack of resources, market developments, globalization, changing consumption patterns, and technological advances in the chain. Sustainability and resilience are some of the most important parameters of the 5.0 industry, along with the approach of artificial intelligence to natural intelligence [4].

In line with the developments of Industry 4.0, Industry 5.0 has created unprecedented changes for all traditional business models and has accelerated the need to redesign and digitize activities. This new industrial revolution offers maximum technological growth, emphasizing sustainability and a forward-looking outlook. So that the lives of future generations are not destroyed. The purpose of this study is to understand better the role and growth potential of the emerging digital technologies of Industry 5.0 and to examine their critical success factors to guide sustainable and resilient innovations in the supply chain of businesses. Advanced robotics, artificial intelligence, big data analysis, Internet of Things, blockchain, and virtual reality/augmented reality are some of the most prominent technologies of the new era that are expected to have the greatest impact on the development of supply chains in the future. Although supply chains have grown rapidly in recent years, focusing only on economic performance to optimize costs or return on investment cannot guarantee development stability, and resilience in the supply chain. Therefore, to benefit from the benefits of digital and smart supply chains in Industry 5.0, it is necessary to examine and evaluate the most important critical success factors in these smart and sustainable chains. Digital technologies in the supply chain offer a wide range of sustainability benefits, from the potential to reduce energy consumption and waste to increasing opportunities for recycling and industrial coexistence. Since sustainability is one of the important drivers of innovation, the recent emergence of disruptive technologies demonstrates the transformative impact of digital innovation on supply chain sustainability. While adopting digital technologies in various industries is still in its early stages, the findings of this research bring more opportunities to promote sustainable results in the future of supply chains. Digital transformation technologies improve efficiency, transparency, and optimization of network logistics resources. This paper provides a way to assess the scope of impact and potential for successful implementation of supply chain digital transformation technologies.

Te structure of this article is as follows. The second section focuses on the supply chain based on AIoE. The third section introduces the most important critical success factors in AIoE-based supply chains in Industry 5.0. The fourth section discusses the investigation approach. The fifth section presents the research findings followed by the conclusion section.

2 Literature Review

In this section, the concepts related to the research are first presented and then an attempt has been made to present and explain the most important critical success factors for a smart and sustainable supply chain based on AIoE.

2.1 AIoE-Based Supply Chain and Industry 5.0

Big data, IoT, blockchain, and robotic process automation are some of the things that make a smart supply chain. Supply chain firms can reduce expenses, shorten item transport times, minimize negative environmental impacts, and accomplish remarkable levels of automation with these innovations. The connection between transformative technologies and the supply chain that gives rise to the intelligent supply chain is that such a supply chain is a self-improving and fully flexible system [5]. Consistency in data sharing and persistent optimization of workflows will be included in a smart supply chain. To better understand the savvy supply chain, it should be said that such a framework can handle numerous things, including deal history, climate conditions, and the sorts of information it gets from its sensors. In this way, giving much superior execution in coordination and supply chain [6]. Hybrid technologies can add extraordinary performance power and efficiency to the supply chain. The combined AIoE technology, created by the combination of IoT and AIoT technologies, was first proposed by Nozari in 2024 [7].

AIoE has shown its potential in processes such as supply chains, besides being a revolutionary technology for all industries. All types of data related to objects, people, and different processes can be extracted, maintained, and analyzed with this technology. Forecasting and monitoring applications can help managers improve distribution operational efficiency and increase decision-making transparency. As a result, the advantages of employing AIoE throughout the supply chain are evident, encompassing all these industries. The framework of AIoE is shown in Fig. 1. As shown in this figure, this powerful technology covers all aspects of intelligent life and turns life into a super-intelligent life with high computing power and speed. People, processes in governments, businesses, and people's daily lives are becoming places to collect and analyze unstructured data. This framework shows that with the growth of this technology, in addition to all the benefits it has, there will be many challenges in life. One of the most important challenges is security and privacy. Therefore, this concept can be modified in the framework of Industry 5.0. Because all parameters of sustainability and resilience of this technology will be considered in a reasonable framework, it will be limited [8].

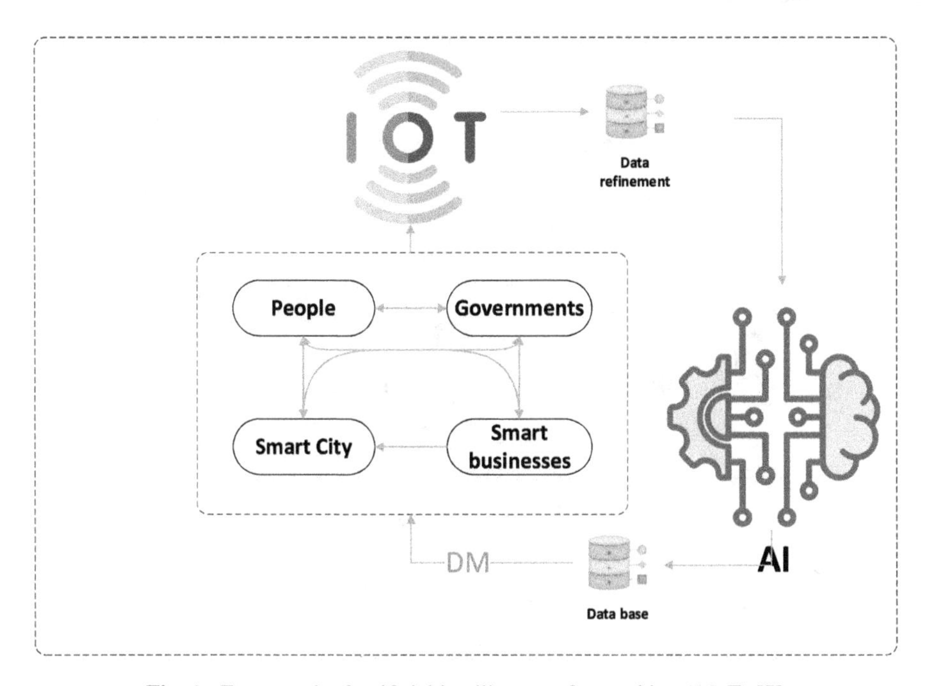

Fig. 1. Framework of artificial intelligence of everything (AIoE) [7].

The role of AIoE in supply chain administration is key and exceptionally imperative because it can offer assistance in decreasing vitality utilization, optimizing different forms, diminishing costs, etc. Subsequently, AIoE and the supply chain nowadays have a close relationship, which can, without a doubt, influence long-term businesses, businesses, and numerous distinctive organizations. This technology shows its increasing power in supply chains because all aspects of the process that this technology can affect exist in the supply chain. Figure 2 shows a conceptual framework for an AIoE-based intelligent supply chain. This framework is derived from the supply chain 6.0 framework provided by Nozari [7].

As shown in Fig. 2, security and privacy issues are always among the most important challenges due to the interlayer interactions under sensor-based networks in this smart supply chain. By intelligentizing the stream of merchandise through sensors, AIoE allows clients to have total item data from crude materials to generation through the Web and utilize them to create obtaining choices. This data can incorporate fixing data, generation preparation data, data related to the fabricating company, wholesaler data, item guarantee, or other data required by the client. This instrument can give ideal choices by utilizing its clever expository control inferred from manufactured insights devices. Instant sharing of information in integrated environments with an emphasis on green processes can increase the stability and resilience of these systems, which is in line with the development of Industry 5.0. They have a robust and easy-to-use database; however, they are not as adept at delivering data in commercial formats and exchanging data among business partners.

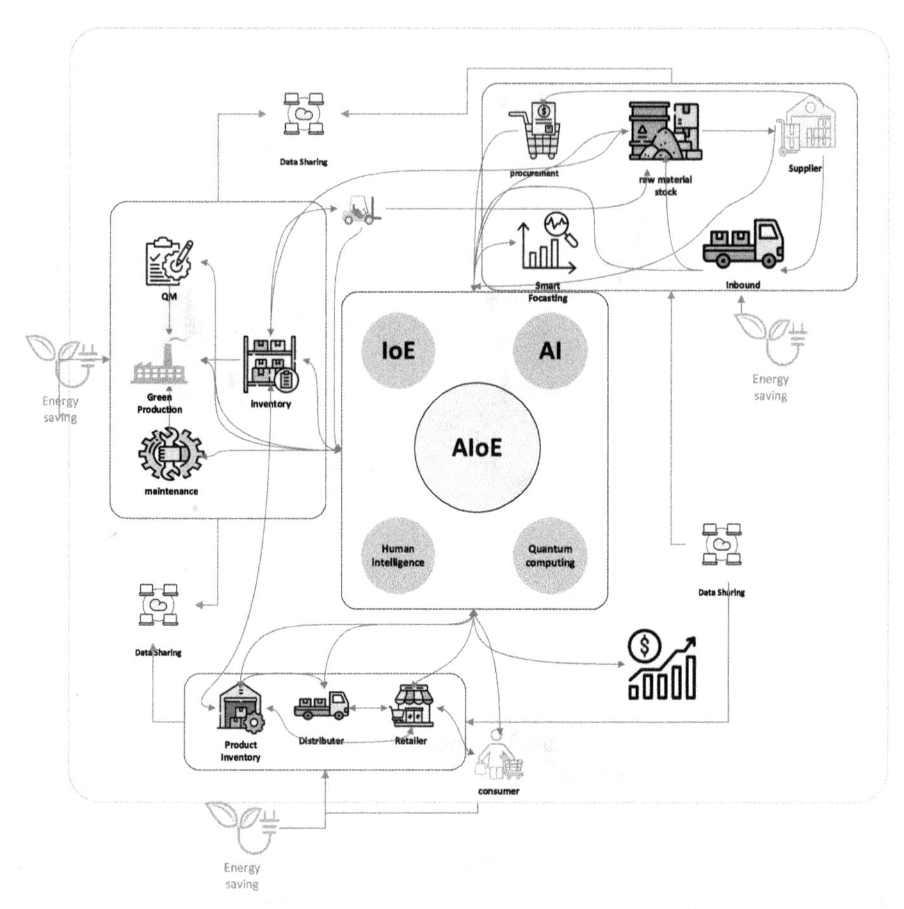

Fig. 2. AIoE-based supply chain [7].

The exchange data can't be analyzed in a minute. Companies can do this by analyzing their rival markets and anticipating their forthcoming commercial strategies to flawlessly capture their merchandise's publicizing share. This may make it easier for businesses to respond to the exhibit. With the intelligentization of items inside the era section, in addition to the precise recognizing confirmation and taking after of rough materials and saving parts in the midst of the era plan, the entire of squandering and breakdowns can be measured. This includes allowing administrators to recognize bottlenecks and weak centers in e-commerce.

2.2 CSF for Sustainable and Resilient Supply Chain Based on AIoE and Blockchain in Industry 5.0

In an AIoE-based economical, intelligent supply chain, communication between things with data innovation is built up through inserted savvy gadgets or through the utilization of one-of-a-kind identifiers and carrier information that can allow intellectuals to communicate with a supporting arranged framework. By labeling things and substances, more data about the status of the workshop and the area of the status of the generation apparatus can be obtained. The valuable data of labels as input information can produce refined programs and move forward coordination [9]. Nearby plan things can distinguish self-organization and shrewd fabricating arrangements. Considering the very large dimensions of intelligent supply chains based on AIoE and the very high capabilities of these intelligent processes, it can be concluded that implementing these systems is very complex and, therefore, requires understanding the full cause-and-effect relationships of all active actors. For this reason, knowing the critical factors of success in AIoE-based smart supply chains can create a deep insight into the effective implementation of these smart systems and help them succeed. Critical success factors (CSFs) are used at all levels of the organization, from management to micro levels of the organization. They can originate from internal initiatives (such as improving the efficiency of the work environment) or external factors (such as changes in technology, laws, or actions of competitors) [10].

In this study, to extract the CSFs in the sustainable and resilient intelligent supply chain based on AIoE, the most important CSFs for the intelligent supply chains were first extracted by reviewing the literature. Available articles were extracted from Google Scholar, Science Direct, and Scopus databases. Then, using experts' opinions, these factors were refined using 5-level Likert scale questionnaires, and for the smart supply chain studied in this research, the experts also considered other cases. Some factors were removed [11]. The experts believed that many critical factors for success in this System are dependent and not independent. This research used the opinions of 15 working industrial and academic experts. The industrial experts had related work records in the supply chain. They were familiar with the field of digital technologies (in the food and pharmaceutical industries, according to the authors' access). Academic experts also had relevant educational and research records in the studied field. 8 of the experts had working and industrial records and 7 of the experts were university professors. Using this expert's opinions, 11 factors were finally considered as the most important CSFs in the intelligent AIoE-based supply chain in Industry 5.0.

Figure 3 shows these critical success factors. The research experts believed that some of the critical success factors presented in this thesis are dependent, but their importance can be examined independently and in different categories. As shown in this figure, with the approval of experts, critical success factors are presented in three general categories: organizational, strategic, and technological. These critical success factors are briefly defined.

- **Skilled manpower:** A capable and expert team with high learning power in transformative technologies and business processes is one of the most important factors affecting the success of smart supply chain systems [12].

- **Organization readiness**: Organizational availability for alter centers on program pioneers and employees' motivation and identity characteristics, organizational resources, and organizational climate. It contrasts with the chance that clients understand the organizational factors related to executing advanced developments in a program [13].

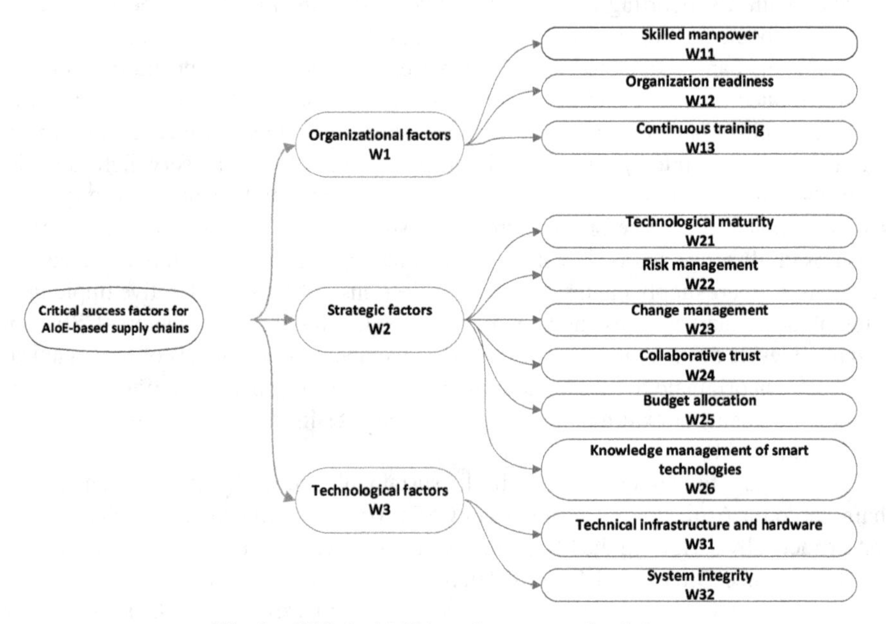

Fig. 3. CSFs in AIoE-based smart supply chain.

- **Continuous training**: The speed of data extraction and calculations in intelligent supply chains requires instant change. When the pace of change is so fast, the only way to maintain lifelong work capacity is to engage in lifelong learning [14].
- **Technological maturity**: Technological maturity refers to the level of the organization that understands the need for technology capabilities. If the organization reaches the desired level of maturity for this understanding, all technological processes will be done on time and with the best quality. Otherwise, the organization will stand in the article of intelligent developments and prevent the growth of technology in the organization [15].
- **Risk management:** As industries grow, some risks can stall or halt progress. Management and controlling supply chain processes become difficult as risk increases. Many of the processes' failures can be attributed to risk and instability in the environment and within the supply chain processes. Therefore, knowing and managing all supply chain risks always helps implement these super-intelligent systems successfully [16, 17].
- **Change management:** Change management involves implementing strategies to control changes and support people's adaptation to them. It is very difficult to manage change successfully [18].

- **Collaborative trust**: Collaborative trust is a personal perception that is a product of one's evaluations, experiences, and inclinations about what one believes and is willing to do, as well as the words, actions, and decisions of others. This can include relying on principles, rules, norms, and decision-making procedures that express collective expectations. This factor can guarantee the organization's success when implementing smart supply chain processes [19].
- **Budget allocation**: Since the development and implementation of transformational technologies are always associated with many financial costs and challenges, allocation is a path-breaking factor in facilitating the effective implementation of smart supply chain processes [20].
- **Knowledge management of smart technologies**: Knowledge management's components (knowledge creation, knowledge storage, knowledge sharing, and knowledge application) significantly impact supply chain processes through the capabilities of transformative technology [21].
- **Technical infrastructure and hardware**: Access to hardware and technical facilities is undoubtedly one of the most important factors in implementing intelligent data-driven systems. Proper budget allocation can always help in this direction [22].
- **System integrity**: System integrity is one of the most important parameters for improving performance and reducing the number of errors in the interconnected network and guarantees the efficiency of operational processes [23].

This research prioritizes these factors using a fuzzy approach to understand better and create insight for implementing these sustainable, intelligent systems.

3 Research Methodology

In this research, to analyze the CSFs in a smart AIoE-based supply chain in Industry 5.0, a fuzzy and non-linear approach is used based on the fuzzy hierarchical analysis method. Precious judgments are required for the conventional diverse-level examination strategy currently in use. The complexity and instability of genuine decision-making issues make it difficult to give precise judgments. Therefore, it would be far more practical and achievable to grant the decision-maker plausibility to make incorrect judgments based on fluffy reasoning instead of logical reasoning. The suggested fuzzy approach for prioritizing requires several cuts and finally aggregating the derived priorities. The proposed non-linear priority method can generate priority values directly from the set of comparison judgments, which are represented as triangular fuzzy numbers, to avoid some of these steps. Fuzzy pairwise comparisons are triangular fuzzy numbers in this method. Priority rates are almost within the range of basic fuzzy judgments when the deterministic vector of weight (priority) is extracted. The weights are determined to establish Eq. (1). The weights are determined to establish Eq. (1)

$$l_{ij} \leq \frac{w_i}{w_j} \leq u_{ij} \tag{1}$$

The above fuzzy inequalities can be measured through the linear membership function of Eq. (2), which applies to each deterministic weight vector (w) with a degree.

$$\mu_{ij}\left(\frac{w_i}{w_j}\right) = \begin{cases} \frac{(w_i/w_j)-l_{ij}}{m_{ij}-l_{ij}} & \frac{w_i}{w_j} \le m_{ij} \\ \frac{u_{ij}-(w_i/w_j)}{u_{ij}-m_{ij}} & \frac{w_i}{w_j} \ge m_{ij} \end{cases} \tag{2}$$

The particular form of the membership functions transforms the fuzzy prioritization problem into a nonlinear optimization problem in the following model.

$$\begin{aligned} & \max\ \lambda \\ & \text{s.t.} \\ & \quad (m_{ij} - l_{ij})\lambda w_j - w_i + l_{ij} w_j \le 0 \\ & \quad (u_{ij} - m_{ij})\lambda w_j + w_j - u_{ij} w_j \le 0 \\ & \quad i = 1, 2, ..., n-1; j = 2, 3, ..., n; j > i \\ & \sum_{k=1}^{n} w_k = 1, w_k > 0, k = 1, 2, ..., n \end{aligned} \tag{3}$$

Unlike the linear method, which can be easily solved using simplex, solving the nonlinear problem (3) above requires a suitable numerical method for nonlinear optimization. Therefore, these non-linear relationships should be solved using software such as Games or Lingo. The optimal value of the consistency index λ, if positive, indicates that all the solution ratios completely satisfy the initial judgments; that is, $l_{ij} \le \frac{w_i}{w_j} \le u_{ij}$. If the compatibility index has a negative value, it shows that the fuzzy judgments are highly inconsistent and the solution ratios almost satisfy them. In the continuation of this research, the CSFs in AIoE-based intelligent supply chain systems will be examined and prioritized.

4 Research Findings

The process of evaluating CSFs in this study involves three distinct phases.

A) Designing a decision tree for the CSFs of the AIoE-based supply chain (Fig. 3).
B) Determining the matrix of pairwise comparisons using linguistic criteria, incorporating experts' assumptions.
C) Investigating and using fuzzy nonlinear mathematical modeling to rank and obtain the weight of the CSfs of the AIoE-based supply chain in the research model.

To prioritize the 11 last success factors extricated in this research, fuzzy questionnaires based on pairwise comparisons utilizing phonetic factors were sent to 15 specialists and college teachers. All specialists replied to the questionnaires; at last, 15 questionnaires were completed and obtained. These tables of pairwise comparison are appeared in Tables 1,2,3 and 4. These comparison tables have been utilized for computation utilizing the nonlinear fuzzy hierarchical examination strategy.

By putting the information obtained from Tables 1,2,3 and 4 within the non-linear show (3) and understanding the demonstration utilizing the Lingo computer program,

Table 1. Pairwise comparison table for overall classification.

	W1	W2	W3
W1	–	–	–
W2	(1, 3.5, 4.75)	–	–
W3	(2, 5.7, 4.2)	(1.5, 3.5, 5)	–

Table 2. Table of paired comparisons for organizational factors.

	W11	W12	W13
W11	–	–	–
W12	(2, 2.7, 3.25)	–	–
W13	(1.5, 2, 5, 5.1)	(1, 3.25, 5)	–

Table 3. Table of paired comparisons for strategic and process factors.

	W21	W22	W23	W24	W25	W26
W21	–	–	–	–	–	–
W22	(2.5, 3, 5.2)	–	–	–	–	–
W23	(2, 3, 4.2)	(1, 3, 5)	–	–	–	–
W24	(1.4, 2.5, 3.8)	(2.2, 3.1, 5)	(2, 2.5, 3.2)	–	–	–
W25	(2, 2, 7, 4.5)	(2.5, 3.2, 5, 5)	(2.1, 3,4)	(3, 3.4, 5)	–	–
W26	(1, 2, 4.1)	(1.5, 2, 325)	(2, 3, 4)	(2.1, 2.7, 2.9)	(1, 2.5, 3.25)	–

Table 4. Table of pairwise comparisons for technological factors.

	W31	W32
W31	–	–
W32	(1, 2.7, 3.7)	–

the weight, and rank of each critical success factor can be obtained. The information shown in these tables are based on phonetic factors and are in the middle value. The calculation comes about for understanding the nonlinear show for common categories and other CSFs within the AIoE-based smart supply chain in Industry 5.0, which appears in Tables 5,6,7 and 8. These tables show that a positive value for the consistency record indicates a worthy consistency of pairwise comparison matrices with a focus on phonetic elements. The weights can be normalized to get the general weight after identifying the common categories and variables within specific categories. In Table 9, the weight of all variables is obtained without considering the category and their general rank, regardless of their category or general rank. This table contains the normalized results.

Table 5. Weight and rank of CSFs in three main sections.

Category	Code	Weight	Rank	The Objective function (λ)
Organizational factors	**W1**	0.1600214	3	0.42431
Strategic factors	**W2**	0.3592142	2	
Technological factors	**W3**	0.4687433	1	

Table 6. Weight and rank of CSFs of AIoE-based smart supply chain in organizational factors.

Category	Code	Weight	Rank	The Objective function (λ)
Skilled manpower	**W11**	0.4651145	1	0.5243
Organization readiness	**W12**	0.2621521	2	
Continuous training	**W13**	0.2621458	3	

Table 7. Weight and ranking of CSFs of AIoE-based smart supply chain in the strategic factors.

Category	Code	Weight	Rank	Objective function (λ)
Technological maturity	**W21**	0.089115	5	0.3245
Risk management	**W22**	0.159152	4	
Change management	**W23**	0.259146	1	
Collaborative trust	**W24**	0.249853	2	
Budget allocation	**W25**	0.178764	3	
Knowledge management of smart technologies	**W26**	0.065436	6	

Table 8. Weight and ranking of the CSFs of AIoE-based smart supply chain in the technological factors section.

Category	Code	Weight	Rank	Objective function (λ)
Technical infrastructure and hardware	**W31**	0.5601145	1	0.3444
System integrity	**W32**	0.4321521	2	

Figure 4 shows the normal weight of all CSFs in AIoE-based smart supply chain systems in Industry 5.0, which is extracted from paired comparison tables and fuzzy hierarchical analysis. This figure illustrates how crucial technical infrastructure and hardware are to the success of AIoE-driven intelligent supply chain networks. As a result, greater attention should be paid to the provision of these facilities. Systems integration and change management are up there. By examining these factors, it can always be seen that many of these factors are dependent and will increase or decrease with the change of other parameters. This figure also shows the ranking of critical success factors in a resilient and sustainable supply chain based on AIoE in Industry 5.0. According to this figure, it can be seen that infrastructural and technological factors are among the most important factors and therefore, special attention should be considered for budget allocation and development of technical infrastructure. System integrity is on the second rank, which has a direct relationship with the first rank factor.

Table 9. Normal weight of CSFs in AIoE-based intelligent supply chain systems in I. 5.0.

Category	Weight	CSF	Weight	Normalized weight	Rank
Organizational factors	0.1600214	Skilled manpower	0.465145	0.074433	5
		Organization readiness	0.262521	0.042009	8
		Continuous training	0.262458	0.041999	9
Strategic factors	0.3592142	Technological maturity	0.089115	0.032011	10
		Risk management	0.159152	0.05717	7
		Change management	0.259146	0.093089	3
		Collaborative trust	0.249853	0.089751	4
		Budget allocation	0.178764	0.064215	6
		Knowledge management of smart technologies	0.065436	0.023506	11
Technological factors	0.4687433	Technical infrastructure and hardware	0.560145	0.262564	1
		System integrity	0.432121	0.202554	2

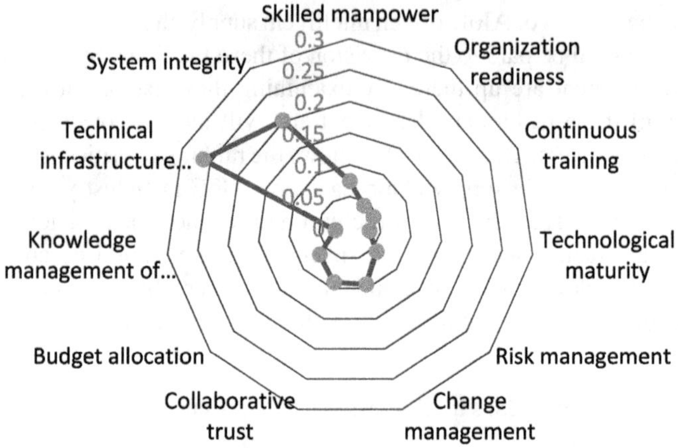

Fig. 4. Normal weight of CSFs in AIoE-based smart supply chain.

5 Conclusion

The increasing growth and expansion of information technology has created a revolution in various aspects of people's lives and the functioning of organizations. This technology has changed the working methods and attitudes of people, organizations, and governments and has created new industries, new jobs, and creativity in doing things. The supply chain, as one of the most important process-oriented parts of large organizations, is also influenced by transformative technologies. Many of these technologies, such as the Internet of Things and artificial intelligence, are so effective in the processes that without them, organizations cannot be superior in the face of competitors. Therefore, understanding the dimensions and critical success factors of smart supply chains will be very important.

Integrating transformative technologies such as AIoE, a combination of the Internet of Things and artificial intelligence of things, in supply chain management can help automate many everyday processes. This feature allows companies and organizations to focus more on their strategic and more effective business activities and not get involved in implementing repetitive activities. Smartening the supply chain and using intelligent software based on AIoE in the supply chain can improve the availability of goods and find the most suitable suppliers. AIoE redefined in this thesis, is a complex yet fascinating topic that can solve several serious issues and challenges in the supply chain and lead to an intelligent supply chain. These smart supply chain systems are always facing many challenges. A wide range of logistics and supply chain management challenges can be solved with supply chain intelligence. Security and privacy are some of the most important challenges. Therefore, the optimal, purposeful, and result-oriented implementation of these systems is of great importance, and the success of those basic steps should be taken. It is always helpful to understand the critical success factors and prioritize them to implement and maintain smart supply chain systems based on transformative technology. The most important critical success factors have been identified and evaluated in this research. As a result, the most crucial elements were identified and refined through a thorough research study incorporating the perspectives of selected experts. Critical success factors were evaluated using a non-linear fuzzy hierarchical analysis method to evaluate critical success factors in smart supply chain systems. Analyses were performed on fuzzy questionnaires based on pairwise comparisons. According to the literature review, this ranking method is an efficient and faster method than the normal hierarchical analysis method. The reason for using the fuzzy method is to consider the uncertainties in the experts' answers and to reach more accurate results. These smart supply chain systems require a special look at technical and hardware infrastructures to succeed, as revealed by the analysis results.

Acknowledgment. Alexandre Dolgui's work on this paper was supported by the European project ACCURATE https://accurateproject.eu, Grant Agreement number 101138269.

Disclosure of Interests. The authors have no competing interests to declare that are relevant to the content of this article.

References

1. Nozari, H.: Investigating key dimensions and key indicators of AIoT-based supply chain in sustainable business development. In: Misra, S., Siakas, K., Lampropoulos, G. (eds.) Artificial Intelligence of Things for Achieving Sustainable Development Goals. LNDECT, vol. 192, pp. 293–310. Springer, Cham (2024). https://doi.org/10.1007/978-3-031-53433-1_15
2. Aliahmadi, A., Nozari, H., Ghahremani-Nahr, J., Szmelter-Jarosz, A.: Evaluation of key impression of resilient supply chain based on artificial intelligence of things (AIoT). arXiv preprint arXiv:2207.13174 (2022)
3. Nozari, H., Ghahremani-Nahr, J., Rahmaty, M., Bayanati, M.: A framework for smart and resilient supply chains based on blockchain and the Internet of Things. In: Blockchain-based Internet of Things, pp. 73–86, Chapman and Hall/CRC (2024)
4. Szmelter-Jarosz, A., Ghahremani-Nahr, J., Nozari, H.: A neutrosophic fuzzy optimisation model for optimal sustainable closed-loop supply chain network during COVID-19. J. Risk Financ. Manag. **14**(11), 519 (2021)
5. Nozari, H., Tavakkoli-Moghaddam, R., Rohaninejad, M., Hanzalek, Z.: Artificial intelligence of Things (AIoT) strategies for a smart sustainable-resilient supply chain. In: Alfnes, E., Romsdal, A., Strandhagen, J.O., von Cieminski, G., Romero, D. (eds.) Advances in Production Management Systems. APMS 2023. IFIPAICT, vol. 691, pp. 805–816. Springer, Cham (2023). https://doi.org/10.1007/978-3-031-43670-3_56
6. Pitakaso, R., Sethanan, K., Worasan, K., Golinska-Dawson, P.: Utilization of data and information flows for enhancing a chambermaid scheduling decision making in a smart hospitality supply chain. Ann. Oper. Res. 1–26 (2024, in Press)
7. Nozari, H.: Supply chain 6.0 and moving towards hyper-intelligent processes. In: Information Logistics for Organizational Empowerment and Effective Supply Chain Management, pp. 1–13. IGI Global (2024)
8. Wang, Z.J., et al.: Smart contract application in resisting extreme weather risks for the prefabricated construction supply chain: Prototype exploration and assessment. Gr. Decis. Negot., 1–39 (2024, in Press)

9. El Mane, A., Tatane, K., Chihab, Y.: Transforming agricultural supply chains: leveraging blockchain-enabled Java smart contracts and IoT integration. ICT Express (2024)
10. Nozari, H., Ghahremani-Nahr, J., Szmelter-Jarosz, A.: AI and machine learning for real-world problems. In: Advances in Computers, vol. 134, pp. 1–12. Elsevier (2024)
11. Rahmaty, M., Nozari, H.: Optimization of the hierarchical supply chain in the pharmaceutical industry. Edelweiss Appl. Sci. Technol. **7**(2), 104–123 (2023)
12. Rady, F.E.E., Saleh, S.A.F., Elshafie, M.H.: A proposed hybrid model for cost management of agility smart supply chains using nanotechnology-case study. Int. J. Account. Manag. Sci. **3**(1), 24–53 (2024)
13. Sakib, S.N.: Blockchain technology for smart contracts: enhancing trust, transparency, and efficiency in supply chain management. In: Achieving Secure and Transparent Supply Chains With Blockchain Technology, pp. 246–266. IGI Global (2024)
14. Nozari, H., Szmelter-Jarosz, A., Ghahremani-Nahr, J.: Analysis of the challenges of artificial intelligence of things (AIoT) for the smart supply chain (case study: FMCG industries). Sensors **22**(8), 2931 (2022)
15. Chien, C.F., Kuo, H.A., Lin, Y.S.: Smart semiconductor manufacturing for pricing, demand planning, capacity portfolio and cost for sustainable supply chain management. Int. J. Logist. Res. Appl. **27**(1), 193–216 (2024)
16. Nozari, H., Fallah, M., Kazemipoor, H., Najafi, S.E.: Big data analysis of IoT-based supply chain management considering FMCG industries. Бизнес-информатика **15**(1), 78–96 (2021)
17. Lee, K.L., Teong, C.X., Alzoubi, H.M., Alshurideh, M.T., Khatib, M.E., Al-Gharaibeh, S.M.: Digital supply chain transformation: the role of smart technologies on operational performance in manufacturing industry. Int. J. Eng. Bus. Manag. **16**, 1–19 (2024)
18. Saruchera, F., Salimi-Zaviyeh, S.G., Vanani, I.R.: Smart supply chain management: The role of smart technologies in the global food industry. In: Building Resilience in Global Business During Crisis, Routledge, India, pp. 152–178 (2024)
19. Nozari, H., Ghahremani-Nahr, J., Szmelter-Jarosz, A.: A multi-stage stochastic inventory management model for transport companies including several different transport modes. Int. J. Manag. Sci. Eng. Manag. **18**(2), 134–144 (2023)
20. Ozbiltekin-Pala, M.: Emerging trends for blockchain technology in smart supply chain management. In: Nozari, H. (ed.) Building Smart and Sustainable Businesses with Transformative Technologies, pp. 52–72. IGI Global (2024)
21. Aliahmadi, A., Nozari, H.: Evaluation of security metrics in AIoT and blockchain-based supply chain by neutrosophic decision-making method. In: Supply Chain Forum: An International Journal, vol. 24, no. 1, pp. 31–42. Taylor & Francis (2023)
22. Shi, T., Lee, S.J., Li, Q.: Smart supply chain management in business education: reflection on the pandemics. Decis. Sci. J. Innov. Educ. **22**(1), 19–32 (2024)
23. Nozari, H., Szmelter-Jarosz, A.: An analytical framework for smart supply chains 5.0. In: Nozari, H. (ed.) Building Smart and Sustainable Businesses With Transformative Technologies, pp. 1–15. IGI Global (2024)

Trading Digital-Valued Assets Within Cyber-Physical Manufacturing Supply Chains: A Scoping Review of Additive Manufacturing and Digital Trade

Kwaku Adu-Amankwa[1(✉)], Andrew Wodehouse[1], Angela Daly[2], Athanasios Rentizelas[3], and Jonathan Corney[4]

[1] Department of Design Manufacturing and Engineering Management, University of Strathclyde, Glasgow, UK
{kwaku.adu-amankwa,andrew.wodehouse}@strath.ac.uk
[2] Leverhulme Research Centre for Forensic Science, School of Science and Engineering, University of Dundee, Dundee, UK
adaly001@dundee.ac.uk
[3] School of Mechanical Engineering, National Technical University of Athens, Athens, Greece
arent@mail.ntua.gr
[4] School of Engineering, University of Edinburgh, Edinburgh, UK
j.r.corney@ed.ac.uk

Abstract. Additive manufacturing, a cyber-physical process that has gained popularity, can revolutionise manufacturing across various industries. By trading digital assets globally, additive manufacturing can support Industry 5.0 principles like sustainability or resilience. Thus, examining the potential for digital trade within cyber-physical manufacturing supply chains is crucial. This paper responds to the research challenge by providing an overview of digital trade (including e-commerce) within the additive manufacturing research landscape. The scoping review technique systematically investigated the extant literature and comprehensively summarised evidence based on their emergent themes and recurring content. The paper enumerates evidence of research efforts made so far within this field. The findings reveal that the topic is still not extensively studied despite this research field's growth. It has also been discovered that existing studies predominately focus on digitally-ordered processes rather than digitally-delivered processes, which primarily involve the trade of physical artefacts rather than cyber artefacts. These knowledge gaps highlight the opportunities for further research that will advance the field and significantly contribute to society's transition into Industry 5.0. It is anticipated that this paper, by providing a comprehensive overview and identifying research gaps, will guide stakeholders (practitioners, researchers, academics, educators, and policymakers) with foundational knowledge on the subject and inspire more studies in this area, ultimately enhancing society's readiness to embrace active trading of digital-valued assets within cyber-physical supply chains using additive manufacturing.

Keywords: Additive Manufacturing · Digital Trade · E-Commerce · 3D Printing · Supply Chain Management · Value Chain · Industry 5.0

Published by Springer Nature Switzerland AG 2024
M. Thürer et al. (Eds.): APMS 2024, IFIP AICT 729, pp. 91–104, 2024.
https://doi.org/10.1007/978-3-031-65894-5_7

1 Introduction

Industrial and technological revolutions have ushered society and enterprises toward an awareness of the great enablement of digitally enabling or digitally transforming one's ability to create and consume artefacts across multiple sectors. A significant resource of the current era is centred around valuable intellectual knowledge on how to make items and send these items to markets that were once impossible due to physical or geographic constraints [1]. One can now use internet technologies to order items of need or want by tapping on their phone or clicking with a computer, and the item can be sent to them digitally or physically in exchange for payment or other valuable items.

This practice of exchanging valued items is what formed the basis of trade, and eventually, nations aligned themselves on regional or international levels based on their trade interests [2]. As we enter Industry 5.0, the digital era has become essential to our daily lives and business practices. This involves using digital technologies to facilitate trade, which is often referred to as Digital Trade.

Digital trade has no clear definition and is sometimes used when referring to electronic commerce (e-commerce), yet it is underpinned by using digital technology to enable or transform existing trade practices internationally and domestically. So, beginning with the OECD's description of digital trade as "all trade that is digitally ordered and/or digitally delivered" [3]; thus, they adopt an alignment with the purchasing mechanisms and the delivery of products. Another perspective considers digital trade as commerce in "products and services delivered via the internet" [4]. However, the U.S. International Trade Commission took a broader perspective to describe digital trade as "domestic commerce and international trade in which the Internet and Internet-based technologies play a particularly significant role in ordering, producing, or delivering products and services" [5]. Undoubtedly, technology and data are inseparable from digital trade, especially as society advances towards Industry 5.0; this complements Industry 4.0, which is mainly centred on interconnectivity among machines and systems, by emphasising sustainable, human-centric and resilient industry practices on manufacturing and societal systems alike [1]. Within the manufacturing sector and beyond, an emergent disruptive technology favoured with the promise of being a crucial contributor to the future of digital transformation is additive manufacturing (popularly referred to as 3D printing). This is mainly due to its hefty reliance on digital technologies and cyber-physical nature, allowing the shift of value-addition within the supply chain between both digital and physical spheres [6]. Simply put, additive manufacturing comprises processes for making a physical component from a digital 3D model data, usually by joining material in layers.

Additive manufacturing is considered a suitable candidate technology to practise distributed manufacturing, simplify complex supply chains, enable on-demand production, and foster near-net shape consolidation, thus contributing to sustainability goals by reducing material usage and physical footprint operations [7]. Accordingly, by adopting the lens of digital trade, we make a case for additive manufacturing's potential to actualise Industry 5.0 principles via some illustrative references. Starting with the influence of evidence about global exchanges on critical digital models during the COVID-19 pandemic to make physical parts close to the consumption point [8]. Moreover, there is a push for holistic digital inventories to build parts on-demand when needed for maintenance or repairs, especially as parts become obsolete [9]. Timely predictions have also been made about manufacturing revolutions to support open trading of intellectual property embedded in digital model files instead of physical components [10]. Ultimately, it boosts hyper-customisation and customer experiences, supports resilience and sustainability, and facilitates co-creation among humans, machines, and data as society moves towards Industry 5.0.

Despite its benefits, the anticipated widespread adoption of additive manufacturing technology to actively trade digital-valued assets for distributed manufacturing practices has not occurred as expected. Industry 5.0 also remains at a nascent stage [11], creating a dilemma between uncertainty and promise. This paper seeks to explore efforts toward trading additive manufacturing's digital assets via systematic searches and synthesises of literature. So, this paper examines literature about characteristics of additive manufacturing that facilitate trading valuable digital assets, such as design files and build instructions or parameters, to produce physical components. The review explores this mode of digital trade—where products are ordered and delivered digitally—within the framework of additive manufacturing supply chains.

The remainder of the paper is structured as follows: Sect. 2 details the PRISMA-ScR methodological step in generating the review protocol, retrieving information, and summarising data. Section 3 presents the emergent descriptive results from the data summaries, while Sect. 4 presents thematic results from the same data summaries. Section 5 synthesises outcomes of the review findings; then, Sect. 6 concludes with a recap summary of all discussions and goals within this paper while highlighting limitations and opportunities for further studies.

2 Research Methodology

The scoping review technique was adopted because it is suitable for exploratory studies that identify and map literature evidence to comprehensively synthesise knowledge beyond traditional academic sources [12]. Tricco et al.'s [13] PRISMA-ScR (Preferred Reporting Items for Systematic reviews and Meta-Analyses extension for Scoping Reviews) methods were used to design, conduct, and report results for this scoping review. The consultation step is excluded as it did not occur before developing and writing this paper. The detailed scoping review methodological steps and justification are described in three parts: Review Protocol, Data Collection, and Data Analysis.

2.1 Review Protocol

A review protocol was developed a priori to establish the review goals in writing and allow agreement on scope within resource constraints. It also served as a guide for reviewers to determine the consistent assessment of candidate literature retrieved from databases based on pre-determined eligibility criteria (inclusion and exclusion), which are presented in Table 1. The Population, Concept, and Context (PCC) acronym for scoping review integration of literature via question development was employed in determining the elements of the study's review protocol, as advised by Peters et al. [12].

Table 1. Inclusion and Exclusion Criteria to Determine Literature Eligibility

Criteria	Inclusion	Exclusion
Document Type	Journal articles, conference papers, reviews, book chapters, dissertation/thesis reports	Books, news items, magazine articles, standards, trade journals, advertisements
Language	English	Non-English
Population	Literature primarily about additive manufacturing or its equivalent searched terms	Literature that fails to demonstrate additive manufacturing as its purpose
Phenomenon	The study focuses on digital trade or e-commerce as its core phenomenon of interest	The study mentions digital trade or e-commerce as illustrations, but it is not the focus of the study
Context	The study describes a focus on additive manufacturing within supply chain settings or context	The study fails to describe an additive manufacturing supply chain context clearly
Timeframe	Unrestricted	N/A
Publication Status	Published and unpublished (pre-prints) were considered	N/A
Country	Unrestricted	N/A

From the content perspective, an initial selection criterion was based on the study population of *literature about additive manufacturing research.* This allowed an oriented focus on retrieving search results within the additive manufacturing research landscape. Also, the selection criterion for the phenomenon of interest was discussions or outputs focused on digital trade or e-commerce channels. This ensured filtered literature upheld the core discussions or outputs aligned with digital trade or e-commerce. Then, the selection criterion based on the context was *discussions or outputs focused on digital trade or e-commerce of additive manufacturing artefacts (digital or physical).* This ensured the selected literature complied with outputs or discussions aligned with the overall study goal. To cover all relevant results, a combination of multidisciplinary database platforms was chosen to cover the fields of engineering, business, management, legal, and computer science. The review team chose *Scopus*, *Web of Science*, *EBSCOhost*, *ProQuest*, and *Westlaw* platforms to query 70 database collections because of their diverse content, extensive document types, access to various academic and non-academic databases, and the ability to filter and export relevant results from their online portals.

The chosen databases were purposefully opened to fulfil the intent of an extensive search scope and outreach to relevant literature that may exist in unpopular or hard-to-reach places; moreover, the concept of digital trade of additive manufacturing assets was considered widely applicable across multiple sectors due to the sector-agnostic applications of additive manufacturing (e.g., food, construction, industrial, home). Search terms employed were grouped into two main categories, *Additive Manufacturing* and *Digital Trade*, where the Boolean operator "OR" was used within categories, whilst the Boolean operator "AND" was used across categories. The emergent search string was applied to the TITLE, KEYWORDS, and ABSTRACT fields of each database as deemed applicable; however, Westlaw was the only result that searched the full text from the onset of the search exercise. The search string was: *("3d print*" OR "additive manufactur*" OR "additive process*" OR "additive technique*" OR "bioprint*" OR "digital fabricat*" OR "digital manufactur*" OR "freeform fabricat*" OR "layer manufactur*" OR "rapid manufactur*" OR "rapid prototyp*" OR "rapid tool*" OR "three dimensional print*")* **AND** *(trade OR commerce).* The wildcard truncated search terms allowed the search term to find different spelling, tense, or word-ending alternatives extensively; however, it also retrieved some good and completely divergent results. Nevertheless, it draws upon the strengths of comprehensively interrogating databases using robust search strategies. The data collection and data analysis are described in the following sections.

2.2 Data Collection

Tricco et al.'s [13] PRISMA-ScR guidelines were used for data collection over three phases: (1) query online databases for data identification; (2) filter through retrieved papers for eligibility screening; (3) enlist eligible papers as finalising to be included for data charting and onward analysis. Figure 1 summarises the reported data collection steps.

Throughout the identification phase, **1,170** papers were identified from the 72 searched databases. Eventually, 471 duplicates were removed from identified papers before being advanced into the screening phase. During the screening phase, **699** papers were screened for relevance by reading their titles and abstracts to determine their suitability for full-text retrieval; 602 papers were excluded after that bibliographic content screening. So, **97** papers were sought for their full-text papers using Google Scholar and SuPrimo (University of Strathclyde Library Search Portal), yet 13 papers were unavailable as full-text online. Eventually, **84** papers advanced to full-text screening by reading each holistically, seeking content alignment with content eligibility criteria, where 71 papers were excluded as ineligible content. At the included phase, 13 papers had data extracted from them to confirm their eligibility for inclusion in further data analysis using descriptive statistics and content analysis techniques.

Peters et al.'s [12] recommendations on quality checks for the data collection approach were performed by collectively piloting the entire screening process with 25 randomly selected papers retrieved from the online databases (5 papers each per co-author). The review team screened them using the eligibility criteria, and discrepancies or required modifications were discussed until mutual agreement was reached. Two authors conducted the full-text screening of papers at the eligibility stage, while the remainder of the review team served to spot-check and resolve disagreements.

2.3 Data Analysis

Included papers were analysed primarily using quantitative methods (content analysis and descriptive statistics) to summarise the extent to which qualifying bibliographic data from retrieval databases and content within each paper. The qualitative method (thematic analysis) was sparingly used but supported the encoding of identified contents into emergent themes in categorising each paper according to this study's review protocol; thus, the results are presented as summaries.

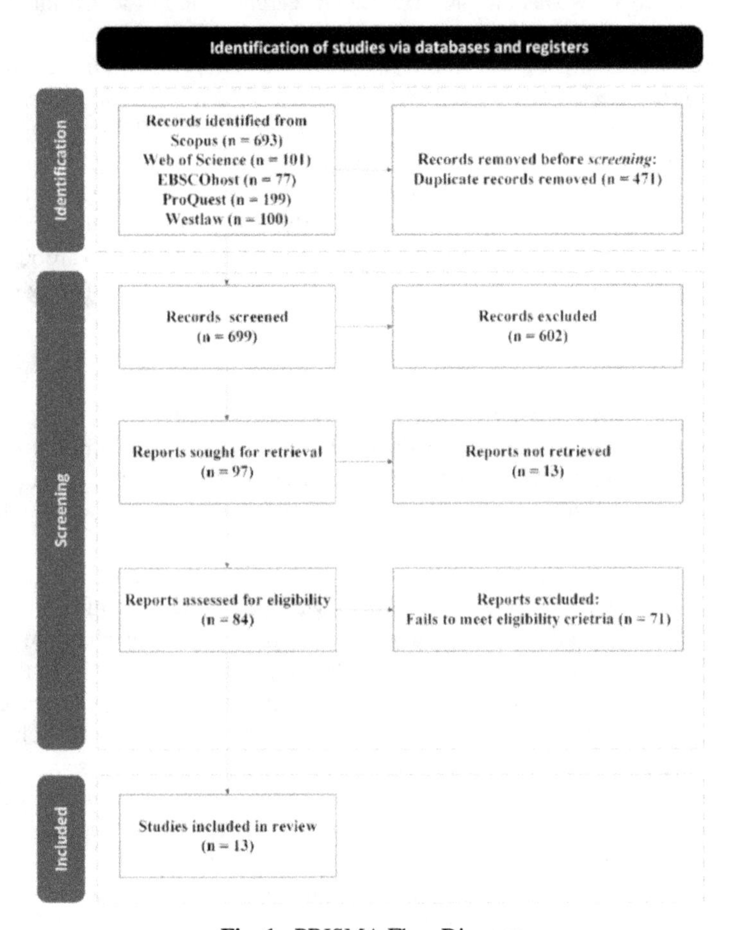

Fig. 1. PRISMA Flow Diagram

3 Findings – Descriptive Summaries

The descriptive summaries provide an overview of the 13 included papers in this scoping review. Most included papers were retrieved from Scopus (92% papers), followed by Web of Science (69% papers), then ProQuest (38% papers), and EBSCOhost (31% papers); there was no representation from Westlaw. Retrieved papers ranged from 2012–2023. A breakdown is presented in the sub-sections below.

3.1 Publication Trends (Years, Authors, Sources)

Analysis of the publication trends revealed that the topic was growing gradually from its initial representation in 2012 with 1 paper to the present state of 2023 with 13 papers.

Still, there were some gap years (2013, 2014, 2017, 2021) with no relevant studies, some peak years (2015, 2019, 2022, 2023) with multiple studies, and regular years (2012, 2016, 2018, 2020) with single studies. The publication sources also varied with no repeated source; however, in terms of document types, Journal Articles (54% papers) were the most prevalent, followed by Conference Papers (38% papers), and then Book Chapters (8% papers). A summary of publication timelines is illustrated in Fig. 2.

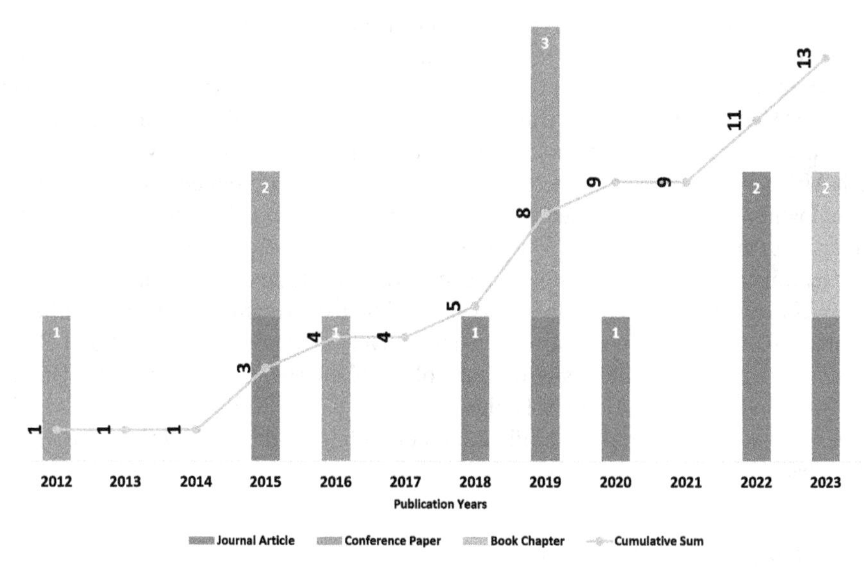

Fig. 2. Publication Distribution Trends by Year

Among the 37 individual authors identified within the included papers, it was discovered that each author had a single document input within this field, so it was challenging to determine the active authors.

3.2 Publication Retrieval Terms (Keywords, Databases)

63 unique Author Keywords and 122 unique Database Index-Terms were identified with the included papers, and their consolidation resulted in 16 Encoded Labels comprising *Additive Manufacturing, Trade, Commerce, Intellectual Property, Supply Chain, Design, Manufacturing, Rules of Origin, Methodology, Business Model, Cloud Computing, Management, Risk, Distributed Manufacturing, European Union*, and *Production.* The most represented terms in Author Keywords were Additive Manufacturing and 3D Printing (7 papers each). Followed by Trade (3 papers), then E-commerce, Innovation, and Supply Chain (2 papers each). All the remaining terms had single-paper representations.

Additionally, when focusing solely on consolidated Author Keywords as labels, it emerged that Trade, Additive Manufacturing, and Management comprised the most represented labels (19%, 18% and 11%, respectively); then *Commerce, Intellectual Property, Design, Supply Chain*, and *Methodology* were adequately represented labels (9%,

6%, 6%, 5%, 5% respectively). Meanwhile, the least represented labels were *Manufacturing*, *Business Model*, *Cloud Computing*, *Risk*, *Distributed Manufacturing*, *European Union*, and *Production* (3% each). Yet it is acknowledged that incorporating Database Index-Terms results in Management and Trade labels obtaining dominance.

3.3 Publication Collaborations (Institutions, Countries)

An analysis of collaboration via country and institutional representation reveals that authors conducted studies within institutions that gave a dispersed global representation by 10 countries comprising Germany, the United Kingdom, Italy, and Spain to represent Europe (55% collectively). The United States alone represented the Americas (15%); then China, Hong Kong, and India represented Asia (15% collectively). Afterwards, South Africa alone represented Africa (10%); Turkey also represented Eurasia (5%). 23 unique institutions were affiliated with the authors: they generally comprise academic (universities) and industry (companies). Among the frequently represented institutions were Central University of Technology (South Africa); FH Aachen (Germany), RWTH Aachen University (Germany), University of Göttingen (Germany), Julius-Maximilians-University (Germany), North Carolina State University (United States), World Bank (United States), Tianjin University of Technology (China), Cardiff University (United Kingdom), and Caritas Institute of Higher Education (Hong Kong) which had multiple representations ranging from 2–4 affiliated authors; whilst all other institutions had single authors each.

4 Findings – Thematic Summaries

The thematic summaries provide an overview of the 13 included papers in this scoping review. The emergent themes were derived from summaries of the charted evidence about goals, outcomes, models, issues, and limitations that were identified within each paper in relation to this review's aim and presented in Table 2. The results are categorised based on the identified cyber-physical sphere in which additive manufacturing trading takes place (i.e. ordering or/and delivery in digital or physical space) and the nature of the traded artefact (i.e. cyber or/and physical) [3, 5, 6].

It is observed that different aspects of digital trade were addressed from both service and product perspectives; these vary from explored e-commerce concepts (e.g. blended shopping, reverse auction) within cyber-physical contexts, proposed models suitable for e-commerce channels, platform development to foster advances in multi-purpose marketplaces that could connect 3D digital content creators (designers) with skilled additive manufacturing production houses (bureaus) for free 3D model exchanges. Meanwhile, other themes examined emerging supply chain reconfigurations, facilitating the coordination of designer needs with manufacturer production capabilities for distributed manufacturing. Finally, the remaining provided insights on the estimated determinants of destination country and transportation cost on the adoption of digital trade, while others explored rule of origin agreements via intellectual property value embedded in digital files, thus supporting consumer-side research, and establishing the befitting utilisation policies associated with selected scenarios for digital trade of additive manufacturing assets. Also, most papers focused on additive manufacturing products over additive manufacturing services, while a similar digitally-ordered trade was closely followed by digitally-delivered trade; then, physical artefacts were popularly represented over cyber artefacts. A blend of these themed labels was represented, and collectively, these formed the 5 cluster groups shown in Table 2.

Table 2. Publication Theme Distribution

Author(s)	Output Summary	Relevance
DIGITALLY-DELIVERED: PHYSICAL ARTEFACT [X4]		
Daduna [14]	An overview of the disruptive nature of AM and its effect on logistics processes in selected e-commerce supply chains to provide insights into modified supply chain configuration	Examined emerging supply chain reconfigurations based on the proximity of production equipment to consumers within the digital trade of additive manufacturing assets (products)
Abeliansky et al. [15]	A proposed economic model that theoretically investigates the impact of additive manufacturing on foreign direct investment and international trade	Provided insights on the estimate determinants of the destination country and transportation cost on the adoption of digital trade of additive manufacturing assets (product)
Freund et al. [16]	An investigation into the effect of additive manufacturing on international trade by examining the case of hearing aids	Revealed a plausible inverse parabolic (U-shaped) effect with global trade exchanges across borders to within borders as local capability increases for the digital trade of additive manufacturing assets (products)
Ekren et al. [17]	An examination of the utilisation policy effects on influencing factors that emerge at the integration of additive manufacturing in e-commerce supply chain networks	Established via theoretical simulations of influencing factors, the befitting utilisation policies associated with selected scenarios for the digital trade of additive manufacturing assets (products/services)
DIGITALLY-ORDERED: CYBER ARTEFACT [X1]		
Pahwa et al. [18]	An algorithm that improves content creators' accessibility to additive manufacturing service providers willing to make the part at an auctioned price	Illustrates the reverse auction mechanism via a designed marketplace to acquire manufacturing-as-a-service as part of the digital trade of additive manufacturing assets (services)
DIGITALLY-ORDERED: PHYSICAL ARTEFACT [X4]		
Fuchs et al. [19]	A toolkit that supports additive manufacturing merchants and producers to organise customer-added value via blended shopping	Explored blended shopping concepts' in-feeding role in the digital trade of additive manufacturing assets (products/services)
Yang et al. [20]	A contemporary platform to trade construction additive manufacturing products and empirically verified with stakeholder needs	Created a marketplace to facilitate the customisation and digital trade of additive manufacturing assets (products)
Stein et al. [21]	A designed marketplace for openly scheduling designer-manufacturer orders with available production capabilities within distributed additive manufacturing	Facilitated coordination of designer needs with manufacturer production capabilities for the digital trade of additive manufacturing assets (service)
Dzogbewu et al. [22]	A bibliometric analysis that provides research directions on the nature of commercialisation of additive manufactured products	Identified the need for research on the consumer side of commercialisation and digital trade of additive manufacturing assets (product)

(*continued*)

Table 2. (*continued*)

Author(s)	Output Summary	Relevance
DIGITALLY-ORDERED/DELIVERED: CYBER ARTEFACT [X2]		
Ng et al. [23]	A cloud-enabled platform system for trading and sharing authentic additive manufacturing ready contents (objects/files)	Developed an online system that fosters free 3D model content exchange amongst creators as part of the digital trade of additive manufacturing assets (products)
Wade [24]	An assessment on factoring the rules of origin into determining the additive manufactured products considering the role of the digital file	Enlightens considerations on determining the digital service inputs have in determining the value based on the origin of products in the digital trade of additive manufacturing assets (products/services)
DIGITALLY-ORDERED/DELIVERED: CYBER/PHYSICAL ARTEFACT [X2]		
Piller et al. [25]	A discussion on the economic and management effects of additive manufacturing business models for commercialisation along its value chain	Uncovered perspectives on the innovation and production locus via business models that facilitate the digital trade of additive manufacturing assets (products/services)
Eyers et al. [26]	A framework that classifies and analyses available e-commerce channels for additive manufacturing management	Proposed models suitable to e-commerce channels for additive manufacturing that harness existing systems development to support the digital trade of additive manufacturing assets (products)

5 Discussion

Based on the emergent findings from the scoping review presented in Sect. 3, 4, this section discusses their collective synthesis. A recall of the study aim is to explore the efforts within the literature about digital trade and additive manufacturing; this had huge expectations due to the varied and broad views about what entails digital trade (see Sect. 1); but the resultant discovery that out of 1,170 retrieved papers from 70 databases, only 13 relevant papers met the inclusion criteria surprisingly revealed a lack of specific attention towards the adopted lens. This could suggest that within the wider context of digital trade research, an additive manufacturing focus barely dominated as compared to the traditionally aggregated trade performance statistics within the larger manufacturing sector or additive manufacturing is often mentioned as an indicative technology facilitating digital trade while the focus remained on something else (this was a familiar observation at full-text screening) or even perhaps the knowledge synthesis method was restrictive that it falsely excluded relevant papers. Regarding the latter, it may be impossible to eliminate interpretation biases or errors; however, by carefully following the guidelines set out in Peters et al. [12], and Tricco et al. [13] it ensured that quality control measures were applied to confirm methodological consistency, on top of doubt clarification on eligibility amongst the review team.

The literature publication trends of the 13 included studies, from a year's range perspective, that revealed relevant literature covered 2012–2023 was not completely surprising, considering that the 3rd Industrial Revolution (Digital Revolution) is recorded to have occurred between the mid-20th to early 21st century (the 1960s-2000s); accordingly, that revolution enabled extensive advancement in internet and computing technology as well as facilitation to develop the maiden additive manufacturing technology within the late 20th century, which was eventually commercialised within the early 21st century alongside e-commerce emergence in late 20th century. Meanwhile, a preliminary review of broader non-academic literature like that of Broadbent et al. [5], Mourougane [3], Jones et al. [27], and Quill et al. [28], provided a core background understanding of the breadth of digital trade development to the point of identification that manufacturing was considered a digitally intensive industry or sector to aid in formulating the study structure and navigating through the academic literature. Moreover, the prevalence of journal articles and conference papers suggests that pending research outputs could be in the publication pipeline. This also indicates high-end contributions to the topic, as authors' works typically receive peer reviews.

It was interesting to discover how diverse contributors were represented across the globe at a continent level, yet the underrepresentation in emerging economies raised questions on speculative inferences about digital technology availability challenges and possible trade integration complexities, perhaps due to uncertainty, lack of trust, or unavailable infrastructure concerns. One would expect that the countries ranked top tier to have the highest number of internet users (top 10 as of 2023 by Statista: China, India, United States, Indonesia, Brazil, Russia, Nigeria, Japan, Mexico, and Philippines in descending order) would have a research focus on digital trade involving additive manufacturing, but rather it looks like advanced economies were at the actualisation stages to consider such digital transformations of manufacturing processes. Furthermore, Wohlers Report 2023 reveals that additive manufacturing equipment/installations are extending into different countries [6], but it is speculated that trade reporting statistics may remain traditional as digital trade is complex.

Coming to the emerging themes, it was not surprising to see that *digitally-delivered: physical artefact* and *digitally-ordered: physical artefact* were the themes dominating the classified papers; this may suggest a classical e-commerce system where the physical artefacts are more of interest to the end-user and the manufacturer or third-party retailer is only able to supply the physical artefacts. Meanwhile, *digitally-ordered/delivered: cyber artefact* being present was evidence of the efforts by scholars about the complete digital trade service within additive manufacturing supply chains to allow an end user to select their preferred place of manufacturing upon receiving the cyber artefact, however, the revelation of the option that *digitally-ordered/delivered: cyber/physical artefact* existed to suggest a classic case of blended ordering and delivery of contents due to lead time or international trade challenges especially for legacy repairs; that way the end user received a combination available physical stock and digital files like what occurred during the COVID-19 pandemic by companies that signed a pledge to make their designs available (for free?) to the additive manufacturing community in collectively fighting the pandemic. Lastly, *digitally-ordered: cyber artefact*, being the least represented group, was still one of the most promising cases that allows an end user to acquire and adapt design files as needed, then also avail their modifications to others; this is quite popular in the hobbyist and prosumer communities.

The overall finding still emphasises that digital trade within additive manufacturing supply chains remains understudied, especially regarding cyber artefacts, as much attention is focused on physical artefacts. As society approaches Industry 5.0, trading digital assets could promote sustainability and human-centric values. This is because production could occur at an approved location near the point of consumption, while the digital file's value could represent the intellectual creations of individuals or organisations involved. However, complex national or international policy issues must be considered, such as determining trade import or export tariffs, applying the WTO's rules, and enforcing intellectual property protection. These issues could exclude users from accessing or using digital content and completely prevent trading amongst themselves. Another big concern is the lack of trust within digital spaces and integrity risks from end-users who may illicitly manufacture products outside their license agreement. Despite these challenges, it is essential to remember the promising potential of trading digital-valued assets within additive manufacturing supply chains for growth towards Industry 5.0. Thus, these considerations make the described challenge VUCA (Volatility, Uncertainty, Complexity, Ambiguity) because it is not simply a "hard system" that needs technology solutions but involves "soft systems" like societal practices and infrastructure to fulfil the promise of such forms of digital trade.

6 Conclusions

Trading digital-valued assets within additive manufacturing supply chains is a practice that shows great promise in extending access to markets that are sometimes geographically inaccessible at the domestic and foreign levels. Especially from a sustainability perspective, it can support keeping end-of-use or end-of-life products longer in service via value retention processes like remanufacturing.

This paper explores efforts within the extant literature on evidence that addresses efforts toward trading additive manufacturing's digital assets. The resultant 13 papers out of 1,170 retrieved papers from 70 databases collectively indicate a research niche and knowledge gap that digital trade within additive manufacturing supply chains remains understudied as few papers were identified as relevant. Thus, further research on this topic is required, and it is proposed that future studies must focus on areas that: (1) aid in understanding the cyber, physical, social, security, and environmental views of this VUCA challenge; (2) provide insights on interventions that address infrastructure and stakeholder concerns; (3) generate empirical evidence that could inform trade policies and organisational practices [27].

Practical efforts towards bridging this identified gap shall include a follow-up study that will consult key stakeholders in academia and industry to extend this study's findings by focusing on how additive manufacturing allows the exchange (trading) of valuable digital assets (design files, build instructions/parameters) to make or remake components in the physical space. Also, a study examining digital trade models within the additive manufacturing supply chain is crucial to building an evidence trail and providing insights on interventions that aid in bridging this topic's gaps.

Limitations identified with this study include language barriers that led to a focus on English language papers, potentially excluding relevant studies in other languages. Also, when writing this paper, the optional step of consulting stakeholders to rationalise key findings and prepare for an extended study has not been conducted. However, the outcomes of this study act as a precursor to follow-up studies by providing researchers and practitioners within the additive manufacturing supply chain with presented evidence from the literature. Despite the promising potential of trading digital assets of additive manufacturing within the supply chain as society nudges gradually towards industry 5.0 and beyond, this topic is underrepresented.

Ultimately, this paper serves as a call to action for practitioners, scholars, policymakers, and governing bodies to primarily understand why, when, which, and how stakeholders would engage or avoid such digital trading practices when it comes to additive manufacturing by investigating multidisciplinary views on the available or needed infrastructure (policies, legislation, technology, etc.) to conduct this practice.

Acknowledgements. This work was supported by the UKRI Made Smarter Innovation Challenge and the Economic and Social Research Council via InterAct [Grant Reference ES/W007231/1]. The first author acknowledges receiving an InterAct Early Career Researcher (ECR) Fellowship.

Disclosure of Interests. The authors have no competing interests to declare relevant to this article's content.

References

1. Xu, X., Lu, Y., Vogel-Heuser, B., Wang, L.: Industry 4.0 and industry 5.0—inception, conception and perception. J. Manuf. Syst. **61**, 530–535 (2021). https://doi.org/10.1016/j.jmsy.2021.10.006
2. Anagaw, B.K., Tabakis, C., Zanardi, M.: The legacy of conflict on trade negotiations (2019)
3. Mourougane, A.: Measuring digital trade (2021). https://doi.org/10.1787/48e68967-en
4. van Rensburg, L.J.J.: The factors influencing ease of doing business with digital trade in countries with strict exchange controls. In: Kolaković, M., Horvatinović, T., and Turčić, I. (eds.) The 6th Business & Entrepreneurial Economics, Plitivice Lakes, pp. 102–108 (2021)
5. Broadbent, M.M., Pinkert, D.A., Williamson, I.A., Johanson, D.S., Kieff, F.S., Schmidtlein, R.K.: Digital Trade in the U.S. and Global Economies, Part 2 (2014)
6. Wohlers Associates: Wohlers Report 2023. Wohlers Associates, Fort Collins (2023)
7. Pei, E., et al. (eds.): Springer Handbook of Additive Manufacturing. Springer, Cham (2023). https://doi.org/10.1007/978-3-031-20752-5
8. Ballardini, R.M., Mimler, M., Minssen, T., Salmi, M.: 3D printing, intellectual property rights and medical emergencies: In: Search of New Flexibilities. IIC - International Review of Intellectual Property and Competition Law, pp. 1–25 (2022). https://doi.org/10.1007/s40319-022-01235-1
9. Zhang, Y., Westerweel, B., Basten, R., Song, J.-S.: Distributed 3D printing of spare parts via IP licensing. Manuf. Serv. Oper. Manag. **24**, 2685–2702 (2022). https://doi.org/10.1287/msom.2022.1117
10. Geissbauer, R., Wunderlin, J., Lehr, J.: The future of spare parts is 3D: a look at the challenges and opportunities of 3D printing (2017)
11. Crnobrnja, J., Stefanovic, D., Romero, D., Softic, S., Marjanovic, U.: Digital transformation towards industry 5.0: a systematic literature review. In: Alfnes, E., Romsdal, A., Strandhagen, J.O., von Cieminski, G., Romero, D. (eds.) APMS 2023. IFIP AICT, vol. 689, pp. 269–281. Springer, Cham (2023). https://doi.org/10.1007/978-3-031-43662-8_20
12. Peters, M., Godfrey, C., McInerney, P., Munn, Z., Tricco, A.C., Khalil, H.: Scoping reviews. In: Aromataris, E., Munn, Z. (eds.) JBI Manual for Evidence Synthesis. JBI (2020). https://doi.org/10.46658/JBIMES-20-12

13. Tricco, A.C., et al.: PRISMA extension for scoping reviews (PRISMA-ScR): checklist and explanation. Ann. Int. Med. **169**, 467 (2018). https://doi.org/10.7326/M18-0850
14. Daduna, J.R.: Disruptive effects on logistics processes by additive manufacturing. IFAC-PapersOnLine **52**, 2770–2775 (2019). https://doi.org/10.1016/j.ifacol.2019.11.627
15. Abeliansky, A.L., Martínez-Zarzoso, I., Prettner, K.: 3D printing, international trade, and FDI. Econ. Model. **85**, 288–306 (2020). https://doi.org/10.1016/j.econmod.2019.10.014
16. Freund, C., Mulabdic, A., Ruta, M.: Is 3D printing a threat to global trade? The trade effects you didn't hear about. J. Int. Econ. **138**, 103646 (2022). https://doi.org/10.1016/j.jinteco.2022.103646
17. Ekren, B.Y., Stylos, N., Zwiegelaar, J., Turhanlar, E.E., Kumar, V.: Additive manufacturing integration in E-commerce supply chain network to improve resilience and competitiveness. Simul. Model. Pract. Theory **122**, 102676 (2023). https://doi.org/10.1016/j.simpat.2022.102676
18. Pahwa, D., Starly, B., Cohen, P.: Reverse auction mechanism design for the acquisition of prototyping services in a manufacturing-as-a-service marketplace. J. Manuf. Syst. **48**, 134–143 (2018). https://doi.org/10.1016/j.jmsy.2018.05.005
19. Fuchs, B., Ritz, T., Stykow, H.: Enhancing the blended shopping concept with additive manufacturing technologies. In: Proceedings of the 8th International Conference on Web Information Systems and Technologies, Porto, pp. 492–501. SciTePress - Science and and Technology Publications (2012). https://doi.org/10.5220/0003933704920501
20. Yang, H., Zhu, K., Zhang, M.: Analysis and building of trading platform of construction 3D printing technology and products. Math. Probl. Eng., 1–11 (2019). https://doi.org/10.1155/2019/9507192
21. Stein, N., Walter, B., Flath, C.: Towards open production: designing a marketplace for 3D-printing capacities. In: Fortieth International Conference on Information Systems (ICIS), Munich, pp. 1–15 (2019)
22. Dzogbewu, T.C., Amoah, N., Fianko, S.K., Afrifa, S., de Beer, D.: Additive manufacturing towards product production: a bibliometric analysis. Manuf. Rev. **9**, 1–21 (2022). https://doi.org/10.1051/mfreview/2021032
23. Ng, K.-C., Pang, W.-M.: A 3D content cloud: sharing, trading and customizing 3D print-ready objects. In: 2016 IEEE Second International Conference on Multimedia Big Data (BigMM), Taipei, pp. 174–177. IEEE (2016). https://doi.org/10.1109/BigMM.2016.82
24. Wade, D.E.: 3D printing, valuation, and service inputs: looking to the future rather than the past to design rules of origin for advanced manufactured products. In: de Amstalden, M., Moran, N., Asmelash, H. (eds.) PEPA-SIEL 2022. European Yearbook of International Economic Law, pp. 119–14. Springer, Cham (2023). https://doi.org/10.1007/978-3-031-41996-6_5
25. Piller, F.T., Weller, C., Kleer, R.: Business models with additive manufacturing—opportunities and challenges from the perspective of economics and management. In: Brecher, C. (eds.) Advances in Production Technology. LNPE, pp. 39–48. Springer, Cham (2015). https://doi.org/10.1007/978-3-319-12304-2_4
26. Eyers, D.R., Potter, A.T.: E-commerce channels for additive manufacturing: an exploratory study. J. Manuf. Technol. Manag. **26**, 390–411 (2015). https://doi.org/10.1108/JMTM-08-2013-0102
27. Jones, E., Kira, B., Sands, A., Garrido Alves, D.B.: The UK and Digital Trade: Which Way Forward? Oxford (2021). https://doi.org/10.35489/BSG-WP-2021/038
28. Quill, P., et al.: Handbook on Measuring Digital Trade, 2nd edn. (2023). https://doi.org/10.1787/ac99e6d3-en

Basic Research on Laborer State Prediction Towards the Realization of Human Digital Twin

Ruriko Watanabe(✉), Yuu Takihara, Kotomichi Matsuno, and Yoshitaka Tanimizu

Faculty of Science and Engineering, Waseda University, 3-4-1 Okubo, Shinjuku-ku, Tokyo 169-8555, Japan
{r.watanabe,sidiwu}@aoni.waseda.jp, ty0114@toki.waseda.jp, tanimizu@waseda.jp

Abstract. In the manufacturing field, the diversification of consumer needs is forcing a shift to variable-mix, variable-volume production, making it increasingly difficult to achieve the conventional uniformity of machine-centered production, the ability to respond to diverse needs, and the improvement of production efficiency. In addition, Japan will become a super-aging society where 40% of the population will be elderly by 2060, and there is an urgent need to restructure the securing and utilization of human resources. To solve these issues, a new manufacturing system in which "people" play a leading role has been proposed. To realize this system, the human-digital twin is attracting attention. The human-digital twin is the reproduction in digital space of an individual's physical, behavioral, and psychological states in the real world, which is thought to enable prediction of laborer fatigue, improvement of work efficiency, and enhancement of laborer safety. This study conducts basic research on predicting laborer fatigue toward the realization of the human digital twin. By conducting demonstration experiments assuming a cell production site, acquiring biometric information, and analyzing it using an autoencoder, it is clarified the prediction of laborer fatigue and the relationship between biometric information and fatigue.

Keywords: Human Digital Twin · Human factor · Cell manufacturing system · Autoencoder

1 Introduction

In recent years, with the diversification of consumer needs, the manufacturing industry has been forced to transform into variable-mix, variable-volume production. However, with conventional, uniform, machine-centered automation investment, it is difficult to respond to diverse needs, produce production, and secure human resources as Japan enters a super-aging society in which 40% of the population will be elderly by 2060. Restructuring of utilization has become an urgent task. In order to solve this problem, a new manufacturing system in which "people" play the leading role is being proposed. In this system, humans and machines collaborate flexibly, taking into account the flexibility of humans and the accuracy of machines. Human-digital twins are attracting attention in order to realize new manufacturing systems where "people" are the main players.

Published by Springer Nature Switzerland AG 2024
M. Thürer et al. (Eds.): APMS 2024, IFIP AICT 729, pp. 105–115, 2024.
https://doi.org/10.1007/978-3-031-65894-5_8

A human digital twin is a digital representation of an individual's physical, behavioral, and psychological state in the real world [1]. Specifically, it is reproduced by measuring physical information, behavioral information, psychological information, etc. and modeling it. Physical information includes 3D models, exercise data, physiological data, etc., behavioral information includes work content, travel route, work time, etc., and psychological information includes cognitive load, stress level, emotional state, etc. By utilizing human digital twins, it is believed that it will be possible to predict laborer fatigue, improve work efficiency, and improve laborer safety. Much of the current research on human digital twins is in the field of medicine, including digital twins of the human body for managing patient health [2] and digital twins of organs for medical services [3]. In comparison to these fields, there has been little in-depth research on worker digital twin (DT) because human needs have not received as much attention as production efficiency and product quality in industry [4]. Human modeling and simulation in human digital twin (HDT) remains a challenge because humans are much more complex systems than machines and other devices, and human behavior is unpredictable [5]. To build human models, some researchers have combined finite element modeling (FEM) and computed tomography (CT) [6]. However, the computational cost is enormous due to the large number of physiological parameters. Simpler and more practical methods are needed for human modeling and simulation in HDT toward the realization of human-centered manufacturing systems [7].

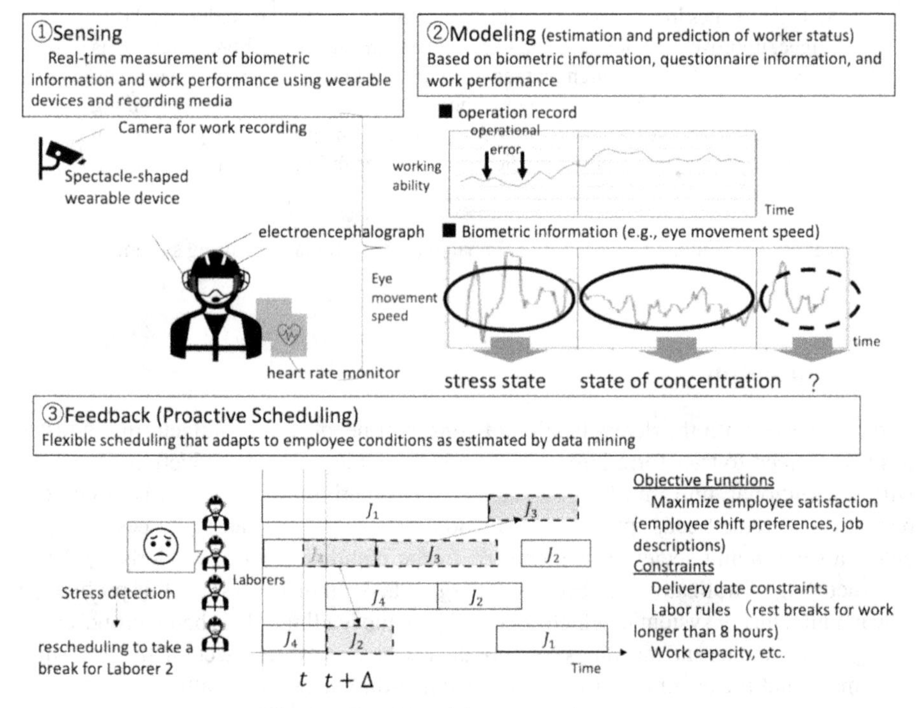

Fig. 1. Overview of the proposed system.

This research aims to construct a laborer support system as shown in Fig. 1, with the aim of realizing a human digital twin. The work support system measures biometric information in real time using a wearable device. Based on the measured biometric information, the system estimates the worker state. Based on the real-time estimation of the worker's state, the system performs work scheduling for the entire workforce, aiming to improve production efficiency and worker wellbeing. As a first step in this research, this paper conducts basic research on predicting laborer fatigue using biological information. Specifically, we conduct a demonstration experiment assuming a cell production site, acquire biological information, and analyze it using an autoencoder to predict laborer fatigue and clarify the relationship between biological information and fatigue.

2 Prediction Method of Human Error Using Biometric Information

This study proposes a method for predicting work errors using waveform analysis and autoencoders based on a laborer's biometric information, assuming light work at a work site.

2.1 Biometric Information

In a previous study of stress prediction using biometric information, Hori et al. [8] stated that autonomic balance can be estimated because the magnitude of the appearance of HF and LF variability waves to heart rate variability varies. In addition, Nozaki et al. [9] evaluated physiological and psychological effects during coloring work through measurements such as cerebral blood flow. The results showed that cerebral blood flow significantly increased during the work compared to that at rest. These studies suggest that cerebral blood flow and heart rate variability are correlated with stress and work performance during work.

In this study, a heart rate monitor and an electroencephalograph (EEG), which are easy to measure and can be employed in general working environments, will be used. The heart rate variability was measured using the my Beat heart rate sensor manufactured by Union Tool Corporation [10], as shown in the Fig. 2(a), and analyzed using the RRI Analyzer manufactured by Union Tool Corporation. The heart rate sensor measures approximately 4 cm.4 cm and is attached to the subject's chest by a belt with electrodes. The information transmitted from the heart rate sensor itself can be captured by a computer in real time through a receiver. The measured heart rate variability consists of a low frequency component (hereinafter referred to as "LF") and a high frequency component (hereinafter referred to as "HF"). The LF range is generally 0.04–0.15 Hz and is considered to reflect both sympathetic and parasympathetic nervous system activity. Sympathetic nerves are autonomic nerves that are active when a person feels stress, and parasympathetic nerves are autonomic nerves that are active when a person feels relaxed and mentally stable. The HF range of 0.15–0.4 Hz is generally considered to reflect the parasympathetic activity.

The HOT-2000 manufactured by NeU [11] shown in Fig. 2(b) is used to measure cerebral blood flow. The HOT-2000 emits light from the forehead that is easily absorbed by hemoglobin in the blood. The wavelength of the light is about 800 nm, and some of the

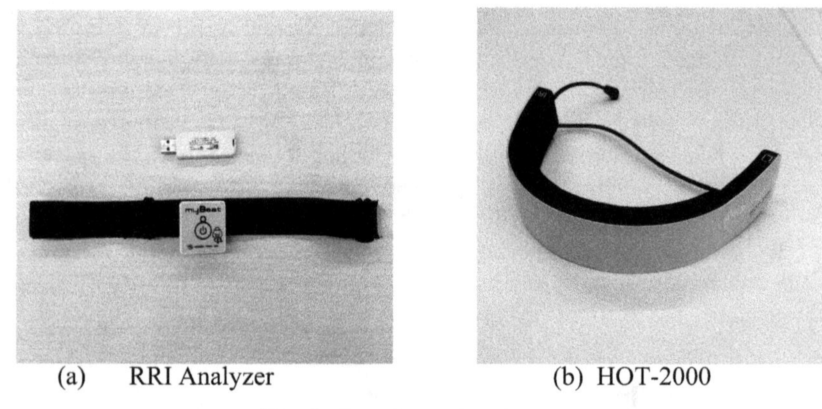

(a) RRI Analyzer (b) HOT-2000

Fig. 2. Experimental equipments

light is absorbed by hemoglobin and some of the light returns to the detector. There are detectors at distance of approximately 3 cm and 1 cm from the irradiated position, and by calculating the difference, the increase and decrease in blood volume can be observed. The change in cerebral blood flow measured by this method is denoted as HbT change (Headbeat Time change). When the sympathetic nervous system becomes dominant due to stress or mental fatigue, a marked increase in cerebral blood flow (activation) is observed in frontal regions such as the medial prefrontal cortex, anterior cingulate cortex and orbitofrontal cortex. These regions play an important role in assessing and coping with stress situations.

2.2 Data Preprocessing

Preprocessing is performed on the time-series data of heart rate variability and cerebral blood flow obtained in the experiment. Figure 3 shows the data intervals extracted by preprocessing.

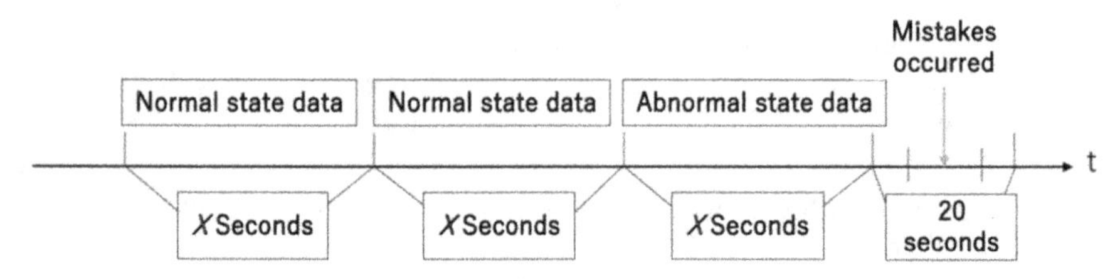

Fig. 3. Data interval to be extracted

Except for the 10 s before and after the error, all working times are labeled at arbitrary intervals. The data before the occurrence of a mistake is considered as the abnormal state data, and the other data as the normal state data. The 10 s before and after a mistake is excluded from the data to be analyzed because it is difficult to prevent a mistake even if it is detected immediately before the mistake occurs, and because the laborer may be upset immediately after a mistake occurs and no longer be in a normal state if he or she is aware of the mistake.

2.3 Feature Extraction Using Waveform Analysis

Waveform analysis is performed on the preprocessed time series data. Since the biological information obtained from experiments is time series data, it is possible to extract a large amount of features by waveform analysis. For each waveform, 778 waveform features are extracted, including statistics such as minimum, maximum, mean, and variance values, as well as waveform features such as frequency. If there are missing values in the obtained feature data, if the feature values are the same for all intervals, or if the values of the features are outliers, the data are removed.

2.4 Predicting Human Error Using Autoencoders

The features extracted by waveform analysis are analyzed using an autoencoder. An autoencoder is a neural network that performs dimensionality reduction [12]. The autoencoder extracts necessary information from the input data and reconstructs data similar to the input data based on that information. The difference between the restored data and the input data is called the reconstruction error. When training with data in the normal state, it is possible to classify normal data from abnormal data because the reconstruction error is larger for data in the abnormal state. Autoencoders, which can create models using only data from normal conditions, are widely used in the field of abnormality detection, where it is more difficult to obtain data from abnormal conditions than from normal conditions.

Let $g_\phi(\boldsymbol{x})$ denote the encoder network and $f_\theta(z)$ denote the decoder network, then the output $\hat{\boldsymbol{x}}$ is represented by Eq. (1).

$$a\hat{\boldsymbol{x}} = f_\theta(g_\phi(x)) \quad (1)$$

In learning, the entire Encoder-Decoder is trained at once based on the loss function in Eq. (2).

$$\frac{1}{n}\sum_{i=1}^{n} \mathcal{L}_{\phi,\theta}(\boldsymbol{x}_i, \hat{\boldsymbol{x}}_i) = \frac{1}{n}\sum_{i=1}^{n} (\boldsymbol{x}_i - f_\theta(g_\phi(\boldsymbol{x}_i)))^2 \quad (2)$$

The reconstruction error is calculated using the created anomaly detection model, and data with a reconstruction error greater than a threshold value is determined to be anomalous. The determination of the threshold value is calculated from Eq. (3). $\overline{r_{train}}$ is the mean of the reconstruction error of the training data, σ_{train} is the standard deviation of the reconstruction error of the training data, and α is the weight coefficient.

$$threshold = \overline{r_{train}} + \alpha\sigma_{train} \quad (3)$$

3 Computational Experiments

The effectiveness of the proposed method will be verified by acquiring biometric information through demonstration experiments and predicting work errors.

3.1 Demonstration Experiments for Data Acquisition

The demonstration experiment is shown in Fig. 4. To acquire biometric information, the subject is asked to insert conductors into a breadboard while wearing a heart rate sensor and an electroencephalograph. The work of inserting conductors into the breadboard is assumed to be the wiring of electronic components at an actual assembly site. The breadboard is an 8 cm × 5 cm plastic board. Numbers from 1 to 30 are written vertically and letters from A to J horizontally, and 10 × 30 holes are provided for each number. Subjects insert 5 conductors per question, and repeat the procedure for 10 sets of 10 questions. At the end of each set, the observer removes the conductors from the breadboard and reinserts them. During this time, the observer takes a break.

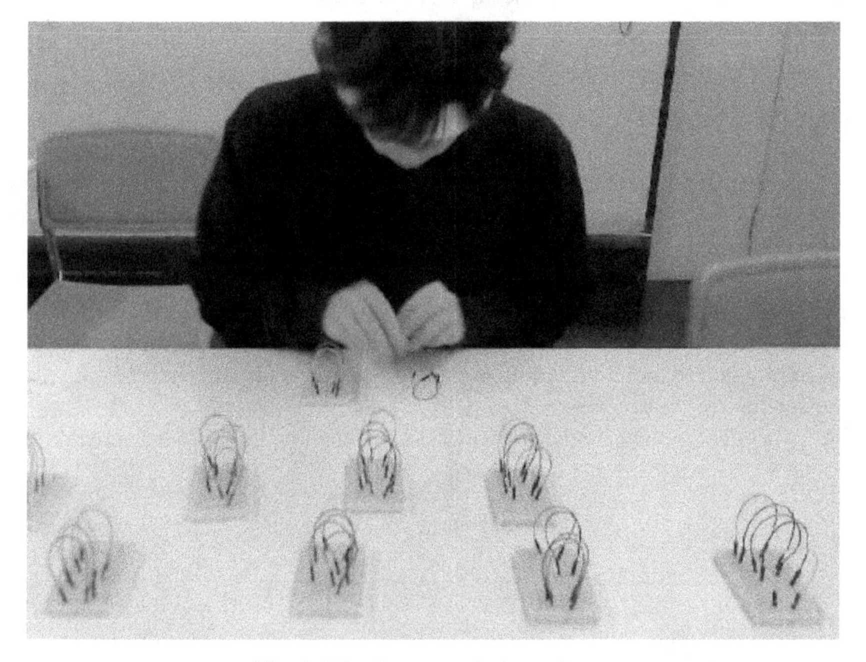

Fig. 4. The demonstration experiment

3.2 Prediction Results Using the Proposed Method

A prediction model was created for each subject using the proposed method, and Table 1 shows the results of predicting work errors and the features that most influence the model. Although there is variation among individuals, it predicts work errors with a maximum prediction accuracy of 100%. The prediction accuracy using data from all subjects is only 59%, but this may be due to differences in biometric information among individuals and insufficient learning of the normal state. In addition, the features with high impact are all related to HbT change; HBT change is a change in cerebral blood flow, which confirms that cerebral blood flow is related to the occurrence of work errors.

Table 1. Subjects' number of work errors and number of data extractions experimental results.

Subjects	Prediction accuracy	Maximum impact feature
A	1.0	Length of HbT change
B	0.7	First difference in HbT change
C	0.45	Sum of HbT change
D	0.5	Length of waveform of HbT change
E	0.72	Sum of squares of HbT change
F	0.63	Sum of HbT change
G	0.56	Length of HbT change
H	0.53	Mean of the second-order derivative of HbT change
I	0.8	First difference of HbT change
J	0.58	Coefficient of variation of HbT change
All	0.59	Length of waveform of HbT change

For subject A, which has the highest prediction accuracy, a histogram of the prediction results is shown in the Fig. 5. In the histogram, the vertical axis represents the number of data, the horizontal axis represents the reconstruction error, blue represents the data for training, orange represents the data for testing under normal conditions, and green represents the data for testing under abnormal conditions. The red line is the threshold value, and if the value of the reconstruction error is greater than the threshold value, it is judged as an abnormal value.

As shown in the Fig. 5, the reconstruction error is small in the normal state, while the data in the abnormal state has a larger reconstruction error and can be identified as being in an abnormal state. In addition, focusing on the two data with large reconstruction errors, it is confirmed that they are the part of the second half of the experiment where work errors were made in succession, and that they are errors of different quality from the other two anomalous data.

To examine in more detail the biometric information that influences the discrimination of abnormality detection, the top 10 influential features for subject A, who has the highest prediction accuracy, are shown in the Table 2. The top 10 most influential features are all related to HbT change.

The second and fifth "Count observed values within the interval" indicate the range of values in the input data, and the value of this feature represents the range of variability in the input data. The larger the value, the greater the variability of the data and the wider the range of values. Conversely, the smaller the value, the smaller the range of variability of the data and the narrower the range of values. The "Number of peaks in HbT change" in the 3rd, 8th, and 10th columns of Table 2 is the number of peaks in the input data and indicates the intensity of the waveform variation in the input data. The "standard error of linear trend in HbT change" in the 4th, 6th, 7th, and 9th columns are the statistics of the standard error of the linear trend in the input data. Both are measures of the variability

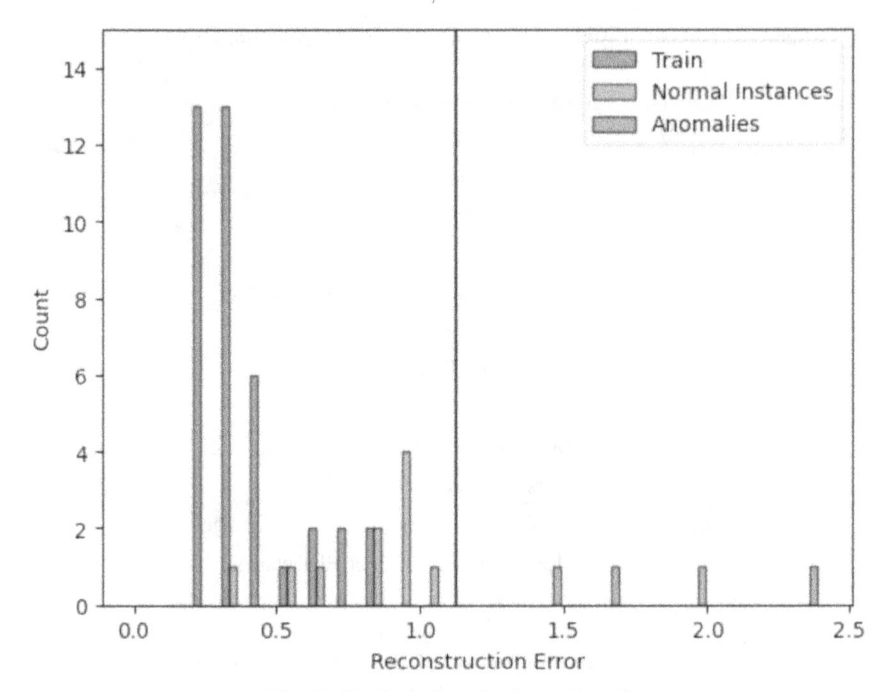

Fig. 5. Prediction results for subject A

of the input data, but the number of peaks focuses on the short-term variability of the data, while the standard error of the linear trend focuses on the long-term variability of the data.

As can be seen from Table 3, the number of values between −100 and 0 is smaller for the data in the abnormal state than for the data in the normal state. It can be said that the change in cerebral blood flow tends to increase in the data from the abnormal state compared to the normal state. This indicates that the amount of change in cerebral blood flow indicates that in the state immediately before a work error occurs, blood flow to the brain increases and the work of the parasympathetic nervous system, which dilates blood vessels and increases cerebral blood flow, is inhibited compared to the normal state. In other words, it is assumed to be a state of stress and reduced concentration.

Next, the same analysis was performed for subject A, who had the highest prediction accuracy, by changing the time interval of the data. The time intervals were 30 s and 2 min. The prediction accuracy at each time interval is shown in Table 2.

From the Table 4, it was found that the best prediction accuracy was achieved by the model with data from a 1-min interval. It is considered that when the time interval of the acquired data was long, it was difficult to distinguish between normal and abnormal conditions, resulting in a decrease in prediction accuracy. On the other hand, when the time interval is short, it is thought that the prediction accuracy decreased because sufficient data could not be acquired for training. From the above, for laborer A, the system of the proposed method can predict 1 min and 10 s before the occurrence of a work error and notify the laborer.

Table 2. The top 10 most influential features for subject A

	Strength of feature influence	Impact feature
1	13.63	Length of HbT change
2	13.63	Count observed values within the interval (−100, 0)
3	9.82	Number of peaks in HbT change (n1)
4	9.27	Maximum standard error of linear trend in HbT change (Chunk length 50)
5	9.04	Count observed values within the interval (−1, 1)
6	8.95	Mean standard error of linear trend in HbT change (Chunk length 50)
7	8.79	Minimum standard error of linear trend in HbT change (Chunk length 50)
8	8.32	Number of peaks in HbT change (n3)
9	7.98	Minimum standard error of linear trend in HbT change. (Chunk length 5)
10	7.98	Number of peaks in HbT change (n5)

Table 3. Average of "Count observed values within the interval (−100, 0)"

Data of normal state	Data of abnormal state
601.54	381.28

Table 4. Prediction accuracy per data interval for subject A

Interval	Prediction accuracy	Maximum impact feature
30 s	0.56	First difference of HbT change
1 min	1.0	Length of HbT change
2 min	0.77	Percentage change in average HbT change

4 Conclusion

This study conducted basic research on predicting laborer fatigue toward the realization of the human digital twin. It is conducted a demonstration experiment assuming a cell production site and proposed a method to predict human errors in real time by acquiring biometric information and analyzing it using an autoencoder.

The results of the computer experiment showed that the model learned for each individual subject was able to predict work errors with high accuracy, although there was some variation among subjects. On the other hand, the prediction accuracy decreased when the data of all subjects were analyzed. This is thought to be due to the influence of individual differences in each subject's biometric information. The feature that had the greatest impact on prediction accuracy was related to cerebral blood flow for all subjects, indicating that cerebral blood flow is related to the occurrence of work errors. The real-time nature of the prediction of work errors was verified by changing the time segment for the subjects with the highest prediction accuracy, and the prediction accuracy was highest when the time segment was 1 min.

As future work, it is necessary to construct a highly versatile prediction model by increasing the amount of data on the subject's normal state and by pre-processing the data in consideration of attributes such as gender. It is also necessary to apply the proposed work error prediction system to reactive scheduling in order to provide appropriate support to workers.

5 Protection of Human Rights and Compliance with Laws and Regulations

This research uses biometric information and falls under the categories of "research that requires the consent and cooperation of the other party, research that requires consideration of the handling of personal information," and "questionnaire surveys that involve personal information. Therefore, in conducting the research, the following considerations were taken into account. Waseda University "Ethics Review Committee for Research Involving Human Subjects" In accordance with the "Code of Ethics for Research Involving Human Subjects" established by Waseda University, we applied to the "Ethics Review Committee for Research Involving Human Subjects" and conducted the research after obtaining their approval. In applying to the Committee, the researcher attended Waseda University's "Theory of Research Ethics" and "JSPS Research Ethics e-Learning Course".

References

1. Miller, M.E., Spatz, E.: A unified view of a human digital twin. Hum. Intell. Syst. Integr. **4**, 23–33 (2022)
2. Liu, Y., Zhang, L., Yang, Y., et al.: A novel cloud-based framework for the elderly healthcare services using digital twin. IEEE Access, 49088–49101 (2019)
3. Martinez-Velazquez, R., Gamez, R., El Saddik, A.: Cardio twin: a digital twin of the human heart running on the edge. In: 2019 IEEE International Symposium on Medical Measurements and Applications (MeMeA), Istanbul, Turkey, June 26–28 (2019)
4. Breque, M., De Nul, L., Petridis, A.: Industry 5.0: towards a sustainable, human-centric and resilient European industry. European Commission, Directorate-General for Research and Innovation, Luxembourg, LU (2021)
5. Wang, B.C., Zheng, P., Yin, Y., et al.: Toward human-centric smart manufacturing: a human-cyber-physical systems (HCPS) perspective. J. Manuf. Syst. **63**, 471–490 (2022)

6. Barricelli, B.R., Casiraghi, E., Gliozzo, J., et al.: Human digital twin for fitness management. IEEE Access, 26637–26664 (2020)
7. He, Q., et al.: From digital human modeling to human digital twin: framework and perspectives in human factors. Chin. J. Mech. Eng. **37**, 9 (2024)
8. Hori, T., Kazuki, A., Suga, K., Yoshimura, R.: An objective stress assessment method. J. Jpn. Soc. Occup. Disaster Med., 330–334 (2018). (in Japanese)
9. Nozaki, M., Miyazaki, T., Nakadate, T.: Effectiveness of occupational therapy: examination of "coloring" in comparison with the Uchida-Kraepelin mental test: using physiological and psychological indicators. J. Showa Gakkai **74**(4), 413–420 (2014). (in Japanese)
10. RRI Analyzer. https://www.uniontool.co.jp/en/
11. NeU brains. https://neu-brains.site/
12. Chandola, V., Banerjee, A., Kumar, V.: Anomaly detection: a survey. ACM Comput. Surv. **41**(3), Article no. 15 (2009)

Resilient Supply Chain Network Planning Method with Two-Stage Stochastic Programming: Extension to Multiple Product Supply Chains

Toshiya Kaihara[1], Hibiki Kobayashi[1], Daisuke Kokuryo[1(✉)], Masashi Hara[2], Yuto Miyachi[2], Dickson Hideki Yamao[2], and Puchit Sariddichainunta[2]

[1] Graduate School of System Informatics, Kobe University, 1-1 Rokkodai-cho, Nada, Kobe, Hyogo 657-8501, Japan
kaihara@kobe-u.ac.jp, hkobayashi@kaede.cs.kobe-u.ac.jp, kokuryo@port.kobe-u.ac.jp

[2] Lion Corporation, 1-3-28 Kuramae, Taito, Tokyo 111-8644, Japan
{h-masashi,y-miya00,yamadeki,puchit}@lion.co.jp

Abstract. In supply chain management, risk management is crucial to ensure a stable product supply while considering economic efficiency. It is essential to design resilient supply chains that can maintain production capacity by preparing for and responding to predictable risks. Our group has proposed strategic planning methodologies for the selection of appropriate material suppliers and the optimization of inventory levels across the supply chain, including suppliers, manufacturers, and wholesalers, with considering potential risks. We have introduced a planning method for resilient supply chain networks using two-stage stochastic programming. This paper extends the proposed method to accommodate supply chains managing multiple products, aiming for more realistic conditions. The effectiveness of the extended method is assessed through computational experiments.

Keywords: Resilient supply chain network · Stochastic programming · Inventory control · Material supplier selection

1 Introduction

Supply chains in the manufacturing industry are becoming increasingly physically longer and routes more complex as products become more complex and diverse due to the development of transportation technology, globalization, and diversification of demand. The impact of risks in the supply chain has increased over the years in the industries [1]. The current situation can be attributed to various changes, including conflicts and a heightened sense of crisis stemming from the global pandemic's impact. Additionally, the supply chain is experiencing increasing destabilization due to tensions between major economic powers and exacerbated by volatile exchange rate fluctuations. These factors collectively influence the stability of the supply chain.

Published by Springer Nature Switzerland AG 2024
M. Thürer et al. (Eds.): APMS 2024, IFIP AICT 729, pp. 116–131, 2024.
https://doi.org/10.1007/978-3-031-65894-5_9

Supply chain risks are categorized into two types: operational risks and disruption risks. Operational risks are associated with normal production activities and may include fluctuations in demand and production timing. Disruption risks arise from unforeseen events that occur irregularly in timing and scale, such as earthquakes, tsunamis, and pandemics. These events are termed 'supply chain disruptions' because they temporarily interrupt the supply chain network due to various external influences. Even large companies can be vulnerable to unexpected risks, and comprehensive risk management that encompasses the entire supply chain is essential for survival. This approach ensures that potential disruptions can be anticipated and mitigated, securing the company's longevity and success in a competitive market [2, 3]. In light of the prevailing supply chain risks, comprehensive risk management that encompasses the entire supply chain is essential to sustain corporate operations. It is crucial to establish a resilient supply chain design that considers a variety of disruption risks to guarantee the supply chain's success [2, 4]. A resilient supply chain refers to a supply chain that can recover from anticipated risks by preparing for and responding to them within an acceptable cost and time frame, and can maintain active steady-state operations thereafter [5]. In general, when economic efficiency of a supply chain is emphasized, the effects of disruptions are more likely to propagate, and when stability is emphasized, efficiency is impaired [4]. The first decision-making process in implementing a supply chain, the design of the supply chain, which determines equipment placement and its capacity, directly influences the second decision-making process, the operational level of the supply chain, such as production volume and delivery routes [6–8].

In order to design an appropriate supply chain in an environment where there is a risk of disruption, research has been conducted focusing on the evaluation of disruption risk by simulation [9, 10] and on the evaluation of disruption risk by probability calculation using a two-stage stochastic programming method [8, 11]. The former is a method that aims to derive an approximate optimal solution while increasing computational efficiency by evaluating disruption risks related to supply chain design through simulation. The latter uses two-stage stochastic programming [12] to view the operational period of the supply chain as a multi-period scenario and simultaneously make decisions at each stage while taking uncertain events into account, thus considering the optimality of decision making after the occurrence of disruption risk.

In our group, we are conducting research on risk management for supply chain disruptions and developing supply chain networks that can effectively respond to such disruptions using two-stage stochastic programming, which can respond probabilistically to the scale and frequency of disruption risk. This approach allows for more robust decision-making by considering various possible future scenarios, thereby enhancing the resilience of supply chains against unforeseen events. Kobayashi et al. proposed a resilient supply chain network planning method with material supplier selection and inventory level determination for supply chains consisting of suppliers, manufacturer including factories and distribution centers, and wholesalers [13]. In order to determine appropriate supplier selection and inventory levels considering future disruption risks, a two-stage stochastic programming was applied. However, previous proposed method focused on the supply chain for single product, and there was a lack of comprehensive studies that took into account the actual supply chain. In this paper, we extend the proposed method so that it can handle multiple products in order to consider more realistic conditions. In computational experiments, the extended proposed method is compared with the traditional exact solution method to evaluate solution accuracy in selecting material suppliers and determining appropriate inventory levels.

2 Target Model and Risks

This chapter offers a comprehensive overview of the targeted supply chain and associated disruption risks.

2.1 Target Supply Chain Network

This paper examines a supply chain network comprising multiple suppliers, a manufacturer with various factories and distribution centers, and wholesalers. The analysis is conducted from the perspective of a manufacturing company. Figure 1 presents an overview of the targeted supply chain network. The objectives are established through discussions with the collaborating company, which specializes in the mass production of daily necessities.

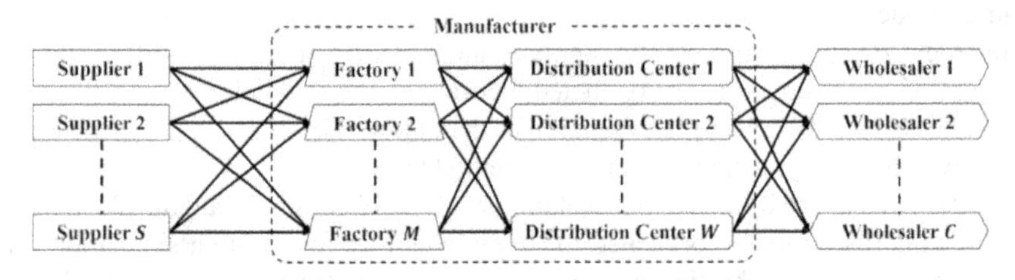

Fig. 1. Targeted Supply Chain Network

The target supply chain network has the following characteristics.

- The target supply chain supplies multiple final products.
- Stock replenishment at each location adheres to a fixed order period system.
- When wholesalers submit orders to the distribution center, any quantity that fails to meet demand is deemed out-of-stock.
- The manufacturer selects material suppliers and sets suitable inventory levels, considering economic viability in response to disruptions.
- Each supplier can supply all types of materials that the supply chain handles.
- Suppliers have a maximum supply capacity, and factories and distribution centers have a maximum storage capacity for materials and products.
- Distance exists between the supply chain locations, and transportation time is taken into account when making decisions.

2.2 Target Risk

In this paper, like the previous paper [13], we consider a multi-period scenario of future events, assuming disruptions that occur at suppliers, factories, and distribution centers. The disruptions and scenarios are established based on reference [10]. The disruptions and scenarios are as follows:

- Any disruption, including demand disruption, will end within the planned period.
- The facility where the disruption occurs will be deactivated.
- The disruption scale is expressed by the length of the stoppage period.
- The longer the stoppage period, the lower the probability of disruption.
- There are two scenarios: a normal scenario in which no disruption occurs and a scenario in which a disruption occurs only once during the planning period, and the length of the stoppage period varies depending on the probability that the disruption occurs.

3 Extending the Proposed Method to Multiple Products

This section describes an extended proposed method for selecting material suppliers and determining appropriate inventory levels using two-stage stochastic programming for planning a resilient supply chain network. Extending previous research [13], we have made it possible to produce multiple products in a factory by applying the symbols and formulations related to the materials and products used to multiple products.

3.1 Notation

The definition of the characters used in the formulation is as follows:

Sets

- $\boldsymbol{S}$: set of suppliers ($s = 1,2, \ldots, S$)
- $\boldsymbol{M}$: set of factories ($m = 1,2, \ldots, M$)
- $\boldsymbol{W}$: set of distribution centers ($w = 1,2, \ldots, W$)
- $\boldsymbol{C}$: set of wholesalers ($c = 1,2, \ldots, C$)
- $\boldsymbol{K}$: set of scenarios ($k = 1,2, \ldots, K$)
- $\boldsymbol{T}$: set of planning periods ($t = 1,2, \ldots, T$)

Parameters

- i: Number of product ($i = 1,2, \ldots, I$)
- j: Number of materials for product i ($j = 1,2, \ldots, J_i$)
- $CAPS_{ij}^{s}$: capacity of material j of product i at supplier s
- $CAPM^{m}$: material inventory capacity at factory m
- $CAPP^{m}$: products inventory capacity at factory m
- $CAPW^{w}$: inventory capacity at distribution center w

- PM^m: production capacity at factory m
- D_i^{ktc}: demand for product i at wholesaler c in scenario k, period t
- CS^{sm}: contract costs between supplier s and factory m
- CB_{ij}^{sm}: purchasing cost per unit of material j of product i from supplier s by factory m
- CTS^{sm}: transportation cost per unit from supplier s by factory m
- CTM^{mw}: transportation cost per unit from factory m by distribution center w
- CTW^{wc}: transportation cost per unit from distribution center w by wholesaler c
- CP_i^m: production cost per unit of product i at factory m
- HM^m: storage cost per unit of material at factory m
- HP^m: storage cost per unit of product at factory m
- HW^w: storage cost per unit of product at distribution center w
- CL_i^c: stock-out loss cost of product i at wholesaler c
- p^k: probability of scenario k
- LSM^{sm}: transportation period from supplier s by factory m
- LMW^{mw}: transportation period from factory m by distribution center w
- $\alpha^{kts}: \begin{cases} 1: \text{ if supplier } s \text{ operates in scenario } k, \text{ period } t \\ 0: \text{ if a disruption occurs at supplier } s \text{ in scenario } k, \text{ period } t \end{cases}$
- $\beta^{ktm}: \begin{cases} 1: \text{ if factory } m \text{ operates in scenario } k, \text{ period } t \\ 0: \text{ if a disruption occurs at facory } m \text{ in scenario } k, \text{ period } t \end{cases}$
- $\gamma^{ktw}: \begin{cases} 1: \text{ if distribution center } w \text{ operates in scenario } k, \text{ period } t \\ 0: \text{ if a disruption occurs at distribution center } w \text{ in scenario } k, \text{ period } t \end{cases}$

Decision Variables at First Stage

- $x^{sm}: \begin{cases} 1: \text{ if supplier } s \text{ is contracted by factory } m \\ 0: \text{ otherwise} \end{cases}$
- lm_{ij}^m: inventory level of material j of product i at factory m
- lp_i^m: inventory level of product i at factory m
- lw_i^w: inventory level of product i at distribution center w

Decision Variables at Second Stage

- pp_i^{ktm}: production quantity of product i factory m in scenario k at period t
- b_{ij}^{ktsm}: quantity of material j of product i purchased from supplier s by factory m in scenario k at period t
- qw_i^{ktmw}: quantity of product i ordered from distribution center w to factory m in scenario k at period t
- qc_i^{ktwc}: quantity of product i ordered from wholesaler c to distribution center w in scenario k at period t

Dependent Variables

- am_{ij}^{ktm}: quantity of materials j of product i arrived of factory m in scenario k at period t
- aw_i^{ktw}: quantity of product i arrived at distribution center w in scenario k at period t
- ac_i^{ktc}: quantity of product i arrived at wholesaler c in scenario k at period t
- im_{ij}^{ktm}: inventory quantity of material j of product i of factory m in scenario k at period t
- ip_i^{ktm}: inventory quantity of product i of factory m in scenario k at period t
- iw_i^{ktw}: inventory quantity of product i of distribution center w in scenario k at period t
- ld_i^{ktc}: stock-out quantity of product i of wholesaler c in scenario k at period t

3.2 Extending Resilient Supply Chain Network Planning Method to Multiple Product Supply Chains

This section outlines a proposed method [13] for planning a resilient supply chain network and its extension to multiple products, involving material supplier selection and inventory level decisions for a supply chain comprised of suppliers, manufacturers, and wholesalers. In the proposed method, two-stage stochastic programming [12] is applied to determine the appropriate supplier selection and inventory levels, taking into account the risk of future disruptions. Additionally, Latin Hypercube Sampling [14] is utilized to uniformly extract scenarios from the entire scenario space, thereby achieving a solution that balances computational time and accuracy effectively. In this paper, when applying the proposed method to a supply chain that handles multiple products, the formulation is extended so that companies in the supply chain can consider multiple materials and products simultaneously. Furthermore, by using the metaheuristic method, simulated annealing, it is possible to derive a solution within a practical time frame.

Figure 2 shows a conceptual diagram of the method expanded to multiple products based on the proposed method. In the event of a disruption, it is assumed that the scenario will diverge from the 'no disruption' scenario to an alternative scenario during the period of the disruption. Considering these potential future scenarios, a two-stage decision-making process will be employed to design a supply chain network. This network aims to minimize the total expected future costs associated with the disruption. The first phase of the decision-making process entails the selection of material suppliers and the determination of suitable inventory levels. This should consider all conceivable scenarios, as depicted in Fig. 2. Thus, the existence of a contract between supplier s and factory m (x^{sm}) and the inventory levels at each facility for each material and each product (lm_{ij}^{m}, lp_i^{m}, lw_i^{w}) are set as the first-stage decision variables, variables common to all scenarios. Next, the second step in the decision process is to determine the production, purchase, and order quantities for each scenario. Here, the amount of product i produced by factory m in period t in scenario k (pp_i^{ktm}), the amount of material j purchased for product i from supplier s (b_{ij}^{ktsm}), the amount of product i ordered from distribution center w to factory m (qw_i^{ktmw}), the amount of product i ordered from wholesaler c to distribution center w (qc_i^{ktwc}) are defined as the second stage decision variables, which are determined for each scenario in the two-stage stochastic programming. Decision variables at each of these stages are considered simultaneously to plan a resilient supply chain network with the aim of minimizing costs in all possible future scenarios.

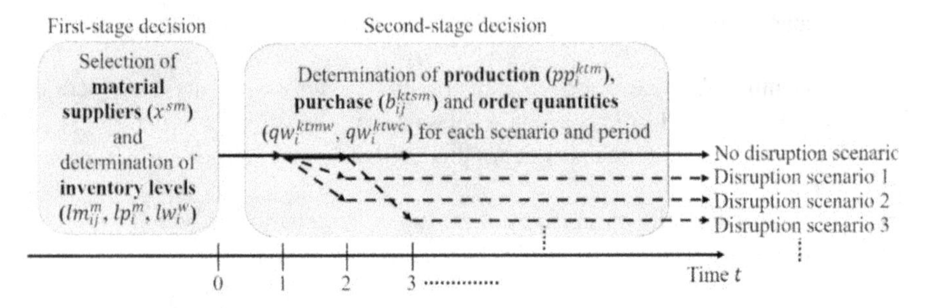

Fig. 2. Conceptual diagram of the extended proposed method

The evaluation of resilience focuses on the overall cost, including the stock-out loss cost. The proposed method is compared with a deterministic approach that overlooks risk factors. Both the total cost under standard conditions and the combined total cost and stock-out loss cost amid disruption risk are assessed. This assessment aids in determining the viability of crafting a resilient supply chain capable of ensuring a consistent product supply, while also factoring in economic viability.

The proposed method employs a comprehensive approach to select material suppliers and determine optimal inventory levels by considering all possible scenarios. However, as the variety of risks and the number of facilities grow, the number of scenarios to be considered increases exponentially, making it impractical to compute an exact solution within a reasonable calculation time. To mitigate this issue while maintaining solution accuracy, the method integrates Latin Hypercube Sampling [14]. This technique guarantees a uniform distribution of samples across the entire sampling space, enabling the selection of varied, suitable scenarios that enhance computational efficiency. As in the previous research [13], the Latin Hypercube Sampling is used to extract effective scenarios in the Second-stage. Moreover, the extended method proposed is augmented with a simulated annealing approach, which expedites the attainment of a quasi-optimal solution.

3.3 Algorithm of Extended Proposed Method

This section describes the algorithm of the extended proposed method. The detailed flow is explained below.

STEP 1: Initialization

A Latin Hypercube Sampling space is generated for each scale of disruption risk. We also initialize the number of repetitions of scenario sampling and the parameters used in the simulated annealing method.

STEP 2: Scenario Sampling

Using Latin Hypercube Sampling, we select scenarios that are consistent in the timing and location of disruption for each scale of disruption risk.

STEP 3: Solving decision variables

The problems of selecting material suppliers and determining appropriate inventory levels are solved using the scenarios selected in **STEP 2**. First, the first stage decision variables, the selection of material supplier x^{sm} and inventory levels lm_{ij}^{m}, lp_{i}^{m} and lw_{i}^{w} are solved using simulated annealing. Then, to solve the rest decision variables, production quantity pp_{i}^{ktm}, purchase quantity b_{ij}^{ktsm}, and order quantities qw_{i}^{ktmw} and qw_{i}^{ktwc} in the second stage, the branch and bound method is used to calculate the objective function value of the candidate solution.

STEP 4: Solution evaluation

The solution obtained in **STEP 3** is evaluated. To evaluate the solution, we fix the first stage decisions and then solve the obtained problem for other scenarios [7]. If the evaluation value of the solution obtained in the second stage of **STEP3** is better than that of the tentative solution, the solution is accepted and update the tentative solution.

STEP 5: Determination of the number of repetitions

Repeat **STEP 2** through **STEP 4** until the set number of iterations is reached.

3.4 Formulation

In this section, the formulation of material procurement source selection and inventory level determination problems that take into account disruption risks for multiple products will be described below:

$$\min \quad COST = CA + \sum_{k \in \boldsymbol{K}} p^{k} \cdot CR^{k} \tag{1}$$

$$\text{where} \quad CA = \sum_{s \in \boldsymbol{S}} \sum_{m \in M} CS^{sm} \cdot x^{sm} \tag{2}$$

$$\begin{aligned} CR^{k} = \sum_{t \in \boldsymbol{T}} \Bigg\{ & \sum_{m \in \boldsymbol{M}} \sum_{i=1}^{I} CP_{i}^{m} \cdot pp_{i}^{ktm} + \sum_{s \in \boldsymbol{S}} \sum_{m \in \boldsymbol{M}} \sum_{i=1}^{I} \sum_{j=1}^{J_i} CB_{ij}^{sm} \cdot b_{ij}^{ktsm} \\ & + \sum_{m \in \boldsymbol{M}} \sum_{i=1}^{I} \left(\sum_{j=1}^{J_i} HM^{m} \cdot im_{ij}^{ktm} + HP^{m} \cdot ip_{i}^{ktm} \right) \\ & + \sum_{w \in \boldsymbol{W}} \sum_{i=1}^{I} HW^{w} \cdot iw_{i}^{ktw} \\ & + \sum_{s \in \boldsymbol{S}} \sum_{m \in \boldsymbol{M}} \sum_{i=1}^{I} \sum_{j=1}^{J_i} CTS^{sm} \cdot b_{ij}^{ktsm} \\ & + \sum_{m \in \boldsymbol{M}} \sum_{w \in \boldsymbol{W}} \sum_{i=1}^{I} CTM^{mw} \cdot qw_{i}^{ktmw} \\ & + \sum_{w \in \boldsymbol{W}} \sum_{c \in \boldsymbol{C}} \sum_{i}^{I} CTW^{wc} \cdot qc_{i}^{ktwc} \\ & + \sum_{c \in \boldsymbol{C}} \sum_{i=1}^{I} CL_{i}^{c} \cdot ld_{i}^{ktc} \Bigg\} \end{aligned} \tag{3}$$

$$am_{ij}^{ktm} = \sum_{s \in \boldsymbol{S}} b_{ij}^{k(t - LSM^{sm})sm}, \forall i, \forall j, \forall m, \forall k, \forall t \tag{4}$$

$$aw_{i}^{ktm} = \sum_{m \in \boldsymbol{M}} qw_{i}^{k(t - LMW^{mw})mw}, \forall i, \forall j, \forall w, \forall k, \forall t \tag{5}$$

$$ac_i^{ktc} = \sum_{w \in \mathbf{W}} qc_i^{ktwc}, \forall i, \forall c, \forall k, \forall t \tag{6}$$

$$D_i^{ktc} = ac_i^{ktc} + ld_i^{ktwc}, \forall c, \forall k, \forall t \tag{7}$$

$$\text{s. t} \quad 0 \le pp_i^{ktm} \le \min\left\{\min_j\left(im_{ij}^{k(t-1)m}\right), \left(lp_i^m - ip_i^{k(t-1)m}\right)\right\}, \forall m, \forall k, \forall t, \forall i \tag{8}$$

$$0 \le \sum_{i=1}^{I} pp_i^{ktm} \le PM^m \cdot \beta^{ktm}, \forall m, \forall k, \forall t \tag{9}$$

$$im_{ij}^{ktm} = im_{ij}^{k(t-1)m} + am_{ij}^{ktm} - pp_{ij}^{ktm} \ge 0, \forall m, \forall k, \forall t, \forall i, \forall j \tag{10}$$

$$ip_i^{ktm} = ip_i^{k(t-1)m} + pp_i^{ktm} - \sum_{w \in \mathbf{W}} qw_i^{ktmw} \ge 0, \forall m, \forall k, \forall t, \forall i \tag{11}$$

$$iw_i^{ktm} = iw_i^{k(t-1)m} + aw_i^{ktm} - \sum_{c \in \mathbf{C}} qc_i^{ktwc} \ge 0, \forall w, \forall k, \forall t, \forall i \tag{12}$$

$$0 \le qc_i^{ktwc} \le \min\left\{\left(D_i^{ktc} - \sum_{w' \in \mathbf{W}'} qc_i^{ktw'c}\right), \left(iw_i^{k(t-1)m} - \sum_{c' \in \mathbf{C}'} qc_i^{ktwc'}\right) \cdot \gamma_{ktw}\right\}, \forall w, \forall c, \forall k, \forall t, \forall i \tag{13}$$

$$0 \le qw_i^{ktmw} \le \min\left\{\left(lw_i^w - iw_i^{k(t-1)w} - \sum_{m \in \mathbf{M}} \sum_{\tau = t - LMW^{mw}+1}^{t-1} qw_i^{k\tau mw} - \sum_{m' \in \mathbf{M}'} qw_i^{ktm'w}\right), \left(ip_i^{k(t-1)m} - \sum_{w' \in \mathbf{W}'} qw_i^{ktmw'} \cdot \beta^{ktm}\right)\right\}, \forall m, \forall w, \forall k, \forall t, \forall i \tag{14}$$

$$0 \le b_{ij}^{ktsm} \le \min\left\{\left(lm_{ij}^m - im_{ij}^{k(t-1)m} - \sum_{s \in \mathbf{S}} \sum_{\tau = t - LSM^{sm}+1}^{t-1} b_{ij}^{k\tau sm} - \sum_{s' \in \mathbf{S}'} b_{ij}^{kts'm}\right), \left(CAPS_{ij}^s - \sum_{m' \in \mathbf{M}'} b_{ij}^{ktsm'}\right) \cdot \alpha^{kts} \cdot x^{sm}\right\}, \forall s, \forall m, \forall k, \forall t, \forall i, \forall j \tag{15}$$

$$x^{sm} \in \{0,1\}, \forall s, \forall m \tag{16}$$

$$0 \le \sum_{i=1}^{I} \sum_{j=1}^{J_i} im_{ij}^m \le CAPM^m, \forall m \tag{17}$$

$$0 \le \sum_{i=1}^{I} lp_i^m \le CAPP^m, \forall m \tag{18}$$

$$0 \leq \sum_{i=1}^{I} lw_i^w \leq CAPW^w, \forall w \tag{19}$$

$$pp_i^{kt'm} = pp_i^{k't'm}, \forall m, \forall k, \forall i, \exists k', \exists t' \tag{20}$$

$$b_{ij}^{kt'ms} = b_{ij}^{kt'sm}, \forall m, \forall s, \forall k, \forall i, \exists k', \exists t' \tag{21}$$

$$qw_i^{kt'mw} = qw_i^{k't'mw}, \forall m, \forall k, \forall w, \forall i, \exists k', \exists t' \tag{22}$$

$$qc_i^{kt'mc} = qw_i^{k't'wc}, \forall m, \forall k, \forall i, \exists k', \exists t' \tag{23}$$

The objective function (1) minimizes the sum of the cost of the first stage (CA) and the expected cost of the second stage (CR).

Equations (2)–(7) are relational expressions. First, Eq. (2) is the contract costs with suppliers at each factory, which is the first stage cost. The second stage cost shown in Eq. (3) is the sum of the expected values of production costs (term 1), purchase costs (term 2), inventory storage costs (terms 3 and 4), transportation costs (terms 5, 6 and 7), and stock-out loss costs (term 8) per product i. Equations (4)–(6) show the quantity of materials and products arrived at each facility in scenario k at period t for each product. Equation (7) is a demand conservation equation that states that the sum of the quantity arrived and the stock-out quantity of wholesaler c in scenario k, period t for product i equals the demanded quantity.

Equation (8)–(23) are constraints. Equations (8) and (9) are constraints on the material j and production capacity of factory m for product i in scenario k and period t. The initial inventory quantity at each facility is set to the inventory level set for each facility. Equations (10)–(12) are constraints on the amount of inventory at the end of the period at each facility for product i in scenario k and period t. Equations (13)–(15) impose constraints on the order and purchase quantities. Downstream facilities place orders or make purchases sequentially from upstream facilities that offer the lowest transportation costs. This ordering strategy ensures cost-efficiency in the supply chain by minimizing transportation expenses. Equation (16) indicates that x^{sm} is a binary variable. Equations (17)–(19) show the range of possible inventory levels for each facility. Equations (20)–(23) are constraints for nonanticipativity [12]. It is shown that the second stage decision variable in period t' (1,2, ..., t) remains consistent with the decision variable in the no disruption scenario (k') even when disruption occurs in scenario k starting from period $t + 1$.

4 Computational Experiments

In this section, we conduct computer experiments to evaluate the usefulness of the extended proposed method. In order to evaluate the extended proposed method, we compare it with the results of a deterministic method that determines material supplier selection and appropriate and inventory levels assuming that no risks occur. The mathematical optimization solver CPLEX12.10 [15] is used to find the exact solution using the branch and bound method.

4.1 Experimental Condition

The experimental conditions are set as follows:

- Number of suppliers (S): 5
- Number of factories (M): 2
- Number of distribution centers (W): 10
- Number of wholesalers (C): 10
- Number of scenarios (K): 29
- Number of planning periods (T): 28
- Number of products (I): 2
- Number of materials for product i (J_i): 2
- Contract costs between supplier s and factory m (CS^{sm}): 100
- Production cost per unit of product i at factory m (CP_i^m): 5
- Purchasing cost per unit of material j of product i from supplier s by factory m (CB_{ij}^{sm}): 3
- Storage cost per unit of material and product at each facility (HM^m, HP^m, HW^w): 3
- Stock-out loss cost at wholesaler c (CL_i^c): 200
- Quantity demanded by wholesaler c in scenario k and period t (D_i^{ktc}): 10
- Number of trail of the extended proposed method: 10

In this experiment, we assume a disruption occurs at a supplier or factory. The stoppage periods at the disruption location are 1, 3, 7, and 14 periods. The occurrence and timing of disruption risks are presumed to be random, and it is assumed that the probability of disruption scenarios occurring, with an equivalent scale of disruption risk, is the same. Table 1 presents the assumed probabilities of disruptions occurring at each location.

Table 1. Probability of occurrence of each disruption risk

Stoppage periods	No	1	3	7	14
Probability	0.50	0.25	0.15	0.075	0.025

Table 2 presents the transportation costs and periods between suppliers and factories (CTS^{sm}, LSM^{sm}).

Table 2. Transportation cost and period between suppliers and factories Values: (CTS^{sm}, LSM^{sm})

	Supplier 1	Supplier 2	Supplier 3	Supplier 4	Supplier 5
Factory 1	(1.0, 1)	(1.5, 1)	(2.0, 1)	(2.5, 2)	(3.0, 2)
Factory 2	(3.0, 2)	(2.5, 2)	(2.0, 1)	(1.5, 1)	(1.0, 1)

Table 3 presents the transportation costs and periods between factories and distribution centers (CTM^{sm}, LMW^{mw}). The transportation costs between distribution centers and wholesaler (CTW^{wc}) are set between 1.0 and 5.5. When a wholesaler places an order with a distribution center, the amount that does not meet the demand is considered stock-out cost, so the distance between the distribution center and the wholesaler is considered only in terms of transportation costs.

Table 3. Transportation cost and period between factories and distribution centers Values: (CTM^{sm}, LMW^{mw})

	Distribution center									
	1	2	3	4	5	6	7	8	9	10
Factory 1	(1.0, 1)	(1.5, 1)	(2.0, 1)	(2.5, 1)	(3.0, 1)	(3.5, 2)	(4.0, 2)	(4.5, 2)	(5.0, 2)	(5.5, 2)
Factory 2	(5.5, 2)	(5.0, 2)	(4.5, 2)	(4.0, 2)	(3.5, 2)	(3.0, 1)	(2.5, 1)	(2.0, 1)	(1.5, 1)	(1.0, 1)

4.2 Results and Discussion

Table 4 shows the materials supplier selection determined using the extended proposed method and the deterministic method, which determines material supplier selection and inventory levels without considering disruption risks. Figures 3 and 4 show the inventory levels of each material and product at factories and distribution centers determined by the extended proposed method and deterministic method. Regarding material supplier selection, in the deterministic method that does not take disruption risk into consideration, each factory only contracts with the one closest supplier, whereas in the extended proposed method, each factory contracts with multiple suppliers. The extended proposed method indicates that within a supply chain, even when a single supplier is capable of procuring the necessary materials, the involvement of multiple suppliers becomes essential when dealing with multiple products. This result suggests that the extended proposed method considers the trade-off between economic efficiency and resilience to disruption risks, similar to what was reported in previous reports [13]. In the extended proposed method, the inventory level for materials was the same as in the deterministic method, but the inventory levels for product at both the factories and distribution centers were set higher as shown in Figs. 3 and 4. This result confirms that the extended proposed method takes the disruption risk into consideration. The results indicate that, with regard to economic efficiency and stability, maintaining a product inventory is preferable to a materials inventory under the present experimental conditions. However, varying inventory levels may be necessary depending on the quantity of products and the associated production costs. Further study is required to fine-tune the parameters of the extended proposed method.

Next, using the results of the selection of material suppliers and the determination of inventory levels by each method, the total cost under normal conditions and the average

Table 4. Combinations and types of material supplier selection for each method in multiple products

Deterministic method	Factory 1: Supplier 1 Factory 2: Supplier 5
Extended proposed method	Factory 1: Supplier 1, 2 Factory 2: Supplier 4, 5

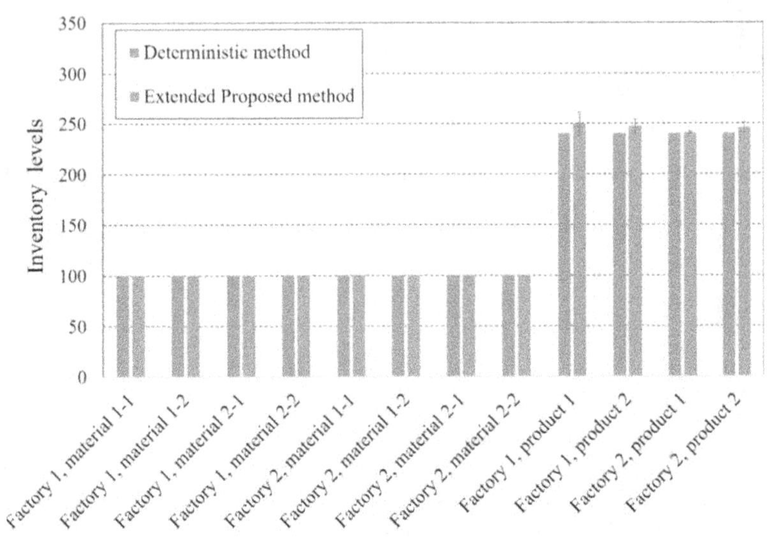

Fig. 3. Inventory level of each material and product in each factory

cost of stock-out losses for each scale of disruption risk are calculated to evaluate the resilience of the supply chain designed by each method. Table 5 presents the results of the total cost, which represents the value of the objective function ($COST$) under normal conditions, for each method. Figure 5 shows the average value of stock-out loss costs under normal times and when multiple disruption risks occur. Regarding the total cost under normal conditions, it has been confirmed that the extended proposed method results in a 0.25% increase compared to the deterministic method. Conversely, it has been observed that the extended proposed method can reduce the cost associated with stock-out losses to about half that of the deterministic method, even in the event of a large-scale stoppage of 7 or more periods. The results indicate that the supply chain, as designed by the extended proposed method, is capable of sustaining a level of resilience in the face of major disruptions. However, within the framework of a multi-product supply chain, it may be difficult to simultaneously maintain economic efficiency and a stable supply of products.

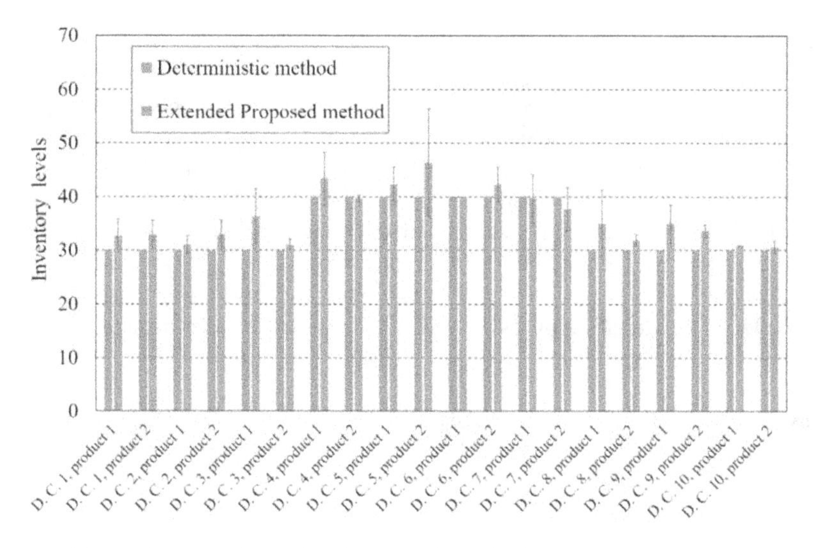

Fig. 4. Inventory level of each product in each distribution center (D.C.)

Table 5. Objective function value (*COST*) under normal condition for each method in multiple products

		COST
Deterministic method		132540.00
Extended proposed Method	Avg.	132875.00
	S.D.	119.11

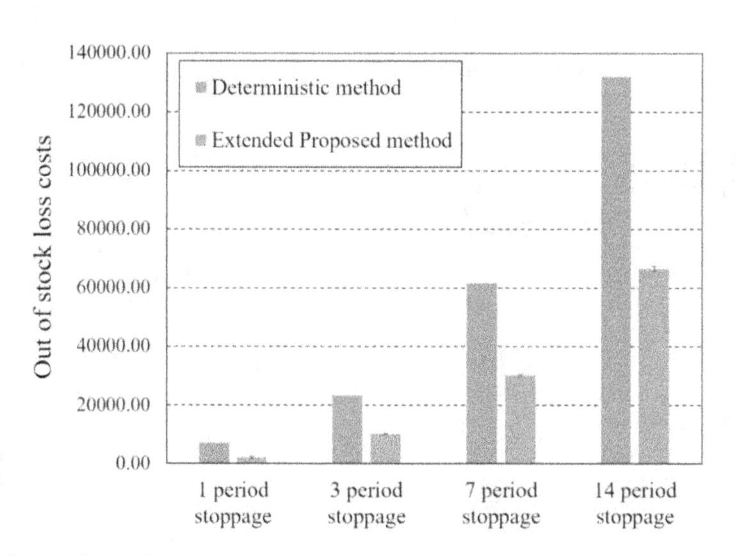

Fig. 5. Costs of Stock-out loss from simulation results for each disruption risk

5 Conclusion

In this paper, we extended the supply chain network planning method, incorporating considerations for disruption risks across multiple products. We assessed its efficacy through computer experiments, comparing it with a deterministic method that disregards risks. Although further improvements such as adjusting parameters are required, the proposed method indicates the feasibility of selecting multiple material suppliers and optimizing inventory levels while factoring in disruption risks. In future studies, it will be necessary to further improve the proposed method for application to actual situation, such as selecting material suppliers for each material and considering setup costs in the production.

Acknowledgement. The authors would like to thank Prof. Nobutada Fujii (Kobe University), Dr. Ruriko Watanabe (Waseda University) and Ms. Rina Tanaka (Lion Corporation) for their valuable advices.

Disclosure of Interests. Toshiya Kaihara, Hibiki Kobayashi and Daisuke Kokuryo have no competing interests to declare that are relevant to the content of this article. Masashi Hara, Yuto Miyachi, Dickson Hideki Yamao and Puchit Sariddichainunta are employees of Lion Corporation.

References

1. Report on the International Economic Survey Project for the Integrated Domestic and International Economic Growth Strategy (Survey for Strengthening Supply Chains in Asia at Large). Ministry of Economy. (in Japanese). https://www.meti.go.jp/meti_lib/report/2020FY/000173.pdf. Accessed 14 Apr 2024
2. Wicaksana, A., Ho, W., Talluri, S., Dolgui, A.: A decade of progress in supply chain risk management: risk typology, emerging topics, and research collaborators. Int. J. Prod. Res. **60**(24), 7155–7177 (2022)
3. Duong, A.T.B., Hoang, T.-H., Nguyen, T.T.B., Akbari, M., Hoang, T.G., Truong, H.Q.: Supply chain risk assessment in disruptive times: opportunities and challenges. J. Enterp. Inf. Manag. **36**(5), 1372–1401 (2023)
4. Golan, M.S., Jenegan, L.H., Linkov, I.: Trends and applications of resilience analytics in supply chain modeling: systematic literature review in the context of the COVID-19 pandemic. Environ. Syst. Decis. **40**, 222–243 (2020)
5. Ribeiro, J.P., Barbosa-Povoa, A.: Supply chain resilience: definitions and quantitative modeling approaches - a literature review. Comput. Ind. Eng. **115**, 109–122 (2018)
6. Fattahi, M., Mahootchi, M., Moattar Husseini, S.M.: Integrated strategic and tactical supply chain planning with price-sensitive demands. Ann. Oper. Res. **242**, 423–456 (2016)
7. Govindan, K., Fattahi, M.: Investigating risk and robustness measures for supply chain network design under demand uncertainty: a case study of glass supply chain. Int. J. Prod. Econ. **183**, 680–699 (2017)
8. Kungwalsong, K., Cheng, C.Y., Yuangyai, C., Janjarassuk, U.: Two-stage stochastic program for supply chain network design under facility disruptions. Sustainability **13**(5) (2021)
9. Silva, A.C., Marques, C.M., de Sousa, J.P.: A simulation approach for the design of more sustainable and resilient supply chains in the pharmaceutical industry. Sustainability **15**(9) (2023)

10. Ohmori, S., Yoshimoto, K.: Optimal risk mitigation planning for supply chain disruption. J. Jpn. Ind. Manag. Assoc. **66**, 12–22 (2015)
11. Tolooie, A., Maity, M., Sinha, A.K.: A two-stage stochastic mixed-integer program for reliable supply chain network design under uncertain disruptions and demand. Comput. Ind. Eng. **148**, 106722 (2020)
12. Shiina, T.: Stochastic Programming. Asakurashoten Co., Ltd., Japan (2015). (in Japanese)
13. Kobayashi, H., et al.: A proposal of resilient supply chain network planning method with supplier selection and inventory levels determination using two-stage stochastic programming. In: Alfnes, E., Romsdal, A., Strandhagen, J.O., von Cieminski, G., Romero, D. (eds.) APMS 2023. IFIPAICT, vol. 692, pp. 714–729. Springer, Cham (2023). https://doi.org/10.1007/978-3-031-43688-8_49
14. Olsson, A., Sandberg, G., Dahlblom, O.: On Latin hypercube sampling for structural reliability analysis. Struct. Saf. **25**, 47–68 (2003)
15. ILOG CPLEX-IBM. https://www.ibm.com/jp-ja/products/ilog-cplex-optimization-studio. Accessed 14 Apr 2024

Study on Developing a Comprehensive Inspection System that Parallel Improves the Accuracy of Manual and Automatic Inspections

Harumi Haraguchi(✉) and Takumi Miyamoto

Ibaraki University, Hitachi 3168511, Japan
harumi.haraguchi.ie@vc.ibaraki.ac.jp

Abstract. This study summarizes a three-year project targeting dental component manufacturing sites. The target inspection departments have always conducted manual inspections twice each. This department wanted to be performed automatically at least once by introducing an automatic inspection machine. We have two problems that need to be solved. The first is to equalize the judgment criteria that differ from one inspection operator to another, and the second is to develop an automatic inspection tool with the same accuracy level as the inspection operator's judgment criteria. However, the target product, a rotary tool for dental treatment (diamond bar), has diamond particles attached to its tip; every part is slightly different. Therefore, creating an inspection tool with a simple threshold setting was impossible. In this study, we developed an automatic inspection tool using machine learning, and at the same time, we developed an inspection training tool to equalize operators' skills. Each tool was repeatedly improved through verification experiments. In addition, we developed feedback rules for the results obtained from the training tools to the training data for the machine learning model to improve the accuracy of the discriminant model. Furthermore, we have proposed a labeling tool that establishes criteria for judging whether a product is quality or defective in consideration of the introduction of new products, thereby realizing the continuous introduction of products and the stabilization of inspection operations.

Keywords: Product inspection · Operator training · Machine learning · Neural network · Image recognition · Medical equipment

1 Introduction

Many studies have been done on automatic inspection using images [1–3]. In recent years, studies have been conducted using machine learning models. The neural network (NN) has been used conventionally in conventional classification recurrences, such as images or statistics. The applicability to the quality assurance check by picture judgment has been observed in recent years because a neural network can recognize the feature quantity and

Published by Springer Nature Switzerland AG 2024
M. Thürer et al. (Eds.): APMS 2024, IFIP AICT 729, pp. 132–145, 2024.
https://doi.org/10.1007/978-3-031-65894-5_10

classify it automatically. A usage example of a collapse measure in a house [4], detection of the flaw of the metal surface [5], and an inspection of a bottle [6], etc., were introduced in a previous study. Dimitri et al. have studied automating the conventional manual inspection of wind turbine blades using CNNs [7]. When inspecting using machine learning such as CNNs, the inspection method depends on the type of inspection target. They are classified into three categories: (1) Supervised, (2) Unsupervised, and (3) Semi-supervised. (1) Supervised Learning is used when the data is labelled. A typical example is the image of a product during inspection [8]. A discriminant model is trained using the labeled data, and binary classifications of normal and abnormal are applied to the unknown data. (2) Unsupervised learning trains models using unlabeled data. This method aims to discover the data's inherent structure and patterns. Typically, methods such as clustering and dimensionality reduction are used. (3) Semi-supervised learning combines a small amount of labeled data with many unlabeled data to train a model. Even when labeled data is scarce, performance improvement can be expected by utilizing unlabeled data. In this study, (1) Supervised machine learning models are selected. This method typically has a problem of an imbalance between quality and defective data. In simple classification problems, unbiased data can often be obtained [9]. However, in anomaly detection, a large bias in data class occurs. In the field of manufacturing, defective products are generally produced infrequently. This leads to the construction of models biased toward one of the classes, making it impossible to build appropriate models [10]. In particular, labels for defective data classes are often only defined as different from quality data, making it difficult to assign labels accurately. To overcome this problem, attempts have been made to resample or create pseudo-anomalous data using generative models. [11, 12].

Even at sites where inspections are automated, manual inspections are often performed to confirm the accuracy of automated inspections. In some cases, manual reconfirmation is performed when the judgment is sensitive. Despite the fact that automatic and manual inspections are used together in some cases in the real site in this way, few studies have integrated the two inspection operations. However, with the proposed integration of automatic inspection tools, machine learning models, and operator training tools, we envision a future where these processes can work in harmony, enhancing the efficiency and accuracy of inspections.

In our previous study, we proposed a system to support the inspection of diamond bars for the inspection work at the diamond bar manufacturing site. We developed an automatic inspection tool using machine learning models and an operator training tool.

About machine learning models, we constructed a machine learning model for the 4-value classification of images using neural networks and convolutional neural networks to visualize the classification distribution from image data [13, 14]. The 4-value classification means the quality of diamond bars was classified into four classes: "Class 0: Clearly quality product," "Class 1: Some operators judged defective, but quality product," "Class 2: Some operators judged quality, but defective product," and "Class 3: Clearly defective product" using machine learning models. However, the accuracy and precision were not sufficient. Therefore, we focused on image processing and proposed a classification model using canny filters and bilateral filters [15]. In previous studies, parameter tuning performed using grid search and other methods [16, 17].

Manual inspection operator needs enough training for accurate determination [18, 19]. Our previous study also developed an inspection training tool to assist operators' judgment and extracted samples with many wrong answers [20, 21]. The inspection training tool was improved, and the change in incorrect response rate and response speed for each inspection operator for samples with many wrong answers was verified [22]. In these previous studies, he found problems with the accuracy of two classifications, "Class 1: Some operators judged defective, but quality product," and "Class 2: Some operators judged quality, but defective product".

However, we have the problems for automatic and manual inspection, respectively.

The problems of machine learning model were data imbalance and accuracy of accompanying it. To solve these problems, data obtained by operator learning tools will be actively feedback into the machine learning model to make balance of data and to improve the accuracy of the automatic inspection tool. In addition, parameter tuning is automated to expand the range of tuning in this study. This update enables the efficient setting of optimal parameters. These comprehensive solutions are designed to address the specific challenges in the current inspection processes, ensuring a more robust and accurate system. The problem with the operator training tool was the variability of training results. While some operators were able to establish the criteria for inspection with the training tools, others did not know where to make the decision. Therefore, in this study, the inspection training tool is further improved so that defects can be judged from the eight segmented images. The results of the training tool are analyzed, and changes are made to the image data of the machine learning model based on the analysis results. Finally, considering introducing new products, we propose a labeling tool to establish criteria for judging whether a product is quality or defective.

The structure of this paper is as follows. Section 2 presents the current status and ultimate goals of the inspection work site that is the subject of this study. Section 3 describes the three inspection support tools proposed in this study. Section 4 presents verification experiments using the tools, and Sect. 5 summarizes the study.

2 Overview of Inspection System

2.1 Current Inspection System

The target model is a manufacturer of precision parts for dental treatment. Figure 1 shows the target products of the inspection, which are called diamond bars. The diameter and length of the diamond bar are 1.8 mm and 20 mm, respectively. The diamond particles were stuck within 10 mm of the point of the bar.

The current inspection rule assigns two inspection operators to each day. The inspection operators perform all product inspections by double-checking. Figure 2 shows the work situation: the first inspection operator removes products from the paper box filled by the production lot and checks the condition of the products using a microscope. The second inspection operator again checks all the products that have been determined to be quality and checks the products in the defective box again. The inspection policy is to avoid mixing defective products with quality products, although it is acceptable for quality products to be mixed in with defective products. Finally, the products judged to be defective have the diamond particles removed and are used again for manufacturing.

While the use of diamond bars is declining due to the decrease in the amount of caries in the Japanese population, the demand for diamond bars is growing significantly, especially in Southeast Asia. The manufacturer has set up a mass production system, but manual inspection cannot keep up with the increased production volume. The problem at this manufacturing site was that the inspection operators were overloaded, and their judgment varied from operator to operator.

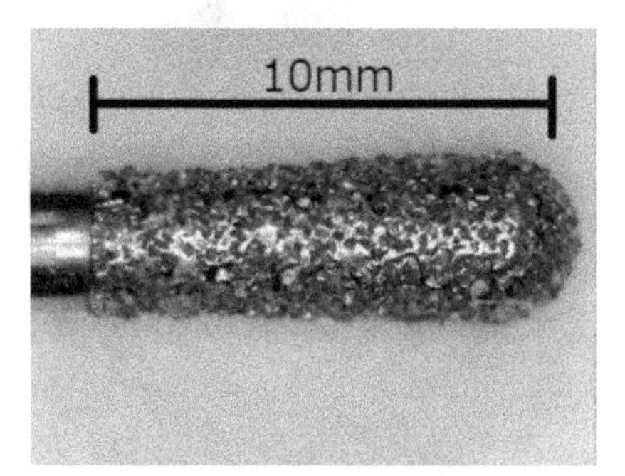

Fig. 1. Diamond bar

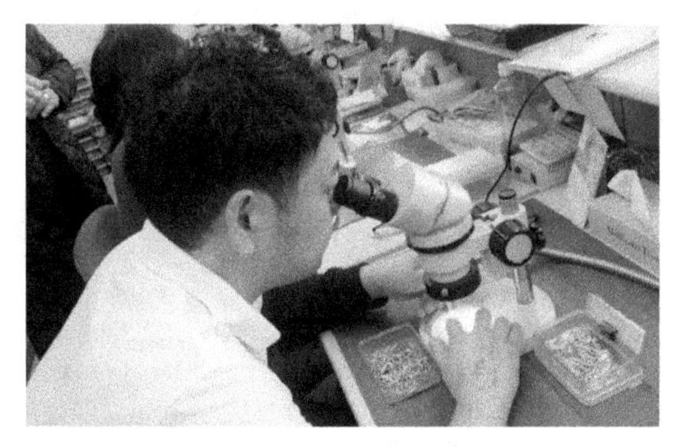

Fig. 2. Manual inspection using a microscope

2.2 Final Goal of the Inspection System

Figure 3 shows the whole image of final goal of the inspection system. We propose three tools for introducing a new product: 1. a labeling tool, 2. an operator training tool, and 3. an auto-discrimination model.

First, the labeling tool sets the criteria for quality and defective products. Based on the data labeled by this tool, sample data for the operator training tool and training data for the automatic discrimination model are created. It is important to note that the data used by the two tools should be the same.

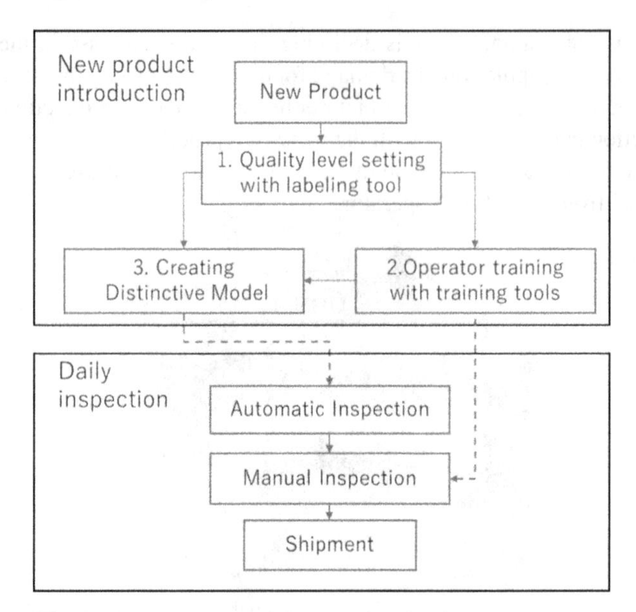

Fig. 3. Overview image of comprehensive inspection system

3 Overview of Inspection Support Tools

This study proposes a new tool, a labeling tool, which proved necessary during the improvement of training and automatic inspection tools.

3.1 Labeling Tool

The Fig. 4 shows the labeling tool screen. In the case of a diamond bar, the condition of diamond adhesion on the tip is used to determine whether the bar is defective or not. Still, the conventional standard was ambiguous: "If there are approximately 1/3 or more areas with no diamond particles attached, the bar is defective.

Therefore, the labeling tool drew a red line on the image of the tip of the diamond bar to divide it into eight sections. The operator checks the boxes below the image for areas of inadequate adhesion of diamond particles. If the number of checks is three or more, the operator determines that more than 1/3 of the diamond bars are defective and presses the "Defective" button. If there are 0 to 2 checked items, press the "Quality" button. In this way, the sample images are labeled as Quality or Defective.

Since the labeling tool sets the quality standard for a new product, only expert inspection operators use it. The labeled sample images are displayed in a list and shared with other expert operators to discuss the patterns of samples that are judged to be defective.

3.2 Inspection Training Tool

The interface of the training tool is similar to that of the labeling tool (Fig. 5). In the previous study, no dividing lines were included in the image. However, after validation

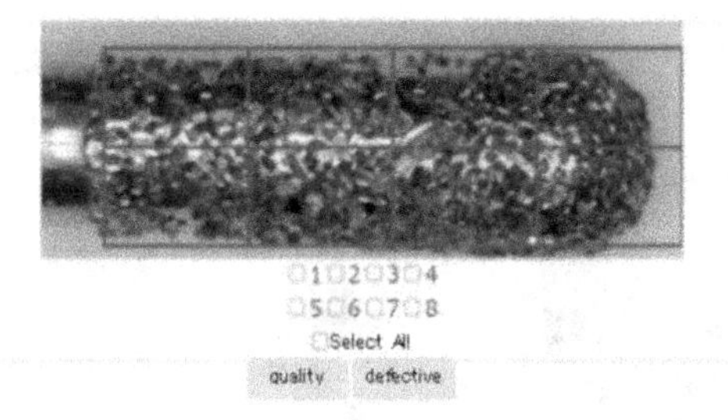

Fig. 4. Labeling tool interface

experiments, it was determined that they were necessary to clarify the basis for the judgment. The difference between the labeling tools is that the correct or incorrect answer is displayed after the quality or defective button is pressed. In addition, the previously judged image and the judgment result are displayed smaller than the current image on the left side. A list is then displayed for every ten image judgments (Fig. 6). This design allows the operator to know the operator's judgment criteria and the percentage of correct answers, which can be used as a reference for subsequent judgments.

The number of images to be judged by the training tool can be set arbitrarily up to the number of images prepared in advance. Images are displayed randomly from the dataset, but the same image is never shown twice.

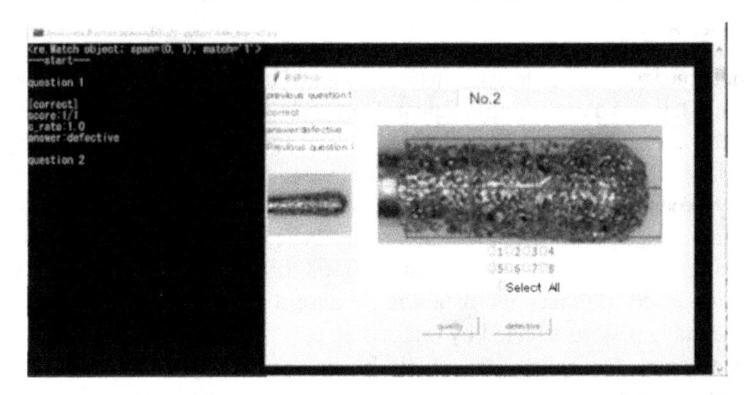

Fig. 5. Tool screen when the user's answer was incorrect in the previous question

3.3 Automatic Inspection Tool

The automatic inspection tool is installed in the inspection machine provided at the end of the diamond bar production line.

In the inspection machine, the diamond bars are set on a fixed stand. The diamond bar is then inspected by cameras installed on both sides of the diamond bar. Products determined to be defective are rejected, and the remaining products are subjected to a final visual inspection (Fig. 7).

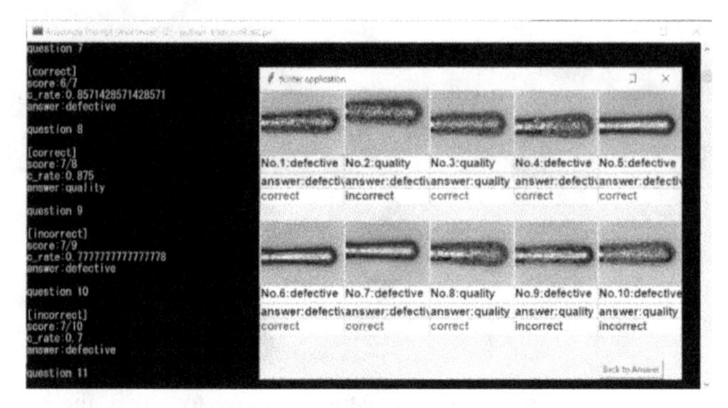

Fig. 6. Tool screen immediately after answering 10 questions

Fig. 7. Diamond bar fixing stand used before automatic inspection

The machine learning model used in the automated inspection tool our previous study [15]. This model uses a convolutional neural network (CNN) to construct the definition model; CNNs are widely used in image recognition and have a convolutional layer and a pooling layer. The parameters of this model were constructed using the results of preliminary experiments and previous studies.

The images used in this tool are the same as those used in the training tool, but the resolution of the images was reduced from 1600×1200 to 128×350 as input data. This resolution was used to reduce the data size because it is known from previous studies that accuracy is largely unaffected by image size [13]. All samples were assumed to be pre-cropped images for template matching. In addition, two types of image processing were used to improve the accuracy of machine learning. The methods used are the Canny filter and the bilateral filter. The canny filter is used for edge detection. The bilateral filter is a type of smoothing filter that smoothes the image while preventing the loss of edge information.

The CNN holds a convolutional layer, a pooling layer, and a combinatorial layer. The network consists of three layers. The activation functions of the layers can be closed in the CNN, and softmax is used for the ReLU and output layer activation functions. The loss function uses cross-entropy, and Adam and batch size are learned 830 epochs into the optimization algorithm. Parameters and hyperparameters are shown in Tables 1 and 2.

The k-division crossing inspection was employed as a model evaluation method. The dataset is divided into k groups with k-division crossing inspection. One group was used as the examination data and the k − 1 group was used as the training data. We obtained the evaluation index from the average by which the k time was repeated. The parameter is estimated as k = 10 in this study.

Table 1. CNN parameters

Layer type	Window	Input	Output
Convolution	3 * 3	256 * 700 * 1	254 * 698 * 16
Max Pooling	2 * 2	254 * 698 * 16	127 * 349 * 16
flatten	–	127 * 349 * 16	709168
Dense	–	709168	16

Table 2. CNN Hyperparameters

Batch size	32
Weight initialization	He's standard distribution
Optimization	Nesterov's Accelerated
Learning rate	0.001
Momentum	0.9
Epoch	100

4 Experiments to Validate the Tools

4.1 Inspection Training Tool

Experiments Condition

The operator needs to judge whether a diamond bar is quality or defective, but different operators judge some bars differently. In addition, some products appear to be quality but are defective; conversely, they are defective but are judged to be quality. Therefore, based on Noda's study, quality sample images are divided into two classes and defect sample images are also divided into two classes. The sample images classified four types as follows:

Class 0: Clearly quality product
Class 1: Some operators judged defective, but quality
Class 2: Some operators judged quality, but defective
Class 3: Clearly defective product

The learning tool has played a crucial role in our study. It has allowed us to verify how operators judge image data, which has been labeled into the four types of our classification system.

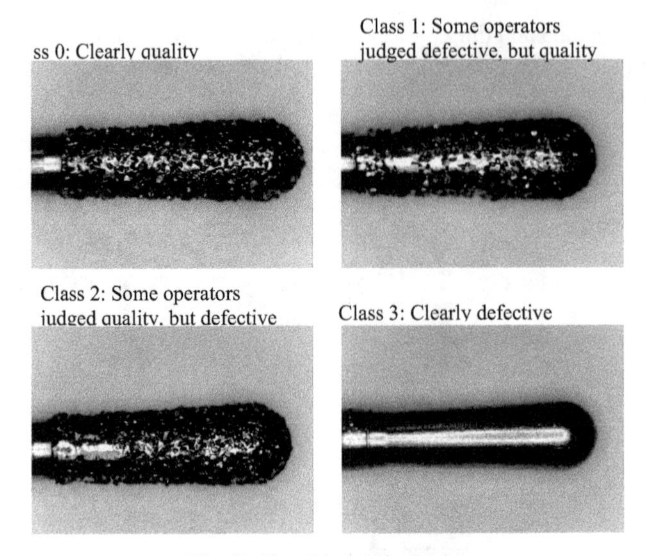

Fig. 8. Four kinds of samples

Examples of sample images classified into four types are shown in the Fig. 8.

In the previous study, the number of subjects and samples were sufficient, but the number of experiments was small (3 to 5 times), so it was not possible to determine the learning effect of the tool. In this study, we conducted a total of 20 training experiments over a 4-week period to verify the learning effect. In 0addition, a one-week interval was left between the third and fourth weeks to examine the forgetting effect.

Five sets of 40 images were prepared as one set. Subjects judged one set of image data once a day during the experiment. Each experiment was conducted with a different set of data, but the final experiment was adjusted so that all sets of data were tested four times each. The Table 3 shows the experimental conditions.

Table 3. Experimental conditions

Number of persons	21
Number of images	200 (40 × 5 sets)
Number of training days	20 (5 days × 4 weeks)

Results

The Table 4 shows the basic statistics for the percentage of correct answers per week. The average percentage of correct answers exceeds 90% for all weeks. The average rate of correct answers increased slightly from week to week. On the other hand, the minimum rate of correct answers varied from week to week, but the standard deviation indicated a highly varied score only in week 2. There was no significant change in the percentage of correct responses after a one-week interval between the third and fourth weeks, and no effect of forgetting was found in this experiment.

Table 4. The basic statistics for the percentage of correct answers per week

	Week 1	Week 2	Week 3	Week 4
Average	0.9191	0.9241	0.9248	0.9368
Max	1.0000	1.0000	1.0000	1.0000
Min	0.8250	0.7750	0.7750	0.8500
S.D	0.0239	0.0350	0.0277	0.0208

The Fig. 9 shows the average weekly rate of correct responses by class of sample images. Class 0, Class 1, and Class 3 had correct response rates of 90% or higher each week. The percentage of correct responses for Class 2 increased slightly from 80% in the first week to 89% in the fourth week.

Compared with previous study, the percentage of correct answers for the training tools was approximately 97% for Class 0, 85% for Class 1, 63% for Class 2, and 97% for Class 3 [15]. It is thought that the operator carefully checks the image by dividing the image into eight sections and checking the image and that the operator decides on the basis of the count of checks whether the image is quality or defective even if the image is difficult to judge, thereby increasing the rate of correct answers.

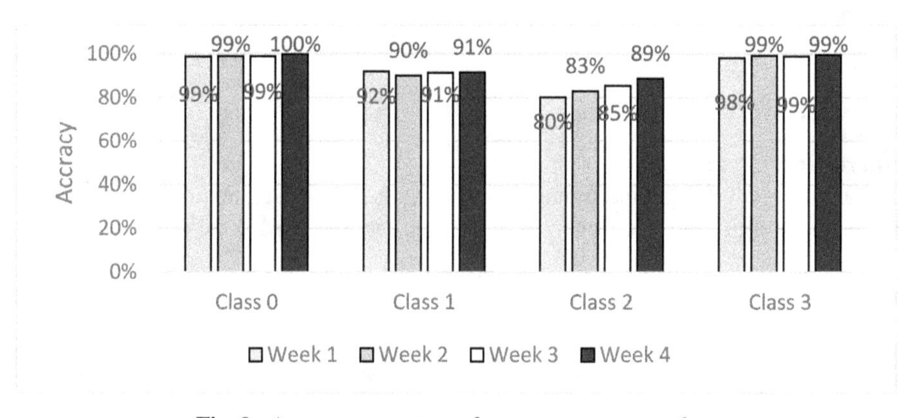

Fig. 9. Average percentage of correct answers per class

A distribution chart of the percentage increase in correct responses for each operator is shown in the Fig. 10, with the mean percentage correct response in week 4 for each Class 2 operator who improved operators correct response rate on the horizontal axis and the percentage correct response of Class 2 in week 4 relative to the work time in week 1 on the vertical axis. The graph shows that 75% of the trainees had an increase in the percentage of correct responses that exceeded 100% (improvement in the percentage of correct responses). These operators were within the range of 93% to 99%. From the above, it can be said that there was a learning effect in Class 2.

Finally, the number of checks on the 8-segmented image, which is a new feature of the training tool, was confirmed. The number of checks is determined to be quality if there are two checks, and defective if there are three or more checks. The average number of checks for class 0 was 0.23, and the average number of checks for class 1 was 0.52. On the other hand, the average number of checks for Class 2 is 3.82 and that for Class 3 is 5.62, indicating that the number of checks for Class 2 and Class 3 is significantly different even for the same defective product. The difficulty of judging Class 2 was also clarified.

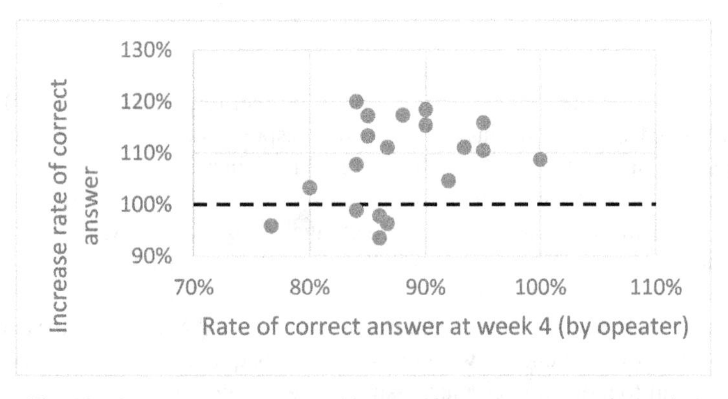

Fig. 10. Correct answers rate and increase in correct answers in week 4

4.2 Automatic Inspection Tool

Experiments Condition

The parameters of the definition model for the automatic inspection tool is shown in the previous section. In this experiment, we prepared the following three types of learning data.

1. Original Data. The first is the data labeled by the labeling tool.
2. Adjusted Data. The sample images with a significantly low percentage of correct responses are removed in the training tool experiment. To compensate for missing data, the remaining data is randomly extracted and flipped upside down to compensate for the number of images deleted as separate data.
3. Relabeling Data. The sample images with a significantly low percentage of correct responses in the training tool experiment are picked up and re-labeled by the expert operator. Since the number of images is the same for all classes in this study, the class with the increased number of images is randomly removed so that it is the same as the other classes, and the class with the decreased number of images is added in the same way as in 2.

Results

Table 5 shows the results of three models. Accuracy was higher for both Adjusted data (98.89%) and Relabeling data (87.69%) than for the original data (84.85%). 92.0% of Relabeling data showed higher Precision than the original data (58.67%). 75.64% of Adjusted data and 89.44% of Relabeling data showed higher Recall than the original data (66.34%).

The inspection policy at the real site is most important to avoid mixing defective products among quality products, and mixing quality products among defective products is not a major problem.

Based on this policy, the most important indicator is precision, which refers to the percentage of data that is actually positive out of those that are predicted to be positive. In other words, model 3 with the highest precision is the best.

Table 5. Results of 3 kinds of models

Model	Accuracy (S.D.)	Precision (S.D.)	Recall (S.D.)
Model 1 (Original Label)	84.85 (±6.50)	58.67 (±8.19)	66.34 (±4.48)
Model 2 (Adjustment Data)	98.89 (±5.51)	65.88 (±8.98)	75.64 (±10.01)
Model 3 (Relabeling Data)	87.69 (±3.96)	92.00 (±5.70)	89.44 (±7.54)

5 Conclusion

In this study, a three-year project to construct an inspection system for dental treatment rotary tools manufacturing sites, we proposed and verified the effectiveness of a tool that aims to improve the accuracy of automatic inspection and equalize the criteria for manual inspection.

The inspection training tool, with its unique feature of dividing the judgment image into eight segments, has proven to be a game-changer. It not only enabled clear judgments of quality and defective items but also achieved a higher rate of correct answers than the rate obtained with the training tool in the previous study. This tool, when used over a sufficient experimental period, has shown to have a learning effect on products in a state where judgment is difficult, thereby enhancing the overall inspection process.

The accuracy of the automatic inspection tool was improved by using the results obtained from the inspection training tool in the machine learning data of the automatic inspection tool.

As part of our proposal, we introduced a labeling tool that plays a crucial role in establishing clear and consistent criteria for judging whether a product is quality or defective. This tool is designed to facilitate inspection operations, particularly when a new product is introduced, thereby ensuring a standardized and efficient inspection process.

This research is limited in scope and may not apply to all inspection tasks. Even at sites where both automatic and manual inspections have been used in current, analyses and improvements related to both inspection tasks have been conducted independently. However, since the purpose of the inspection is to control product quality, we have shown that they should not be separated by method and that synergistic effects can be expected by bringing together the knowledge of both inspection methods.

Future issues include proposals for generically improving inspection quality

Acknowledgments. We would like to thank Sun-Techno Corporation for their useful discussions and advice.

Disclosure of Interests. Authors have no competing interests to declare that are relevant to the content of this article.

References

1. Timothy, S.N., Anil, K.J.: A survey of automated visual inspection. Comput. Vis. Image Underst. **61**(2), 231–262 (1995). https://doi.org/10.1006/cviu.1995.1017
2. Huang, S.-H., Pan, Y.-C.: Automated visual inspection in the semiconductor industry: a survey. Comput. Ind. **66**, 1–10 (2015). https://doi.org/10.1016/j.compind.2014.10.006
3. Roland, T.C., Carles, A.H.: Automated visual inspection: a survey. IEEE Trans. Pattern Anal. Mach. Intell. **PAMI-4**(6), 557–573. IEEE (1982). https://doi.org/10.1109/TPAMI.1982.4767309
4. Rin, T., Hiroyuki, I., Masato, Y., Masashi, F., Azuma, O.: Detecting collapsed buildings using convolutional neural network for estimating the disaster debris amount. J. Inf. Process. **56**, 1565–1575 (2016). (in Japanese)
5. Satorres, M., Ortega, V., Gámez, G., Gómez, O.: Quality inspection of machined metal parts using an image fusion technique. Measurement **111**, 374–383 (2017)
6. Wang, J., Fu, P., Gao, R.X.: Machine vision intelligence for product defect inspection based on deep T learning and Hough transform. J. Manuf. Syst. **51**, 52–60 (2019)
7. Dimitri, D., Benjamin, S., Michael, L., Michael, F.: Automatic optical surface inspection of wind turbine rotor blades using convolutional neural networks. Procedia CIRP **81**, 1166–1170 (2019). https://doi.org/10.1016/j.procir.2019.03.286
8. Paul, B., Michael, F., David, S., Carsten, S.: MVTec AD–a comprehensive real-world dataset for unsupervised anomaly detection. In: Proceedings of the IEEE/CVF Conference on Computer Vision and Pattern Recognition, pp. 9592–9600. IEEE (2019). https://doi.org/10.1109/CVPR.2019.00982
9. Jia, D., Wei, D., Richard, S., Li-Jia, L., Kai, L., Li, F.: ImageNet a largescale hierarchical image database. In: 2009 IEEE Conference on Computer Vision and Pattern Recognition, pp. 248–255. IEEE (2009). https://doi.org/10.1109/CVPR.2009.5206848
10. Chawla, N.V.: Data mining for imbalanced datasets: an overview. In: Maimon, O., Rokach, L. (eds.) Data Mining and Knowledge Discovery Handbook, pp. 853–867. Springer, Boston (2005). https://doi.org/10.1007/0-387-25465-X_40
11. James, P.T., Michael, C.: Resampling approach for anomaly detection in multispectral images. In: Algorithms and Technologies for Multispectral, Hyperspectral, and Ultra special Imagery IX, vol. 5093, pp. 230–240. SPIE (2003)
12. Hironori, M., Kenji, F.: ALGAN anomaly detection by generating pseudo anomalous data via latent variables, vol. 10, pp. 44259–44270. IEEE (2022)
13. Yuki, N., Harumi, H., Jun, M.: A study of the inspection support tool development using the neural network. In: Proceeding of Ibaraki Area Division Conference 2019, pp. 1001–1004. The Japan Society of Mechanical Engineers (2019). (in Japanese)

14. Harumi, H., Yuki, N.: A study of developing the parts inspection support tool using the neural network. In: Proceeding of Manufacturing Systems Division Conference 2020, pp. 205–206. The Japan Society of Mechanical Engineers (2020). (in Japanese)
15. Riku, A., Harumi, H.: Study on improving the performance of inspection support tools using image processing. In: Proceeding of Manufacturing Systems Division Conference 2023. The Japan Society of Mechanical Engineers (2023)
16. Hussain, A., Simone, A.L.: Hyperparameter optimization comparing genetic algorithm against grid search and Bayesian optimization. IEEE Congress on Evolutionary Computation (CEC) (2021)
17. Nurshazlyn, M.A., Dhanapal, D.D.P.S.: Hyperparameter optimization in convolution neural network using genetic algorithms. Int. J. Adv. Comput. Sci. Appl. **10**(6) (2019)
18. Kota, F., Noriyoshi, K., Daisaku, N., Shunpei, K., Eiji, K.: Application of estimated proficiency to machining technology education. In: Proceedings CIEC 2019 PC Conference, pp. 148–149 (2019). (in Japanese)
19. Taisei, K., Takahito, T.: Using error-based simulation (EBS) and concept maps development and evaluation of a system to promote abstraction operations in metacognitive activities. JSiSE Res. Rep. **34**(6), 199–204 (2020). (in Japanese)
20. Riku, A., Harumi, H.: A study on the sample extraction for a quality inspection tool and operator training, proceeding of manufacturing systems division conference 2021. The Japan Society of Mechanical Engineers (2021). (in Japanese)
21. Riku, A., Harumi, H.: Sample extraction of a quality inspection tool for dental parts manufacturing industry. In: IEEM 2021, pp. 843–847. IEEE (2021)
22. Shingo, K., Masatsuki, S., Riku, A., Harumi, H.: Improvement of inspection training tools and validation of the accuracy of machine learning discriminant models using the results. In: IEEM 2022, pp. 1–5. IEEE, December 2022

Autonomous Vehicles as a Way to Mitigate Traffic Accidents: A Literature Review of Obstacles of Its Implementation

Daniele dos Santos Ramos Xavier[1,2](✉), João Gilberto Mendes dos Reis[1,3], Daniel Laurentino de Jesus Xavier[1,2], and Gabriel Santos Rodrigues[1]

[1] RESUP – Research Group in Supply Networks and Logistics – Postgraduate Program in Production Engineering, Universidade Paulista – UNIP, Dr. Bacelar. 1212, Sao Paulo 04026002, Brazil
{daniele.xavier4,daniel.xavier3, gabriel.rodrigues119}@aluno.unip.br, joao.reis@docente.unip.br

[2] Centro Paula Souza – Faculdade de Tecnologia da Zona Leste, Av. Águia de Haia, 2983 - Cidade Antônio Estêvão de Carvalho, São Paulo - SP 03694000, Brazil

[3] Social and Applied Sciences Center, Mackenzie Presbyterian University—MPU, R. da Consolação, 930 - Consolação, São Paulo - SP 01302-907, Brazil

Abstract. Autonomous vehicles have been announced as innovative solutions for transport with the potential for changes in the transport of cargo and passengers. The objective of this paper is to analyze the challenges of vehicle automation adoption to mitigate traffic accidents. To do so, a literature review was conducted, and the implications were verified in the light of the Brazilian case. The results suggest that the country has great potential to benefit from the implementation of autonomous vehicles since it has a large number of traffic accidents caused by human error, such as distraction, tiredness, and recklessness, among others. However, the literature indicated that the barriers to the implementation of autonomous cars refer to; the need for improvements in obstacle detection by autonomous vehicles; users' perception regarding the adoption of this technology; questions about the interaction between pedestrians and drivers; cost competitiveness of vehicles; as well as the need of improvement in road safety and reduction of accidents.

Keywords: Autonomous vehicles · Technological evolution · Automation · Vehicle accidents

1 Introduction

Autonomous vehicles have been announced as a groundbreaking transport solution [1]. However, there is much debate about whether they can be considered safe and potentially reduce traffic accidents. Fagnant and Kockelma [2] suggest that autonomous vehicles have this potential and can also reduce travel time and improve energy efficiency.

M. Thürer et al. (Eds.): APMS 2024, IFIP AICT 729, pp. 146–157, 2024.
https://doi.org/10.1007/978-3-031-65894-5_11

In 2016, the United States National Highway Traffic Safety Administration reported that human errors are involved in about 94–96% of all motor vehicle accidents [1]. That same year, it was predicted that autonomous vehicles could eliminate 90% of traffic collisions due to the elimination of human error [1, 2].

Nonetheless, this optimistic view ignores the potential additional risks that these technologies can bring. It is important to carefully consider the challenges and safety concerns that may arise with autonomous vehicles to ensure responsible implementation and minimize any new risks introduced by this new form of transportation [3].

Questions of safety in autonomous vehicles started to be a concern after the first pedestrian death by an autonomous vehicle in its testing phase occurred in Tempe, AZ, in 2018 [4]. Nevertheless, the potential benefits far outweigh the potential risks associated with this new technology.

Indeed, the number of accidents and death rates on roads around the world has increased significantly in recent years [3, 4]. In Brazil, for instance, 64,447 accidents were recorded in 2022 on federal highways, 52,948 of which resulted in victims (dead or injured) [5]. In France, "traffic accidents are the leading cause of hospitalization in adolescence, with the 18–24-year-old age group accounting for 23% of deaths by traffic accidents" caused by traffic environment and human factors [6].

Autonomous driving has emerged in recent years as an alternative to conventional driving mode, offering great expectations in reducing these numbers of traffic accidents and providing increased safety, comfort, air quality, and road capabilities. Although autonomous technology is ready to be implemented on the roads, several barriers prevent this phenomenon, including ethical, moral, legislative, and regulatory issues, that represent the main challenges of introducing autonomy in driving [7].

In this paper, we aim to analyze the challenges of implementing autonomous vehicles as a way to mitigate traffic accidents considering the presented in the literature and evaluate them from a Brazilian perspective.

This article is divided into sections. After this introduction, Sect. 2 presents figures of traffic accidents in Brazil, Sect. 3 the methodology approach, and Sect. 4 discusses the implications of the study based on the literature. Finally, Sect. 5 presents our final remarks.

2 Traffic Accidents in Brazil

Traffic accidents are one of the most important justifications for implementing autonomous vehicles. Litman [4] points out that the progress of autonomous vehicles promises to positively transform the way people move, bringing essential benefits to society and the environment. One of the reasons for this progress is the reduction in the number of accidents. Figure 1 shows the number of accidents in Brazil between 2007 – 2022.

As can be seen in Fig. 1, despite the reduction in the number of victims observed since 2011 on Brazil's federal highways, there is still a large number of accidents and victims per year [5].

Brazil's economy is highly dependent on road transport, Fig. 2.

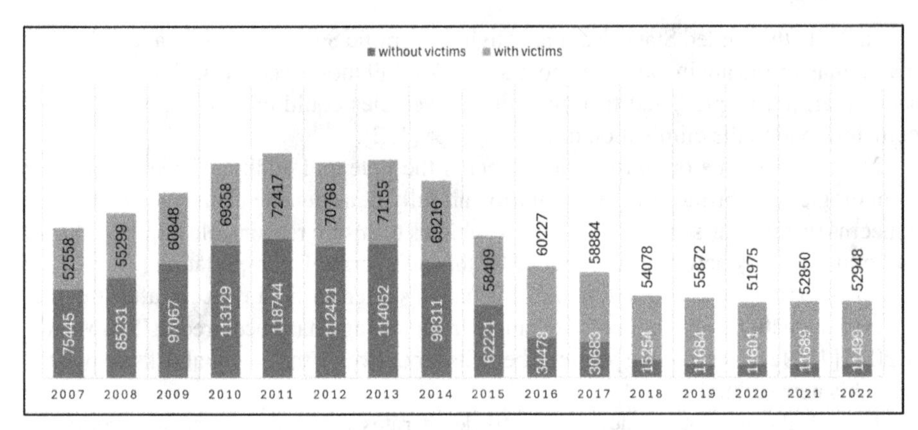

Fig. 1. Accidents by type (with victims and without victims) in Brazil (2007 – 2021). Source: Author using data from [5]

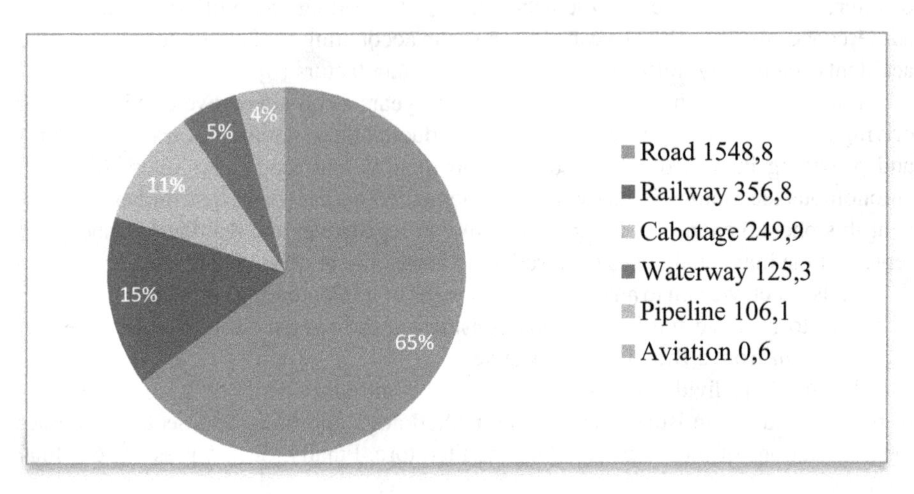

Fig. 2. Cargo transportation by highways in Brazil in billions of TKUs. Source: The authors with data from the National Logistics Plan Report, 2025 [9]

Around 65% of cargo transport in Brazil passes through highways, according to the Executive Report of the National Logistics Plan 2025. This high dependence compromises the competitiveness of Brazilian products, especially in the international market, since transport expenses can represent up to 7% of national Gross Domestic Product (GDP) [8].

The Federal Highway Police which is responsible for safety on federal roads in Brazil, indicates that the biggest cause of accidents is the lack of attention or reaction from drivers, motorcyclists, and pedestrians, contributing to 36% of cases. Inappropriate behavior such as disregarding traffic rules, speeding, and alcohol consumption are also significant factors, totaling 29.4% of occurrences. Head-on collisions are the most common type of accident, resulting in almost 40% of road fatalities [10].

Ministry of Health, between 2010 and 2019, stated that approximately 392 thousand people lost their lives in land transport accidents in Brazil. These accidents included pedestrian accidents, accidents involving bicycles, motorcycles, cars, pickup trucks, trucks, buses, and other types of land vehicles, such as service vehicles [10].

Table 1. Deaths by mode in Brazil from 2000 to 2019

Mode	Deaths 2000–2009	Percentual 2000–2009	Deaths 2010–2019	Percentual 2010–2019	Variation (%)
Pedestrian	96,917	28	75,648	19.3	−21.9
Bicycle	13,717	4	13,785	3.5	0.5
Motorcycle	58,310	16.8	118,720	30.2	103.6
Car/Light vehicle	72,462	20.9	91,940	23.4	26.9
Truck	6,598	1.9	8,111	2.1	22.9
Bus	1,650	0.5	1,802	0.5	9.2
Others	96,497	27.9	82,923	21	−14.1
Total	392,929	100	392,929	100	13.5

Source: Authors using data from the Datasus [10] and IPEA[11]

In Table 1, it is observed that there was a 13.5% increase in the total number of deaths in traffic accidents compared to the previous decade (2000–2009). However, in terms of the mortality rate per 100,000 inhabitants, the rates remained practically the same as in the previous decade, with a small increase of 2.3%.

Therefore, there is an expectation that autonomous vehicles can contribute to the reduction/mitigation of accidents on Brazilian roads.

3 Methodology

To answer the objective of this study of investigating the challenges faced in the process of evolution of vehicle automation, bibliographic research was used [12]. To this end, the following methodological steps were adopted.

1. **Identification of Bibliographic Sources:** Initially, the Scopus database was searched. From the search generated by Scopus, we obtained 36 results. Subsequently, the most relevant articles for the research were selected, resulting in 12 scientific articles, as presented in the references, along with government databases and other web research.
2. **Selection of Articles, Theses, and Materials:** Using keywords related to the topic, such as "vehicle automation", "vehicle automation challenges", and "advanced automotive technologies", among others, the relevant and updated materials available were selected. The selection was based on the relevance of the content, the quality of the sources, and the date of publication.

3. **Material Organization:** The materials were organized by themes relevant to the study, such as vehicle automation technologies, technical and safety challenges, regulations and legal aspects, and social and environmental impacts, among others. This organization allowed for a more in-depth and structured analysis of the different aspects involved in the process of evolution of vehicle automation.
4. **Analysis and Synthesis of Results:** After organizing the materials, a critical and comparative analysis was carried out of the different points of view and approaches found in the literature. Patterns, knowledge gaps, and emerging trends related to challenges faced in the development and implementation of vehicle automation were identified.
5. **Interpretation and Discussion of Results:** Finally, the results were interpreted, and their theoretical and practical implications were discussed. The main conclusions and recommendations for future research and the development of public policies and strategies in the field of vehicle automation were highlighted.

4 Results and Discussions

4.1 Automation Stages

Vehicle automation is one of the most significant technological advances of the 21st century, promising to revolutionize the way we travel and interact with vehicles. This complex process involves a series of steps, ranging from conception and development to implementation and maintenance of automated systems.

Cars with full automation, capable of traveling safely and without human intervention, do not exist, as stated by Zaparolli [13]. However, autonomous cars that are capable of interacting with city infrastructure, generating improvements in traffic, and at the same time comfort and safety for users and pedestrians are the future of mobility, but it is not a reality for all countries.

To understand this transformation, it is necessary to present the stages of vehicle automation. The classification of autonomous vehicles is essential to understand their capabilities and limitations. In addition to guiding the development and regulation of these innovative systems.

Autonomous vehicles are categorized according to their level of automation, determined by the National Highway Traffic Safety Administration (NHTSA), which developed a five-level scale describing the degree of control that both the driver and the vehicle have over driving functions. Figure 3 presents how autonomous vehicles are classified.

The levels of automation in Fig. 3 can be defined as follows:

- **Level 0 – No Automation:** The driver maintains full and exclusive control of the vehicle's main controls - brake, steering, and accelerator - at all times. The driver is solely responsible for supervising the road and ensuring the safe operation of the vehicle.
- **Level 1 - Function-specific automation:** At this level, one or more specific control functions can be automated, but they operate independently of each other. The driver maintains overall control and is responsible for safe operation. Moreover, he/she may cede limited authority over a primary control, as in the case of adaptive cruise control.

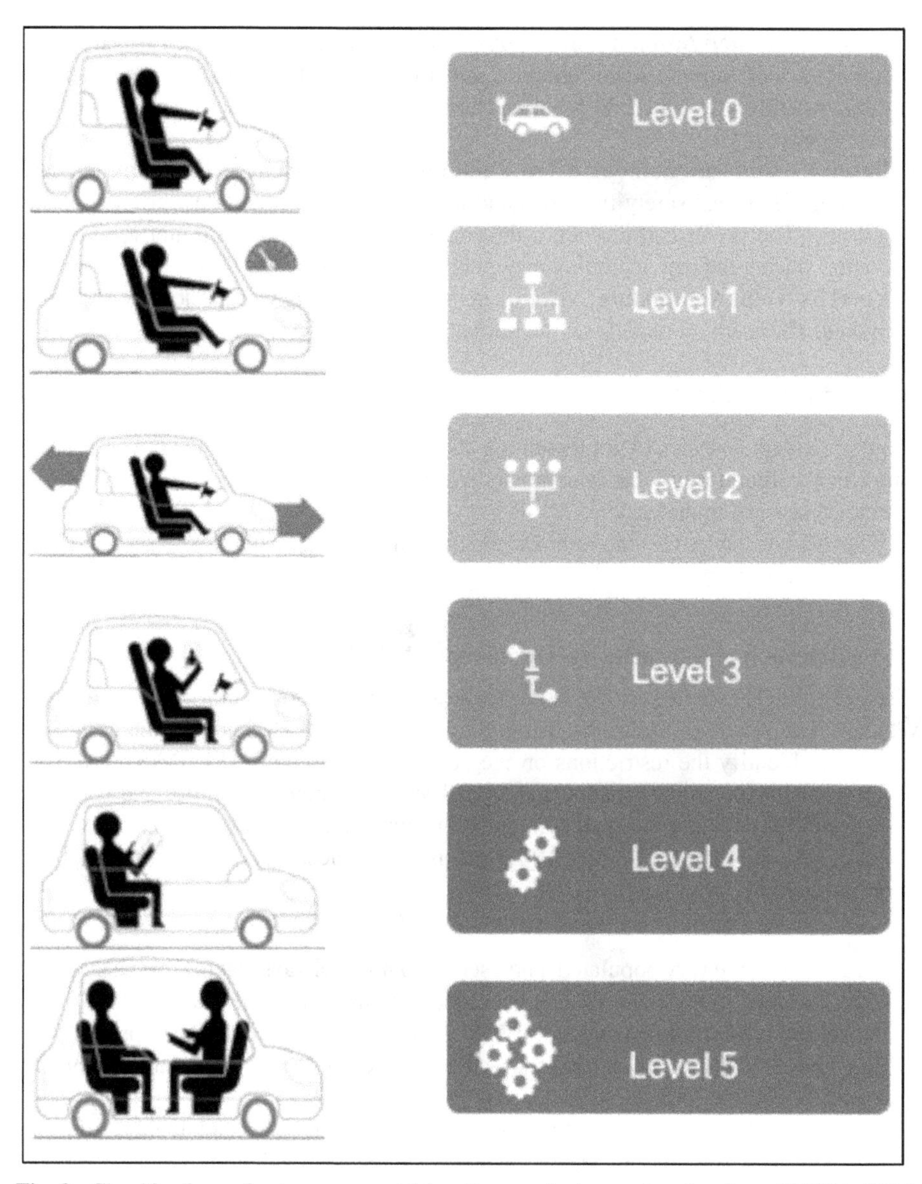

Fig. 3. Classifications of autonomous vehicles. Source: Authors using data from NHTSA, 2024 [13]

- **Level 2 - Combined Function Automation:** Involves the automation of at least two primary control functions that work together to relieve the driver of control of these functions. The driver is still responsible for monitoring the road and safe operation, even though the vehicle may assume limited authority in certain driving situations.

- **Level 3 - Limited Autonomous Automation:** At this level, the driver can relinquish full control of safety-critical functions under certain traffic or environmental conditions, trusting the vehicle to monitor changes and transition back to driver control when necessary.
- **Level 4 - Full autonomous driving automation:** The vehicle is capable of performing all critical safety functions and monitoring road conditions throughout the journey. The driver can provide a destination or navigation but is not required to control during the trip, including occupied and empty vehicles [14].
- **Level 5 – Total Automation:** Vehicles do not need steering wheels, accelerators, and brakes. There are no models on the market yet [13].

4.2 Main Obstacles

There are many obstacles to implementing autonomous vehicles on a commercial basis found in the literature. Table 2 summarizes the main studies regarding autonomous vehicles discussed in this article.

The main obstacles for autonomous vehicles will be discussed following considering the Brazilian scenario.

4.2.1 Detection of Obstacles by Autonomous Vehicles

Xiaoyan et al. [15] studied critical issues regarding the detection of objects by automated vehicles. The safety of the traffic using automated vehicles depends on the ability of vehicles to identify the restrictions on the path. The authors tested LIDAR, RADAR, vision cameras, ultrasonic sensors, and IR and reviewed their capabilities and behavior in several different situations and concluded the current developments appear to be not sophisticated enough to guarantee 100% precision and accuracy, hence further valiant effort is needed.

Considering Brazil‘s case some aspects should be considered:

- Large centers are very populated, and users tend not to always respect the basic traffic rules, meaning vehicles need to detect objects in any area of the road;
- Urban and rural traffic conditions in Brazil are comprehensive and may present different needs to detect obstacles, such as the presence of pedestrians or animals on the roads, in addition to population density;
- Lack of standardization of road infrastructure, as well as poor road quality;
- Legislation in Brazil is in the process of developing regulations for autonomous vehicles. Issues such as safety, responsibilities in cases of accidents, or systems to detect obstacles have not yet been defined;
- The cost of adopting technologies that can detect obstacles may vary and require supplier availability and investment.

Table 2. Main studies regarding autonomous vehicles

Author	Research	Methodology	Results
Yu, et al. [15]	Study on object detection by autonomous vehicles	Testing of LIDAR, RADAR, vision cameras, ultrasonic sensors, and IR in various situations	Conclude that current developments do not guarantee 100% accuracy, requiring further effort
Yuen et al. [16]	Examination of factors influencing users' behavioral intention to adopt autonomous vehicles	Application of an integrated model based on the diffusion of innovation theory and the technology acceptance model	Indicate that perceived usefulness and perceived ease of use positively influence users' intention to use autonomous vehicles
Rouchitsas and Alm [17]	Research on pedestrian-driver interaction in an autonomous vehicle environment	Review of empirical work on external human-computer interfaces	Identification of the need for a communication interface to enhance interaction between pedestrians and autonomous vehicles
Chen, T. Donna; Kockelman, Kara M.; Hanna, Josiah P [18]	Study on the cost competitiveness of shared autonomous electric vehicles	Use of an agent-based simulation model of a fleet	Discovery that cost competitiveness depends on the automation of recharging and there are synergies between shared autonomous vehicle fleets and electric vehicle technology
Canis [19]	Analysis of the potential of autonomous vehicles to improve road safety	Review of discussions in the US Congress on autonomous vehicle regulation	Highlight the interest and challenges faced by Congress regarding the regulation and implementation of autonomous vehicles, including safety issues, legislation, and data access

4.2.2 User Perception of the Adoption of Autonomous Vehicles

Yuen et al. [16] examined the factors that influence a user's behavioral intention to use autonomous vehicles. To this end, an integrated model was applied based on the innovation diffusion theory (IDT) and the technology acceptance model (TAM). The authors' results indicated that perceived usefulness (PU) and perceived ease of use (PEOU) positively influence the behavioral intention of autonomous vehicle users.

Considering these aspects for Brazil, some points can be highlighted:

- The perception of Brazilians regarding technology and innovation. This perception can be influenced by cultural and educational factors for the acceptance and adoption of autonomous vehicles;
- Safety is a concern for users of autonomous vehicles and in Brazil, considering the high levels of traffic safety, the road infrastructure can be challenging and influence the perception of safety and reliability of autonomous vehicles;
- User experience with autonomous vehicles through pilot tests, simulations or interactions with semi-autonomous systems can influence the perception and acceptance of the technology.

4.2.3 Pedestrian-Driver Interaction

Rouchitsas and Alm [17] researched the new traffic environment with the implementation of semi-autonomous and fully autonomous vehicles, in which there is no visual communication between drivers and pedestrians. Through a comprehensive account of empirical work in the field of external human-computer interfaces, it identified that autonomous vehicles have benefited from the presence of a communication interface for interaction between pedestrians and vehicles.

When considering Brazil about the interaction of pedestrians and drivers of autonomous vehicles, it can be considered that:

- The interaction between pedestrians and drivers is influenced by the culture and traffic behavior in the country. There are different road users and
- It is important to consider how these factors affect the interaction between them;
- Brazil has high rates of accidents involving pedestrians, so it is important to ensure that autonomous vehicles are capable of interacting safely with pedestrians and minimizing the risk of accidents;
- The use of communications technology may be necessary to facilitate safe interaction, including visual, auditory, and tactile signaling systems that warn pedestrians about the actions of autonomous vehicles.

4.2.4 Cost Competitiveness of Autonomous Electric Vehicles

Chen et al. [18] when studying the cost competitiveness of shared autonomous electric vehicles for non-electric vehicles, used an agent-based model to simulate a fleet and identified that the cost competitiveness of shared autonomous vehicles compared to non-electric vehicles depends on recharging automation. There are natural synergies between shared autonomous vehicle fleets and electric vehicle technology. Autonomous vehicle fleets solve the practical limitations of current non-autonomous electric vehicles, including travelers' range anxiety, access to charging infrastructure, and time management charging.

When considering Brazil to the cost competitiveness of autonomous electric vehicles, it is important to take into account the following aspects:

- The cost of purchasing autonomous electric vehicles can be a relevant obstacle, especially in a market where electric vehicles are still relatively new and may have higher prices than traditional vehicles that use fossil fuels;
- Charging infrastructure is essential for autonomous electric vehicles and in Brazil, this infrastructure is still under development and presents high installation and maintenance costs;
- The tax incentives under discussion could help promote the adoption of autonomous electric vehicles, making them more cost-competitive with traditional vehicles;
- Incentives and partnerships between automotive companies can help access and develop more efficient and economical technological solutions, reducing the costs associated with autonomous vehicles.

4.2.5 Improvement in Road Safety and Accident Reduction

Canis [19], when studying the potential of autonomous vehicles to bring improvements in road safety and reduce accidents, presented that the American Congress showed interest in federal supervision of the testing and implementation of autonomous vehicles due to the number of accidents and fatalities. Autonomous vehicle legislation and a bill have been passed, but neither has been enacted. Key issues discussed include the traditional division of vehicle regulation between the federal government and states, the number of autonomous vehicles allowed for testing on the road, legislation related to cybersecurity threats, and access to data generated by autonomous vehicles. The research presents the interest and challenges facing Congress and regulatory agencies regarding the regulation and implementation of autonomous vehicles, as well as the potential impact on transportation and highway safety policies.

When considering Brazil concerning the possible improvement of road safety and reduction of accidents with the adoption of autonomous vehicles, it is important to take into account the following aspects:

- Brazil presents challenges in the development and implementation of legislation and regulations on autonomous vehicles, making it necessary to define government responsibilities, the number of vehicles allowed for testing on the roads, and legislation on cyber security;
- Definition of safety policies for the introduction of autonomous vehicles, considering whether they fit into existing policies in addition to the development of strategies to improve road safety with the prevention and possible reduction of accidents and victims;
- Brazil must evaluate the effects of autonomous vehicles on road safety and the eventual reduction of accidents, with simulation and modeling studies to analyze possible impacts of proposed policies and regulations.

5 Final Remarks

This research studies vehicle automation and its barriers considering a Brazilian perspective. The results suggest that Brazil has great potential to benefit from the implementation of autonomous vehicles since it has a large number of traffic accidents, 40% of which are caused by human error, such as distraction, tiredness, and recklessness, among others.

The literature search carried out indicated that the barriers to the implementation of autonomous cars refer to the need for improvements in obstacle detection by autonomous vehicles, users' perception regarding the adoption of this technology, questions about the interaction between pedestrians and drivers, cost competitiveness of vehicles, and potential improvement in road safety and reduction of accidents.

Also considering that the country mainly uses road transport, the implementation of this new technology has a greater potential impact on reducing logistics costs than countries that have a greater balance between modes.

It is important to consider that many aspects need to be built or improved. Autonomous vehicles tend to be more expensive than other vehicles and this aspect can impact and hinder large-scale adoption, especially in countries where purchasing power is limited. Large-scale adoption of autonomous vehicles could lead to unemployment in traditional sectors such as taxi drivers and truck drivers, requiring reskilling and transition programs to mitigate these impacts.

These aspects require not only technological advances, but also the formulation of effective policies, the development of adequate infrastructure, and efforts to promote the acceptance and adoption of this innovation by society.

The research expectation is to provide relevant information about the potential of autonomous vehicles for improving transport safety in Brazil, as well as identify the main obstacles and opportunities for their adoption.

The limitation of this study lies in its bibliographic nature. However, since we are dealing with an emerging technology and the study's primary aim was exploratory, this limitation was not considered a significant issue. The exploratory nature of the study was appropriate for providing an initial overview and identifying areas for more in-depth future research.

The implications of the study are connected to the fact that examining how different levels of automation impact accident rates and identifying both the benefits and potential vulnerabilities associated with each level of automation is crucial and will lead to a more comprehensive understanding of the effects of automation on traffic safety. Moreover, it contributes to the development of policies and technologies that mitigate risks and enhance the safety of autonomous vehicles in cities.

As future research, this investigation suggests analyzing the socioeconomic impacts of the introduction of autonomous vehicles in cargo transport in Brazil; measuring possible effects on employment in the transport sector, and the impact on logistics companies and the supply chain; in addition suggests research on income distribution and accessibility to transportation services in a autonomous environment.

Acknowledgments. This study was financed in part by the Coordenação de Aperfeiçoamento de Pessoal de Nível Superior - Brasil (CAPES). Finance Code 001.

References

1. Robinson-Tay, K.: The role of autonomous vehicles in transportation equity in Tempe. Arizona. Mobilities , 1–17 (2023). https://doi.org/10.1080/17450101.2023.2276100
2. Bonnefon, J.-F., Shariff, A., Rahwan, I.: The social dilemma of autonomous vehicles. Science **352**, 1573–1576 (2016). https://doi.org/10.1126/science.aaf2654
3. Kok, I., Zou, S.Y., Gordon, J., Mercer, B.: The Disruption of Transportation and the Collapse of the Internal-Combustion Vehicle and Oil Industries (2017)
4. Litman, T.: Autonomous Vehicle Implementation Predictions: Implications for Transport Planning. https://www.vtpi.org/avip.pdf
5. Painel CNT de Acidentes Rodoviários. https://cnt.org.br/painel-acidente. Accessed 08 Apr 2024
6. Gicquel, L., Ordonneau, P., Blot, E., Toillon, C., Ingrand, P., Romo, L.: Description of various factors contributing to traffic accidents in youth and measures proposed to alleviate recurrence. Front. Psychiatry. **8** (2017). https://doi.org/10.3389/fpsyt.2017.00094
7. Lousa, A.J.M.: Veículos Autónomos e Conectados - Tecnologia e Identificação de Possíveis Alterações na Infraestrutura de Transporte. https://estudogeral.uc.pt/bitstream/10316/84931
8. CNT: Confederação Nacional do Transporte. https://cnt.org.br/home
9. Empresa de Planejamento e Logística: Plano Nacional de Logística. https://portal.epl.gov.br/plano-nacional-de-logistica-pnl
10. Costa, R.R. da: Instituto de Pesquisa Econômica Aplicada. https://www.ipea.gov.br/portal/categorias/45-todas-as-noticias/noticias/13899-estudo-aponta-aumento-de-13-5-em-mortes-no-transito. Accessed 07 Apr 2024
11. DATASUS. https://datasus.saude.gov.br/mortalidade-desde-1996-pela-cid-10/
12. Lakatos, E.M., Marconi, M. de A.: Fundamentos de metodologia científica. Atlas, São Paulo (2017)
13. Zaparolli, D.: O futuro da mobilidade com carros autônomos. https://revistapesquisa.fapesp.br/o-futuro-da-mobilidade-com-carros-autonomos/
14. Rocha, P.A.M.C.: Implicações do Veículo Autónomo no Futuro das Cidades. https://revistapesquisa.fapesp.br/o-futuro-da-mobilidade-com-carros-autonomos
15. Yu, X., Marinov, M.: A study on recent developments and issues with obstacle detection systems for automated vehicles. Sustainability **12**, 3281 (2020). https://doi.org/10.3390/su12083281
16. Yuen, K.F., Cai, L., Qi, G., Wang, X.: Factors influencing autonomous vehicle adoption: an application of the technology acceptance model and innovation diffusion theory. Technol. Anal. Strateg. Manag. **33**, 505–519 (2021). https://doi.org/10.1080/09537325.2020.1826423
17. Rouchitsas, A., Alm, H.: External human-machine interfaces for autonomous vehicle-to-pedestrian communication: a review of empirical work. Front. Psychol. **10**, 2757 (2019). https://doi.org/10.3389/fpsyg.2019.02757
18. Chen, T.D., Kockelman, K.M., Hanna, J.P.: Operations of a shared, autonomous, electric vehicle fleet: Implications of vehicle & charging infrastructure decisions. Transp. Res. Part Policy Pract. **94**, 243–254 (2016). https://doi.org/10.1016/j.tra.2016.08.020
19. Canis, B.: Issues in Autonomous Vehicle Testing and Deployment. https://sgp.fas.org/crs/misc/R45985.pdf

Analysis of People's Continental Behavior Regarding Cycling in Light of the Cyclability Index

Izolina Margarida de Souza[1,2](✉), João Gilberto Mendes dos Reis[1,3], Alexandre Formigoni[4], and Lucas Santos de Queiroz[4]

[1] RESUP – Research Group in Supply Networks and Logistics – Postgraduate Program in Production Engineering, Universidade Paulista – UNIP, Dr. Bacelar, Sao Paulo 1212, 04026002, Brazil
guidariana@hotmail.com, joao.reis@docente.unip.br
[2] Centro Paula Souza – Faculdade de Tecnologia da Zona Leste, Avenue Águia de Haia, 2983 - Cidade Antônio Estêvão de Carvalho, São Paulo - SP 03694000, Brazil
[3] Social and Applied Sciences Center, Mackenzie Presbyterian University—MPU, R. da Consolação, 930 - Consolação, São Paulo - SP 01302-907, Brazil
[4] Mestrado Profissional Em Gestão E Tecnologia Em Sistemas Produtivos. Rua Dos Bandeirantes, 169, São Paulo 01124-010, Brazil
alexandre.formigoni@fatec.sp.gov.br

Abstract. The excessive use of motor vehicles in urban mobility has caused many negative impacts on both society and the environment in several cities around the world. Issues such as traffic congestion, noise, and air pollution caused by the excessive release of CO_2 into the atmosphere, and inefficient use of public spaces make people more stressed, with cardiovascular and respiratory problems. In recent years, there has been much discussion about the use of bicycles as an alternative form of active and sustainable mobility, which not only does not contribute to pollution but also promotes physical activity. After categorizing cities considered bicycle-friendly by continent, the objective of this study is to evaluate if there are significant differences in the bike-friendliness index presented by these cities when compared by continent groups. Data were collected from the Global Bicycle Cities Index - GBCI for the year 2022. Indicators such as safety, infrastructure, events, critical mass, and availability of shared bicycles are considered, and the social network analysis tool using graphs from NetDraw® software is utilized. It is concluded that there are no significant differences among the available indices except related to the availability of shared bicycles.

Keywords: Active Mobility · Cyclability · Sustainability

1 Introduction

In recent years, there has been a global discussion about sustainable urban displacements, with a focus on active mobility: cycling and walkability. In this regard, the bicycle plays an essential role, whether conventional or electric (e-bikes). However, studies show that

M. Thürer et al. (Eds.): APMS 2024, IFIP AICT 729, pp. 158–171, 2024.
https://doi.org/10.1007/978-3-031-65894-5_12

the perception of bicycles as a sustainable mode of transportation still faces a lot of prejudice, as this mode of transport is still associated with poverty while the use of motor vehicles represents a certain social status [1, 2].

The use of bicycles in large urban centers has been highly encouraged in many countries, especially in more developed ones, because it not only reduces CO_2 emissions levels in the atmosphere but also promotes physical activity. Moreover, it improves people's quality of life through exercise and social interaction, helps reduce noise pollution, and optimizes spaces by occupying smaller areas [3].

On the other hand, the increase in bicycle use in urban areas faces some obstacles such as cultural issues, especially in the East, which has almost no female cyclists due to religious prohibitions, being subordinate to their husbands, and also facing the discomfort of traditional clothing such as the sari and salwar kameez [1]. It also faces social, economic, environmental, geographical, and mainly governmental obstacles [1].

Many cities are oriented towards the use of motor vehicles, which have individual trips above 43%, present high levels of pollutants, long congestion routes, and a lack of parking space. When deciding on sustainable changes by implementing cycling infrastructures, they do so with low investment, without effective planning, implementing disaggregated, poorly signaled, and low-quality structures. These structures, besides not attracting new cyclists, are rejected by experienced ones [4].

The car-centric culture is extremely harmful to global society, due to the constant demand for resources. There is a significant environmental impact, the degradation of people's health caused by stress and a sedentary lifestyle. Moreover, the high number of accidents caused by cars results in the loss of 1.3 million lives on all continents yearly. Finally, it provokes a disparity of in-car infrastructure, to the detriment of other active mobility alternatives, which encourages even more car use [5].

Indeed, car use implies many conflicts due to dangerous intersections, regions with excessive potholes, roads with high volumes of trucks and motorcycles at very high speeds, incompatible traffic rules, deficient signaling, conflict with motorized vehicles at the point of viaduct convergence, high rear-end collision rates followed by run-over and death and so on [1].

Implementing segregated lanes for cycling is not always an easy task and certain that infrastructure will be used by cyclists. The idea is to provide greater safety and tranquility to them, however, in some regions, these segregation structures are stolen, and cars end up invading the cyclist's lane, exposing them to risk again [4].

In addition to all the infrastructural fragility evidenced, cyclists still endure the rude and aggressive behavior exhibited by other road users through constant verbal harassment and persecution by drivers [1]. These are some of the challenges faced by cyclists who do not live in a bike-friendly city and do not prioritize cyclists in traffic.

Considering that many cities, especially in emerging countries, have their social development devices far from the outskirts and have expensive and inefficient public transport, active mobility promotes social inclusion and accessibility. In addition to favoring the reduction of polluting gas emissions and reducing the consumption of non-renewable energy [6]. The practice of physical exercises stimulated by active mobility can reduce the number of people with respiratory and cardiovascular diseases, consequently, it allows the reduction of public spending on diseases that develop due to lack of physical activity [7].

Considering the importance of the dissemination of active mobility, this study is important once it highlights the behavior of people who share urban spaces on each continent and allows us to identify which factors are inhibiting the spread of active mobility. There is an evident difference in the structure offered to cyclists in various regions of the world.

Based on a research database generated by the 2022 Global Bicycle Cities Index (GBCI) [8], after categorizing cities considered bike-friendly by continent, the objective of this study is to evaluate if there are significant differences in the bike-friendliness indices presented by bike-friendly cities when compared by continent groups. For this purpose, indicators such as safety, infrastructure, events, critical mass, and availability of shared bicycles are considered. The data analysis is made using graph theory (social network analysis) plotting graphically the results of the research investigated based on the indicators selected. Graphics were performed using NetDraw® 2.178 software.

The article is divided into sections. After this introduction, a short literature review is presented. Section 3 indicates the methods adopted in this research. Section 4 compares and discusses the results and finally, Sect. 5 presents the remarks of our study.

2 Literature Review

2.1 The Challenge of Sustainable Mobility

The challenge of sustainable mobility is one of the main environmental and social problems in the world today. With the growth of the world population and accelerated urbanization, passenger traffic is expected to increase by 50% by 2030, while freight volume will grow by 70% in the same period [9]. This growth places significant pressure on transport systems, requiring innovative and sustainable solutions.

The main challenges to be faced when implementing sustainable mobility are [10]:

- **Accessibility and Inclusion:** sustainable mobility must ensure that everyone, regardless of their geographic location or socioeconomic situation, has access to efficient and safe transport systems. Inequality in access to transport is an ongoing challenge, especially in developing countries, where infrastructure is often inadequate or non-existent;
- **Environmental Impact:** The transport sector is one of the largest contributors to the emission of greenhouse gases. To mitigate these impacts, it is essential to promote the use of clean technologies and reduce dependence on fossil fuels. Policies that encourage the use of electric vehicles and improve energy efficiency are fundamental to achieving environmental sustainability;
- **Efficiency and Innovation:** Transport efficiency can be improved through the use of intelligent technologies, such as traffic management systems, ITS (Intelligent Transport Systems), and autonomous vehicles. These innovations have the potential to reduce congestion and improve road safety;
- **Safety:** Ensuring the safety of transport systems is a priority. This includes the implementation of adequate infrastructure, awareness campaigns on traffic safety, and the development of technologies that increase the safety of public road users.

2.2 The Bicycle as a Sustainable Way of Mobility

Active transport is a movement that depends exclusively on the person moving, whether on foot, scooter, tricycle, or bicycle. Therefore, active transport, or non-motorized transport, directly results in improving the quality of life of the inhabitants who practice it. Mostly, because of practicing physical activities regularly, and reducing the emission of noise or atmospheric sources usually produced by motor vehicles [11].

Bicycling is a sustainable alternative that promotes physical activity, improves mental health, and contributes to reducing carbon emissions. Studies show that cities that invest in cycling infrastructure, such as Copenhagen and Amsterdam, can achieve significant improvements in air quality and public health [12].

On the other hand, in other regions in which those responsible for policies related to active mobility are economically backward, lack competitiveness, and are socially backward, investments in infrastructure for the use of motor vehicles persist. Concomitant to this movement, groups that depend on structures for non-motorized transport, such as bicycles, in some regions, are not organized and do not have the strength to articulate in favor of their needs [11].

For those who cannot invest in a vehicle or do not want to incur maintenance costs, there is the alternative of using shared bicycles, a fundamental service for encouraging active mobility. The use of shared bicycles, generally used in the initial and final sections of users' journeys, encourages the use of public transport by promoting intermodality between bus, subway, or train terminals. Furthermore, apps show stations with available services, cycle lanes, or cycle paths, allowing cyclists to travel more safely. The bicycle-sharing service offers the customer a vehicle (bike) in a virtual environment, where it is also possible to pay for the rental using cards [13]. The shared bicycle service is an extremely relevant criterion in evaluations of the most cycling cities in the world due to its importance in promoting the practice.

Based on those aspects is extremely relevant to provide studies that discuss the adoption of bike use in different parts of the world and compare them.

3 Methodology

This research analyzes the cyclability index for cities worldwide and compares them based on geography and location. To do so, we collected the data from Global Bicycle Cities Index that ranked the 90 cities bike friendly worldwide based on a methodology created by the French company insurance Luko [8].

Among the 90 cities evaluated and their - indices presented by the Global Bicycle Cities Index, the top five cities in each continent were chosen: Europe, Asia, Africa, North America, South America - except Oceania, which only presented four cities. These data are described in Table 1.

Table 1. Cities analyzed separated by continent according to the Global Bicycle Cities Index.

City	Continent	Safety	Infra structure	Sharing	Critical Mass	Event
Utrecht	Europe	4.412	4.052	2.833	1.991	5.634
Munster	Europe	4.482	3.936	3.434	3.171	6.099
Antwerp	Europe	4.344	3.535	4.489	2.648	7.061
Copenhagen	Europe	4.419	4.012	2.708	4.305	5.387
Amsterdam	Europe	4.353	4.024	3.497	3.119	6.927
Hangzhou	Asia	4.383	3.488	3.871	1.188	6.167
Montreal	N. America	4.472	3.827	1.792	2.177	6.894
Melbourne	Oceania	4.571	3.857	1.099	0.884	5.162
Wellington	Oceania	4.436	4.081	2.303	1.550	7.080
Vancouver	North Am	4.385	3.912	3.584	2.159	7.283
Beijing	Asia	4.467	3.459	2.398	2.385	5.953
San_Francisco	North Am	4.219	3.937	3.401	3.070	7.858
Sydney	Oceania	4.555	3.686	0.693	2.286	5.846
Portland	North Am	4.291	4.126	3.466	2.248	7.434
Casablanca	Africa	4.258	3.413	0.000	1.552	6.904
Auckland	Oceania	4.549	3.518	0.000	0.392	6.117
Seattle	North Am	4.321	4.042	2.079	1.442	7.431
Shanghai	Asia	4.489	3.476	2.944	0.122	5.736
Singapore	Asia	4.466	4.032	3.434	2.313	5.827
Santiago	Latin Am	4.418	3.700	1.386	1.522	5.894
Cairo	Africa	4.353	3.062	0.000	3.299	7.565
Sao_Paulo	Latin Am	4.123	2.286	0.693	4.097	6.164
Johannesburg	Africa	3.452	3.364	1.609	3.215	5.858
Nairobi	Africa	4.451	3.306	0.000	2.791	6.131
Bogota	Latin Am	4.049	3.338	0.000	3.009	6.341
Cali	Latin Am	4.008	2.119	0.000	2.456	6.433
Buenos_Aires	Latin Am	4.427	2.550	1.609	4.110	6.728
Tokyo	Asia	4.400	3.982	1.099	1.047	7.107
Lagos	Africa	4.300	0.000	0.000	3.080	6.264

The Global Bicycle Cities Index highlights five main criteria, which are [8]:

- **Safety:** considers deaths in bicycle accidents, accidents, and bicycle theft scores. Each item is always in the ratio of 1,000 per 100,000 cyclists per city;
- **Events:** events that occur in cities such as "car-free day," bike tours, and races;

- **Critical mass:** for this criterion, the number of participants or the size of the critical mass is evaluated. It is called critical mass when a group of people organizes through social networks and takes up space on the streets with the aim of celebrating the bicycle as a means of transportation. This event always takes place on the last Friday of the month in over 300 cities worldwide;
- **Infrastructure:** for infrastructure, the total number of bike shops per 100,000 cyclists is considered; as the length of bike lanes per population, road quality, and quality of investment;
- **Shared bike stations:** for evaluation, the number of stations per city and the number of bikes available per station per 100,000 inhabitants are considered.

For the analysis, the Social Network Analysis (SNA) method was used, which is a research method based on graph theory, probability theory, and geometry [14].

The data were processed using the NetDraw® 2.178 software, creating a VNA file showing the cities and their respective scores. The VNA file is a way to organize data for analysis in social network software. The networks were generated for each of the evaluated criteria, facilitating the analysis of the generated figure and the interpretation of the phenomenon by visually identifying it.

4 Results and Discussion

Figure 1 presents the network of selected cities in the study, grouped according to the scores obtained for each measured indicator.

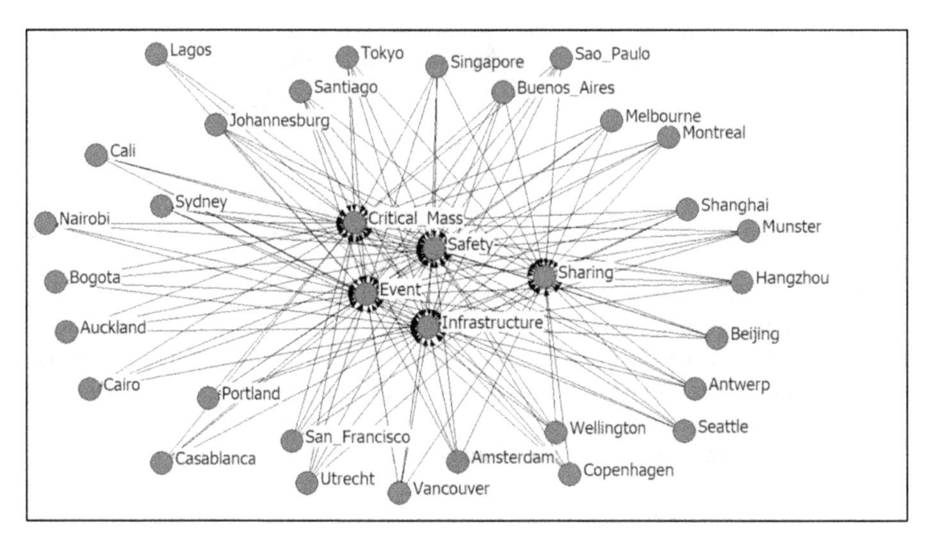

Fig. 1. Network of evaluation indicators of bicycle-friendly cities and cities from the selected sample

Observing Fig. 1 note that all selected cities according to each continent exhibit a similar behavior among the evaluated criteria. However, regarding the "Bike Sharing" criterion, a differentiated behavior is observed in the graph analysis, which places the node referring to this criterion distant from the others. This is justified by the fact that not all selected cities offer bike-sharing services. The cities that do not have connections with the "Sharing" node are Casablanca, Auckland, Cairo, Nairobi, Bogotá, Cali, and Lagos. It is noteworthy that, except for Auckland, all cities are part of developing countries. According to Prieto-Curiel and Ospina this occurs due to the higher the per capita Gross Domestic Product (GDP) of the analyzed region, that generates a higher dependence on car mobility [5]. In this case, there is a need for a strategy that discourages car use.

Figure 2 the K-core analysis presents the network of evaluated cities and highlights in red nodes all those that meet all indicators.

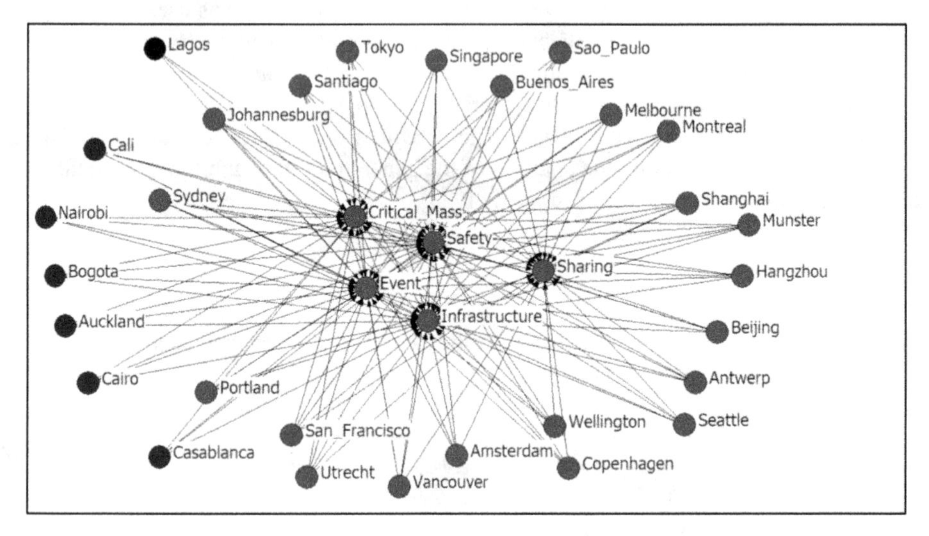

Fig. 2. K-core

All European cities connect with nodes from all evaluated criteria and occupy the top positions in the ranking of the most bike-friendly cities, highlighting their efforts and high investments in sustainable transportation and infrastructure [15].

Auckland, Oceania, has a strong culture of car use, with car trips representing 74.8% of all trips in the country, while in Auckland, they represent 78.8%. The bicycle mode share for the entire country is 2.3%, while in Auckland, it is only 0.9%. The factor contributing the most to the reduction of bicycle users on the streets of Auckland is safety, which is directly related to the volume of cars and dangerous intersections in its urban area [15, 16].

Auckland is the only evaluated city from a developed country that does not meet all the criteria evaluated by the GBCI. Analyzing Fig. 2 - K-Core, it is observed that all cities that do not meet all evaluated criteria, represented by nodes in blue and black, except for Auckland, belong to developing countries. This phenomenon may be related to unfavorable social formation, weakened educational foundations, a set of beliefs and values, lifestyle, and environmental awareness that complete a framework of characteristics of cities that have expanded very rapidly, in a short period, with little or no planning and are part of economically weaker countries [1].

The cities Lagos, Nairobi, Bogota, Cairo, Cali, and Auckland do not have bike-sharing services, as shown in Fig. 3 – Sharing (equal colors means same continent/area). The African cities of Lagos, Nairobi, and Cairo have practically the same scenario, with a worsening situation for Nairobi, where vehicles used by the poorest people living in precarious areas near work and school are vehicles that have already been discarded from countries like Japan and Europe, with their catalyzers losing efficiency, increasing the layer of pollution in the region. There is little investment in public transportation, which is precarious, and investments in non-motorized mobility infrastructure have been underestimated, representing little or no safety for active mobility users, with poor signaling, and large cracks in the roads, increasing the risk for users. Such a scenario ends up impacting the availability of bike-sharing services, which depend on infrastructure for service provision [17].

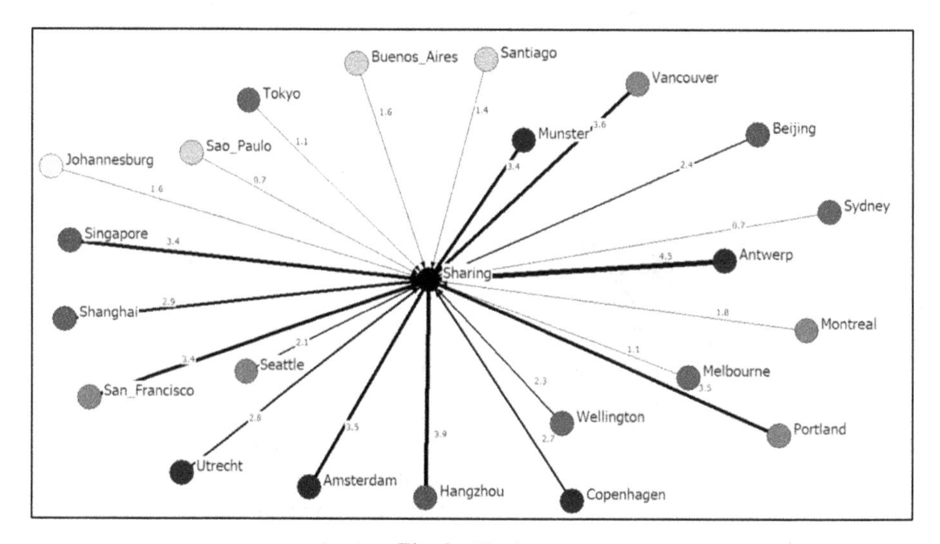

Fig. 3. Sharing

The GBCI evaluates both the number of sharing stations and the number of bicycles offered per station [8]. The Bike Sharing Service (BSS) is of fundamental importance for the expansion of active mobility and promotes sustainability both environmentally and economically as it eliminates the cost of owning/maintaining a bicycle, which can end up generating maintenance and storage costs. On the other hand, the BSS faces spatial constraints regarding the service offering and the distances to be traveled [18].

Still analyzing Fig. 3, differences in the sizes of the vertices are noted, representing greater or lesser BSS availability in the analyzed cities. In this regard, European cities are also highlighted, favored by their geographical structure characterized by flat spaces and short distances to be traveled by bicycles, as well as well-elaborated planning and high investment in cycling structures [18].

The implementation of BBS is an innovative type of business that, in addition to providing a positive economic impact, emerged through the development of Information Communications and Technologies (ICTs) and is of great importance when seeking to develop bicycle culture in certain regions that are still adapting to the mode [18].For the infrastructure analysis, in Fig. 4 (equal colors means same continent/area), segregated roads for cyclist use are considered, as the investment made by the evaluated regions, and the store structure that meets the demand for products and services generated by cyclists [1].

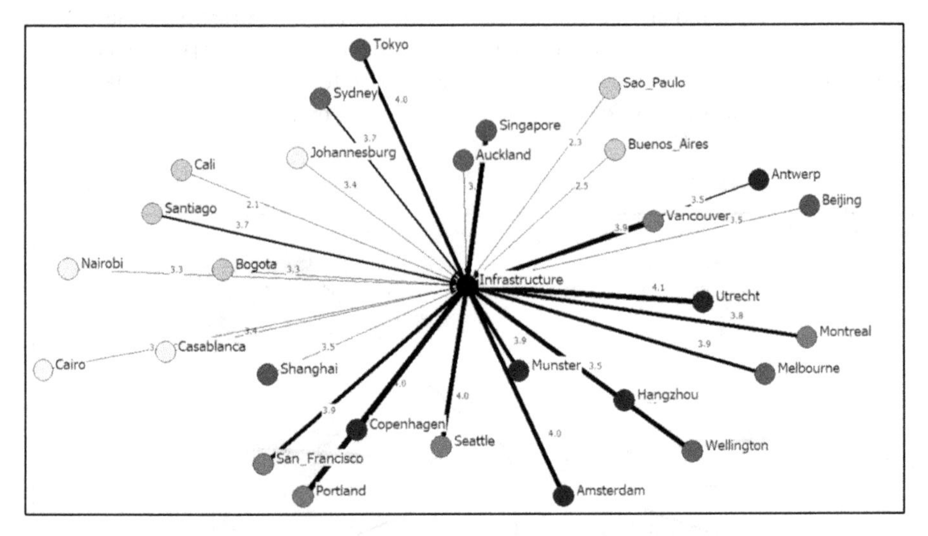

Fig. 4. Infrastructure

Note in Fig. 4 that São Paulo and Buenos Aires are Latin American cities that, like other parts of the world, have undergone industrialization and suffered strong impacts from the automotive industry. They grew rapidly and without planning in movements of conurbation and suburbanization. For some years, these cities prioritized mobility by motor vehicle and suffered the consequences of heavy traffic, long kilometers of congestion, high levels of noise and air pollution, and lack of space for people in the cities [2, 19].

It is noted that after many investments in favor of structures for cars, attempts are being made to make adjustments for the implementation of structures for cyclists. There are many common points between Buenos Aires and São Paulo. São Paulo's Sustainable Mobility Law in 2012 and Buenos Aires in 2009 provided opportunities for greater investments in structures for cyclists, both in bike lanes, intermodal connection structures, and bike parking. It is worth noting that this readjustment becomes much more challenging since the structures for vehicles have already occupied all the spaces, leaving cyclists to share space with pedestrians [19], and often presenting low quality.

Despite São Paulo's low index, the system has 260 stations and over 2,700 bicycles strategically spread throughout the city.

In terms of Events, highlighted in Fig. 5 (equal colors means same continent/area), there is a certain similarity among all the cities analyzed, considering that events are always important and also attract the attention of non-cyclists. Many cities end up hosting international events such as the World Cup, Whoop UCI Mountain Bike World Series, among others, which arouse interest in viewers and the first step towards cycling practice [20]. There is a slight decrease in the indices presented in the event criterion by the leading cities in the ranking such as Utrecht, Melbourne, and Copenhagen, which are cities that already have a cycling culture.

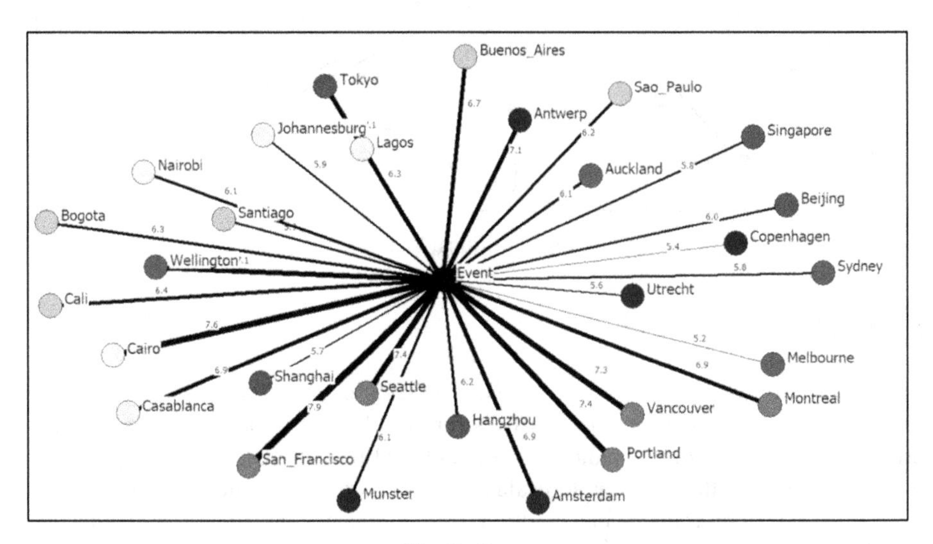

Fig. 5. Event

São Paulo also hosts a series of weekly events such as non-competitive group rides, as well as the "Night Riders," which is a free night ride also held in other Brazilian cities. These are important events that increase visibility and attract new enthusiasts to the mode [21].

Figure 6 presents critical mass network (equal colors means same continent/area).

Analyzing Fig. 6, critical mass movements, the cities of Buenos Aires and São Paulo stand out, cities in Latin America that have car culture ingrained in their mobility structures and that for many years prioritized motor vehicles in urban spaces. Critical mass proposes to carry out coercion so that these spaces become accessible again to other mobility alternatives with the strength of joint action. Cities in which bicycle users are excluded from public roads tend to have higher rates of critical mass. It is interesting to highlight that, although European cities already have a strong structure of inclusive active mobility, critical mass rates are not yet low.

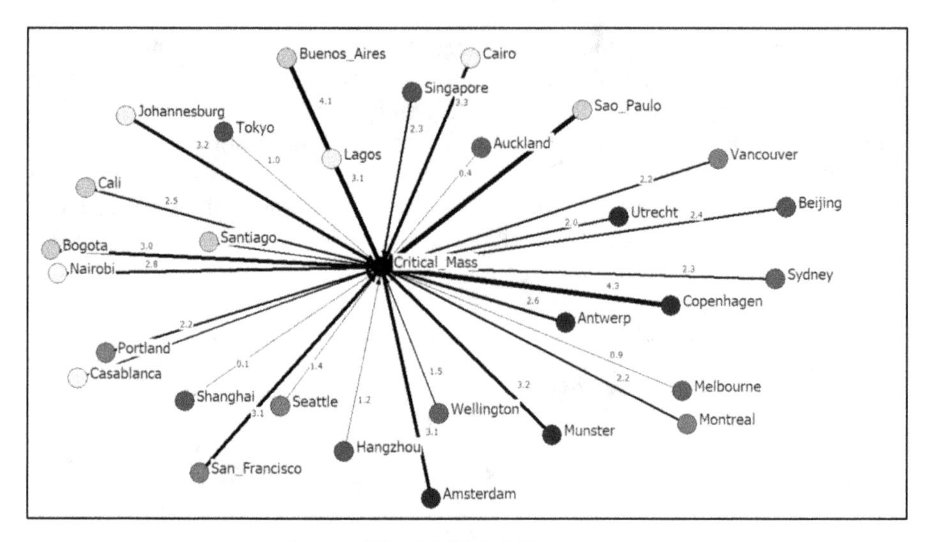

Fig. 6. Critical Mass

5 Final Remarks

The objective of this research was to investigate the differences between countries from different continents regarding the initiatives for the adoption of bicycles as sustainable urban transport based on the indicators of the Global Bicycle Index. After segregating them by continent, the aim was to evaluate if there are significant differences in the cyclability index presented by bicycle-friendly cities when compared across continents.

The objectives were achieved as it was noted that cities exhibit similar behaviors when belonging to the same continent. European cities stand out, with a history of planned cities, significant investments, culture, and education focused on bicycle use, recognizing its importance as active and sustainable mobility, with indices of 40% or more of people using bicycles as a mode of transport.

Cities in South America are characterized by belonging to developing countries that prioritized cars in their infrastructures for a long time but are now seeking to make active mobility more inclusive after legislation and environmental awareness. Cities in North America, despite investing well in improving bicycle infrastructure, still have a strong car culture, with a significant issue related to the social prestige associated with cars. It's worth noting that this prestige is observed in other continents as well, but per capita income is a limiting factor for cities belonging to continents with weaker economies.

Asian cities also maintain a certain behavioral pattern, investing in infrastructure and seeking to meet the demand generated by bicycle users among the evaluated criteria. Cities in Africa, due to their economic and social fragility, high population density, and low investments in mobility, whether in public transport or active mobility, have low scores in the evaluated criteria. Lagos stands out with low investment in infrastructure supporting BSS.

Cities in Oceania, although well-scored in GBCI analyses, have a different behavior in terms of BSS. Auckland, for instance, offers good infrastructure for cyclists but has a strong car culture, lacking in traffic safety, which reduces interest in active mobility.

Given this scenario, it can be suggested that there are no significant differences in the indices presented by cities when separated by continent because they exhibit similar behaviors when they share similar economic, cultural, and social characteristics. Studying and discussing urban mobility is crucial regardless of culture, belief, economic power, or region. The vulnerabilities of individuals in enduring high levels of pollution and stress, the loss of time and money imposed by kilometers of congestion, are relentless, regardless of continental origin or affiliation.

It's important to note that investments and public policies alone are not enough without environmental awareness and the development of thinking toward more inclusive and sustainable urban mobility. This is observed in cities in North America and Auckland, where there is investment in infrastructure, but people are still trapped in values that go against sustainable mobility.

There are still those who advocate for the use of electric vehicles, a mode that does not exempt society from congestion, pollution caused by tire friction with the ground, and financial loss caused by time wasted in traffic jams. The advancement of technology favors the development of autonomous cars; however, this alternative will further highlight the economic vulnerability of developing countries, strengthening exclusion and social distancing.

Achieving sustainable mobility is a multifaceted goal that requires co-operation between governments, the private sector, and civil society. Addressing the challenges of accessibility, environmental impact, efficiency, and safety holistically is crucial to creating transport systems that meet current and future needs without compromising the environment and social equity.

The contributions of the study are connected to the development of public policies to motivate the use of bicycles in urban areas. The perception of that bicycle use is a sustainable mode of displacement with benefits to the environment and personal health. The need for bicycle supply chains to adapt their systems to produce affordable and efficient equipment that motivates bicycle use and makes the experience pleasurable which means going beyond the competitive cycling and sportive performance for add value markets.

As future research, we suggest a deeper investigation regarding the reasons that leave to adopt or not adopt cycling as a way of displacement in cities. Investigate why industries are disconnected of this implicit demand and how is it possible to modify their models of market and production to increase bicycle use. Analyze the role of bike infrastructures in their adoption and how to mitigate the effects of car industry preference.

Acknowledgements. Coordenação de Aperfeiçoamento de Pessoal de Nível Superior-Brasil (CAPES)- Code 001 for funding granted to Author 1.

References

1. Sarker, R.I., Morshed, G., Sikder, S.K., Sharmeen, F.: Trends in active and sustainable mobility: experiences from emerging cycling territories of Dhaka and Innsbruck. In: Urban Ecology, pp. 163–183. Elsevier (2020). https://doi.org/10.1016/B978-0-12-820730-7.00010-0
2. Malichova, E., Tokarcikova, E.: Does personality influence interest in bike sharing In: 72nd International Scientific Conference on Economic and Social Development, p. 116. Varazdin (2021)
3. Foletta, N.: Europe's Vibrant New Low Car(bon) Communities - Houten: case study (2010). https://www.itdp.org/publication/europes-vibrant-new-low-carbon-communities/
4. Khajehpour, B., Miremadi, I.: Assessing just mobility transitions in the global south: the case of bicycle-sharing in Iran. Energy Res. Soc. Sci. **110**, 103435 (2024). https://doi.org/10.1016/j.erss.2024.103435
5. Prieto-Curiel, R., Ospina, J.P.: The ABC of mobility. Environ. Int. **185**, 108541 (2024). https://doi.org/10.1016/j.envint.2024.108541
6. Schneider, N., Helen, L., Costa, F., Cipullo, G.M.: Caso118 - Estrategia Itau Unibanco Mobilidade Urbana. https://archivo.cepal.org/pdfs/bigpushambiental/Caso118-EstrategiaItauUnibancoMobilidadeUrbana.pdf
7. Dekoster, J., Schollaert, U.: Cidades para bicicletas, cidades de futuro. EUR-OP, Luxemburgo (2000)
8. Luko, S.: Pontuações dos índices das cidades mais cicláveis do mundo (2023). https://www.hellogetsafe.com/en-de/adieu-luko-hello-getsafe
9. Mohieldin, M., Vandycke, N.: Mobilidade Sustentável para o Século XXI (2017). https://www.worldbank.org/en/news/feature/2017/07/10/sustainable-mobility-for-the-21st-century
10. Gallo, M., Marinelli, M.: Sustainable mobility: a review of possible actions and policies. Sustainability **12**, 7499 (2020). https://doi.org/10.3390/su12187499
11. Pojani, D., Stead, D.: Policy design for sustainable urban transport in the global south. Policy Des. Pract. **1**, 90–102 (2018). https://doi.org/10.1080/25741292.2018.1454291
12. Mueller, N., et al.: Health impact assessment of cycling network expansions in European cities. Prev. Med. **109**, 62–70 (2018). https://doi.org/10.1016/j.ypmed.2017.12.011
13. Carvalho, J.G.A., Rosa, D.M.M., Camargo, M.E.P., Aquino, G.P., Boas, E.C.V.: Vá de Bike: Plataforma IoT para Aluguel de Bicicletas. In: Anais do XLI Simpósio Brasileiro de Telecomunicações e Processamento de Sinais. Sociedade Brasileira de Telecomunicações (2023). https://doi.org/10.14209/sbrt.2023.1570916139
14. Zaw, T.N., Lim, S.: The military's role in disaster management and response during the 2015 Myanmar floods: a social network approach. Int. J. Disaster Risk Reduct. **25**, 1–21 (2017). https://doi.org/10.1016/j.ijdrr.2017.06.023
15. Ermagun, A., Erinne, J., Maharjan, S.: Equity of bike infrastructure access in the United States: a risky commute for socially vulnerable populations. Environ. Res.: Infrastruct. Sustain. **3**, 035001 (2023). https://doi.org/10.1088/2634-4505/ace5cf
16. Jahanbani, R.: Safer Cycling Networks in Auckland (2020). https://www.researchbank.ac.nz/bitstreams/c7caa57a-4aba-40ec-848e-a1edd8c848fd/download
17. Rajé, F., Tight, M., Pope, F.D.: Traffic pollution: a search for solutions for a city like Nairobi. Cities **82**, 100–107 (2018). https://doi.org/10.1016/j.cities.2018.05.008
18. Giuffrida, N., Pilla, F., Carroll, P.: The social sustainability of cycling: assessing equitin the accessibility of bike-sharing services. https://www.sciencedirect.com/science/article/abs/pii/S0966692322002137

19. Instituto Cidades Sustentáveis: Buenos Aires tornou-se referência em práticas de mobilidade sustentável. https://www.cidadessustentaveis.org.br/boaspraticas/detalhes/45
20. Aliança Bike. https://aliancabike.org.br/
21. Ativo: Night Riders 2023: passeio ciclístico noturno está de volta a São Paulo. https://www.ativo.com/bike/night-riders-2023-passeio-ciclistico-noturno-esta-de-volta-sao-paulo/

Analytical and Computational Models for In-Store Shopper Journeys

Henprasert Korrawee, Prakash Chakraborty, and Vittaldas Prabhu(✉)

Pennsylvania State University, University Park, PA 16802, USA
{kvh5781,prakashc,vittal.prabhu}@psu.edu

Abstract. Retailing is an important part of the supply chain, the link at which money is transferred from the consumer and flows to upstream trading partners. However, retailing has received very little attention from researchers in terms of developing quantitative models that are used in the rest of the supply chain. This paper proposes analytical and computational models based on queuing theory and discrete-event simulation, respectively, for in-store shopper journeys. Specifically, based on empirical data from retail stores, we propose a $M/G/\infty$ and $M_t/G/c$ queues in tandem to model the shopping process and the checkout process to characterize key aspects of shopper journeys. Such an analytical model is probably the very first of its kind proposed to mathematically characterize retailing and offers prospects for enabling better understanding and decision-making in this multi-trillion-dollar industry. A key advantage of such analytical models is that they can provide quick answers to questions such as average customer waiting times at checkout for different levels of staffing, which is an important issue in customer service vs staffing cost. A shortcoming of such analytical models is that they are based on assumptions therefore their output has limited accuracy, and these models cannot predict variance in the output. To overcome these shortcomings, this paper presents ShopperSim, which is a stand-alone discrete-event simulation software being developed using SimPy, a Python library, to simulate shopper journeys in a variety of store formats. To ensure practical relevance of these developments, ShopperSim has been programmed to statistically reproduce empirically observed shopping time, basket size, and store area covered – key attributes for shopper journeys. Tests indicate that ShopperSim can reproduce empirically observed statistics of key attributes for shopper journeys. Data generated from ShopperSim indicates that $M/G/\infty$ and $M/G/c$, a much simpler and more tractable tandem queuing model may be adequate for the tested case of a convenience store. In the future, analytical and computational models developed here can be leveraged in several ways for education, training, and operational decision-making in stores, including gamification of these decisions.

Keywords: Retailing · shopper journeys · analytical models · simulation models

Published by Springer Nature Switzerland AG 2024
M. Thürer et al. (Eds.): APMS 2024, IFIP AICT 729, pp. 172–186, 2024.
https://doi.org/10.1007/978-3-031-65894-5_13

1 Introduction

Retailing is an important part of supply chain, the link at which money is transferred from the consumer and flows to upstream trading partners. In the United States retailing is about $7 trillion in revenue annually. However, retailing continues to be a very competitive business with relative low profit margins typically 0.5% to 3.5% and return on assets typically below 10%. Considerable effort goes into understanding shopper behavior and product preferences [1]. While omnichannel retailing has gained considerable importance, in-store shopping remains a significant channel [2]. Much of academic research in retailing has been from a marketing perspective [3]. However, after the COVID pandemic, there is an increased impetus for strategizing all aspects of retailing and to leverage technology advances [4]. Retailing has received very little attention from researchers in terms of developing quantitative models that are used in rest of the supply chain. Past research has successfully used technology such as RFID on shopping carts to understand paths taken by shoppers in stores [5]. A noteworthy study by Sorenson et al. [6] is one of the few in open literature that collected extensive data to characterize shopper behavior in stores across several formats and several countries. The key contribution of this study is empirical measurements used to characterize shopper journey as statistical distributions of the following attributes:

1. Trip Duration (X_t) – lognormal distribution, in units of time
2. Store Coverage (X_c)– logit normal distribution, in units of area
3. Basket Size (X_b) – Poisson lognormal distribution, in number of items purchased.

One of the operational challenges in retailing is to reduce staffing costs while maintaining appropriate levels of customer service quality [7]. In-store shopper footfalls vary considerably during the day and across the days, weeks, and months. To maintain acceptable customer service, store managers need to adapt to this dynamically varying demand by adjusting staffing levels. Some of the changes that have been adopted in stores include express checkout and self-checkout counters. Much of the research studies investigating checkout process have relied on simulation modeling [8, 9]. Miwa and Takakuwa (2008) used simulation to model entire shopper journey through retail stores [10]. Terano et al. (2009) used an agent-based simulator for analyzing customer behavior in supermarkets [11]. Such simulation approaches have also been proposed for checkout staffing decisions [12] and real-time staffing-decisions [13].

In this paper we propose a $M/G/\infty$ and $M_t/G/c$ queues in tandem to model the shopping process and the checkout process in Sect. 2. Section 3 describes ShopperSim developed using SimPy, a Python library, to simulate shopper journeys. Computational techniques used to statistically reproduce empirically observed shopping time, basket size, and store area covered are also discussed. Section 3 also discusses computational logic used in ShopperSim to generate higher resolution dynamics of detailed shopper's movements among store aisles during a journey. Section 4 details the tests and results of ShopperSim generated results compared to empirically observed statistics. Section 5 concludes with a discussion and several possible future research directions.

2 Analytical Models for Shoppers Journey

The shopper's journey through a retail store is modelled using a two-station tandem queue as shown in Fig. 1. Shoppers enter the store according to some arrival process, and then shop according to a distribution as specified in [6], at the end of which they arrive at the checkout counters. We model the shoppers' entry to the store as Poisson arrivals, which is usually a good approach to model such exogenous independent arrivals. In stores with dedicated entrances, it would be relatively easy to measure arrival times using infrared sensors. Similarly point-of-sale (POS) devices at checkout can readily provide departure times from the checkout queue. More sophisticated but expensive movement tracking systems can also be used [14].

The general service times here is used to characterize time spent in the store by different types of shoppers moving across different aisles in the store to pick different items they need. In retail stores there are usually considerably large number of items on shelves compared to shoppers, and usually shoppers do not need to wait to pick items on shelves that they want to purchase. From a modeling perspective, this can be viewed as having "infinite" resources for shoppers because they do not have to incur any waiting time to pick items on shelves. Therefore, the shopping stage can be modeled as an infinite server system. The departure process from the shopping stage queue is the arrival process for the checkout stage queue. Characterizing this departure process is important because it influences the waiting times experienced by the customers at the checkout stage. We model the second queue with finite servers in which the number of servers here represents the number of checkout counters open at a particular time. The checkout time at this queuing stage is modeled to as general service times that varies randomly based on the basket size of the shopper and on factors such as physical size and weight of the items in the basket and reliability with which UPC codes on these items can be scanned. Additional source of variability in this service can arise from the payment mode used (e.g. NFC, credit card, cash, food stamps, cash, etc.).

We now provide a full description of the two queues mentioned above. Assume the arrival process of shoppers to the store follow a homogeneous Poisson process A_t with intensity or rate given by λ. Retail stores may experience varying arrival intensities over its operating time, and in the sequel, we provide staffing decision for high, medium and low arrival intensities. In Fig. 2 below we provide the difference between the two models for time-varying intensities differing in their period. Here intensity represents the number of shoppers that arrive per unit time. For fixed service time distribution, the difference is pronounced when the intensity exhibits periodic behavior with large periods. However, Poisson arrivals with constant intensity is a good model for short time periods and as we will see below provides a simpler approximation to our tandem queue. This leads to

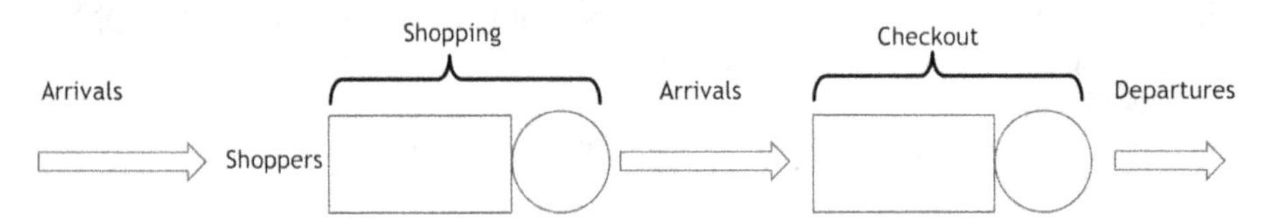

Fig. 1. A tandem queue model for shopper journeys.

simplified staffing decisions which is inherently a short time horizon decision problem based on the arrival intensity of customers.

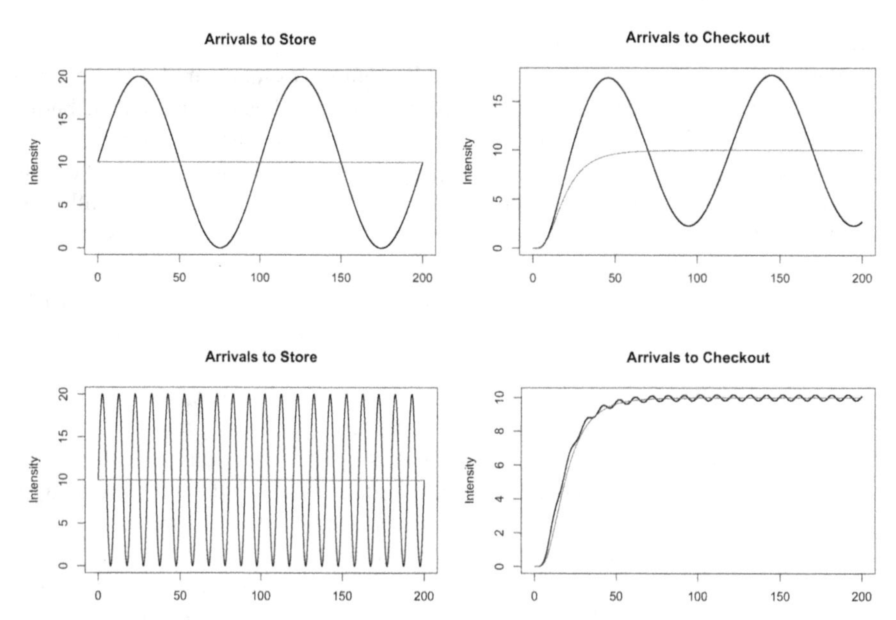

Fig. 2. Time-varying intensity (black) and constant averaged intensity (blue) with corresponding arrival processes to checkout.

We assume that the store has enough retail space to accommodate all incoming shoppers to the store arriving according to this rate. After arriving to the store, a shopper spends a random amount of time X_t to shop, while exploring a random proportion X_c of the store and placing a random number of items X_b to their basket.

As mentioned before, these three variables have been empirically measured and analyzed by Sorensen et al. [6] for various types of retail stores in different geographical locations. Their statistical analysis and data fitting reveals that for every shopper the random variables X_t, X_c, X_b are well approximated by independent and identical draws from pairwise associated Lognormal, Logit-normal, and Poisson Lognormal distribution, respectively, where the parameters of each distribution may vary depending on store type, size, and location. Association is measured through correlation coefficients, which is provided for each pair of random variables. It is important to note that the joint distribution of (X_t, X_b, X_c) is not specified. Instead, we have the respective marginal distributions and the pairwise correlations. Given this incomplete picture, one may create a joint distribution through what is known as a copula [15].

Let F denote the joint distribution function for (X_t, X_b, X_c) with respective marginal distributions being F_1, F_2 and F_3 respectively. Then Sklar's theorem [15] says that.

$F(x_1, x_2, x_3) = C(F_1(x_1), F_2(x_2), F_3(x_3)),$

where C is a distribution function over $[0,1]^3$ and referred to as a copula. Note that if the random variables admit marginal densities, which is true in our case, the joint density function is related to the marginal densities through the following relation

$$f(x_1, x_2, x_3) = c(F_1(x_1), F_2(x_2), F_3(x_3))f_1(x_1)f_2(x_2)f_3(x_3),$$

where c is the density of the copula C.

In the following when the joint distribution is required, as for example during simulations, we will use a Gaussian copula to create a fully described tri-variate distribution such that the prespecified marginals and correlations are maintained.

After spending X_t amount of time shopping, the shoppers move to the checkout counters, with c servers. The checkout times X_{co} are independent and identically distributed according to a positive valued random variable depending on the basket size of the shopper. For simplicity and analytical reasons, we assume that conditional on the basket size X_b, the checkout times are Exponential random variables with mean ηX_b where η is some non-negative parameter fixed for all shoppers. However, note that such a model for the checkout distribution is easily modified to match data. Further note that presence of finitely many servers creates queues before the checkout counters, and we assume that shoppers are checked out FCFS. In this model queue "jumping" by shoppers and other behaviors while waiting in a queue is ignored for tractability [16]. Also, the model proposed can be extended to self-checkout process, which may have a larger variance in service time because skilled staff are replaced by shoppers themselves.

With the above prescription, we find that indeed we have a tandem queue with two systems. The first queuing system is a $M/G/\infty$ queue whose departure process feeds into the second queue. It is well known that for the $M/G/\infty$ queue, the departure process is not homogeneous Poisson anymore. Instead, the departure process is given by a non-homogeneous Poisson Process with time-varying intensity δ_t, $t \geq 0$. In our case this intensity function is given by [17]

$$\delta_t = \lambda F_1(t),$$

where recall λ is the intensity of arrivals to the store and F_1 is the shopping time distribution. Since the departure of the first queue is the arrival process of the second queue, the second queue is now a $M_t/G/c$ queue, where c denotes the number of servers operating the checkout counters. Time varying queues like our second queue is notoriously difficult to analyze. In the following we adopt two approximations to the checkout queue.

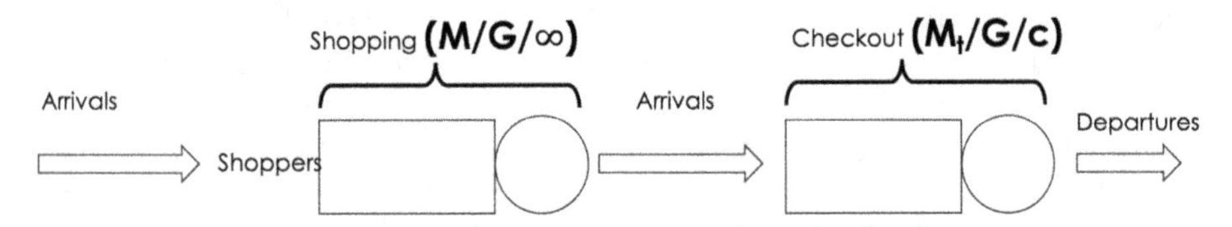

Fig. 3a. A tandem queue model with non-homogenous Poisson arrival at checkout.

Fig. 3b.A tandem queue model with homogenous Poisson arrival at checkout.

Note that as t gets large $F_1(t)$ approaches 1. This implies that $\delta_t \approx \lambda$ for large t. That is, the departures from the first queue tend towards a homogeneous Poisson Process in

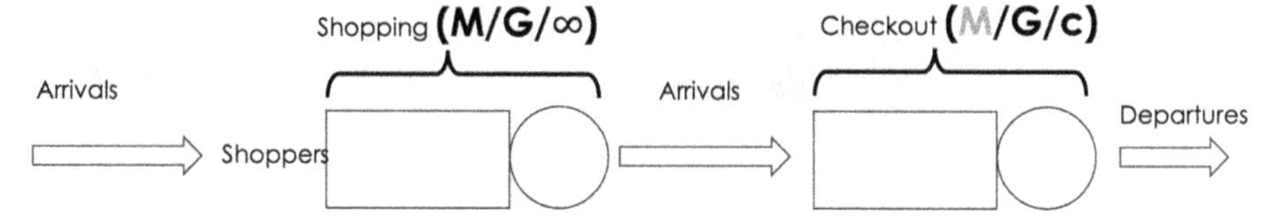

Fig. 3b. A tandem queue model with homogenous Poisson arrival at checkout.

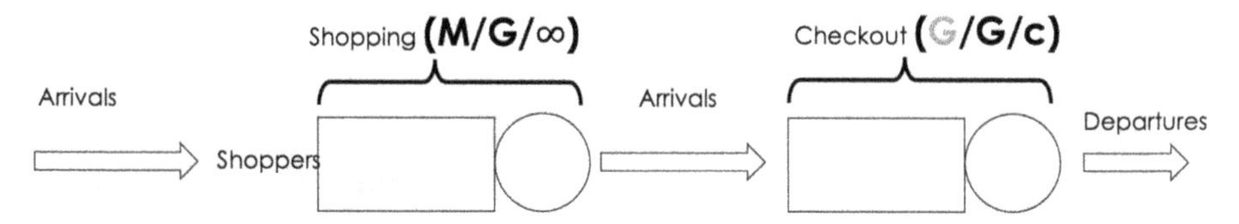

Fig. 3c. A tandem queue model with general arrival process at checkout.

contrast to a non-homogeneous one. So, for large t one may approximate our $M_t/G/c$ queue by a $M/G/c$ queue; we call this Case A.

Alternatively, we may approximate the $M_t/G/c$ queue by a $G/G/c$ queue – we call this Case B. Here we hope general inter-arrival times would account for the transient state of the system before it reaches steady-state behavior for large t.

For both these cases note that the interarrival times are assumed to be identically distributed, which is not true if the arrivals are truly non-homogeneous as in a $M_t/G/c$ queue. Let A denote a generic inter-arrival time to the second queue and S a generic service or checkout time. Furthermore, denote

$$\rho = \frac{E(S)}{cE(A)}$$

to be the system load or utilization rate. Then the Allen/Cunneen-formula [18] can be used for approximating the mean wait time:

$$E(W) \approx \frac{P_c E(S)}{c(1-\rho)} \frac{c_A^2 + c_S^2}{2},$$

where c_A^2 and c_S^2 denote respectively the squared coefficients of variation for inter-arrival and service times respectively, and P_c is the probability that all servers are blocked which is given by the Erlang C formula.

With regards to the above formula, Cases A and B differ in the values of $E(A)$ and $Var(A)$. In Case A we use $E(A) = \sqrt{Var(A)} = \frac{1}{\lambda}$ since the arrivals there are approximated by a Poisson process with constant rate λ. In Case B, however, we use ShopperSim (described in the next Section) to generate arrival times and consequently inter-arrival times to obtain sample estimates of $E(A)$ and $Var(A)$. The results are presented in Fig. 4 for three arrival intensity levels $\lambda_h = 20$ (High),$\lambda_m = 2$ (Medium) and $\lambda_l = 0.2$ (Low) for the two cases Case A: $M/G/c$ and Case B: $G/G/c$ mentioned before. From these results we can observe that when there are adequate checkout counters open, i.e., number of servers, expected waiting times for shoppers at the checkout queue will be quite similar regardless of the model. Importantly, there is no difference in expected wait times at low to medium intensities, which is what stores would experience most of the time (Fig. 4).

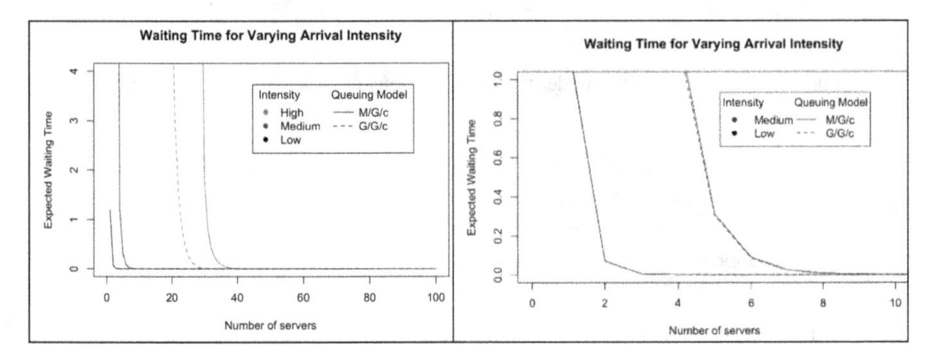

Fig. 4. Expected waiting time at the checkout as a function of number of servers.

Retail staffing has been analyzed in the literature with and without simulation [9, 22–24]. Here the arrivals to the checkouts are modeled exogenously from past historical data. However, shopper arrivals to checkout counters have not been well studied in the literature while also considering the shopping behavior in the store. To the best of our knowledge ours is the first work using a tandem queuing model for shopper behaviors in stores.

3 Computational Models for Shopper Journeys

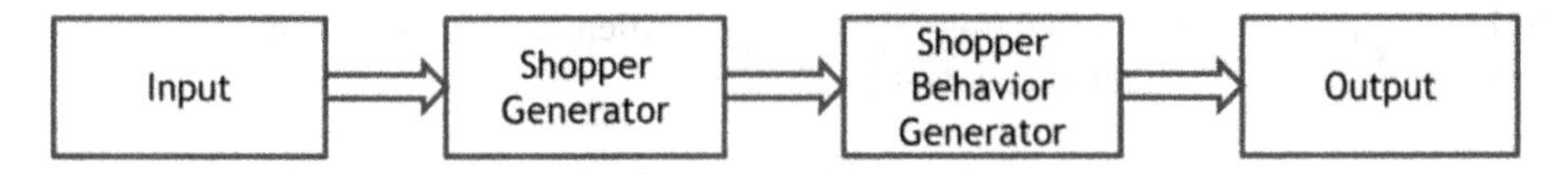

Fig. 5. ShopperSim simulation architecture.

As shown in Fig. 5, the ShopperSim simulation architecture is based on SimPy and makes use of SimPy resources and containers to facilitate access to statistical data such as queue lengths and waiting times [19]. Moreover, SimPy's event tracking capabilities are harnessed to monitor and record key simulation metrics, enabling comprehensive analysis and evaluation of various retail strategies and scenarios. Store layouts can be configured as a grid in which elements of the grid that shoppers can walk, and shelves are appropriately encoded. For a given store layout, ShopperSim is designed to accommodate two principal inputs: the staffing level at checkout counters and the customer arrival rate, denoted as λ. Figure 6 shows some screenshots of ShopperSim to illustrate the software user interface. Currently ShopperSim is in alpha-testing, and planned to be used in teaching during the next academic year.

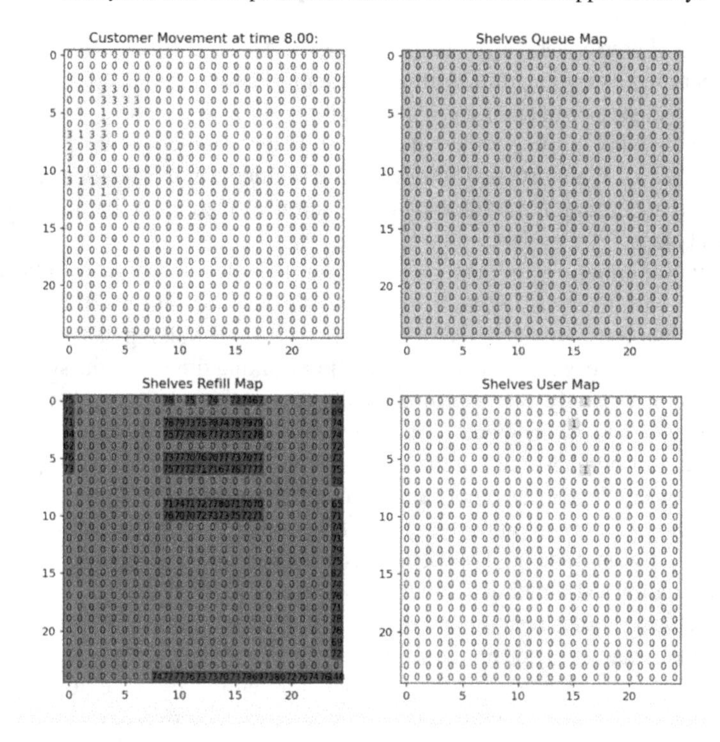

Fig. 6. ShopperSim screenshots to illustrate the software.

The simulation methodology encompasses several key stages integrated to statistically reproduce shopper journeys through stores. At the outset, the customer arrival process is simulated using a Poisson process, with interarrival times following an Exponential Distribution. Upon customer arrival, a function is invoked to dynamically assign key parameters X_t, X_c, and X_b. These parameters, drawn from respective probability distributions empirically obtained from retail stores by [6]. The key function of ShopperSim is to generate shopper journeys that are statistically consistent with these distributions. However, these distributions aggregate shopper journeys at a high level and cannot characterize higher resolution dynamics such as shopper's movements in a particular store layout. ShopperSim fills such gaps by including logic for detailed shopper movements among store aisles during a journey, checkout process, and entering/exiting the store. These additions can be expected to result in different statistics but the shapes of various distributions in ShopperSim should be consistent with those of [6].

For ShopperSim, correlated random variable generators are implemented using the NORTA (Normal to anything) method, facilitated through the utilization of the Julia API to access the BigSimR and distribution libraries. NORTA transforms marginal normal distributions into alternative marginal distributions. Specifically, in our investigation, we utilize NORTA to convert the marginal normal distribution to Lognormal for variable (X_t), Logit-Normal for variable (X_c), and Poisson Log-Normal for variable (X_b). These transformations are conducted while preserving the specified correlations between each marginal distribution, as provided by [6]. Of particular importance is the conversion of the proportion of area coverage, X_c, to the number of grids to be traversed by the sample customer, denoted as υ. This transformation establishes a concrete spatial framework for subsequent customer interactions by multiplying X_c with the area of the store. Shelf positions are encoded in the grid layout of the store environment being simulated. The sample of basket X_b, is apportioned across the grid segments in υ. Specifically, the basket size is divided according to multinomial distribution with equal probability among the grid segments in υ.

The temporal dimension of customer behavior is modeled through the allocation of shopping time within each grid area to determine the walking time and the buying time. The walking time is calculated based on walking speed and walking distance to cover υ segments. Buying time is calculated by subtracting walking time from the shopping time and is apportioned equally and deterministically to each item in the basket. Additional shopper behavior can be included such as window shopping wherein shoppers only browse but do not purchase, or purchase a fraction of the items browsed, or purchase every item browsed. When customers engage in purchases, the allocation of time within each grid area is contingent upon the proportion of items to be purchased relative to the total items, accommodating both scenarios where all and some items are purchased. This approach allows simulating a variety of shopper journey types.

For simulating higher resolution dynamics of shopper movement among the store aisles, ShopperSim uses logic based on a nearest neighbor heuristic in which Manhattan distance is used as the distance metric. Distinct store entrances and exits are specified which in turn serve as starting points and ending points, respectively, of simulated shopper journeys on the grid layout of the given store. At each location, the nearest neighbor from the remaining locations is selected as the next location to move and the path for the movement is also generated. Once all the locations are visited the shopper proceeds to the cashier in the checkout area and then exit the store. This approach allows simulating a variety of store layouts and retail settings.

To illustrate this concept, consider a store represented by a 5×5 grid from Fig. 7. Suppose the movement to be simulated to cover area υ is by traversing path connecting grid coordinates $[(0, 4), (2, 4), (2, 3), (3, 0), (4, 4)]$, where the shopper could be purchasing items for their basket from the last four locations on this list. From the entrance, starting point, $(0, 4)$, path generated using nearest neighbor search to reach $(2, 4)$ is: $(0, 4)$, $(1, 4)$, $(2, 4)$. Time elapsed is tracked based on distance and walking speed. Shopper movement to the next grid coordinate from $(2, 4)$ to $(2, 3)$ is simulated, and $(2, 4)$ is removed from the list. After visiting the four locations the shopper proceeds to the cashier in the checkout area indicated by a green triangle in and then exit the store through the designated exit, marked with a black star in Fig. 7d.

It should also be noted that shopper journeys in ShopperSim may include waypoints where the shopper is neither browsing nor purchasing, covering intermediary areas in the store. In the above illustration (1,4) is such a waypoint. ShopperSim tracks the area covered in each simulated journey. It should be emphasized that the resulting area covered may differ from the X_c generated for a particular customer and is discussed in Sect. 4. Another noteworthy point is that while generating X_c, X_t, and X_b, it is possible that in some instances $X_b = 0$ while X_c and X_t are some positive numbers. This can

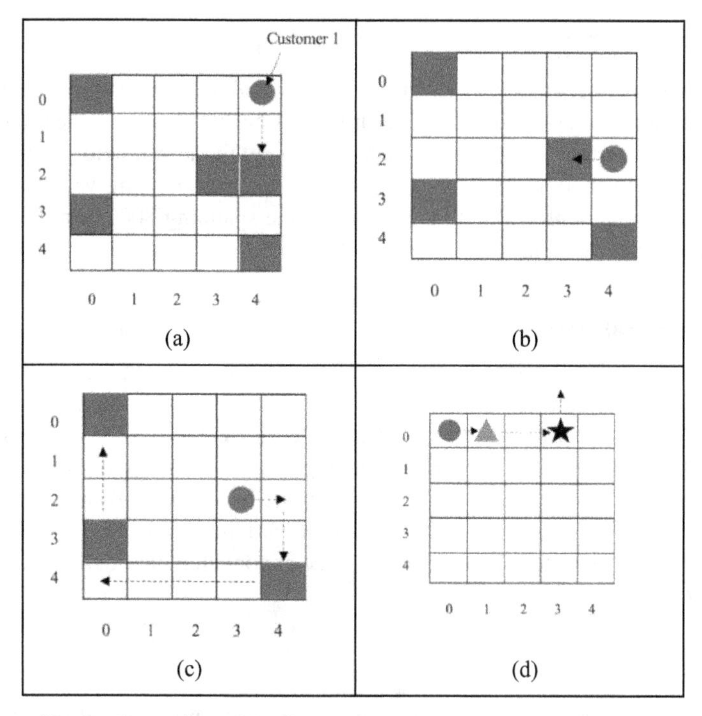

Fig. 7. Illustration of detailed movement during a shopper journey.

be interpreted as ShopperSim of having generated a "window shopper" who does not purchase anything but spends some time in the store.

4 Experimental Result

The experimental setup involved conducting simulations on a MacBook Air featuring an M1 chip, with 8 GB of memory, 256 GB SSD. The simulations were executed using SimPy version 1.12.1rc1 and Python version 3.9.18. The computational time for each simulation run varied depending on the complexity of the model and the desired level of precision. Furthermore, the simulation was conducted based on a grid area of 25 × 25 grid using the parameters of the Convenience Store/Drug Store from [6]. To test the accuracy and reliability of the simulation results, rigorous validation procedures have been conducted. The validation process involved comparing the simulated output generated by the NORTA generator with empirical data or theoretical expectations where applicable. This comparison allowed for the assessment of the simulation model's ability to replicate the empirically obtained distributions under different scenarios. Details of the validation methodology and the results obtained are discussed next.

4.1 Result Validation

First, we assess the accuracy of four key random variables generated by ShopperSim: interarrival time, area covered, shopping time, and basket size, variables that characterize shoppers' journey. The following subsections detail the validation methodologies employed for each random variable and present the corresponding validation results. A single simulation employed with a simulated time spanning 8000 min, taking actual time of approximately 50 min.

4.1.1 Interarrival Time

In ShopperSim, customer arrivals follow a Poisson process with a rate parameter 2, resulting in an exponential distribution of interarrival times with rate λ. In these tests, it is set to 2 to reflect a moderate arrival rate. To assess the fidelity of the simulated interarrival time distribution, a goodness-of-fit analysis is conducted by comparing it to the theoretical exponential distribution with a rate of 2. The analysis reveals a close match, with a sample mean of 0.499 and sample variance of 0.251, indicating strong alignment with the expected characteristics of the exponential distribution with a rate of 2.

4.1.2 Area Covered

The distribution for area covered is generated using a logit-normal distribution. Specifically, the NORTA generator is used to produce logit-normal distribution, with a mean of 0.36 and a variance of 0.005. Assessing the goodness of fit of a logit-normal distribution is a two-step process: first, is to transform the output of area covered using a logit transform; and second, examining the normality of the transformed using a goodness-of-fit test. Histogram of ShopperSim generated area covered is present in Fig. 8. The area covered has a mean of 0.71 and a variance of 0.006, which is higher than the mean of 0.36 but consistent with the variance of 0.005. This inflation is caused by factors in ShopperSim such as additional movement, discretization of the store layout using grids and using Manhattan distance. This suggests that Shopper Sim generated data maintains the characteristics of a logit-normal distribution and future work should investigate the discrepancy between the mean rigorously and develop ways to reduce this discrepancy.

4.1.3 Shopping Time

Shopping time is found to have a lognormal distribution, and its goodness-of-fit can be readily assessed. Using NORTA a lognormal distribution is generated with a mean of 8.89 and a variance of 23.10. The resulting output has a mean of 9.01 and a variance of 23.63, indicating a close match between the simulated and specified distributions. Here the difference between the two means is lower compared to the means of the area covered because ShopperSim ensures that the sum of walking time and buying time is equal to the shopping time. The results of this analysis are presented in Fig. 9, which is acceptable.

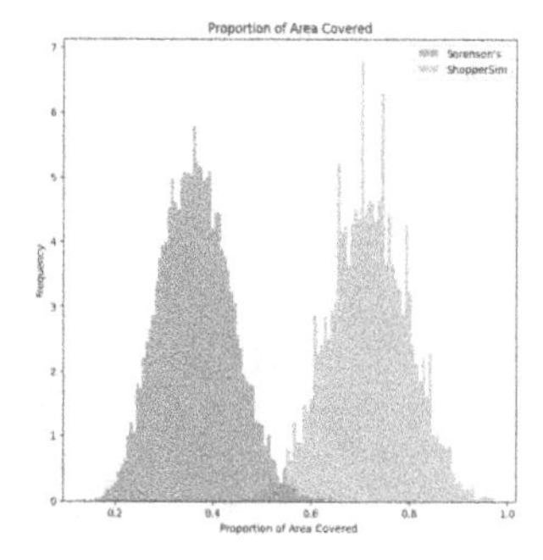

Fig. 8. Histogram of the percentage of store area covered during a shopper journey.

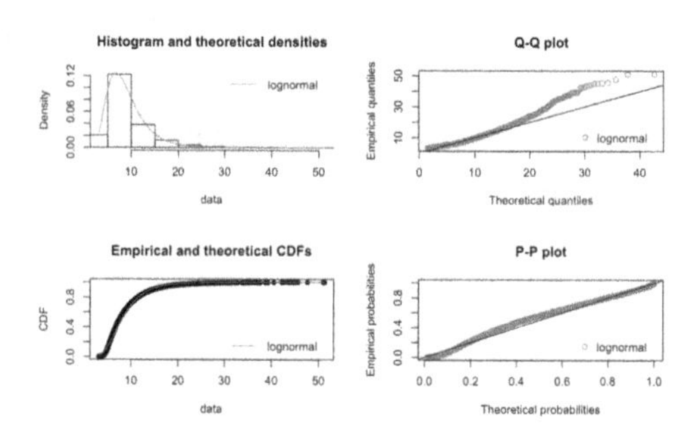

Fig. 9. Goodness of fit test for shopping time distribution.

4.1.4 Basket Size

The basket size distribution follows Poisson-lognormal distribution, which involves two generating steps: first, generating the rate from a lognormal distribution, and then substituting the rate into a Poisson distribution. As before, the NORTA generator is used in ShopperSim. Using NORTA a Poisson lognormal distribution is generated with a mean of 1.44 and a variance of 3.04. The resulting output has a mean of 2.15 and a variance of 2.09. Goodness of fit tests for Poisson lognormal is not straightforward and left for future work.

4.2 Departures from Shopping Stage

As discussed in Sect. 2, shopper journeys can be modeled as a tandem queue in which the first stage is shopping, and the second stage is the checkout process. One important modeling issue is to ascertain the departure process from the shopping stage which in turn becomes the arrival process to the checkout. The ability of ShopperSim to simulate the departure process from the shopping stage can be used to fill an important gap in developing more effective models for retail stores. A convenience store is used as the test case, and 30 replications, each for 500 min of simulated time, are run using ShopperSim. Each replication takes about 5 min of computer time.

Figure 10 illustrates the departure process from the 30 replications, alongside the expected number of departures derived from the $M/G/\infty$ analytical model, depicted by

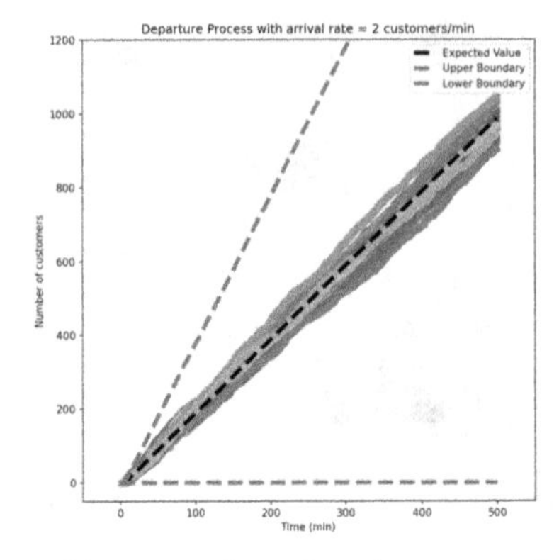

Fig. 10. Number of departures from shopping stage arriving at checkout.

a dashed line. Given the good agreement between the analytical model and computational results from ShopperSim's, these results can be leveraged in several ways for education, training, and operational decision-making in stores. Detailed analysis of data generated from ShopperSim has revealed that after a brief transient period that can be regarded simulation warmup time, the interarrival time distribution is a good fit for exponential distribution. This implies that Case A of $M/G/c$ would be a good model for the checkout queuing stage.

5 Discussion and Future Work

Even though retailing is a large industry in terms of its economic value and its role in modern supply chains, there has been little work to mathematically characterize key aspects of stores. This paper presented analytical model based on queuing theory in which shopper journeys are models as a tandem queuing system with $M/G/\infty$ and $M_t/G/c$ queues to model shopping and checkout stages, respectively. A simplified model of this $M/G/\infty$ and $M/G/c$ queues in tandem has been found to be suitable. Our model is suitable for short time periods when arrivals to the store can be assumed to have approximately constant intensity. The study of the case where the shopping stage is modeled by $M_t/G/\infty$ or $G_t/G/\infty$ queue is left for future work. The paper also presented a discrete-event simulation software called ShopperSim being developed using SimPy. ShopperSim can be used to simulate a variety of store formats to get a better understanding of shopper journeys. Empirically observed distributions for shopping time, basket size, and store area covered during shopper journeys have been programmed into ShopperSim. Statistical tests indicate that ShopperSim produces output that is consistent with empirical distributions.

Tests that have been conducted with ShopperSim are for Convenience Store/Drug Store formats, which is a limitation of the current work. Additionally appropriate tests need to be developed to efficiently test Poisson-lognormal distributions, which has not been researched much. Ongoing work is focused on extensive testing to include a variety of store formats from multiple countries around the world. One future direction would be to enhance the nearest neighbor logic currently used to other techniques that may be a more natural fit for shopper's movement such as influence maps [20] and potential fields [21]. In the future, analytical and computational models developed here should also be tested for arrival rate that vary over time, making the queuing system non-stationary.

Gamification using ShopperSim for training students and store managers to better understand the impact of store layout, tradeoffs between staffing cost and customer service level in terms of waiting time experienced. One of the major motivations of developing ShopperSim is to use it to gamify operational decision-making in a retail services engineering course taught in our university as an elective for undergraduate and master's students. The software is in its "alpha" stage of development and focus until now has been to rigorously test its ability to faithfully reproduce specified statistics. Hands-on learning projects would include students collecting arrival, shopping, and checkout processes data from real retail stores, and specifying these in ShopperSim. Students would then interactively change staffing levels to develop more intuition about queuing dynamics in the context of shopper journeys. Teams of students would act as "consultants" to retailers in helping to improve the operations of the store where the data is collected. Store managers could continue to "play" with ShopperSim after these student projects to continuously improve their operations. In the future every such project will effectively become a detailed case study providing us constant feedback for improving features and user interface. Such engaged learning experiences would complement other courses that cover probability, statistics, stochastic process, and simulation and synthesize various facets of the curriculum, and help students to understand the strengths and shortcomings of analytical and computational models in the context of real world.

References

1. Korgaonkar, P.A., Karson, E.J.: The influence of perceived product risk on consumers' e-tailer shopping preference. J. Bus. Psychol. **22**, 55–64 (2007)
2. Mishra, R., Singh, R.K., Koles, B.: Consumer decision-making in Omnichannel retailing: literature review and future research agenda. Int. J. Consum. Stud. **45**(2), 147–174 (2021)
3. Lilien, G.L., Rangaswamy, A.: Marketing engineering: computer-assisted marketing analysis and planning. DecisionPro (2004)
4. Grewal, D., Gauri, D.K., Roggeveen, A.L., Sethuraman, R.: Strategizing retailing in the new technology era. J. Retail. **97**(1), 6–12 (2021)
5. Larson, J.S., Bradlow, E.T., Fader, P.S.: An exploratory look at supermarket shopping paths. Int. J. Res. Mark. **22**(4), 395–414 (2005)
6. Sorensen, H., et al.: Fundamental patterns of in-store shopper behavior. J. Retail. Consum. Serv. **37**, 182–194 (2017)
7. Dabholkar, P.A., Thorpe, D.I., Rentz, J.O.: A measure of service quality for retail stores: scale development and validation. J. Acad. Mark. Sci. **24**, 3–16 (1996)
8. Kwak, J.K.: Analysis on the effect of express checkouts in retail stores. J. Appl. Bus. Res. (JABR) **33**(4), 767–774 (2017)
9. Antczak, T., Weron, R., Zabawa, J.: Data-driven simulation modeling of the checkout process in supermarkets: insights for decision support in retail operations. IEEE Access **8**, 228841–228852 (2020)
10. Miwa, K., Takakuwa, S.: Simulation modeling and analysis for in-store merchandizing of retail stores with enhanced information technology. In: 2008 Winter Simulation Conference, pp. 1702–1710. IEEE (2008)

11. Terano, T., Kishimoto, A., Takahashi, T., Yamada, T., Takahashi, M.: Agent-based in-store simulator for analyzing customer behaviors in a super-market. In: Velásquez, J.D., Ríos, S.A., Howlett, R.J., Jain, L.C. (eds.) Knowledge-Based and Intelligent Information and Engineering Systems. KES 2009. Lecture Notes in Computer Science(), vol. 5712, pp. 244–251. Springer, Berlin, Heidelberg (2009). https://doi.org/10.1007/978-3-642-04592-9_31
12. Williams, E. J., Karaki, M., Lammers, C., Verbraeck, A., Krug, W.: Use of simulation to determine cashier staffing policy at a retail checkout. In: Proceedings 14th European simulation symposium, pp. 172–176 (2002)
13. Mou, S., Robb, D.J.: Real-time labour allocation in grocery stores: a simulation-based approach. Decis. Support. Syst. **124**, 113095 (2019)
14. Cho, S.I., Kang, S.J.: Real-time people counting system for customer movement analysis. IEEE Access **6**, 55264–55272 (2018)
15. Nelsen, R.B.: An Introduction to Copulas. Springer (2006)
16. Dahm, M., Wentzel, D., Herzog, W., Wiecek, A.: Breathing down your neck!: the impact of queues on customers using a retail service. J. Retail. **94**(2), 217–230 (2018)
17. Eick, S.G., Massey, W.A., Whitt, W.: The physics of the Mt/G/∞ queue. Oper. Res. **41**(4), 731–742 (1993)
18. Allen, A.O.: Queueing models of computer systems. Computer **13**(04), 13–24 (1980)
19. Matloff, N.: Introduction to discrete-event simulation and the simpy language. Davis, CA. Dept. Comput. Sci. Univ. Calif. Davis. **2**, 1–33 (2008)
20. Mark, D.: Modular tactical influence maps. In: Game AI Pro 360: Guide to Tactics and Strategy, pp. 103–124. CRC Press (2019)
21. Ontanón, S., Synnaeve, G., Uriarte, A., Richoux, F., Churchill, D., Preuss, M.: A survey of real-time strategy game AI research and competition in StarCraft. IEEE Trans. Comput. Intell. AI Games **5**(4), 293–311 (2013)
22. Fisher, M., Gallino, S., Netessine, S.: Setting retail staffing levels: a methodology validated with implementation. Manuf. Serv. Oper. Manag. **23**(6), 1562–1579 (2021)
23. Chuang, H.H.C., Oliva, R., Perdikaki, O.: Traffic-based labor planning in retail stores. Prod. Oper. Manag. **25**(1), 96–113 (2016)
24. Williams, E.J., Karaki, M., Lammers, C., Verbraeck, A., Krug, W.: Use of simulation to determine cashier staffing policy at a retail checkout. In Proceedings 14th European Simulation Symposium, pp. 172–176 (2002)

Human-centred Manufacturing and Logistics Systems Design and Management for the Operator 5.0

A Meta-heuristic Approach for Industry 5.0 Assembly Line Balancing and Scheduling with Human-Robot Collaboration

Jingyue Zhang(✉), Jinshu Zhou, and Shigeru Fujimura

Graduate School of Information, Production and Systems, Waseda University, Fukuoka, Japan
jingyue625@fuji.waseda.jp, xolles_ryo@ruri.waseda.jp, fujimura@waseda.jp

Abstract. Throughout the development of Industry 5.0 towards a paradigm prioritizing human-centricity and sustainability, the potential of assembly lines with human-robot collaboration (HRC) is substantial. In HRC environments, human and robot operators share the workplace and can perform tasks simultaneously and collaboratively, amplifying work efficiency and operators' welfare. This research investigates solutions to the assembly line balancing problem (ALBP) with HRC, where multiple human and robot operators work together. Using an adaptive simulated annealing (SA) framework for addressing ALBP with HRC, two innovative mechanisms are introduced-a new fitness value calculation method for roulette wheel selection and a pioneering heuristic approach. These mechanisms are devised to establish an innovative meta-heuristic approach based on SA for enhancing task allocation and resource management, improving productivity and operators' well-being through strategic workload balancing between human and robot operators, and minimizing cycle times and the total number of operators required. The computational results using actual production data show that these mechanisms significantly enhance the solution quality, particularly in the large-size case study involving collaboration between multiple humans and robots in each workstation.

Keywords: Assembly line balancing · Human-robot collaboration · Meta-heuristic

1 Introduction

The assembly line balancing problem (ALBP) is vital in modern manufacturing, focusing on maximizing the output of a production process at a given resource or minimizing the cost of a required resource at a fixed output level. In the context of Industry 4.0, automotive machines and robots with high precision, repeatability, and strength play a crucial role. Robotic assembly lines have sig-

M. Thürer et al. (Eds.): APMS 2024, IFIP AICT 729, pp. 189–204, 2024.
https://doi.org/10.1007/978-3-031-65894-5_14

nificantly increased productivity and exposed skill differences between robots and the human workforce. However, with the evolution of the industrial environment and technological advancement, ALBP faces new challenges. Many manufacturers found that while robots are advantageous in some ways, many tasks still rely on human workers at the level of intelligence and skill. Manual work is an indispensable and critical component of modern industrial production because of its flexibility and adaptability.

Towards the vision of Industry 5.0, human-centric development will further drive the use of assembly lines based on human-robot collaboration (HRC). Traditional solution approaches face significant challenges when dealing with complex assembly tasks and diverse collaboration modes in this context because they are mainly designed for purely manual or fully automated production environments and are insufficient for handling HRC scenarios. Additionally, although mathematical programming and exact algorithms can potentially obtain optimal solutions in ALBP, the computation time can be overly long, especially for complex and large-scale assembly operations. The extended computation time poses significant challenges to meet real-time decision-making requirements. This is particularly problematic in industries such as automotive manufacturing, as the increasing complexity of product types and production processes have exceeded the capabilities of traditional ALBP solutions. Consequently, more intelligent and flexible methods are required to manage task allocation and meet the production demand. Additionally, with the aging global population, more innovative solutions are necessary to improve productivity while ensuring the health and safety of workers [1]. The research and application of ALBP-HRC can enhance the flexibility and efficiency of the assembly process. Furthermore, close cooperation between humans and robots contributes to the well-being of human operators, avoiding them from injuries and occupational diseases and giving them a higher sense of job satisfaction and empowerment. However, there is insufficient exploration of ALBP-HRC. Most existing studies focus on workstation configurations considering a single human or robot operator, with a relatively simple problem setup and fewer solutions adapted to such problems. A more careful consideration of problem settings and further investigations of solution methods are essential for enhancing the efficiency of solving ALBP-HRC and ensuring alignment with real-world production scenarios. This paper aims to fill the research gap with a more comprehensive consideration. Inspired by the research of Nourmohammadi et al. [2], two main contributions are presented in this study. A new method is introduced to determine the fitness value of each task when using the roulette wheel selection to arrange task sequence, and a heuristic mechanism is proposed and integrated into the improved meta-heuristic algorithm based on simulated annealing (SA) to solve ALBP-HRC involving multiple human and robot operators. The proposed mechanisms were tested with actual production data through computational experiments, discovering more effective solutions to minimize cycle time (ct) and the number of operators.

The structure of the paper is as follows: Sect. 2 introduces the literature review and research significance of ALBP-HRC. Then, Sect. 3 elaborates on the problem definition and prerequisites, and Sect. 4 introduces the proposed methods. Section 5 presents the experimental settings and computational results of actual cases in the automotive manufacturing industry. Finally, the contribution of this study and the future research prospects are summarized.

2 Literature Review

ALBP is a classic problem extensively studied in literature since its first mathematical formula was proposed in 1955. It seeks to effectively allocate tasks to workstations on assembly lines constrained by operations [3]. The fundamental simple assembly line balancing problem (SALBP) allocates the workload of the entire assembly process to multiple workstations, each of which executes a series of priority-constrained tasks during the serial production process [4]. It attempts to maximize efficiency by minimizing the number of required workstations, limiting the total amount of work performed by each workstation, and ensuring that specific tasks are executed before others under priority constraints between tasks. If a task precedes another, it must be operated before the subsequent task.

In the progression of Industry 4.0, the deployment of fully automated robotic systems on production lines has been explored. For example, the most suitable method for assigning assembly tasks and robot types to each workstation was studied, with a proposed optimization method for the robotic welding ALBP considering switch time [5]. Chutima [6] summarized the development of robot assembly lines, in which robots and automation equipment replace the roles of human operators. However, the paper also pointed out that HRC assembly lines would become a new direction, and further research is necessary to cover practical industrial needs [6]. The importance of robotics technology in production is increasing, but it still has limitations. Human operators' intelligence and creativity are beyond the reach of robots. Tasks involving complex decision-making and delicate manual skills are difficult for current robotics technology to replicate. In addition, the flexibility of human operators to handle unexpected situations and urgent tasks makes them crucial in rapidly changing production environments. They can respond and adjust immediately based on the situation, a capability that highly automated systems often lack. At the same time, the participation of human operators also brings a humanized touch to the production process, improving the diversity and adaptability of the workplace.

An assembly line incorporating HRC can integrate the advantages of human and robot operators, which leads to a natural shift from Industry 4.0 to Industry 5.0. Such an assembly line conforms to the core vision of Industry 5.0 to achieve a harmonious symbiosis between humans and robots on the production line [1]. Industry 5.0 emphasizes the re-integration of human creativity and flexibility on the foundation of high automation and intelligence to achieve more comprehensive and sustainable industrial progress. This emerging framework is designed to adapt to the latest industrial revolution developments and the application of cutting-edge technologies to enable more efficient, adaptable, and humane production processes. Different forms of HRC in assembly lines have been studied, such as collaboration technologies and organizational and economic considerations, which emphasize the importance of flexibility and variability in the assembly process and point out the advantages of robots as intelligent assistants to improve the efficiency of complex assembly processes [7]. Applying HRC in assembly lines has brought new challenges and opportunities to production processes. The emergence of the HRC environment aims to address the limitations of traditional manual or robotic assembly lines while improving their productivity, flexibility, and reconfigurability. In an HRC assembly line, different tasks can be shared based on the abilities of humans and robots to achieve more efficient execution, which traditional single manual or robot lines cannot attain. Advancements in collaborative robotics, featuring embedded interaction, sensors, and security measures, enable robots to work closely with humans in the shared workspace. Demiralay and Kara [8] integrated the collaborative robot (Cobot) into ALBP, focusing on the gripper allocation for the Cobot. A mixed integer linear programming (MILP) was proposed and solved to present how a Cobot with an appropriate gripper can enhance assembly line optimization and improve production efficiency. Trends and future research directions for ALBP-HRC, particularly new challenges in resource selection, task allocation, line balancing, and scheduling decisions, were analyzed. These challenges include allocating tasks efficiently, managing HRC, and optimizing production processes, providing new possibilities for improving assembly lines' flexibility, efficiency, and human operators' working conditions [9].

When incorporating HRC into assembly lines, the complexity of solving the ALBP increases compared to traditional approaches. Traditional exact algorithms may require excessive computation time and often struggle to find optimal solutions due to the difficulty of coordinating tasks between humans and robots and balancing work efficiency. These methods are inadequate in real-world production because of the increased complexity of various constraints and advanced methods are necessary to address this issue [10]. New solutions for ALBP have focused on heuristic methods. Comparisons of 20 constructive heuristic and meta-heuristic algorithms have shown that these methods are often more effective in finding solutions than traditional precise methods [11]. At present, some studies apply meta-heuristic algorithms to ALBP-HRC. The integration of Cobots in manual assembly lines was studied, and a hybrid solution based on the genetic algorithm (GA) was proposed to address the high complexity of the problem. The productivity benefits of the proposed method were verified through computational experiments using generated instances [12]. GA with Methods-Time Measurement (MTM) was also used for multi-objective optimization, considering human and robot capabilities based on predefined processes and constraints to determine subprocess times [13]. A MILP model was introduced for solving small-scale problems, with an adaptive simulated annealing (ASA) algorithm for large-scale instances, including customized solution representation, feasible solution generation, and adaptive neighborhood search. The proposed ASA utilized the historical performance of the algorithm to dynamically update the selection probability of neighborhood search to adapt to the characteristics of ALBP-HRC [2]. Based on practical case studies in the automotive industry, ALBP-HRC with heterogeneous operations was also solved with an algorithm based on GA using custom parameters and features. The calculation results showed that the developed GA can provide decision-makers with efficient human-robot heterogeneous solutions [14].

From the literature review and the authors' best understanding, existing research on optimizing assembly line configuration in the HRC environment mainly studies simplified problem settings and rarely considers real cases. Moreover, most research has not addressed the scheduling aspect of task sequence in assembly lines. However, with the increase in production demand and the complexity of work environments, more comprehensive solutions are needed to effectively arrange assembly line operations that involve collaboration between multiple humans and robots. The novelty of this study mainly reflects the solution methods for ALBP-HRC with various numbers of human and robot operators assigned in each station. Based on the work by Nourmohammadi et al. [2], an innovative approach is introduced to calculate the fitness value for different tasks when using roulette wheel selection. Also, a heuristic mechanism is proposed to enhance the efficiency of the improved SA. The effectiveness of the enhanced algorithms using actual data from the automotive industry is validated through computational experiments. These improvements provide a new perspective and approach to solving ALBP-HRC.

(a) Collaboration of One Human and One Robot.

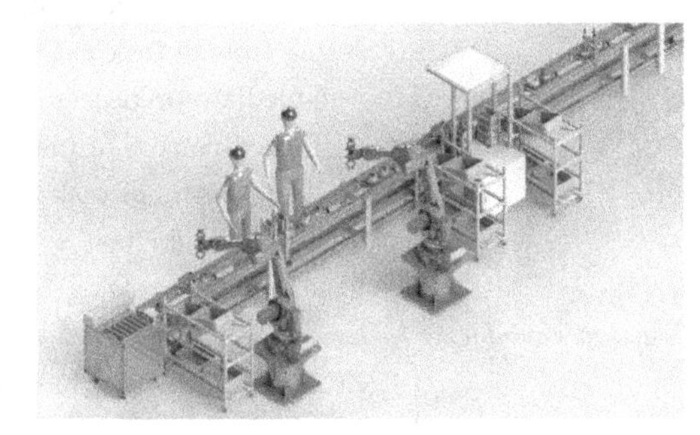

(b) Collaboration of Multiple Human and Robot Operators.

Fig. 1. Human and Robot Collaboration on Assembly Lines

3 Problem Description

With the market demand diversification, assembly line flexibility becomes the key to improving competitiveness. HRC combines the creativity and adaptiveness of humans and the efficiency of robots, which offers a revolutionary way to achieve this goal. This study focuses on assembly lines in automotive manufacturing environments, with four real cases aimed at assembling mass balance systems (MBS). In these cases, humans and robots work independently and collaboratively on specific tasks. Each workstation is configured with a different number of human and robot operators, considering their respective skills and characteristics. Figure 1a and Fig. 1b show the HRC workspace in an assembly line with different numbers of human and robot operators assigned.

Table 1. Notations for the ALBP-HRC (derived from Nourmohammadi et al. [2])

Notation	Definition
Indices:	
i, k	Set of tasks ($i, k = 1, \ldots, NT$)
j	Set of stations ($j = 1, \ldots, NS$)
h	Set of humans ($h = 1, \ldots, NH$)
r	Set of robots ($r = 1, \ldots, NR$)
Parameters:	
NT	Number of tasks
NS	Number of stations
NH	Maximum number of human operators for each station
NR	Maximum number of robot operators for each station
th_i	Processing time of task i if operated by human
tr_i	Processing time of task i if performed by robot
p_i	Set of immediate predecessors of task i
$pall_i$	Set of all predecessors of task i
$sall_i$	Set of all successors of task i
jt_i	Set of joint tasks for task i
f_i	The fitness value of task i
Decision variables:	
hu_{jh}	$\begin{cases} 1, & \text{if human } h \text{ is used in station } j \\ 0, & \text{otherwise} \end{cases}$
ro_{jr}	$\begin{cases} 1, & \text{if robot } r \text{ is used in station } j \\ 0, & \text{otherwise} \end{cases}$
ct	Cycle time
t_i	Processing time of task i after being assigned to either a human or robot
o_{ij}	Assigned operator for task i in station j
j_i	Assigned station for task i
st_i	The start time of task i
co_i	The completion time of task i
SCT_j	The completion time of station j

Implementing HRC on an assembly line entails significant challenges in task allocation and distribution of human and robotic resources between workstations. These challenges directly affect the assembly line's overall efficiency and output quality. Additionally, task scheduling complexity increases with HRC involvement. In line with the approach of Nourmohammadi et al. [2], this study aims to minimize the overall ct as the primary objective and the total number of operators as the secondary objective. The same set of constraints used in the referenced work is applied in this study. Table 1 presents the notation in this paper, and the objective function is detailed in Eq. (1). The value of the secondary objective is always less than 1.

$$\text{Minimize } z = ct + \frac{\sum_{j=1}^{NS} \sum_{h=1}^{NH} hu_{jh} + \sum_{j=1}^{NS} \sum_{r=1}^{NR} ro_{jr}}{(NH + NR) \times NS + 1} \tag{1}$$

The assumptions and constraints of the ALBP-HRC are outlined as follows:

- A single product model is assembled on a straight line.
- Human and robot operators have different skills and capabilities, addressed by different task processing times.
- In each workstation, the allocation of human and robotic resources remains within the prescribed limits for NH and NR.
- The maintenance and failure of robots are not considered.
- Each task is performed by one human or one robot.
- A task cannot be processed until all of its predecessors are completed.
- When solving task i and its joint tasks jt_i, human and robot operators perform them at the same station simultaneously, with the same start and completion times.
- ct is the maximum completion time among all stations, measured in seconds.
- The completion time of a station is determined by when its final task is completed.

4 Algorithm Design

SA is a typical representative of meta-heuristic algorithms, yielding relatively high-quality solutions and presenting notable flexibility [15]. It possesses excellent potential in solving complex problems such as ALBP-HRC. Based on adaptive simulated annealing (ASA) proposed by Nourmohammadi et al. [2], this study first addresses a new fitness value calculation method of roulette wheel selection to determine the weights of different tasks. After that, the selected task i is assigned to station j with operator o_{ij}, and the st_i and co_i are determined. Then, a heuristic mechanism is developed to reduce the maximum completion time at each workstation. Such a method randomly moves tasks that do not have predecessors or with all predecessors assigned to complete at previous workstations to an idle period to balance the workload. Both mechanisms are integrated into the ASA framework to form a novel meta-heuristic algorithm to minimize ct and the number of operators by optimizing task assignments to improve overall production efficiency.

4.1 Adaptive Simulated Annealing with New Fitness Value (ASA_nf)

When arranging tasks in an assembly line, the sequence of tasks is critical, and the completion times of tasks directly determine *ct*. Therefore, reducing the completion time of each workstation is the key to improving the overall assembly line efficiency. When generating initial and feasible solutions, the roulette wheel selection is applied to select a candid task [2]. This method assumes that the probability of selection is proportional to the individual's fitness [16]. The fitness value of each task is determined by its characteristics. Then, the probability p_k of selecting task k is:

$$p_k = \frac{f_k}{\sum_{i=1}^{NT} f_i} \tag{2}$$

An effective strategy to determine the fitness value is considering the relationships between tasks in the scheduling process. In the initial and feasible solution generation, the task to be assigned next is selected from the candidate list of tasks [2]. The candidate list contains only tasks that do not have a predecessor or whose predecessors are already assigned. Therefore, from the precedence relationship perspective, each task in the candidate list has an equal chance of being selected. However, the number of successors may vary from task to task. Suppose the selection process tends to arrange the tasks with a longer successors' total completion time in a more backward position in the workstation. In that case, it may result in more tasks being processed in the later stations, increasing *ct* and the number of operators required, which is not conducive to the overall objective.

Conversely, suppose the roulette wheel selection expands the spectrum of opportunities to tasks with a longer total completion time for all successors. In that case, it may help improve the overall workstation efficiency. In the proposed ASA_nf, A new fitness calculation method for each task is addressed to determine the weight of different tasks during the scheduling process. The fitness value of task i is defined as the sum of all the minimum processing time by its successors and itself. Tasks with larger fitness values have higher probabilities to be selected.

$$f_i = \sum_{k \in salli} \min(th_k, tr_k) + \min(th_i, tr_i) \tag{3}$$

4.2 Heuristic Adaptive Simulated Annealing (HASA)

As described in Algorithm 1, the heuristic mechanism aims to improve task allocation to be more tightly arranged, thereby reducing the completion time per workstation and improving overall efficiency. Firstly, the mechanism calculates the station completion time of each workstation. Next, tasks in the current workstation without any predecessors or with all the predecessors assigned to previous stations are recognized as unbound tasks (ut). These tasks are unrestricted, permitting their assignments to available time slots not already occupied by other tasks. Then, in the current station j, the idle time slots between the start time 0 and station completion time (SCT_j) are identified. Afterward, one of the tasks from ut is randomly selected and moved to an idle time slot. After a task is moved, the algorithm updates all tasks and operator schedules for the current station j and re-calculates SCT_j. If the task allocation within the station is optimized to become more compact and efficient, resulting in a reduced SCT_j compared to the original SCT_j, the adjustment is accepted; otherwise, the adjustment is rejected.

Algorithm 1. Heuristic Mechanism

```
1: Input: solution, NS, NT, NH, NR, th_i, tr_i, p_i, pall_i, jt_i
2: SCT = set of the completion times for stations;
3: station = 0; initialize the station counter;
4: while station < NS do
5:     station = station + 1; starting from station 1
6:     ut = tasks in station with no predecessors and tasks in station which all prede-
       cessors already assigned to previous stations;
7:     idle_time_slots = set of idle time slots for operators in station starting from
       time 0 ending at SCT_station;
8:     Randomly move a task from ut to a time slot in idle_time_slots regarding
       station;
9:     Update solution of all tasks and operators for station in a new copy;
10:    new_SCT_station = the station completion time for station in the new copy;
11:    if new_SCT_station < SCT_station then
12:        Accept solution for station in the new copy;
13:    else
14:        Reject solution for station in the new copy
15:    end if
16: end while
17: Output: updated solution
```

Figure 2 illustrates the adjustment of scenario 1 from case study 2 by Algorithm 1 in a single iteration. One human operator H_1 and one robot operator R_1 are assigned in each station. Each horizontal rectangle with different colors represents different tasks, and the number in the center signifies the task's index. Figure 2a represents the current solution before applying Algorithm 1. There are 19 tasks in total, each with a varying number of predecessors. The predecessor of Task 10 is Task 9, which is processed by operator 1 (human) in station 2 at time $0 - 7$s, making Task 10 an unbounded task. Algorithm 1 identifies and relocates Task 10 to the idle time slot between Task 9 and Task 17. The updated and accepted arrangement is shown in Fig. 2b, reducing the completion time in station 2 from 88s to 59s, consequently decreasing the ct from 88 s to 73s.

Figure 3 presents the HASA flowchart that implements both mechanisms based on ASA.Firstly, the problem and SA parameters are input into the HASA.

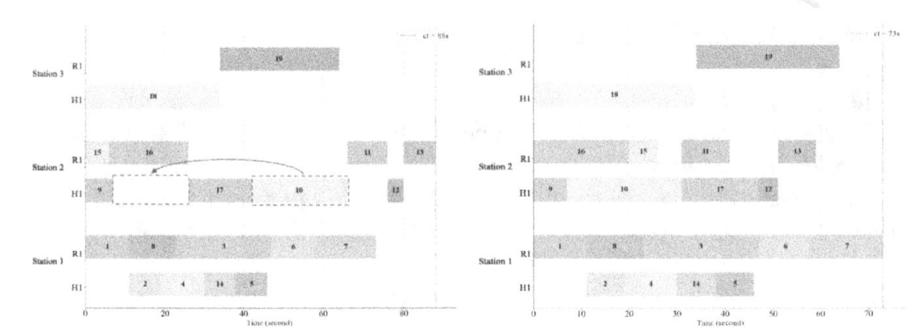

(a) Before implementation of Algorithm 1 (b) After implementation of Algorithm 1

Fig. 2. Algorithm 1 Implementation: Case Study 2, Scenario 1

Next, it generates an initial solution (iso) and sets it as the current solution (cso), the best solution (bso), and the worst solution (wso) simultaneously. In the main loop of HASA, the algorithm initializes a neighborhood selection matrix and sets the initial neighborhood selection probability. Then, it iteratively loops from the initial temperature T_0 and randomly selects a neighborhood search strategy based on the current neighborhood selection probability [2]. Then, a feasible solution is generated based on the changes from the neighborhood search strategy. In the process of generating initial and feasible solutions, the roulette wheel selection is used to arrange tasks sequentially. The probability of different tasks being selected in the roulette wheel selection is calculated using the proposed mechanism.

After the feasible solution is generated, the proposed heuristic mechanism further optimizes it as the neighborhood solution (nso). If the objective function value of the neighborhood solution ($z(nso)$) is superior to $z(cso)$ or is accepted according to specific probability criteria, that is, if $\Delta z = z(nso) - z(cso) \leq 0$ or $rand < \exp(-\Delta/T)$, the algorithm accepts nso and updates cso with nso [2]. Then, the algorithm checks and determines whether to update the bso and wso. After each iteration, the neighborhood selection matrix is updated, and the temperature is reduced [2]. When the temperature drops to the final temperature T_f, the algorithm stops and outputs the bso found.

5 Experiments and Results

5.1 Experiment Settings

Experimental data in this study are derived from Nourmohammadi et al. [2], as shown in Table 2. The dataset includes four real cases in the automotive industry, each with two scenarios differing in the maximum number of human and robot operators in each station. The total number of tasks and their precedence relationships vary in each case. Additionally, the processing times for human and robot operators differ for each task. Some tasks require HRC in each case.

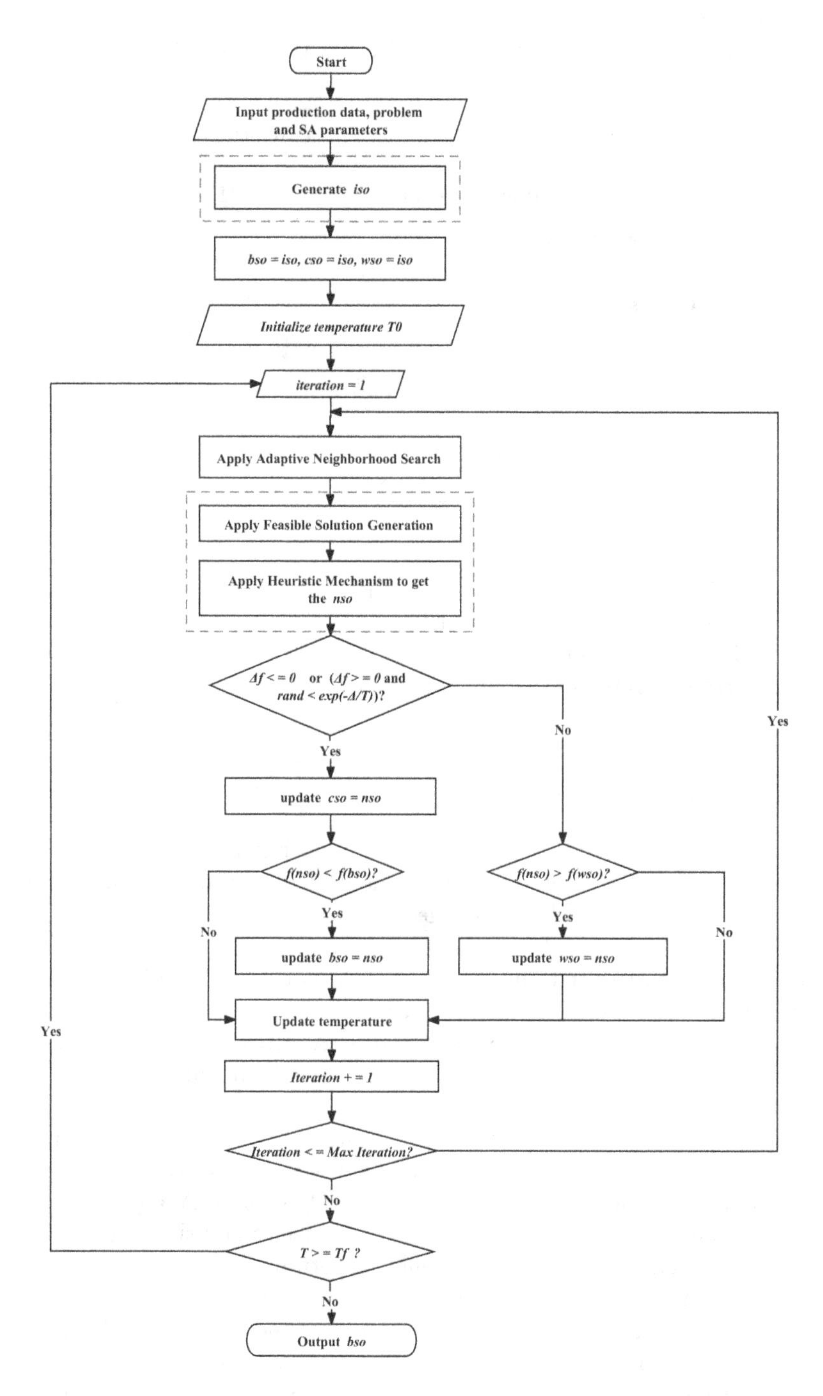

Fig. 3. Flow Chart of HASA

For each scenario of the four cases, five runs of computational experiments were performed, utilizing three mechanisms: ASA, ASA_nf, and HASA.

Table 2. Experimental data for case studies

Problem	Problem Scenarios	NT	NS	NH	NR
Case study1	S1	28	2	1	1
	S2	28	2	2	2
Case study2	S1	19	3	1	1
	S2	19	3	2	2
Case study3	S1	41	4	1	1
	S2	41	4	2	2
Case study4	S1	88	9	1	1
	S2	88	9	2	2

The experiments were conducted using Python 3.9 on a PC with an AMD Ryzen 7 5700X 8-Core Processor, clocked at 3.40 GHz and supported by 16.0GB of RAM. The SA parameters were set to $T_0 = 100, T_f = 0.1, cr = 0.99$ with $Max\_Iteration = 100$, consistent with those used by Nourmohammadi et al. [2].

5.2 Computational Results

Firstly, the ASA is coded and replicated. Next, based on replicated experiments, this study implemented the new calculation method of fitness value in roulette wheel selection and then carried out the ASA_nf experiments. Based on ASA_nf, the proposed heuristic mechanism is integrated to perform the HASA experiments. Considering the inherent randomness of SA and the influence of different processor, memory, and system configurations, the OFV and CPU time obtained by the five runs of the ASA experiment presented in this paper marginally differs from the results in Nourmohammadi et al.[2].

The best results of ASA, ASA_nf, and HASA are presented in Table 3, which includes the objective function values (OFV) and CPU times (measured in seconds) for four real cases, each with two scenarios. The GAP calculations for ASA_nf and HASA are derived based on the OFV of ASA replicated in this study. Since the objective function aims to minimize *ct* and the number of operators, a negative GAP in Table 3 represents an optimization over the ASA result. ASA_nf and HASA achieve smaller OFV values to varying degrees, demonstrating task scheduling and resource allocation enhancements. Specifically, ASA_nf reduces the OFV, while HASA shows a more significant decrease. Also, ASA_nf and HASA yield better results quickly for all cases, even for the large-scale problem referred to as case study 4. For the two scenarios in case study 2, the

Table 3. Computational Results of ASA, ASA_nf, and HASA

No.	Problem	Scenario	ASA		ASA_nf			HASA		
			OFV	CPU	OFV	CPU	GAP	OFV	CPU	GAP
1	case study 1	s1	60.0	33.1	59.6	31.9	**−0.67%**	58.8	72.0	**−2.04%**
2		s2	37.3	30.1	35.1	27.7	**−6.27%**	35.1	73.8	**−6.27%**
3	case study 2	s1	56.6	23.4	56.6	19.5	0.00%	56.6	48.4	0.00%
4		s2	56.3	24.3	56.3	20.9	0.00%	56.3	57.5	0.00%
5	case study 3	s1	55.3	48.5	54.9	46.9	**−0.73%**	53.7	114.0	**−2.98%**
6		s2	32.8	49.6	32.8	42.0	0.00%	31.5	127.0	**−4.12%**
7	case study 4	s1	64.9	120.0	63.9	106.7	**−1.56%**	62.1	342.5	**−4.59%**
8		s2	56.9	97.7	49.0	107.3	**−16.21%**	46.4	193.8	**−22.65%**
Avg.			52.5	53.3	51	50.4	**−2.92%**	50.1	128.6	**−4.91%**

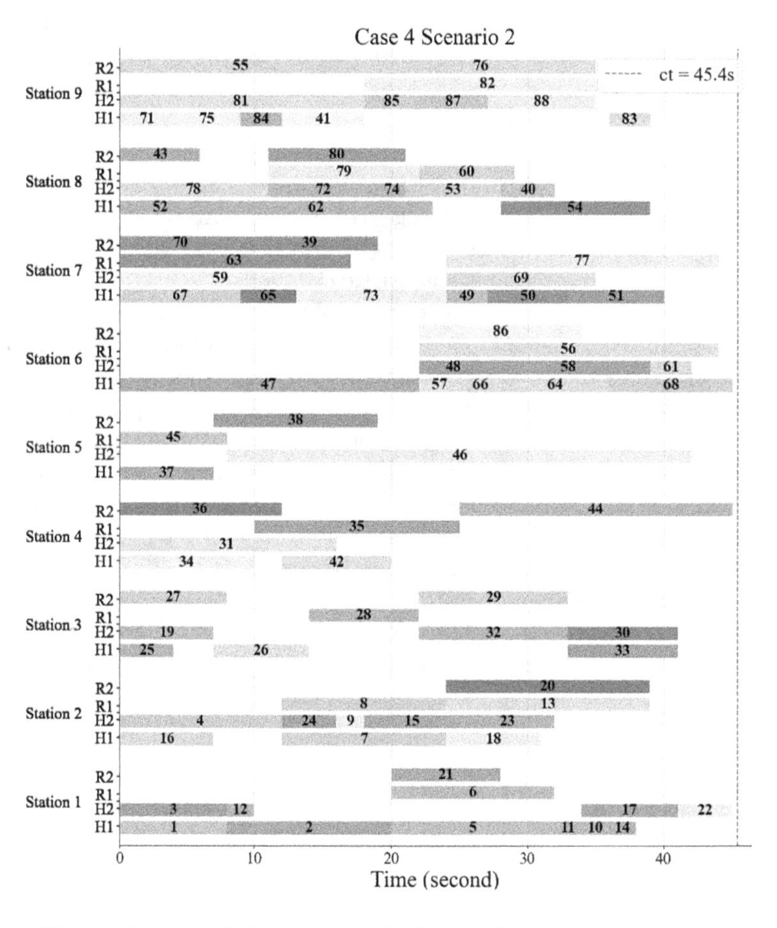

Fig. 4. The Best Solution from HASA for Case Study 4 Scenario 2

OFV results for ASA, ASA_nf, and HASA are consistent with those in Nourmohammadi et al. [2], where these solutions have been verified as optimal. In scenario 2 of case study 4, a large-scale production instance involving 88 tasks, with each workstation equipped with two human and two robot operators, the HASA result with a GAP of −22.65% presents the most efficient improvement, and the solution is shown in Fig. 4.

Overall, the proposed ASA_nf and HASA obtain better results for ALBP-HRC to different extents, effectively minimizing *ct* and the number of operators to achieve better task allocation and resource management in assembly lines within a relatively short CPU time.

6 Conclusion

From Industry 4.0 to Industry 5.0, the harmonious coexistence of humans and robots in the workplace is becoming an essential trend for future development. Based on the work of Nourmohammadi et al. [2], this study for ALBP-HRC introduces significant advancements by developing and incorporating new solution methods of task allocation of the workstation, minimizing the cycle time and the total number of operators in scenarios where workstations are equipped with varying numbers of human and robot operators.

This paper introduces two innovative mechanisms integrated into ASA to create ASA_nf and HASA for solving ALBP-HRC. ASA_nf incorporates a new calculation method for fitness values to improve task allocation. Based on ASA_nf, HASA further integrates a heuristic mechanism to better balance workstation workloads. Actual production data are used to validate these mechanisms in computational experiments. The experimental results indicate that ASA_nf and HASA present varying degrees of advanced task allocation and resource arrangement. Significant improvements are observed in large problem instance tests, with both algorithms showing substantial enhancements and short CPU times when workstations are equipped with multiple human and robot operators.

While ASA_nf and HASA demonstrate better performance in ALBP-HRC, opportunities remain for further improvement. This study has not comprehensively addressed the maintenance schedules, the potential failure of robots, and the ergonomic risks of human operators. Future research could prioritize formulating solutions that more thoroughly integrate these factors to better address ALBP-HRC. Also, developing more advanced meta-heuristic algorithms in this context is a challenging and valuable research direction.

Disclosure of Interests. The authors have no competing interests to declare that are relevant to the content of this article.

References

1. Demir, K.A., Döven, G., Sezen, B.: Industry 5.0 and human-robot co-working. Procedia Comput. Sci. **158**, 688–695 (2019). https://doi.org/10.1016/j.procs.2019.09.104
2. Nourmohammadi, A., Fathi, M., Ng, A.H.: Balancing and scheduling assembly lines with human-robot collaboration tasks. Comput. Oper. Res. **140**, 105674 (2022). https://doi.org/10.1016/j.cor.2021.105674
3. Salveson, M.: The assembly line balancing problem. J. Ind. Eng. **77**, 18–25 (1955)
4. Scholl, A., Becker, C.: State-of-the-art exact and heuristic solution procedures for simple assembly line balancing. Eur. J. Oper. Res. **168**(3), 666–693 (2006). https://doi.org/10.1016/j.ejor.2004.07.022, balancing Assembly and Transfer lines
5. Zhou, B., Wu, Q.: A novel optimal method of robotic weld assembly line balancing problems with changeover times: a case study. Assem. Autom. **38**, 376–386 (2018). https://doi.org/10.1108/AA-02-2018-026
6. Chutima, P.: A comprehensive review of robotic assembly line balancing problem. J. Intell. Manuf. **33**(1), 1–34 (2022). https://doi.org/10.1007/s10845-020-01641-7
7. Krüger, J., Lien, T.K., Verl, A.: Cooperation of human and machines in assembly lines. CIRP Ann. Manuf. Technol. **58**, 628–646 (2009). https://doi.org/10.1016/j.cirp.2009.09.009
8. Demiralay, Y.D., Kara, Y.: Considering gripper allocations in balancing of human-robot collaborative assembly lines. In: Alfnes, E., Romsdal, A., Strandhagen, J.O., von Cieminski, G., Romero, D. (eds.) Advances in Production Management Systems. Production Management Systems for Responsible Manufacturing, Service, and Logistics Futures. APMS 2023. IFIP Advances in Information and Communication Technology, vol. 689, pp. 691–701. Springer, Cham (2023). https://doi.org/10.1007/978-3-031-43662-8_49
9. Kheirabadi, M., Keivanpour, S., Chinniah, Y.A., Frayret, J.M.: Human-robot collaboration in assembly line balancing problems: Review and research gaps. Comput. Ind. Eng. **186**, 109737 (2023). https://doi.org/10.1016/j.cie.2023.109737
10. Becker, C., Scholl, A.: A survey on problems and methods in generalized assembly line balancing. Eur. J. Oper. Res. **168**, 694–715 (2006). https://doi.org/10.1016/j.ejor.2004.07.023
11. Fathi, M., Fontes, D.B.M.M., Moris, M.U., Ghobakhloo, M.: Assembly line balancing problem: a comparative evaluation of heuristics and a computational assessment of objectives. J. Model. Manag. **13**, 455–474 (2018). https://doi.org/10.1108/JM2-03-2017-0027
12. Weckenborg, C., Kieckhäfer, K., Müller, C., Grunewald, M., Spengler, T.S.: Balancing of assembly lines with collaborative robots. Bus. Res. **13**, 93–132 (2020). https://doi.org/10.1007/s40685-019-0101-y
13. Raatz, A., Blankemeyer, S., Recker, T., Pischke, D., Nyhuis, P.: Task scheduling method for HRC workplaces based on capabilities and execution time assumptions for robots. CIRP Ann. **69**, 13–16 (2020). https://doi.org/10.1016/j.cirp.2020.04.030
14. Nourmohammadi, A., Fathi, M., Ng, A.H.: Balancing and scheduling assembly lines with human-robot collaboration tasks. Compute. Oper. Res. **140**, 105674 (2022). https://doi.org/10.1016/j.cor.2021.105674

15. Dréo, J., Siarry, P., Pétrowski, A., Taillard, E.: Metaheuristics for Hard Optimization: Methods and Case Studies. Springer Berlin, Heidelberg, 1 edn. (2006). https://doi.org/10.1007/3-540-30966-7, published: 24 November 2005 (Hardcover), 12 February 2010 (Softcover), 16 January 2006 (eBook)
16. Goldbeg, D.E.: Genetic algorithms in search, optimization, and machine learning. Addison-Wesley Pub. Co, Reading, Mass (1989)

Game-Based Design of a Human-Machine Collaboration Monitoring System

Mónika Gugolya[1], Tibor Medvegy[1], János Abonyi[2,3], and Tamás Ruppert[1,2](✉)

[1] Department of System Engineering, University of Pannonia, Egyetem str. 10, Veszprém, Hungary
ruppert@abonyilab.com

[2] HUN-REN-PE Complex System Monitoring Research Group, University of Pannonia, Egyetem str. 10, Veszprém, Hungary

[3] Department of Process Engineering, University of Pannonia, Egyetem str. 10, Veszprém, Hungary

Abstract. In a human-machine collaboration scenario, identifying a specific use case can be challenging due to the wide range of potential applications and interactions. Additionally, effective monitoring of the behavior of both human and machine agents during this collaboration poses significant challenges. The developed game-based process enables the analysis of the behaviours, leading to improved efficiency and collaboration. Indicators such as agent utilization and waiting times serve as valuable metrics to represent the quality of collaboration. This paper presents a setting in the Industry 5.0 laboratory, where monitoring and evaluation of humans and robots is possible. An experimental design is described and executed based on the developed game-based scenario, and exploratory analyses are performed based on the measured data.

Keywords: Human-machine · Collaboration · Evaluation · Experiment · Productivity

1 Introduction

The manufacturing workforce faces a severe shortage of human labor globally and in Europe, driven by an aging population and rapid technological changes. A WHO report predicts that by 2024, those over 65 will outnumber those under 15 in the WHO European Region. This demographic shift poses social, economic, and health challenges, affecting the European economy by slowing growth and reducing competitiveness [19]. Consequently, the need for solutions becomes apparent, and while some automation is possible, establishing a shared space for close interaction between humans and machines is crucial.

The interactions between operators and machines have changed with the new industrial revolution [15]. However, even in the so-called smart factories, where there is data sharing between intelligent robots, workers are still seen as specta-

Published by Springer Nature Switzerland AG 2024
M. Threr et al. (Eds.): APMS 2024, IFIP AICT 729, pp. 205–219, 2024.
https://doi.org/10.1007/978-3-031-65894-5_15

tors, not an active part of the system. To change this and bring human-centered manufacturing to the forefront, robots should not replace the jobs of human workers but work together and assist the workers in helping with efficiency and productivity. The interaction between operators and machines has evolved over the years. The focus is on a system where automation further enhances human worker abilities [14]. Operator 4.0 aims to create a partnership between workers and machines that involves trust and cooperation. In this way, smart factories can combine the powers of intelligent robots and help their qualified and competent operators with new skills to take full advantage of every opportunity in the created system [16]. The operator can be supported in many forms; depending on that, it improves one of its skills. As a result, workers can be supported in many ways, improving their skills and abilities. As Industry 4.0 enhances the capabilities of manufacturing systems, it should also empower the workforce. The concept of Operator 5.0, particularly the "Resilient Operator 5.0," envisions a skilled individual using creativity and technology to overcome challenges, ensuring sustainable operations and the well-being of the workforce in difficult conditions [12].

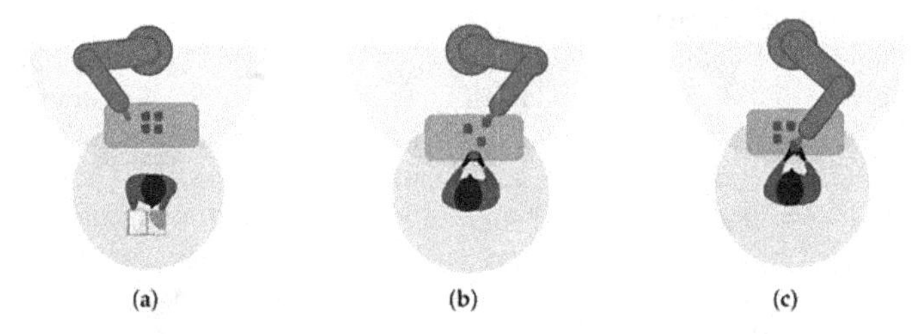

Fig. 1. Collaboration levels: (a) Coexistence; (b) Cooperation; (c) Collaboration [4]

A new generation of robots called collaborative robots - cobots [9] was introduced. These industrial robots have been set up to replace human assistants or help in varied repetitive manufacturing tasks that are hazardous and monotonous to the worker. In the case of using collaborative robots, humans and robots are sharing a workspace. Thus, new human-robot interaction systems have been created for such systems to be able to enhance the abilities of both humans and robots [7]. Cobots are designed to work alongside humans, using advanced safety technologies to ensure the appropriate level of automation. However, the systems become more complex as the collaboration between humans and cobots increases. This complexity has limited the industrial application of cobots in labor-intensive processes such as assembly [11].

Human-robot collaboration can be classified into levels depending on how humans and robots work together in their respective spaces (see Fig. 1). These levels are closely linked to how they interact during their work [4]. The coexistence level involves robots kept separate from humans; assistive robots do not have an independent task space and are designed to work alongside humans.

Cooperation is the next level of human-robot interaction, where robots and humans share the same workspace and resources but sequentially work on separate tasks. There is an intervention zone where tasks may be shared, but there is no physical contact between the two. Collaborative robots can be helpful in tasks that require human and robot input, such as assembly lines. The final level is collaboration, where collaborative robots work alongside humans, sharing the same workspace and tasks. These robots can make decisions based on human and robot input and may involve physical contact or communication, such as gestures or voice commands [5]. Collaborative robots are helpful when tasks require close cooperation between humans and robots, such as in advanced manufacturing settings [13]

The performed tasks should be monitored to create the Operator 4.0 scheme, and the data should support the operator's activities. Creating an intelligent space [10] could be possible. While working, it should guide the operator when help is needed. The movements of the operators could be followed by a camera system or real-time locating system (RTLS) to obtain and analyze data on the situation. Incoming data can be processed using artificial intelligence and machine learning solutions [17]. With these, improving process performance and quality is achievable [14]. The tic-tac-toe game has been used in several research studies [6,8,18,20]. Ambsdorf et al. investigate the effects of explainable artificial intelligence (XAI) on human perceptions in human-robot interactions. In their study, two simulated robots played a tic-tac-toe game; one robot explained its moves while the other did not. The results showed that explanations made the robot appear more lively and human-like but did not significantly impact perceived competence, intelligence, likeability, or safety. This suggests that while XAI can enhance the perceived humanity of robots, it does not necessarily influence other critical user perceptions [3]. The need for resilient operators and the challenge of the upcoming or even ongoing demographic situation highlight the need for more efficient and human-centred human-machine collaboration. Ahmad et al. highlight the significance of adaptive robots in enhancing human-robot interaction across various social environments. Adaptive robots adjust their behaviors based on user emotions, personalities, and past interactions. It is highlighted, that there is a need for more research into different adaptive behaviors, standardized evaluation metrics, and ethical concerns related to data privacy and user comfort [2]

To frame our study within the existing body of knowledge, we aim to address the following research questions derived from the current state of the art in human-robot collaboration: How can we design an approachable user study setup for human-robot collaboration? How does adjusting task difficulty influence waiting time in human-robot collaboration scenarios? By exploring these questions, we intend to contribute to the development of effective and practical methodologies for studying and enhancing human-robot interactions. Our goal is to create an experimental setup for monitoring and analysing robot-human collaboration. With the developed game-based process and monitoring framework, an experimental design is proposed. The novelty of our work is the experimental design and the simple smart method of measuring and labeling the experimental data. The developed game-based collaboration process simulates the manufacturing situation where the human and the robot work sequentially within the same process. In our interpretation, understanding the collaboration behaviour of this sequential work is crucial for a better and "deeper" collaboration and interaction between human and machine.

2 The Developed Game-Based Process and Experiment Design

Analyzing and understanding human behavior plays a crucial role in designing more efficient human-machine systems. We have developed a game-based process to simulate the manufacturing scenario with a cobot and an operator. A monitoring system is proposed to observe the behavior and interactions between the two agents. The process represents a sequential work between the human and the cobot, but the main point of our experiments is to observe and analyse the behaviour and dependencies of the two agents during the process. We designed an experiment to see how the effects of different factors (such as main and additional task difficulty) on collaboration behaviour.

2.1 Designed Process and the Workstation Setup

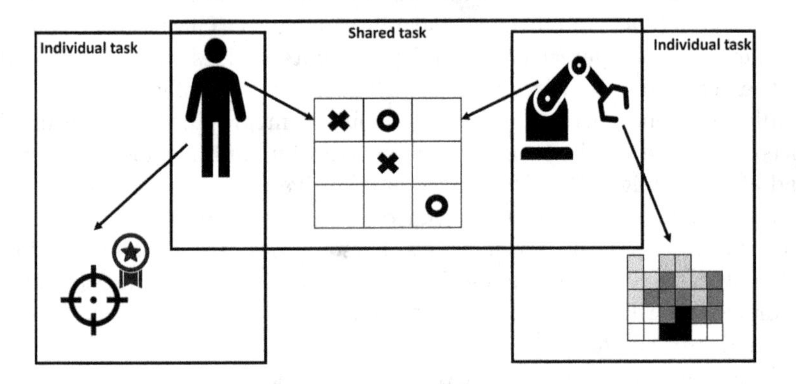

Fig. 2. The developed game-based process.

To explore human-robot collaboration, a designed use case is presented, centred around a simple tic-tac-toe game, a two-player board game where the players are the human and the cobot. The tic-tac-toe game was chosen because it is an easy and well-known game, or it can be learned in a short time. It was also beneficial because it was easy to create pieces that the robot can grab and play the game with. This scenario includes three tasks (see Fig. 2): the tic-tac-toe game between the human and the cobot, the human's reaching for the next tile (through another dexterity game - see details in Sect. 2.3), and the robot's independent pattern-building activity. Each task is unique, making it possible to observe and measure the collaborative dynamics between the human operator and the cobot. As the Fig. 2 shows, both the human and the cobot have additional individual tasks. The common task (the tic-tac-toe game) where they depend on each other. During the analyses this sequencing process will be the interesting part, how they behave during the process and what are the productivity indicators. This process is related to the real industrial scenarios where agents are interdependent next to the production line or even within a U-cell workstation.

The defined workspace represents the collaborative space, where the cobot and operator share a workstation and interact with each other. The robot arm used is a Universal Robots UR5e model, which allows safe interaction with humans. To set up the play area (see Fig. 3), a board and game pieces are required, which must be suitable for the cobot to pick up and place the pieces smoothly on the board. The game setup includes 3D printed pieces (see Fig. 4) that are designed to be easily picked up and manipulated by the robot arm. These circular pieces, with X and O shapes, are hollow on the inside to facilitate stacking and save space. In addition, a specially designed board ensures consistent part placement, with slots for parts and storage areas.

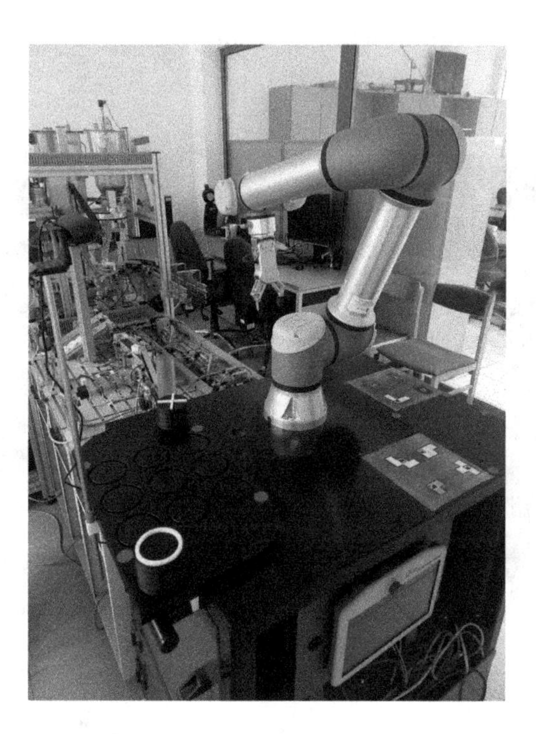

Fig. 3. The developed workspace

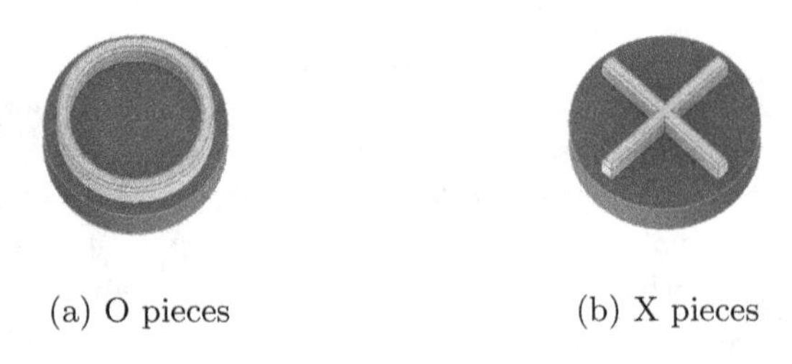

(a) O pieces (b) X pieces

Fig. 4. The model of the pieces for the game

The cobot is controlled by a custom Python program based on the simple tic-tac-toe solution implementation, the recognition of the board's current state by video-based object recognition. The previous state of the board is stored in a matrix containing positions and information about whether the step was made by a human or a cobot. The algorithm can calculate what the cobot's next step should be, and the algorithm's output is the trajectory of the robot arm's next action, which is positional data of where the robot will go next. The pre-programmed pattern-building trajectories are handled the pattern building secondary task between two tic-tac-toe steps.

2.2 The Proposed Monitoring System

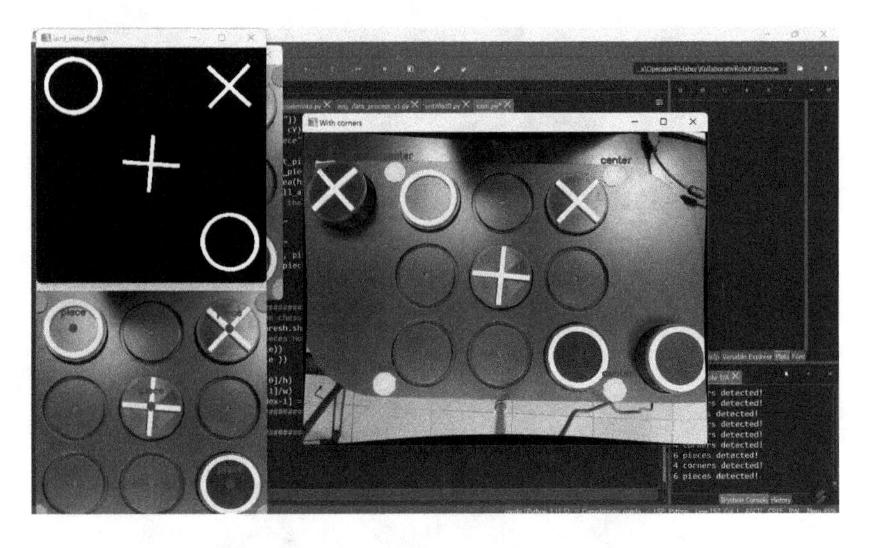

Fig. 5. The tic-tac-toe game's state recognition program

When robots and humans work in the same workspace, their movements should be monitored and observed as they provide valuable information for future use.

Depending on what the robot arm is doing, the operator's actions may vary, and a camera system can be used to observe these changes. To do this, a camera is positioned above the work area to record what is happening, capturing images of the work area and detecting changes in the environment, such as the location of objects. As we have described, the control of the cobot depends on the current state of the game, which is detected by the camera vision using OpenCV (see in Fig. 5). This is done with the help of four red dots on the corner of the board. The picture is then reduced to the area between the red dots. By dividing this image into three parts and looking at the contours, the places and types of the pieces can be determined. This recognition was done until the program recognized the new piece, and the other recognized piece numbers was what it was supposed to be. Figure 6 shows the structure of the developed monitoring system. In order to evaluate and analyse the efficiency of the collaboration, we measure both human and cobot times in terms of work and waiting times. As we already stated a camera is used to recognise the situation within the workspace. An observer (with a mobile application) measured the times of the operator during the experiments. More details about the measured data are given in Sect. 2.3.

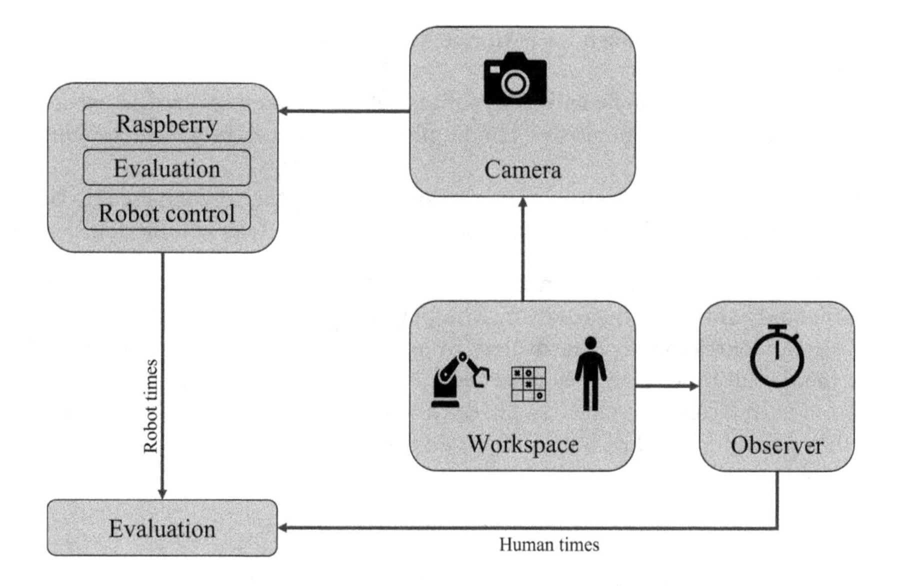

Fig. 6. The developed monitoring system

Through a series of tasks, including robot playing tic-tac-toe, human playing tic-tac-toe, robot building pattern and human playing with the additional hand dexterity game this paper aims to explore the efficiency, adaptability, and collaborative synergy between the human and the robot. Furthermore, including time as a pivotal metric adds a quantitative dimension to the evaluation. Timing each task provides insights into human and robot collaborators' efficiency and analyzes their strengths and areas for improvement. The emphasis on time also aligns with the dynamic nature of collaborative work, where swift and precise actions often define success.

2.3 Design of Experiment

The measured variables provide insights into task efficiency, engagement levels, and the impact of different factors on human-robot collaboration. The proposed experimental design provides a basis for conducting large-scale measurements and obtaining comprehensive data on human-robot collaboration. To design an experience, two key factors can be considered regarding the individual and shared task between the human and the cobot:

1. **Additional task difficulty:** In this factor, different difficulty levels are considered - medium and difficult.
2. **Robot's algorithm efficiency:** This factor explores the efficiency of the robot's decision-making algorithm at three levels - low, medium and hard.

We preliminary defined the variables which should be measured:

1. **Time taken:** Time taken for each task (tic-tac-toe, pattern building) is measured in seconds.
2. **Idle time:** Total idle time during collaborative tasks, measured in seconds.
3. **Frequency of idle periods:** The number of instances in which the human or the robot is idle.
4. **Reasons for idle states:** Categorize and measure the reasons for idle time, such as waiting for instructions, processing delays or task completion.

Table 1 provides a comprehensive overview of the measured data, corresponding tools and units of measurement. The player's time, which is the time taken by a human participant, is measured by a mobile phone application. On the other hand, the robot's time is recorded by the computer. It was stored as the robot made different movements. These time-based metrics provide insight into the efficiency and speed of both human and machine components.

Table 1. The measured data

Measured data	Tool	Unit
Player's time - tic-tac-toe	Phone application	seconds
Player's time - waiting	Phone application	seconds
Player's time - dexterity tester	Phone application	seconds
Robot's time - tic-tac-toe	Computer log	seconds
Robot's time - waiting	Computer log	seconds
Number of idle period	Manually logged	-
Number of winning	Manually logged	-

To setup our experiment, two factors were selected; the first factor involved varying the difficulty of the tic-tac-toe game-solving algorithm. Two difficulty levels were implemented: an easy level where the robot's moves were randomly selected and a difficult level where the robot strategically chose moves to increase the chances of winning. In both cases, humans could defeat the cobot. The second factor focused on the difficulty of the task performed by the human while the cobot made its moves or built the pattern. This task involved a skill-based challenge using a dexterity tester tool see in Fig. 7. The objective was to touch the metal rod to the sensor at the bottom of a hole without letting it touch the walls. Different hole sizes were used and the difficulty levels varied. Before the experiment, each participant tried the game (during the trial session).

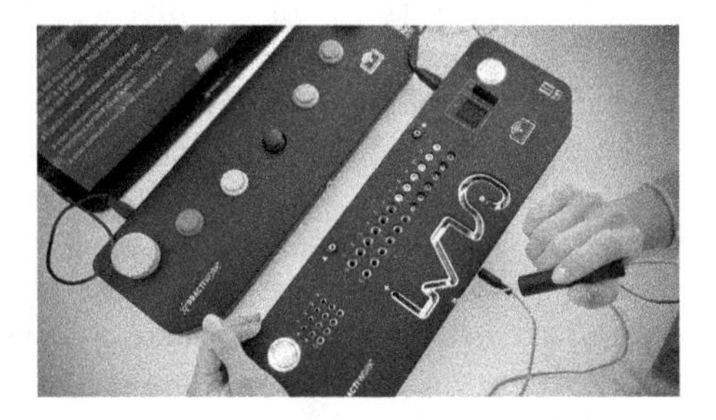

Fig. 7. The dexterity tester [1]

Participants had to complete the task perfectly once, without the screen flashing red, in the easy difficulty, while in the hard difficult level, they had to do it twice. In cases where participants struggled significantly, a time limit of 90 seconds was set, after which they would receive their tic-tac-toe piece and could continue with their move in the game.

Our contribution includes conducting user studies with collaborative robots and testing quantitative objective measurements, complementing the extensive simulation work prevalent in the field.

3 Experiments and Results

The experiment involved playing a game of tic-tac-toe against a robot. In addition to playing against the participant, the robot had to perform other tasks to increase its utilisation, and the participants also had to perform a hand dexterity game to reach the next piece for the tic-tac-toe game. The experimental design took into account the factors outlined in the previous experimental design. Three measurements were taken for each scenario, resulting in 12 measurements for one participant. Four individuals participated in the experiment, each completing the entire process. Experiment 1 (E_1) was when both difficulties were easy, Experiment 2 (E_2) was when the dexterity tester task was hard and the tic-tac-toe easy, Experiment 3 (E_3) was when the dexterity tester task was easy and the tic-tac-toe hard, and lastly, Experiment 4 (E_4) was when both difficulties were hard. Each participant had an ID. The robot and the person were also distinguished; Agent 1 (A_1) was the robot, and Agent 2 (A_2) was the participant.

The measured variables included the time of individual actions in each task: the moves made in tic-tac-toe by both the robot and the participant, the time of the steps taken by the robot during pattern building, the amount of time the player spent on the dexterity tester task, and the waiting time for both the player and the robot. Analysing these variables provides a comprehensive understanding of the dynamics of the experiment. The moves in the tic-tac-toe game reflect the strategic decisions of both the human player and the robot, allowing the effectiveness of the game-solving algorithms to be assessed at different levels of difficulty. In addition, the waiting times for the player and the robot provide valuable information about the interaction dynamics, highlighting any delays or periods of inactivity during the experiment. Table 2 shows the average waiting times (t_{avg}) for the different measurements for each person and the robot, as well as the normalised waiting times (t_{norm}) for each participant and the robot.

An exploratory analysis of the progress of each game was carried out but due to the limited number of participants, complex statistical analysis could not be conducted. The results were plotted on Gantt charts - one for each level of experimental difficulty, as shown in Fig. 8. The measured waiting times are of particular importance, as they are a crucial indicator of how efficiently the human and robot managed their allocated time. The different Gantt charts vividly illustrate how time allocation unfolded for specific individuals over the four unique experiments. These visual representations provide valuable information on the temporal dynamics of each interaction, showing how participants navigated their tasks. The graphs illustrate examples where the robot and human had to wait, providing insight into the dynamics of the interaction. The graphs also show that the games ended with different numbers of moves. With regard to tasks, human participants experienced particularly long periods, especially in the waiting and dexterity tests. At the same time, the robot spent most of its time on pattern building or waiting.

Figure 9 shows the waiting times for each scenario. The notation represents the waiting times for the participants as P_1, P_2, P_3 and P_4. Notably, the robot experienced the shortest waiting time with the first participant, while the longest was with the second participant. For humans, most of the waiting occurred in E_3 when the dexterity test was easy but the tic-tac-toe game was difficult. The values could be normalised to the maximum. This was done for the human participants and shown in Fig. 10. This makes it easier to compare the waiting times of the individual participants. It is also shown that the maximum waiting times were observed in E_3. It is noticeable that P_1 did not experience any significant changes in waiting times across the E_4. In contrast, P_2 had to wait much less in the E_1, indicating a longer duration for the manual dexterity test. It can also be seen that waiting times were longer when the task was easier.

Table 2. The average and normalized waiting times in all of the experiments

Experiment	Participant	Agent	Code	t_{avg} [s]	t_{norm} [s]
1	1	1	11	6	0,915
		2	12	36	0,866
	2	1	21	27	1,000
		2	22	17	0,488
	3	1	31	8	1,000
		2	32	32	0,880
	4	1	41	8	0,582
		2	42	33	0,966
2	1	1	11	7	1,000
		2	12	33	0,788
	2	1	21	24	0,858
		2	22	29	0,819
	3	1	31	6	0,778
		2	32	29	0,809
	4	1	41	14	1,000
		2	42	28	0,833
3	1	1	11	4	0,654
		2	12	42	1,000
	2	1	21	11	0,399
		2	22	35	1,000
	3	1	31	6	0,733
		2	32	36	1,000
	4	1	41	10	0,717
		2	42	34	1,000
4	1	1	11	6	0,854
		2	12	36	0,870
	2	1	21	11	0,388
		2	22	29	0,815
	3	1	31	6	0,763
		2	32	27	0,748
	4	1	41	11	0,801
		2	42	25	0,724

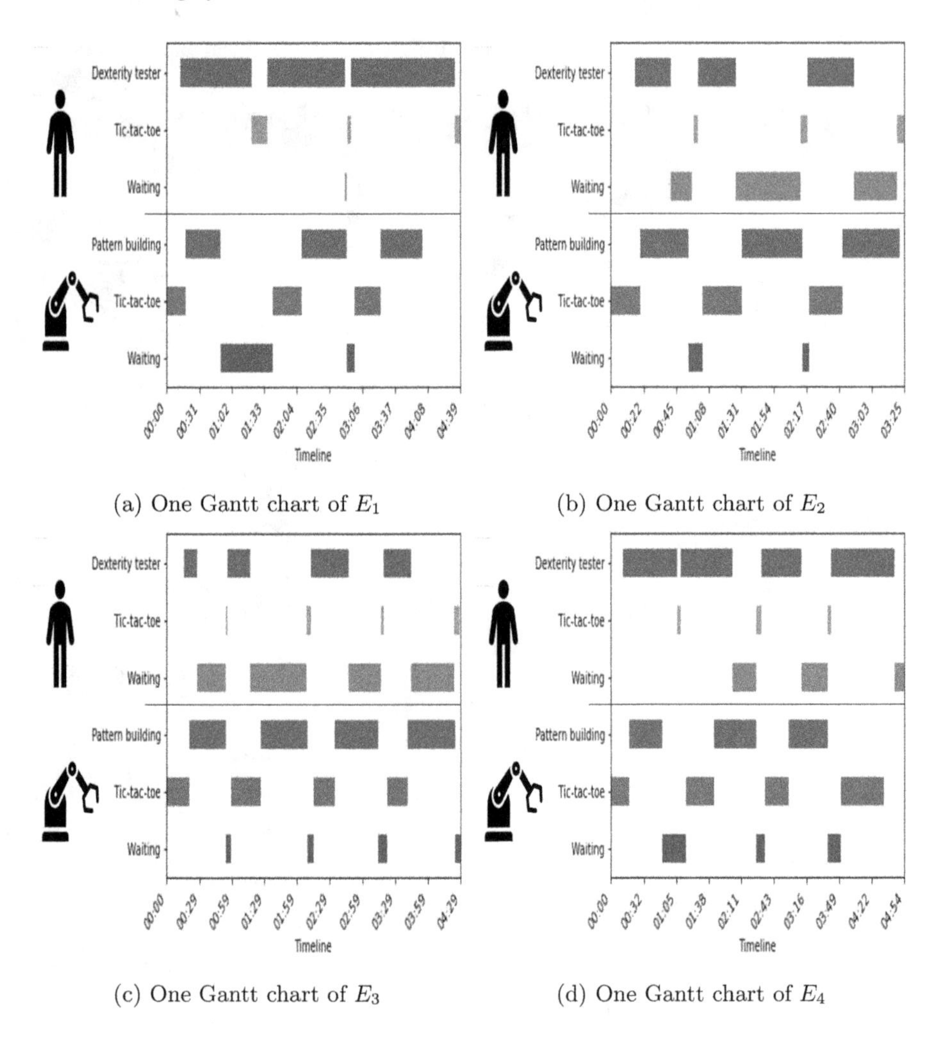

(a) One Gantt chart of E_1

(b) One Gantt chart of E_2

(c) One Gantt chart of E_3

(d) One Gantt chart of E_4

Fig. 8. Gantt charts for each experiment

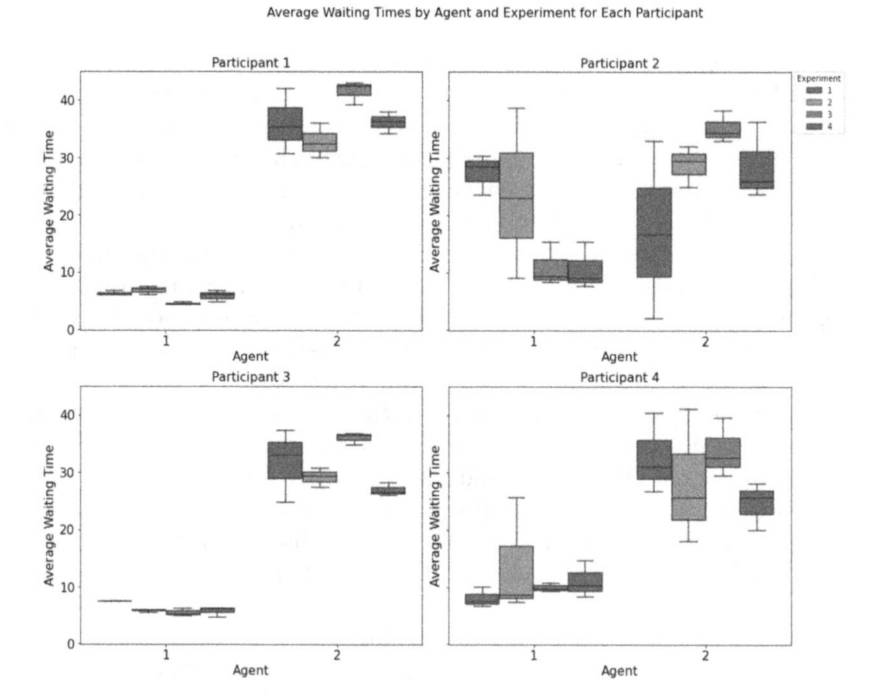

Fig. 9. The waiting times of the cobot (Agent 1) and the human (Agent 2).

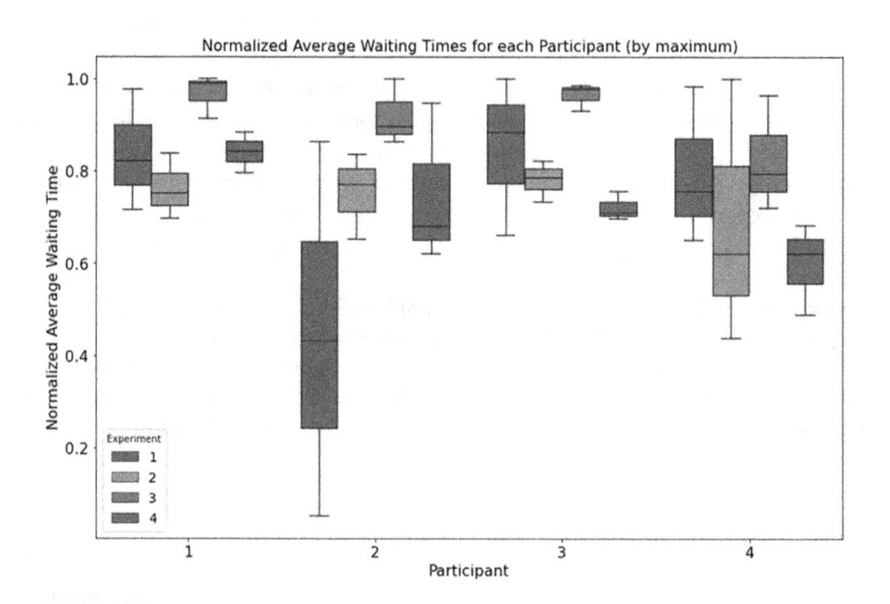

Fig. 10. The normalized waiting times

4 Conclusion

We have developed a game-based process to monitor and analyse human-robot collaboration. The proposed process, monitoring system, and experimental design provide a good basis for further analysis of this type of collaborative behaviour. Our first exploratory experiments show valuable information about the efficiency of these sequentially dependent tasks between humans and cobots. The joint task shows the importance of this dependency between the paired agents and their individual tasks. To answer our research questions, we can state, that in the experiments with the harder task for the human, the robot's waiting time was less than the easy task. In addition, the proposed experimental design is based on specific tasks that incorporate various factors to monitor collaboration. These factors include the sequence of tasks, the efficiency of the robot-control algorithm, the configuration of the workspace, and the familiarity of the operator. We measured several variables, including the time taken, idle time, the frequency of idle periods, and the reasons for idle states. These experiments aim to manipulate these factors, collect data, and analyse collaborative performance. Taking into account factors and variables, we conducted experiments that will be useful to make human-robot collaboration more effective and measure collaboration.

The designed experiment is a good basis for further analysis, where we can explore human behavior during the different factors of the experiments. The developed monitoring system gives a certain digitalized way of measuring the key values during the experiments. Physiological signals play a crucial role in understanding human behaviour, and using the insights from my results can be valuable for optimisation and planning in the future. In our future studies, we aim to concentrate on the following areas to support our comprehension of human-robot collaboration: further refining our hypotheses to encompass a broader spectrum of factors influencing collaboration dynamics; enlarging and diversifying the participants to enhance the generalizability of our findings; incorporating subjective measures and physiological sensors to capture participants' subjective experiences and physiological responses and exploring methods to mitigate learning effects and enhance the validity of our experimental results.

Acknowledgment. This work has been implemented by the TKP2021-NVA-10 project with the support provided by the Ministry of Culture and Innovation of Hungary from the National Research, Development and Innovation Fund, financed under the 2021 Thematic Excellence Programme funding scheme. The work of Tamás Ruppert is supported by the János Bolyai Research Scholarship of the Hungarian Academy of Science.

References

1. Practiwork. https://practiwork.hu/content/4/4_0.jpg. Accessed 06 June 2024
2. Ahmad, M., Mubin, O., Orlando, J.: A systematic review of adaptivity in human-robot interaction. Multimodal Technol. Interact. **1**, 3 (2017)

3. Ambsdorf, J., et al.: Explain yourself! Effects of explanations in human-robot interaction. In: 2022 31st IEEE International Conference on Robot and Human Interactive Communication (RO-MAN), pp. 393–400. IEEE (2022)
4. Arents, J., Abolins, V., Judvaitis, J., Vismanis, O., Oraby, A., Ozols, K.: Human-robot collaboration trends and safety aspects: a systematic review. J. Sens. Actuator Netw. **10**(3), 48 (2021)
5. Bauer, A., Wollherr, D., Buss, M.: Human-robot collaboration: a survey. Int. J. Humanoid Rob. **5**(01), 47–66 (2008)
6. Dell'Ariccia, A., Bremers, A.W., Lee, W.Y., Ju, W.: "ah! he wants to win!"': Social responses to playing tic-tac-toe against a physical drawing robot. In: Proceedings of the Sixteenth International Conference on Tangible, Embedded, and Embodied Interaction, pp. 1–6 (2022)
7. Hentout, A., Aouache, M., Maoudj, A., Akli, I.: Human-robot interaction in industrial collaborative robotics: a literature review of the decade 2008–2017. Adv. Robot. **33**(15–16), 764–799 (2019)
8. Junior, J.M., de Paula Caurin, G.A., Camolesi Jr., L.: Collaborative rules operating manipulators. INTECH Open Access Publisher (2010)
9. Knudsen, M., Kaivo-Oja, J.: Collaborative robots: frontiers of current literature. J. Intell. Syst. Theory Appl. **3**(2), 13–20 (2020)
10. Lee, J.H., Hashimoto, H.: Intelligent space-concept and contents. Adv. Robot. **16**(3), 265–280 (2002)
11. Malik, A.A., Brem, A.: Digital twins for collaborative robots: a case study in human-robot interaction. Robot. Comput.-Integr. Manuf. **68**, 102092 (2021)
12. Mourtzis, D., Angelopoulos, J., Panopoulos, N.: Operator 5.0: a survey on enabling technologies and a framework for digital manufacturing based on extended reality. J. Mach. Eng. **22** (2022)
13. Mukherjee, D., Gupta, K., Chang, L.H., Najjaran, H.: A survey of robot learning strategies for human-robot collaboration in industrial settings. Robot. Comput.-Integr. Manuf. **73**, 102231 (2022)
14. Peruzzini, M., Grandi, F., Pellicciari, M.: Exploring the potential of operator 4.0 interface and monitoring. Comput. Ind. Eng. **139**, 105600 (2020)
15. Romero, D., Bernus, P., Noran, O., Stahre, J., Fast-Berglund, Å.: The operator 4.0: human cyber-physical systems & adaptive automation towards human-automation symbiosis work systems. In: Nääs, I., et al. (eds.) APMS 2016. IAICT, vol. 488, pp. 677–686. Springer, Cham (2016). https://doi.org/10.1007/978-3-319-51133-7_80
16. Romero, D., et al.: Towards an operator 4.0 typology: a human-centric perspective on the fourth industrial revolution technologies. In: proceedings of the International Conference on Computers and Industrial Engineering (CIE46), Tianjin, China, pp. 29–31 (2016)
17. Ruppert, T., Jaskó, S., Holczinger, T., Abonyi, J.: Enabling technologies for operator 4.0: a survey. Appl. Sci. **8**(9), 1650 (2018)
18. Thormann, C., Winkler, A.: Playing tic-tac-toe with a lightweight robot. In: Lepuschitz, W., Merdan, M., Koppensteiner, G., Balogh, R., Obdržálek, D. (eds.) RiE 2022. LNNS, vol. 515, pp. 150–160. Springer, Cham (2022). https://doi.org/10.1007/978-3-031-12848-6_14
19. Weber, T., et al.: Measures to tackle labour shortages: lessons for future policy. Eurofound (2023)
20. Yang, Y., Williams, A.B.: Improving human-robot collaboration efficiency and robustness through non-verbal emotional communication. In: Companion of the 2021 ACM/IEEE International Conference on Human-Robot Interaction, pp. 354–356 (2021)

Assessing Trustworthy Artificial Intelligence of Voice-Enabled Intelligent Assistants for the Operator 5.0

Alexandros Bousdekis[1(✉)], Gregoris Mentzas[1], Dimitris Apostolou[1,2], and Stefan Wellsandt[3]

[1] Information Management Unit (IMU), Institute of Communication and Computer Systems (ICCS), National Technical University of Athens (NTUA), Athens, Greece
{albous,gmentzas}@mail.ntua.gr

[2] Department of Informatics, University of Piraeus, Piraeus, Greece
dapost@unipi.gr

[3] BIBA - Bremer Institut Für Produktion Und Logistik GmbH at the University of Bremen, Bremen, Germany
wel@biba.uni-bremen.de

Abstract. The concept of Trustworthy Artificial Intelligence (TAI) focuses on the establishment of trust in AI systems' development, deployment, and use. In this realm, the European Commission (EC) developed the Assessment List for Trustworthy Artificial Intelligence (ALTAI) in order to enable the assessment of trustworthiness in the AI systems under development. Since this is an emerging topic, there is little evidence on how to apply ALTAI. In this paper, we present the application of ALTAI on a Digital Intelligent Assistant (DIA) for manufacturing. In this way, we aim at contributing to the enrichment of ALTAI applications and to the drawing of remarks regarding its applicability to diverse domains. We also discuss our responses to the ALTAI questionnaire, and present the score and the recommendations derived from the ALTAI web application.

Keywords: Trustworthy AI · voice assistant · AI ethics · Assessment List for Trustworthy Artificial Intelligence · ALTAI · Industry 5.0

1 Introduction

To maximize the benefits of Artificial Intelligence (AI), while at the same time mitigating its risks and dangers, the concept of Trustworthy AI (TAI) promotes the idea that individuals, organizations, and societies will be able to achieve the full potential of AI if trust is established in its development, deployment, and use [1]. The TAI concept has been studied in several works [2–6]. The increasing literature implies that ethics have been put at the core of the development of AI technologies [7], especially of those that incorporate predictive capabilities [4, 8]. In European Commission (EC)'s strategy, published in 2018, AI must be lawful, ethical, and robust [1, 9], and is defined as

Published by Springer Nature Switzerland AG 2024
M. Thürer et al. (Eds.): APMS 2024, IFIP AICT 729, pp. 220–234, 2024.
https://doi.org/10.1007/978-3-031-65894-5_16

"systems that display intelligent behaviour by analyzing their environment and taking actions – with some degree of autonomy – to achieve specific goals". In this context, EC developed the Assessment List for Trustworthy Artificial Intelligence (ALTAI), "a practical tool that helps business and organizations to self-assess the trustworthiness of their AI systems under development" [10]. Since this is an emerging topic, up to now there is little evidence on how to apply ALTAI [11].

In this paper, we present the application of ALTAI to a Digital Intelligent Assistant (DIA) for manufacturing. DIAs represent a new type of interaction between operators and machines in the context of Industry 5.0 aiming at establishing mutually beneficial relationships between smart technologies, and the Operator 5.0 [12]. The DIA was developed in the COALA ("COgnitive Assisted agile manufacturing for a LAbor force supported by trustworthy Artificial Intelligence") EU research project which provides a proactive and pragmatic approach to support operative situations characterized by high cognitive load, time pressure, and zero tolerance for quality issues. For more details on the COALA concepts and technologies, the reader may refer to [13–15].

The rest of the paper is organized as follows: Sect. 2 briefly reviews the main frameworks for TAI. Section 3 reviews the related works on applications of ALTAI. Section 4 discusses our responses to the ALTAI questionnaire, and presents the score and the recommendations derived from the ALTAI web application. Section 5 presents the main concluding remarks on the applicability of ALTAI, while Sect. 6 concludes the paper.

2 Review of Frameworks for Trustworthy AI

During the last years, several frameworks have arisen referred to as Beneficial AI [16], Responsible AI [17, 18], Ethical AI [2, 19], and Trustworthy AI [1, 20]. An overview of some of the most representative works is presented in Table 1. Despite their value for TAI realization, they exhibit two main limitations [6]: (i) Several TAI principles may conflict with each other, depending on the application cases; (ii) They are general and they do not provide sufficient guidance on how they are transferred into practice.

Table 1. Description of the main TAI frameworks.

Framework	Description
Asilomar AI Principles [16]	23 principles of beneficial AI, organized into three categories: research issues, ethics and values, and long-term issues
Montreal Declaration of Responsible AI [17]	10 ethical principles that promote the fundamental interests of people and groups, and 8 recommendations for responsible AI
UK AI Code [19]	5 principles for an ethical AI code, intended to position the UK as a future leader in AI
AI4People [2]	Synthesis of 6 frameworks, which resulted in 5 foundational principles for ethical AI, and a set of 20 action points
OECD Principles [20]	5 principles for the responsible stewardship of trustworthy AI

(*continued*)

Table 1. (*continued*)

Framework	Description
Governance Principles for the New Generation AI [18]	A framework and action guidelines for the governance of AI, based on 8 principles for the development of responsible AI
EU Ethics Guidelines for TAI [1]	4 principles and 7 key requirements for achieving TAI. An assessment list for the operationalization of the requirements

3 Related Works on Applications of the ALTAI Tool

Although the Ethics Guidelines for TAI [1] have received some criticism [21, 22], they emerge to be influential in EU, since they try to achieve an inclusive consensus of how societies can deal with the opportunities and challenges of AI. They serve as the basis for the regulation of AI, the EU AI Act [23], and the principles for TAI in EU [24]. Whilst the ALTAI list is not the only example of an AI impact assessment [25–27], its visibility benefits from the central role it plays in EU AI policy [28].

Table 2. Remarks from ALTAI applications.

Ref	Main Remarks
[11]	ALTAI variants should be developed for the various software lifecycle phases Domain-specific adaptations of ALTAI should evolve ALTAI should be reorganized to support readability
[29]	ALTAI should entail definitions of widely used terms ALTAI should have specific versions for different domains Overlaps and redundancies were found within/throughout the 7 requirements ALTAI should consider what questions are relevant to which development stage
[30]	The weighing of each question to the final scores is not clear to the user The ALTAI Polar diagram presents the results in an unappealing way for early-stage organizations who might score poorly to present their results publicly ALTAI does not consider the appropriate level of governance There is no consideration to the relative risk of AI systems in the assessment process The consequence of the prerequisite Fundamental Rights Impact Assessment (FRIA) failing would negatively affect the ALTAI score Some consideration should be given to the merits and demerits of the nature of the tool Organizations need to understand how they compare to their peers as well
[31]	ALTAI do not publish how each answer contributes to the final scores The ALTAI recommendations seem extensive, repeated, and difficult to understand The ALTAI score is difficult to be interpreted in terms of guidelines for improvement
[32]	ALTAI lacks a clear strategy toward various entities that can be affected by unintended harmful consequences of the AI systems Relevant ethical issues can be mapped out on how data moves across the AI systems The ALTAI risk-based approach can be reinforced to categorise the ethical risks and countermeasures in relation to the specific stakeholders
[28]	ALTAI considers AI as an ethical issue instead of a technology, or family of techniques ALTAI creates questions of applicability that are independent of the actual ethical and social consequences of the specific AI system under examination The focus on trustworthiness does not fully represent all the ethical and social concerns Limitations exist when applying an ex-ante instrument (i.e. ALTAI) at the research stage

Applications of ALTAI include: Advanced Driver-Assistance System [11], Early Warning System in Education [4], AI-supported Air Traffic Controller Operations [29], AI-based technologies for ageing and healthcare [30], neuroinformatics [28]. Table 2 presents the main derived remarks from such applications.

4 Application of ALTAI in Digital Intelligent Assistants for Manufacturing

4.1 Analysis and Discussion

The Ethics Guidelines for TAI [1], published by the High-Level Expert Group on Artificial Intelligence (AI HLEG), contains an Assessment List to help assess whether the AI system adheres to the seven requirements of TAI: (1) Human agency and oversight; (2) Technical robustness and safety; (3) Privacy and data governance; (4) Transparency; (5) Diversity, non-discrimination and fairness; (6) Societal and environmental well-being; (7) Accountability. ALTAI is a method to drive the self-assessment and requires interdisciplinary expertise as well as a continuous evaluation procedure. In this section, we present and discuss our responses to the ALTAI questionnaire for each requirement in order to reflect the discussion and the concluding remarks of the workshop among the technical partners. We pursued an intermediate and a final workshop in which we used the ALTAI prototype web-based tool. The participants were representatives of the technical partners of the COALA consortium who had been developing the technological solution.

Human Agency and Oversight. AI systems should support human autonomy and decision-making, as prescribed by the principle of respect for human autonomy. In this section, ALTAI asks to assess the AI system in terms of respect for human agency, as well as human oversight.

Human Autonomy. Human autonomy deals with the effect AI systems that are aimed at guiding, influencing or supporting humans in decision making processes. It also deals with the effect on human perception and expectation when confronted with AI systems that 'act' like humans and with the effect of AI systems on human affection, trust and (in)dependence. The COALA solution has been designed to interact, guide and support decisions by human end-users that affect humans. Through the DIA, it interacts with the operators, provides insights, supports the decisions as well as the training process of novice operators. The users know from the beginning that they interact with an AI system; thus, COALA could not generate confusion for the end-users on whether a decision, content, advice or outcome is the result of an AI algorithm. Moreover, COALA can potentially affect human autonomy by interfering with the end-user's decision-making process in an unintended way, since the outcomes of its functionalities are communicated to the user through the voice-enabled interface in a non-intrusive way. Although the risk is low, we have set up monitoring mechanisms in order to ensure that it will not cause over-reliance, since training is foreseen beforehand (e.g. the didactic concept and on-the-job training). In this sense, it is unlikely to manipulate human behavior. We have also taken measures to mitigate the risk of manipulation by also protecting the technology stack of the solution.

Human Oversight. This subsection helps to self-assess necessary oversight measures through governance mechanisms such as human-in-the-loop (HITL), human-on-the-loop (HOTL), or human-in-command (HIC) approaches. COALA is overseen by a mix of HITL and HIC approach. More specifically, the producer (manager or supervisor or administrator) controls how the employee should or can use the assistant. The user has control under these conditions. The manager has a complete control over the system. The employee (end user) has not a complete control but some degrees of freedom, depending also on their experience level. The humans that are involved in the use of the DIA have been given specific training on how to exercise oversight through the concept of AI-focused didactic concept [33]. We have also established detection and response mechanisms for undesirable adverse effects.

Technical Robustness and Safety. A crucial requirement for achieving TAI systems is their dependability and resilience. Technical robustness requires that AI systems are developed with a preventive approach against risks and that they behave reliably and as intended while minimising unintentional and unexpected harm. The questions in this section of ALTAI address four main issues: 1) security; 2) safety; 3) accuracy; and 4) reliability, fall-back plans and reproducibility. The COALA solution has not been certified for cybersecurity but it is compliant with various security standards that are inherent to the technology stack. This is achieved by the use of state-of-the-art technologies which are based on well-established technology standards. We use Keycloak in order to add authentication and secure services of the COALA components. Keycloak is based on standard protocols and provides support for OpenID Connect, OAuth 2.0, and SAML. In addition, COALA includes an anonymization component to fulfil privacy requirements. Therefore, it is not exposed to cyber-attacks, as it has also been verified by the IT departments of the use case partners. However, we have put measures in place to ensure its integrity, robustness and overall security. Finally, the users are informed about the duration of security coverage and updates.

General Safety. The COALA solution may potentially have adversarial, critical or damaging effects in case of risks or threats such as design or technical faults, defects, outages, attacks, misuse, inappropriate or malicious use. However, the probability of such cases is low because COALA does not automatically implement actions. The decision is finally taken by the operator. In each COALA business case, risks and risk levels have been defined from the very beginning of the project. These risks are continuously assessed throughout the evolution of the project and the progress of the development and deployment activities, and a specific process has been put in place in order to facilitate the consistent continuous risk assessment. Moreover, we have identified the possible threats to the AI system (design faults, technical faults, environmental threats) and the possible resulting consequences. We have also assessed risks related to possible malicious use, misuse or inappropriate use as well as on the safety criticality levels of their possible consequences. We have also assessed the dependency of critical system's decisions on its stable and reliable behavior.

Accuracy. COALA, as most of the AI systems, can potentially have critical, adversarial or damaging consequences. One cause of this can be a low level of accuracy which can lead to wrong guidance, misleading predictions, as well as inappropriate recommendations and advice. We have put in place measures to ensure that the data (including training data) used by the Data Analytics component is up to date, of high quality, complete and representative of the environment. This is also ensured by the data sources of the COALA use cases, as well as by the Data Management component which acquires and structures the data. We have also put in place a procedure to monitor and document COALA's accuracy by implementing mechanisms for evaluating the embedded algorithms and for providing interpretability insights. Further, these mechanisms consider whether the system's operation can invalidate the data or assumptions it was trained on, and how this might lead to adversarial effects. The results of the aforementioned accuracy evaluation mechanisms can be communicated to the end-users upon request either through the voice interface or through the Graphical User Interface (GUI).

Reliability, Fall-Back Plans and Reproducibility. COALA could potentially cause critical, adversarial or damaging consequences in case of low reliability and/or reproducibility. However, the probability of such cases is low. In any case, we have put in place procedures to monitor if the system meets the goals of the intended applications and whether specific contexts or conditions need to be taken into account to ensure reproducibility. We have also put in place verification and validation methods and documentation to evaluate and ensure different aspects of the system's reliability and reproducibility. Processes for the testing and verification of the reliability and reproducibility have been documented and operationalized. There have also been defined tested failsafe fallback plans to address COALA errors; they have been covered during the integration and have been validated in the context of the evaluation procedure. In addition, the Data Analytics component has embedded internal mechanisms for handling the cases where the system yields results with a low confidence score, while the voice interface has been subject to a UX study for chatbot breakdown assessment. All these activities were performed in close collaboration with the COALA use cases. COALA incorporates online continual learning in the sense of accommodating new knowledge while retaining previously learned experiences. In general, this is crucial for agents operating in changing environments and required to acquire, fine-tune, adapt, and transfer increasingly complex representations of knowledge. COALA tackles with this challenge in the following ways: (i) The didactic concept and the change management process teach and guide workers competencies when they collaborate with AI. It demonstrates how AI-specific worker education can help building trust in AI systems [33]; (ii) the Knowledge Management component captures best practices on the factory shop floor and facilitates knowledge acquisition, representation and inference [34]; (iii) the Data Analytics component incorporates ML algorithms that are capable of being updated, taking into account new data that are recorded [35].

Privacy and Data Governance. Closely linked to the principle of prevention of harm is privacy, a fundamental right particularly affected by AI systems. Prevention of harm to privacy also necessitates adequate data governance that covers the quality and integrity of the data used, its relevance in light of the domain in which the AI systems will be deployed, its access protocols and the capability to process data in a manner that protects privacy. In COALA, we have considered the impact of the AI system on the right to privacy, the right to physical, mental and/or moral integrity and the right to data protection.

Depending on the use case, we have established mechanisms that allow related flagging issues. There is the data anonymization service to protect the privacy of the workers and achieve General Data Protection Regulation (GDPR) compliance. In addition, in order to thoroughly implement the GDPR, we have defined a Data Protection Officer (DPO) role in the consortium from the very beginning of the project so that he is involved in all the phases of the COALA lifecycle. We have also adopted measures to enhance privacy by design and default (e.g. encryption, anonymisation), which are continuously assessed throughout the development phases. It should be noted that COALA does not use or process personal data (including special categories of personal data) when being trained and developed. Where applicable, we have adopted a policy for data minimization. We have also taken into account the right to withdraw consent, the right to object and the right to be forgotten in the COALA solution. We have considered the privacy and data protection implications of data collected, generated or processed as well as the privacy and data protection implications of non-personal training-data or other processed non-personal data.

Transparency. A crucial factor for achieving TAI is transparency which encompasses three elements: 1) traceability, 2) explainability and 3) open communication about the limitations of the AI system. Technical robustness requires that AI systems be developed with a preventive approach to risks and in a way that they reliably behave as intended while minimising unintentional and unexpected harm, and preventing unacceptable harm. This should also apply to potential changes in their operating environment or the presence of other agents (human and artificial) that may interact with the system in an adversarial manner. In addition, the physical and mental integrity of humans should be ensured.

Traceability. This subsection helps to self-assess whether the processes of the development of the AI system, i.e. the data and processes that yield the AI system's decisions, is properly documented to allow traceability, increase transparency and, build trust in AI in society. In COALA, we have put in place measures to continuously assess the quality of the input data to the AI system.

Explainability. This subsection helps to self-assess the explainability of the AI system. The questions refer to the ability to explain both the technical processes of the AI system and the reasoning behind the decisions or predictions that the AI system makes. Explainability is crucial for building and maintaining users' trust. AI-driven decisions should be explained and understood to those directly and indirectly affected, in order to allow contesting of such decisions. An explanation as to why a model has generated a particular output or decision is not always possible. These cases are referred as "black boxes" and require particular attention. In those circumstances, other explainability measures may be required, provided that the AI system as a whole respects fundamental rights. The degree to which explainability is needed depends on the context and the severity of the consequences of erroneous or otherwise inaccurate output to human lives. COALA explains the decisions and all its outcomes to the users through an explainability engine. Moreover, it incorporates a Large Language Model (LLM) which segments from a pdf a recommendation allowing the user going back where the answer comes from. COALA continuously surveys the users to ask them whether they understand the decisions of the AI system taking, at the same time, into account that the operator should not be disturbed by unnecessary detailed explanations.

Communication. This subsection helps to self-assess whether the AI system's capabilities and limitations have been communicated to the users in a manner appropriate to the use case at hand. This could encompass communication of the AI system's level of accuracy as well as its limitations. Since COALA is based on chatbot and voice-enabled technologies, in order to facilitate the interaction between humans and AI, the users are explicitly informed that they interact with an AI system and not with a human. We have established mechanisms to inform users about the purpose, criteria and limitations of the decisions generated by COALA. To do this, we use a "capabilities" intent explaining what the assistant can do. We use a "FAQ" intent pointing users at a learning nugget with basics about digital assistants. We use an "out of scope" intent to indicate when the assistant cannot answer/help the user (because training data contains example utterances that are out of scope). Regarding the LLM in particular, if a question is out of context (i.e. if semantic similarity is below a threshold), it does not provide answers. We communicated the technical limitations and potential risks of the AI system to end-users, since the results about the level of accuracy and/or error rates are available to be exposed to the user upon request. Moreover, we provided appropriate training material and disclaimers to users on how to adequately use the COALA system as part of a didactic concept. The evaluation of the COALA solution incorporated user tests for the aforementioned findings.

Diversity, Non-discrimination and Fairness. In order to achieve TAI, inclusion and diversity should be enabled throughout the entire AI system's life cycle. AI systems (both for training and operation) may suffer from the inclusion of inadvertent historic bias, incompleteness, and bad governance models. Identifiable and discriminatory bias should be removed in the collection phase where possible. AI systems should be user-centric and designed in a way that allows all people to use AI products or services, regardless of their age, gender, abilities or characteristics. Accessibility to this technology for persons with disabilities, which are present in all societal groups, is of particular importance.

Avoidance of Unfair Bias. In COALA, we have established a set of procedures to avoid creating or reinforcing unfair bias, both regarding the use of input data as well as for the algorithm design [37]. We have considered diversity and representativeness of end-users and subjects in the data. We research and use mostly open-source state-of-the-art technical tools in order to improve understanding of the data, model and performance. We assessed and put in place processes to test and monitor for potential biases during the entire lifecycle of COALA (e.g. biases due to possible limitations stemming from the composition of the used data sets - lack of diversity, non-representativeness). We have put in place educational and awareness initiatives to help system designers and developers be more aware of the possible bias they can inject in designing and developing the AI system. In this context, we have created an Ethics Board and we performed an ethics survey. Depending on the use case, we ensure a mechanism that allows for the flagging of issues related to bias, discrimination or poor performance. Moreover, we have established ways of communicating on how and to whom such issues can be raised yet, while we have also identified the subjects that could potentially be (in)directly affected by the AI system, in addition to the end-users. In COALA, we use the widely used generic definition of "fairness" without having elaborated on its instantiation to the requirements of the project. However, the COALA ethics survey provides the means to do this.

Accessibility and Universal Design. Particularly in business-to-consumer domains, AI systems should be user-centric and designed in a way that allows all people to use AI products or services, regardless of their age, gender, abilities or characteristics. Accessibility to this technology for people with disabilities, which are present in all societal groups, is of particular importance. In this context, we have ensured that Universal Design principles are taken into account during every step of the planning and development process, while we have also taken into account the impact on the potential end-users. We have also assessed whether the team of developers are engaged with the possible target end-users. The COALA project, as a European collaborative project, dictates the close collaboration between the technical partners and the use cases by its nature. Therefore, following an agile software development methodology, the use case partners were continuously interacting with the technical partners. The rest of the ALTAI questions are not applicable to the COALA solution; however, there is not such an option in the alternative responses. Consequently, our responses were given in a way to maintain their neutrality and to avoid affecting the resulting ALTAI recommendations to the degree this is feasible.

Stakeholder Participation. In order to develop TAI, it is advisable to consult stakeholders who may directly or indirectly be affected by the AI system throughout its life cycle. COALA, as a European collaborative research project, by nature includes the close collaboration between the use case partners and the technical partners in its design and development. In addition, the project included tasks dedicated to ethics and human-centric AI, while its governance included an Ethics Board. However, the stakeholders were mainly high-skilled industrials and developers, something which is not a representative case in AI software development.

Societal and Environmental Wellbeing. In line with the principles of fairness and prevention of harm, the broader society, other sentient beings and the environment should be considered as stakeholders throughout the AI system's life cycle. Ubiquitous exposure to social AI systems in all areas of our lives may alter our conception of social agency, or negatively impact our social relationships and attachment. While AI systems can be used to enhance social skills, they can equally contribute to their deterioration. This could equally affect peoples' physical and mental well-being. The effects of AI systems must therefore be carefully monitored and considered. AI systems should serve to maintain and foster democratic processes and respect the plurality of values and life choices of individuals.

Environmental Wellbeing. This subsection helps to self-assess the positive and negative impacts of the AI system on the environment. Measures to secure the environmental friendliness of an AI system's entire supply chain should be encouraged. COALA has not any potential negative impacts system on the environment taking into account the Context of the ALTAI questions. In this sense, the respective questions of the ALTAI questionnaire in this section were not applicable. We answered positively in order not to affect the resulting recommendations.

Impact on Work and Skills. This subsection helps self-assess the impact of the AI system and its use in a working environment on workers, the relationship between workers and employers, and on skills. The COALA solution has a major impact on human work and work arrangements, since it is aimed at supporting the operations on the shop floor in manufacturing environments. Furthermore, COALA adopts the didactic concept as well as learning nuggets in order to support a training approach for introducing changes through the advice of dialogs by the voice interface [33]. This approach results in measurable changes in workers behavior. Moreover, COALA implements an assistant function to support novice workers in their learning and working activities while reconfiguring and operating production lines [15, 34]. In order to prepare the implementation, deployment, and evaluation of the COALA concepts and technologies, we have informed and consulted the impacted workers and their representatives in advance in order to ensure that the work impacts are well understood. This is also part of the change management procedure that we have adopted [37]. We use a didactic concept to teach workers about challenges, capabilities, and risks of DIAs. We use a change management process focused on AI to prepare managers and workers for human-AI collaboration. To that extent, we have taken measures to counteract de-skilling risks. Workers keep focusing on knowledge-intensive tasks while the assistant takes over repetitive tasks (that are subject of de-skilling).

Accountability. The principle of accountability requires that mechanisms are put in place to ensure responsibility for the development, deployment and/or use of AI systems. This topic is closely related to risk management, identifying and mitigating risks in a transparent way that can be explained to and audited by third parties.

Auditability. This subsection helps to self-assess the existing or necessary level that would be required for an evaluation of the AI system by internal and external auditors. In applications affecting fundamental rights, including safety-critical applications, AI systems should be able to be independently audited. We have established mechanisms that facilitate the AI system's auditability.

Risk Management. Both the ability to report on actions or decisions that contribute to the AI system's outcome, and to respond to the consequences of such an outcome, must be ensured. Identifying, assessing, documenting and minimising the potential negative impacts of AI systems is especially crucial for those (in)directly affected. We have established an "AI ethics review board" to discuss the overall accountability and ethics practices, including potential unclear grey areas. We have also established processes for third parties (e.g. suppliers, end-users, subjects, distributors/vendors or workers) or workers to report potential vulnerabilities, risks or biases in the AI system. In addition, we have not put in place redress by design mechanisms in cases COALA significantly adversely affect individuals.

4.2 Results and Recommendations

The outcomes of the ALTAI tool are: (i) A visualisation of the self-assessed level of adherence of the AI system with the 7 requirements for TAI; and, (ii) Recommendations based on the answers to the questionnaire. Figure 1 depicts the results of the Assessment List in the form of a Polar diagram for the 7 requirements of ALTAI. Table 3 presents the resulting recommendations per ALTAI requirement. It should be noted that some questions are not applicable to the COALA solution or the alternative responses that are provided do not represent accurately the opinion of the COALA consortium. These issues inevitably affect the scores for the 7 ALTAI requirements. This fact dictates the addition of one more step in this self-assessment procedure, a validation of the resulting recommendations with regards to the scope of the AI system under examination.

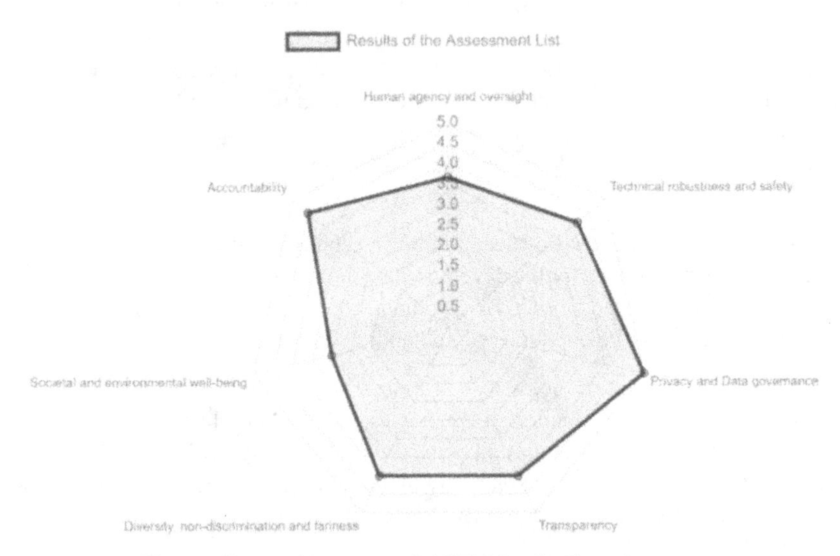

Fig. 1. The resulting score of ALTAI for the 7 requirements.

Table 3. ALTAI recommendations per requirement.

	Resulting ALTAI Recommendations for each out of the 7 Requirements
1	**Human agency and oversight**
No recommendation for this requirement.	
2	**Technical robustness and safety**
No recommendation for this requirement.	
3	**Privacy and Data Governance**
i.	Whenever possible and relevant, align the AI-system with relevant standards (e.g. ISO, IEEE) or widely adopted protocols for data management and governance.
4	**Transparency**
No recommendation for this requirement.	
5	**Diversity, non-discrimination and fairness**
i.	Your definition of fairness should be commonly used and should be implemented in any phase of the process of setting up the AI system.
ii.	Consider other definitions of fairness before choosing one.
iii.	Consult with the impacted communities about the correct definition of fairness.
iv.	Ensure a quantitative analysis to measure and test the applied definition of fairness.
v.	Establish mechanisms to ensure fairness in your AI system.
vi.	You should assess whether the AI system's user interface is usable by those with special needs or disabilities or those at risk of exclusion.
vii.	You should assess the risk of the possible unfairness onto the end-user's communities.
6	**Societal and environmental well-being**
No recommendation for this requirement.	
7	**Accountability**
i.	3rd party auditing can contribute to generate trust in the technology and the product itself. Additionally, it is a strong indication of adhering to industrial standards.
ii.	If AI systems are increasingly used for decision support or for taking decisions themselves, it has to be made sure these systems are fair in their impact on people's lives, that they are in line with values that should not be compromised and able to act accordingly, and that suitable accountability processes can ensure this.
iii.	A risk management process should include new findings since initial assumptions about the likelihood of occurrence for a specific risk might be faulty.
iv.	Acknowledging that redress is needed when incorrect predictions can cause adverse impacts to individuals is key to ensure trust.

5 Conclusions and Remarks on the ALTAI Tool

In this paper, we presented ALTAI's application on a DIA for the Operator 5.0, but we also conclude to some remarks:

1. ALTAI has been designed for end products; it does not address the various phases of software development lifecycle. This could ensure the compliance of the AI system with ethics guidelines during its development, but also enable early improvements in the context of agile software development. Therefore, there is the need to create ALTAI variants for the various development phases or to connect some questions to different phases to treat them differently (e.g. reducing their scoring weight). In addition, the assessment could outline potential risks if the system developers do not address a shortcoming.
2. ALTAI incorporates generic questions aiming at addressing every AI system. However, for example, in the case of COALA, the AI system refers to a manufacturing environment, i.e. a professional environment with expert and qualified users. In contrast, an AI system referring to a different application domain or even more a generic audience of end-users, may have different requirements. Therefore, there is the need for domain-specific adaptations of ALTAI with context-specific questions.
3. ALTAI considers as "AI system" the software and does not treat it as a socio-technical system, potentially leading to disregard of unforeseen challenges. Moreover, the way artificial agents learn may not be understandable to humans, making uncertainty and unpredictability present to a higher degree than in traditional systems.
4. No alternative response is accurate for some questions. Some responses should have been "not applicable", "not yet", or "to some extent" instead of the options "yes / no / don't know". Given the limitation (1), options such as "not yet" would indicate that something has not been implemented yet, but it has been planned. Given the limitation (2), ALTAI in each current form could have provided options such as "not applicable". These affect the resulting assessment score and the recommendations, some of which may not be applicable. To this end, these results need further validation during the development activities. In order to provide different response options, ALTAI could adopt a Likert-scale approach.
5. While ALTAI covers the seven key requirements of TAI, the current structure hinders its applicability. Several sections of ALTAI are unbalanced, since some of them cover large parts of the assessment list, while there are overlapping and redundant questions. ALTAI could be reorganized to support readability.
6. In the long-term, the tool could be enhanced with generative AI, e.g. generating stories about a hypothetical market introduction of the AI system with a shortcoming or good rating resulting in story part to explain it. E.g., a shortcoming in "privacy and data governance" could lead to a story part where personal data is misused.

Acknowledgements. This work is funded by the European Union's H2020 project COALA (https://www.coala-h2020.eu/) (Grant agreement No 957296). The work presented here reflects only the authors' view and the European Commission is not responsible for any use that may be made of the information it contains.

References

1. AI HLEG.: Ethics Guidelines for Trustworthy AI. Brussels: European Commission (2019)

2. Floridi, L., et al.: AI4People-An ethical framework for a good AI society: opportunities, risks, principles, and recommendations. Mind. Mach. **28**(4), 689–707 (2018)
3. Wiens, J., et al.: Do no harm: a roadmap for responsible machine learning for health care. Nat. Med. **25**(9), 1337–1340 (2019)
4. Baneres, D., Guerrero-Roldán, A.E., Rodríguez-González, M.E., Karadeniz, A.: A predictive analytics infrastructure to support a trustworthy early warning system. Appl. Sci. **11**(13), 5781 (2021)
5. Vilone, G., Longo, L.: Notions of explainability and evaluation approaches for explainable artificial intelligence. Inf. Fusion **76**(1), 89–106 (2021)
6. Thiebes, S., Lins, S., Sunyaev, A.: Trustworthy artificial intelligence. Electron. Mark. **31**(2), 447–464 (2021)
7. Georgieva, I., Lazo, C., Timan, T., van Veenstra, A.F.: From AI ethics principles to data science practice: a reflection and a gap analysis based on recent frameworks and practical experience. AI Ethics, 1–15 (2022)
8. Kazim, E., Koshiyama, A.: AI assurance processes (2020). SSRN 3685087
9. Smuha, N.A.: The EU approach to ethics guidelines for trustworthy artificial intelligence. Comput. Law Rev. Int. **20**(4), 97–106 (2019)
10. Ala-Pietilä, P., et al.: The assessment list for trustworthy artificial intelligence (ALTAI). European Commission (2020)
11. Borg, M., et al.: Exploring the assessment list for trustworthy AI in the context of advanced driver-assistance systems. In: 2021 IEEE/ACM 2nd International Workshop on Ethics in Software Engineering Research and Practice (SEthics), pp. 5–12. IEEE (2021)
12. Romero, D., Stahre, J.: Towards the resilient operator 5.0: the future of work in smart resilient manufacturing systems. Procedia CIRP **104**, 1089–1094 (2021)
13. Freire, S.K., et al.: Lessons learned from designing and evaluating CLAICA: a continuously learning AI cognitive assistant. In: Proceedings of the 28th International Conference on Intelligent User Interfaces, pp. 553–568 (2023)
14. Bousdekis, A. et al.: Human-AI collaboration in quality control with augmented manufacturing analytics. In: Dolgui, A., Bernard, A., Lemoine, D., von Cieminski, G., Romero, D. (eds.) Advances in Production Management Systems. Artificial Intelligence for Sustainable and Resilient Production Systems. APMS 2021. IFIP Advances in Information and Communication Technology, vol. 633, pp. 303–310. Springer(2021). https://doi.org/10.1007/978-3-030-85910-7_32
15. COALA project - Deliverable 2.5. Digital intelligent assistant core for manufacturing demonstrator– version 2. https://ncld.ips.biba.uni-bremen.de/s/S8W8dmm2ei4RbGw. Accessed 03 June 2024
16. Future of Life Institute: Asilomar AI Principles (2017). https://futureoflife.org/ai-principles/. Accessed 14 Mar 2024
17. Université de Montréal: Montreal Declaration for a Responsible Development of AI (2017). https://www.montrealdeclaration-responsibleai.com/the-declaration. Accessed 14 Mar 2024
18. Chinese National Governance Committee for the New Generation Artificial Intelligence.: Governance Principles for the New Generation Artificial Intelligence–Developing Responsible Artificial Intelligence (2019). https://www.chinadaily.com.cn/a/201906/17/WS5d07486ba3103dbf14328ab7.html. Accessed 14 Mar 2024
19. UK House of Lords. AI in the UK: ready, willing and able? (2017). https://publications.parliament.uk/pa/ld201719/ldselect/ldai/100/10002.htm. Accessed 14 Mar 2024
20. OECD: OECD Principles on AI (2019). https://www.oecd.org/going-digital/ai/principles/. Accessed 14 Mar 2024
21. Metzinger, T.: Ethics washing made in Europe. Der Tagesspiegel (2019)
22. Veale, M.: A critical take on the policy recommendations of the EU high-level expert group on artificial intelligence. Euro. J. Risk Regul. **11**(1) (2020)

23. European Commission: Proposal for a Regulation on a European approach for Artificial Intelligence (No. COM (2021) 206 final). European Commission, Brussels (2021)
24. Stix, C.: The ghost of AI governance past, present and future: AI governance in the European Union. arXiv preprint arXiv:2107.14099 (2021)
25. Reisman, D., Schultz, J., Crawford, K., Whittaker, M.: AI Now institute: algorithmic impact assessments: a practical framework for public agency accountability algorithmic impact assessments: a practical framework for public agency. AI Now **9** (2018)
26. IEEE: IEEE 7010–2020—IEEE Recommended Practice for Assessing the Impact of Autonomous and Intelligent Systems on Human Well-Being (Standard). IEEE (2020)
27. Zicari, R.V., et al.: Z-Inspection®: a process to assess trustworthy AI. IEEE Trans. Technol. Soc. **2**(2), 83–97 (2021)
28. Stahl, B.C., Leach, T.: Assessing the ethical and social concerns of artificial intelligence in neuroinformatics research: an empirical test of the European Union assessment list for trustworthy AI (ALTAI). AI Ethics **3**(3), 745–767 (2023)
29. Stefani, T., et al.: Applying the assessment list for trustworthy artificial intelligence on the development of AI supported air traffic controller operations. In: 2023 IEEE/AIAA 42nd Digital Avionics Systems Conference (DASC), pp. 1–9. IEEE (2023)
30. Rajamäki, J., Gioulekas, F., Rocha, P.A.L., Garcia, X.D.T., Ofem, P., Tyni, J.: ALTAI tool for assessing ai-based technologies: lessons learned and recommendations from SHAPES pilots. Healthcare **11**(10), 1454, MDPI (2023)
31. Radclyffe, C., Ribeiro, M., Wortham, R.H.: The assessment list for trustworthy artificial intelligence: a review and recommendations. Front. Artif. Intell. **6**(1), 1020592 (2023)
32. Gavornik, A., Podroužek, J., Mesarcik, M., Solarova, S., Oresko, S., Bielikova, M.: Utilising the Assessment List for Trustworthy AI: Three Areas of Improvement. ceur-ws.org (2022)
33. COALA project - Deliverable 3.6. AI-focused Didactic Concept for Factory Workers – final. https://ncld.ips.biba.uni-bremen.de/s/Wy7ywFBFjLw8oKx. Accessed 03 June 2024
34. COALA project - Deliverable 3.4. Cognitive Advisor Service – Version 2. https://ncld.ips.biba.uni-bremen.de/s/LkgM8BLiMmmgn7q. Accessed 03 June 2024
35. Fikardos, M., Lepenioti, K., Bousdekis, A., Bosani, E., Apostolou, D., Mentzas, G.: An automated machine learning framework for predictive analytics in quality control. In: Kim, D.Y., von Cieminski, G., Romero, D. (eds.) Advances in Production Management Systems. Smart Manufacturing and Logistics Systems: Turning Ideas into Action. APMS 2022. IFIP Advances in Information and Communication Technology, vol. 663, pp. 19–26. Springer, Cham (2022). https://doi.org/10.1007/978-3-031-16407-1_3
36. COALA project - Deliverable 7.6. Report on the application of ethical principles for AI in manufacturing – Final. https://ncld.ips.biba.uni-bremen.de/s/9PQPpMTgSpHrtqQ. Accessed 03 June 2024
37. COALA project - Deliverable 5.3. Change management process for human – AI Collaboration – Final. https://ncld.ips.biba.uni-bremen.de/s/s6nRtrNbGKsTppa. Accessed 03 June 2024

A Bibliometric Perspective of Integrating Labor Flexibility in Workload Control

Alireza Ahmadi[1,2](✉), Alessandra Cantini[1], and Alberto Portioli Staudacher[1]

[1] Department of Management, Economics, and Industrial Engineering, Politecnico Di Milano, Piazza Leonardo da Vinci 32, 20133 Milan, Italy
Alireza.ahmadi@polimi.it

[2] Escuela Técnica Superior de Ingenieros Industriales, Universidad Politécnica de Madrid, C. de José Gutiérrez Abascal, 2, Chamartín, 28006 Madrid, Spain

Abstract. In an era assessed by quick technological advancement and shifting work paradigms, the integration of labor flexibility (LF) within Workload Control (WLC) systems presents a critical yet underexplored facet of industrial operation. This research embarks on a comprehensive bibliometric and systematic literature network analysis crossing a decade of studies from 2014 to 2024, uncovering pivotal contributions and identifying prevalent themes in LF and WLC. Although our inquiry reveals an intensifying interest in this field, particularly during the COVID-19 pandemic, it discloses a noticeable research shortfall in empirical explorations of LF's role within various order release methodologies. Addressing this gap, the study brings to light the growing importance of human dynamics, such as learning curves and worker heterogeneity, in optimizing WLC. Synthesizing the most significant scholarly works, the paper points out the urgency of adopting cross-disciplinary approaches to enrich future research attempts.

Keywords: Labor Flexibility · Workload Control · Bibliometric Analysis

1 Introduction

LF is essential in management and organizational studies. It means a company can adapt its workforce to changing business conditions, market demands, and technology. LF helps companies operate more efficiently, cut costs, and stay competitive in today's fast-paced business environment [1]. Integrating cyber-physical systems, IoT, and AI has transformed manufacturing processes in intelligent manufacturing. Smart factories manage interconnected operations in real time, enhancing productivity and adaptability. This shift requires a workforce capable of managing complex systems and adapting to new technologies. Continuous training programs and agile workforce planning are essential for labor management in intelligent manufacturing [2–6].

Modern industries require LF beyond task handling, including rapid skill acquisition and role adaptability. Broad-skilled workers can quickly adapt to new technologies and product changes, improving operational resilience and agility. Companies with a

Published by Springer Nature Switzerland AG 2024
M. Thürer et al. (Eds.): APMS 2024, IFIP AICT 729, pp. 235–250, 2024.
https://doi.org/10.1007/978-3-031-65894-5_17

broadly trained workforce are more flexible than those with highly specialized skills, which improves resource utilization, manufacturing times, and customer service. Japan's manufacturing sector found that LF is necessary for flexible manufacturing systems, emphasizing the need for multi-skilling in the workforce [7, 8]. However, a comprehensive bibliometric analysis that integrates various aspects of LF within WLC frameworks is lacking. The literature shows poorly integrated LF and WLC strategies to improve workforce allocation and production. This gap requires research highlighting the benefits of LF, such as increased adaptability, reduced production lead times, and improved operational performance, and explaining how to fully leverage these benefits [7–11].

Fundamental theories and models, such as the Theory of Constraints (TOC), WLC, and the Lancaster University Management School (LUMS) methodology, form the underpinnings for comprehending LF within WLC frameworks [11, 12]. These paradigms traditionally visualize LF with workflow and workload structuring but do not offer a comprehensive model that summarizes the dynamic interaction among LF, production planning, and WLC [2, 3, 12]. Furthermore, LF goes above just task handling; it captures the workforce's capacity for rapid skill acquisition and role adaptability, which is crucial in modern industries. Firms with broad-skilled workers can quickly respond to new technologies and product changes, underscoring the importance of LF for operational resilience and agility [13–15]. Historically, firms focusing on a broadly trained workforce have demonstrated greater flexibility than those with highly specialized skills [7]. This flexibility aids in efficient resource utilization, reduces manufacturing time, and improves customer service. Notable research from Japan's manufacturing sector indicates that LF is vital for the success of flexible manufacturing systems, highlighting the significance of multi-skilling in the workforce [8]. To put it in a nutshell, this study not only offers a foundational insight into the role of LF within WLC but also provides the first comprehensive overview of the state of the art literature on this analyzed topic, highlighting a significant research gap that needs addressing. There is a pressing need to create multidisciplinary frameworks that effectively implement LF. This approach is crucial to ensure that businesses are not just surviving but actively thriving in the dynamic and competitive arena of the manufacturing sector. [2, 3, 9, 12].

Moreover, to the best of the authors' knowledge, there is still a lack of a comprehensive overview in the literature concerning the initial stages of LF integration within WLC systems and a bibliometric review on this specific subject. Understanding the existing studies on LF in WLC, mainly focusing on the early stages of integrating LF strategies, could significantly contribute to two primary areas. Firstly, it allows for the organization and recognition of the main contributions, influential authors, journals, and countries in this research area. Secondly, it aids in identifying both current and emerging research trends within this realm, laying a solid foundation for future studies. Therefore, based on a Systematic Literature Network Analysis (SLNA), this paper presents a bibliometric review focused on LF in WLC. This review explores the critical themes, methodologies, and dialogues that have shaped the field from 2014 to 2024. The review aims to address these research questions (RQ):

RQ1: Who are the most productive and influential countries, journals, and authors, and what are the pivotal contributions in the literature on LF within WLC?

RQ2: What are the primary themes and research streams driving the development of LF research within WLC?

The goals of this paper are threefold. First, the existing literature on LF in WLC should be collected, emphasizing the initial phase of LF strategy implementation. Second, the key contributors in this field will be identified by analyzing publication and citation counts and proposing an innovative graphical descriptive tool. Lastly, the authors examine past and current research themes related to LF in WLC, focusing on author keywords and their co-occurrences. This study will provide descriptive metrics on the reviewed literature, offering an overview of the knowledge landscape in this area. The paper aims to set the stage for future research endeavors in this domain, equipping researchers with insights to identify potential gaps and suggest new study avenues. The paper is sorted out as follows: Sect. 2 outlines the theoretical background to understand this study. Specifically, it summarized the results of a preliminary SLNA, which confirms the literature gap addressed in the present study, thereby justifying how this work can contribute to the extant research. Section 3 provides the method for conducting the bibliometric review. Section 4 displays the results of the bibliometric review. Section 5 closes the paper with a discussion of the findings.

2 Theoretical Background

To confirm the literature gap addressed in this study, a preliminary SLNA was conducted to search for existing literature contributions, providing a comprehensive overview of the research on LF in WLC. The SLNA was conducted on March 2nd, 2024, by searching scientific contributions on the Scopus database, considered the best search engine for scientific journal coverage [18]. Initially, contributions including (in the title, abstract, or keywords) keywords related to LF were investigated utilizing the following search query:

(TITLE-ABS-KEY ("labo allocation*" OR "labo* assignment*" OR "work* allocation*" OR "work* assignment*" OR "flexible staffing" OR "staffing flexibility" OR "employee scheduling" OR "work scheduling" OR "shift assignment*" OR "shift flexibility" OR "resource* allocation*" OR "allocation* resource*" OR "dynamic staffing" OR "dynamic labo* allocation*" OR "skill-based assignment*" OR "skill-based allocation*" OR "cross-training" OR "demand-driven staffing" OR "demand-driven labo* allocation*" OR "lab* flexibility" OR "work force flexibility" OR "work* flexibility" OR "manpower movement*" OR "manpower movement*" OR "flexible staff*" OR "staff* flexibility" OR "staff scheduling" OR "flexible employee" OR "employee flexibility" OR "work* flexibility" OR "flexible work arrangements" OR "operational flexibility" OR "human resource flexibility" OR "multi-skilling" OR "labor adaptability" OR "rotational shifts" OR "adaptive staffing" OR "staff adaptability" OR "workforce adaptability" OR "rotating assignments").*

This search query yielded 166,301 documents. To estimate better the studies offering literature reviews on LF, our second attempt was using the keywords related to the literature review using the following search query:

AND (TITLE-ABS-KEY ("Bibliographic Analysis" OR "systematic analysis" OR "Review" OR "Systematic literature Analysis" OR "Bibliometric Analysis")).

This combination of search queries yielded 11,615 documents. Aiming to extract all existing contributions in the analyzed domain, no filter was used on the papers' publishing date, and the subject area was inserted. Then, only Articles, Conference Papers, and Reviews were included in the final Publication Stages, considering only English-written documents. In this way, 10,339 results were found. Finally, we used different queries related to WLC to exclude all the papers that were not related to LF in WLC (which is the topic of this research) using the following query:

AND (TITLE-ABS-KEY ("workload control" OR "workload-control" OR "work-load control" OR "work load control" OR "workload balancing" OR "workload management" OR "material flow control" OR "task management" OR "capacity management" OR "load leveling" OR "load balancing")).

The papers that were filtered obtained 150 results. The number of first-sight contributions identified may appear high. However, by consulting the title, abstract, and keywords of these papers, it emerged that, in contrast to the integration of LF and WLC, few studies focused on a literature review. For example, Costa et al. [16] research primarily focuses on the efficiency of production control methods in hybrid MTO-MTS through simulation. It discusses bottleneck-based versus load-based release models without addressing LF or the integration of such flexibility within WLC frameworks. The paper's emphasis on mechanical aspects of production control, such as bottleneck management and load-based release models, does not align with the specific inquiry into LF, hence its exclusion. Bertolini et al. [17] investigate the performance of hybrid production planning systems using discrete event simulation but do not delve into labor aspects of integrating LF within these systems. The focus is strictly on the technical comparison of production systems, which diverges from the labor-centric analysis required for understanding LF within WLC.

Moreover, other research topics emerged using synonyms of main keywords unrelated to the topic of interest (e.g., load leveling, load balancing, workload management, workload balancing, material flow control, task management, and capacity management). Therefore, it was considered more appropriate to manually select the collected papers based on the consulted titles, abstracts, and keywords, thus removing the documents not concerning the topic of our interest. After the manual selection, 2 documents remained (excluding 148 papers), showing a greater interest of researchers in LF within WLC but not related to providing any literature review on this topic. Figure 1 summarizes the following screening process based on 3 exclusion criteria. This bibliometric analysis addresses the gaps in existing studies by expanding the scope beyond specific settings like Industry 4.0 and pure flow shops to include diverse manufacturing environments. It integrates theoretical insights and empirical applications to overview LF and WLC comprehensively. This approach enhances the accuracy of research findings and identifies under-researched areas and emerging trends. By utilizing bibliometric methods, this analysis ensures robust and reliable findings that bridge existing research gaps and guide future investigations into effectively integrating LF with WLC across various industrial contexts.

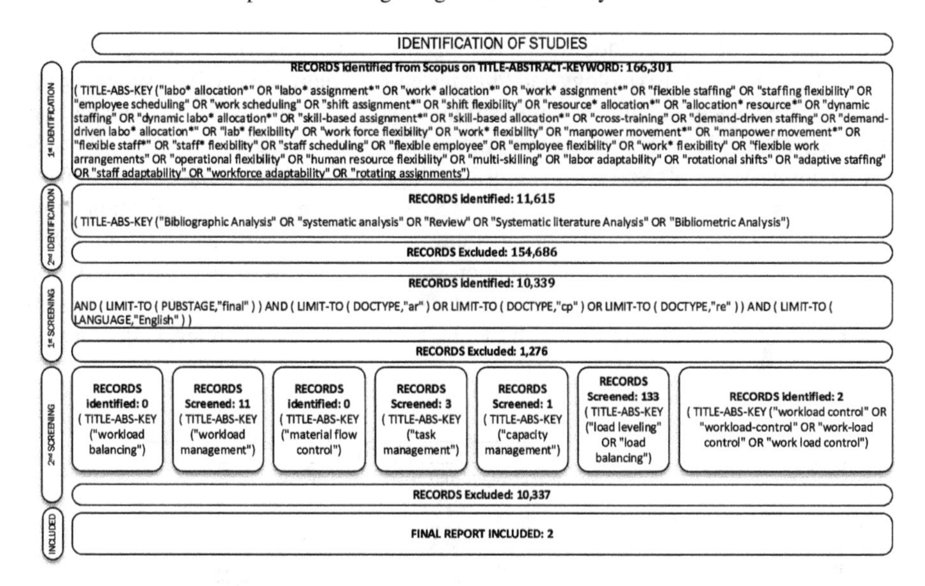

Fig. 1. Preliminary SLNA to justify the literature gap addressed in this study.

3 Methodology

A systematic bibliometric analysis was performed to fill the WLC LF literature gap. Scopus, known for its extensive scientific journal coverage, was used for the systematic search on April 1st, 2024. A comprehensive WLC literature search started. [18]. The primary search query employed was:

(TITLE-ABS-KEY("workload control" OR "workload-control" OR "work-load control" OR "work load control" OR "lab flexibility" OR "wrok force flexibility" OR "work* flexibility" OR "manpower movement" OR "manpower movement").*

This extensive query captured all WLC-related documents for diverse terms. A staggering 2,103 documents were found in the initial search, indicating a large literature base. To target WLC LF, a refined search query was used:

AND TITLE-ABS-KEY("labo allocation*" OR "labo* assignment*" OR "work* allocation*" OR "work* assignment*" OR "flexible staffing" OR "staffing flexibility" OR "employee scheduling" OR "work scheduling" OR "shift assignment*" OR "shift flexibility" OR "resource* allocation*" OR "allocation* resource*" OR "dynamic staffing" OR "dynamic labo* allocation*" OR "skill-based assignment*" OR "skill-based allocation*" OR "cross-training" OR "demand-driven staffing" OR "demand-driven labo* allocation*" OR "lab* flexibility" OR "work force flexibility" OR "work* flexibility" OR "manpower movement*" OR "manpower movement*" OR "flexible staff*" OR "staff* flexibility" OR "staff scheduling" OR "flexible employee" OR "employee flexibility" OR "work* flexibility" OR "flexible work arrangements" OR "operational flexibility" OR "human resource flexibility" OR "multi-skilling" OR "labor adaptability" OR "rotational shifts" OR "adaptive staffing" OR "staff adaptability" OR "workforce adaptability" OR "rotating assignments")).*

This tailored search yielded 1,371 documents, indicating significant LF in this domain. Pharmacology, Toxicology and Pharmaceutics, Neuroscience, Veterinary, Earth and Planetary Sciences, Chemical Engineering, Chemistry, Biochemistry, Genetics and Molecular Biology, Health Professions, Agricultural and Biological Sciences, Physics and Astronomy, Materials Science, Energy, Nursing, Arts and Humanities, Environmental Science, Psychology, Mathematics, and Medicine were excluded. Publication dates from 2014 to 2024 were restricted to ensure relevance and focus. Articles and Conference Papers were included in the final publication stage, reviews were not considered, and English-language documents were considered. This reduced the entries to 440. This step was essential to align search results with LF and WLC, yielding 440 relevant documents (Fig. 2).

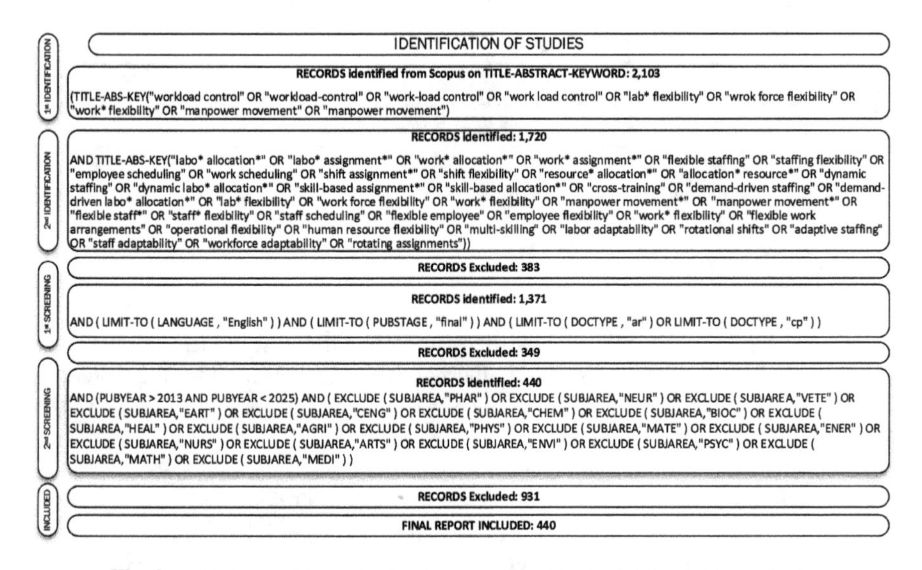

Fig. 2. SLNA to achieve the database analyzed in the bibliometric analysis.

The authors collected WLC LF research for bibliometric analysis using this SLNA. The bibliometric analysis identified major themes, contributors, and research trends without searching for review-specific literature. This refined biblio-metric analysis would lay the groundwork for a systematic and informed bibliographic literature review. This study used bibliometric network analysis to examine WLC LF trends and research areas. RQ1 was answered by examining article publication and citation rates. This showed us which countries, journals, and authors write most about this topic and are most influential [19]. The most productive ones have the most publications and are good starting points for learning about WLC LF. Other researchers frequently use and cite the most influential ones. The authors used Cantini et al.'s unique graphical tool, Qualitative Authors' Relevance Assessment (QARA), to better understand these influential authors and their key papers [20]. The authors answered RQ2 with co-word network analysis [21]. This involved examining authors' keywords and common word combinations. This clarifies research themes and trends. The authors analyzed data using Microsoft Excel™, Bibliometrics (an R-tool), and VOSviewer. We utilized Microsoft Excel™ and Bibliometrics to gather publication and citation statistics for RQ1. VOSviewer and Bibliometrics examined authors' keywords, relationships, and research trends (RQ2).

4 Results

4.1 Outcome Related to RQ1: Productivity and Influence in LF and WLC

The database contains 440 articles (87% articles and 13% conference papers) by 1233 authors published in 316 journals between 2014 and 2024. The average paper contains 13.7 citations. Figure 3 shows the annual citation count (blue line) and publication temporal distribution (gray histogram), indicating that researchers have been aware of WLC and LF for a decade. Contrary to expectations, the literature on this topic is scarce, particularly now, with an annual growth rate of −4.59%. After peaking in 2022 (nearly 60% of papers published), the publication trend has slowed over the last three years, indicating waning scientific interest. Since the search query was conducted in April, only 20 articles for 2024 have been displayed, indicating that more publications may appear by the end of the year. The citation curve peaked in 2018, indicating that significant contributions occurred that year.

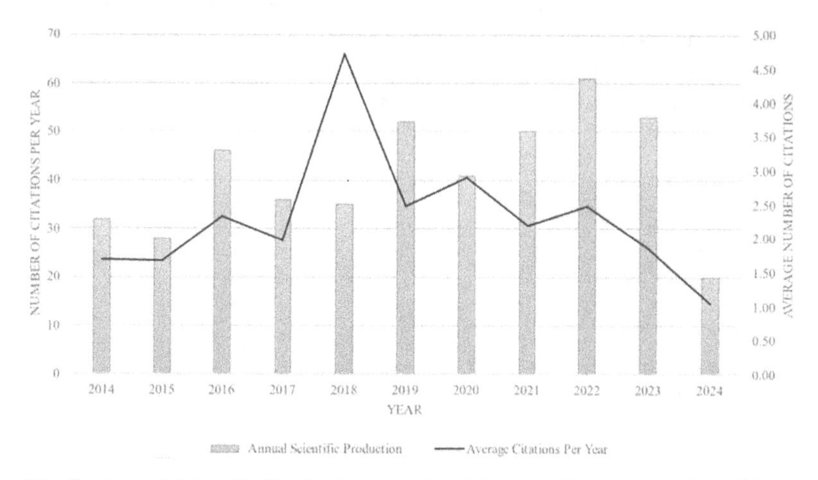

Fig. 3. Annual Scientific Production (gray) and Average Citations Per Year (blue).

RQ1 was answered by studying the geographical distribution of publications and citations to identify the most productive and influential countries. Figure 4 shows countries' productivity by WLC LF publications. The most productive countries, the USA, China, Italy, Australia, India, and the UK, have darker colors and 254, 121, 101, 99, 81, and 78 publications. Figure 3 shows that Eastern and Western countries published the analyzed field, while Africa contributed little to research development. However, the USA, UK, China, Italy, India, and the Netherlands are the most influential countries, with

688, 524, 494, 440, 365, and 326 citations, respectively. Five of six countries' leaders in both fields are on the list of most productive and influential. In this case, countries with many publications provide the most interesting scientific contributions, according to other researchers. This consideration does not necessarily affect authors and journals, so this study examined their productivity and influence. It showed the importance of not limiting literature analysis to the most productive countries in the field, and investigating other countries could be interesting.

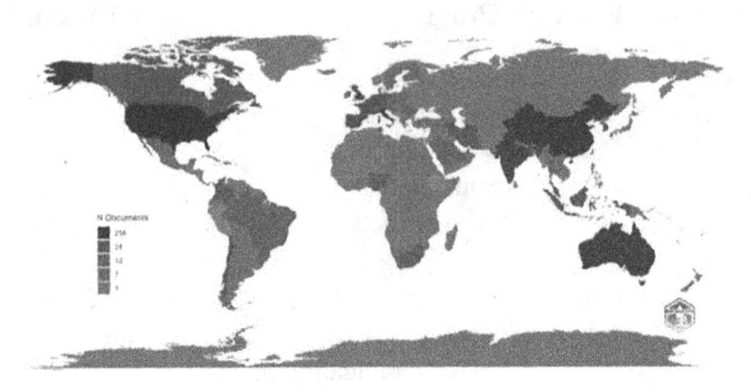

Fig. 4. Total number of publications per country.

Three analyses determined the most productive and influential journals. Bradford's Law identified relevant journals [22]. Sort topical journals by publication count in descending order. If successive journal zones have the same number of papers on a topic, the geometric series 1: *n*:*n*2:*n*3:... Forms.

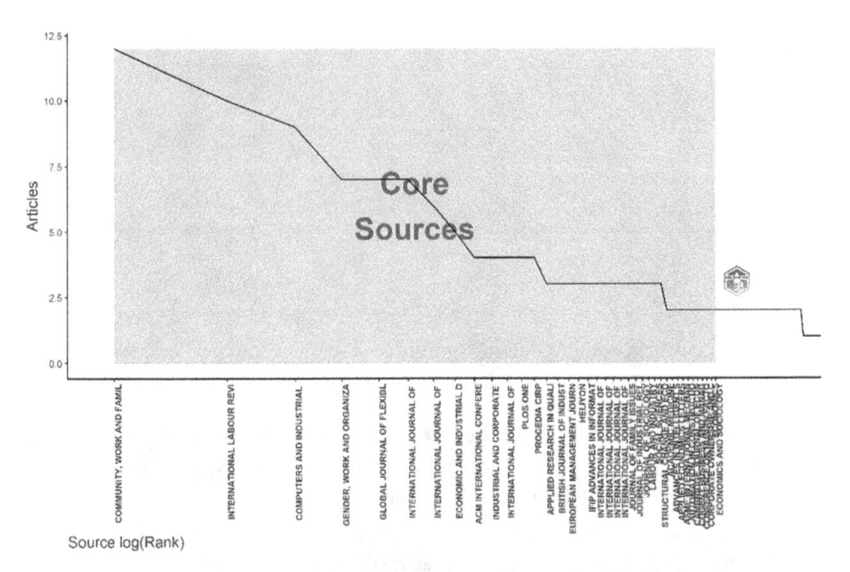

Fig. 5. Most productive journals, according to Bradford's Law.

Figure 5 demonstrates that Bradford's first zone (gray rectangle) contains the most productive journals (core sources): Community, Work, and Family (CW&F) (12 publications), International Labor Review (ILR) (10 publications), Computers and Industrial Engineering (C&IE) (9 publications), Gender, Work, and Organization (GWO) (7), Global Journal of Flexible Systems Management (GJFSM) (7), International Journal of Human Resource Management (IJHRM) (7), International Journal of Production Research (IJPR) (6), Economic and Industrial Democracy These core sources (63 out of 440 journals) account for one-seventh (14%) of the analyzed database. In a second analysis, after the most productive journals were published, the publication trend of the top six core sources was determined (Fig. 6). It shows no high-persistence journal with consistent publication over the years. However, CW&F, ILR, C&IE, and IJHRM confirmed their importance, indicating a significant interest in the analyzed topic over the last decade. In contrast, GWO demonstrated a more recent interest, producing only 7 papers in the last three years. Because Bradford's Law identifies the most productive journals but not the most influential ones, a third analysis was performed to determine the journal's average number of Citations Per Publication (CPP, Eq. 1).

$$CPP = \frac{Total Number of Citations}{Total Number of Publications} \tag{1}$$

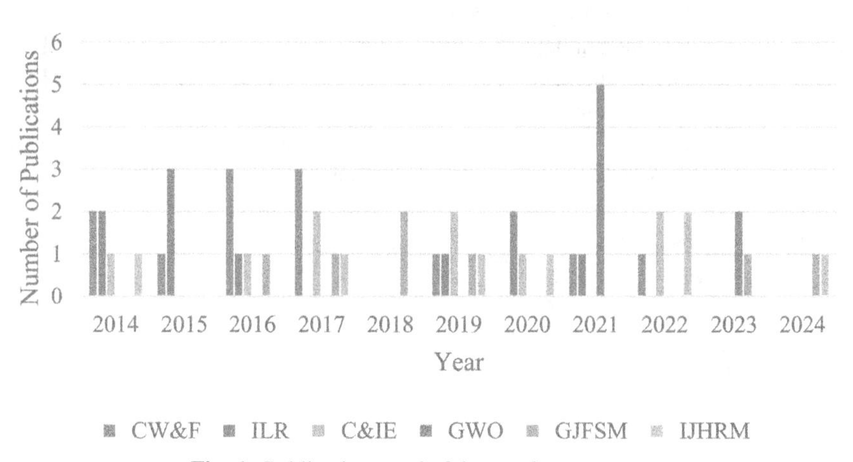

Fig. 6. Publication trend of the top 6 core sources.

As a result, the top ten most influential sources were identified as having the highest CPP. According to Table 1, two aspects are worth noting. First, the CPP analysis allowed us to identify the field's most influential journals and papers. For example, EMJ's contribution was highlighted, with 199 citations over ten years, and the peak in citations was confirmed in Fig. 3 in 2018. Second, none of the journals listed in Table 1 appear in Figs. 4–5, demonstrating that the most significant literary contributions were not published in the core sources and emphasizing the distinction between the most productive and influential journals.

Finally, the most productive and influential authors were identified by considering their publications and CPPs and proposing the novel graphical descriptive tool QARA

Table 1. Most influential journals based on CPP.

Element	Number of citations	Number of publications	CPP
CW&F	247	12	21
C&IE	213	9	24
GJFSM	173	7	25
IJHRM	113	7	16
IJPR	168	6	28
ILR	119	10	12
EID	88	5	18
GWO	38	7	5
BJIR	50	3	17
EMJ	199	3	66

developed by Cantini et al. [20], which allows for summarizing key information on authors' productivity and influence. Figure 7 depicts the QARA based on the top 15 authors in terms of CPP. However, it could be extended to include all authors. In the QARA, a dot indicates the publications each author provides each year. Specifically, the dots' size can be small or large according to the number of annual documents published yearly by each author (in this case, 1 or 2 publications, respectively); the dots' color intensity follows a chromatic scale based on the total number of citations received yearly by each author (here, dark blue corresponds to 30 citations, while light blue corresponds to 10); and, finally, the authors' names are listed on the y-axis, placing the author's name at the top.

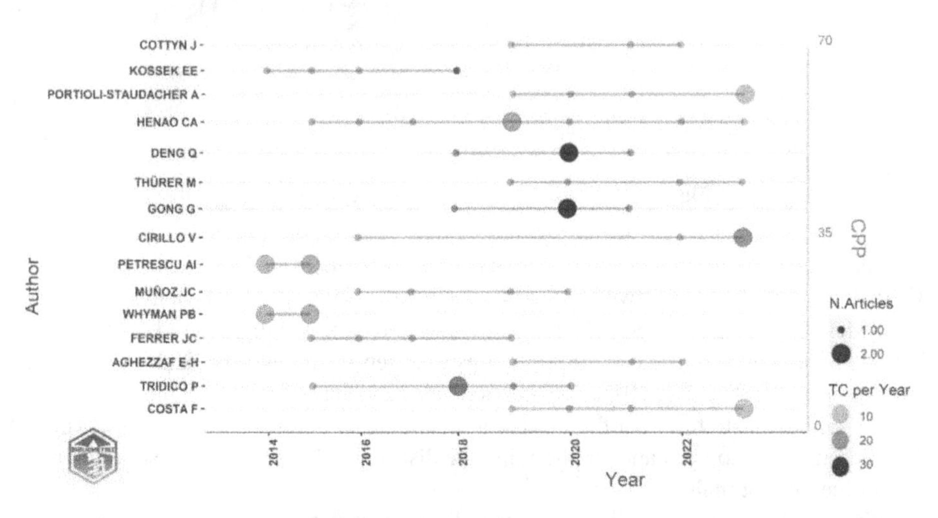

Fig. 7. Qualitative Authors' Relevance Assessment.

From the QARA (Fig. 7), four useful considerations emerge for answering RQ1. First, the most productive authors in the field appear based on the total number of publications (number and size of dots). In particular, the most productive author is Henao, with 8 publications. Secondly, the most influential authors are identified based on the highest CPPs, recognizing Cottyn and Kossek as the top 2 authors in the y-axis (with CPPs equal to 69 and 62, respectively). This result shows the difference between the most productive and influential authors. Thirdly, by looking at the dots' distribution and size, it is possible to check each author's temporal publication trend, observing the publication cadence and the date of the first publication. Finally, it is possible to identify the most influential papers in the existing literature on the analyzed topic.

4.2 Outcome Related to RQ2: Primary Themes and Research Streams in LF and WLC

The co-occurrence of authors' keywords in VOSViewer was used to answer RQ2 about workplace, work, labor, worker, and workforce flexibility. Results in Fig. 8 show keywords with a minimum of two co-occurrences and their reciprocal links. The colors and keywords in Fig. 8 identified five major research themes related to the analyzed topic, which were confirmed by reviewing the database paper abstracts: Core concepts of flexibility relate to labor and overall flexibility; work environment and conditions relate to workplace and work flexibility; human resource and personnel management affects labor, worker, and workforce flexibility; and technical and organizational issues affect all types of flexibility. Workers and LF prioritize employee satisfaction and well-being. How companies plan and deploy their employees to meet changing demands, such as cross-training and dynamic team formation. (vii) Leadership and Cultural Impact: research on how supportive leadership and corporate culture affect flexible work arrangements and transformation. Labor laws, industry regulations, and economic policies affect how firms implement flexible work practices.

RQ2 was answered by creating a Thematic Map of author keywords[23] and using Bibliometrics. Figure 9 confirmed Fig. 8's findings and highlighted two driving (motor) themes that dominated the research: workplace flexibility, LF, overall flexibility, workforce flexibility, work flexibility, and employee flexibility. Motor themes are dense and central. They are well-developed and research-relevant. Figure 9 shows "flexibility," "gender," "COVID-19," "employment," and "productivity." Basic themes are central but sparse. Though less developed than motor themes, they are crucial to the field. Figure 9 shows the terms "workplace flexibility," "flexible work arrangements," "employee engagement," "labour flexibility," "occupations," "inequality," "working conditions," "future of work," "work flexibility," "labour market," "gig economy," "human resource management," ore "remote working." Low centrality, high density for niche themes. These are well-developed but not field-tested, often representing specialized research. There are "workforce flexibility," "multiskilling," "labour productivity," "personnel scheduling," "europe," "retail services," "non-standard work," "panel data," "product innovation," and "childcare" in Fig. 9. Low centrality and density indicate new or declining themes that are underdeveloped or losing traction in the field. Figure 9 shows "job satisfaction," "work-life balance," "remote work," and "women". This perspective

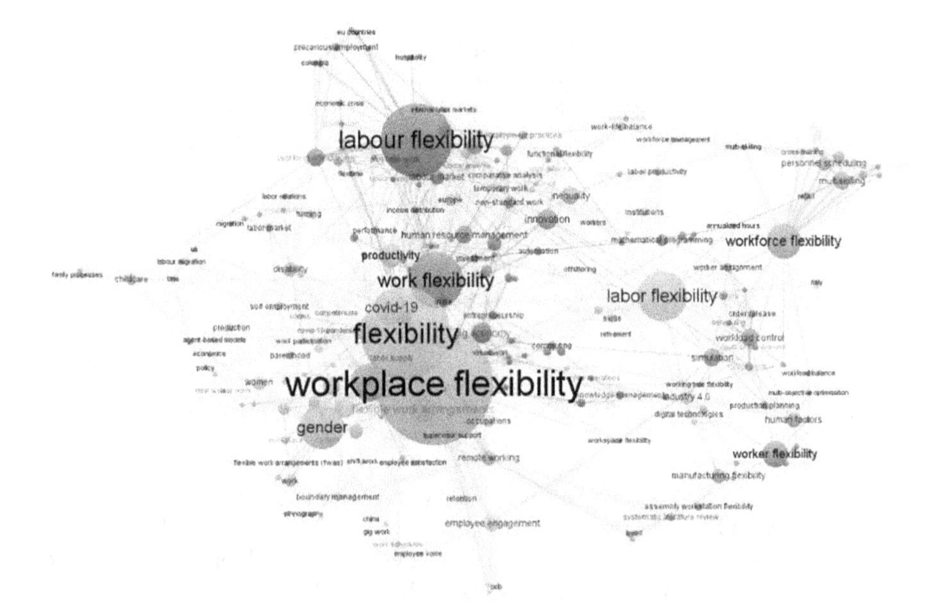

Fig. 8. Co-occurrence of authors' keywords.

emphasizes flexibility management human factors over operational logistics. It emphasizes employee well-being and job satisfaction while building an adaptable workforce that can handle changing workloads. This theme balances operational efficiency with employee needs, using technology to improve communication, scheduling, and personal development. It promotes cultural and organizational changes like workforce-focused leadership and policies. This approach is crucial in today's workplace, where employee well-being is increasingly valued for organizational success.

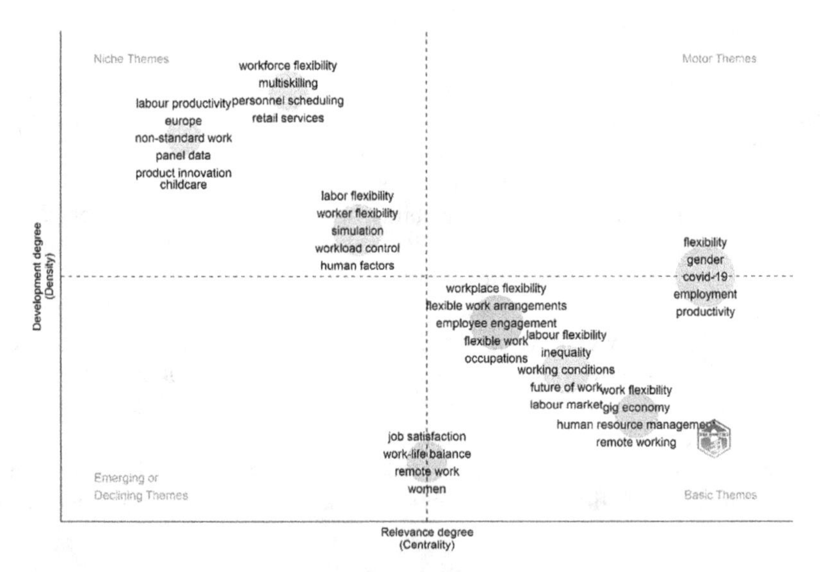

Fig. 9. Thematic Map of authors' keywords

5 Conclusions

This paper reviews the existing literature on LF in WLC. The field's most productive and influential countries, journals, and authors were identified using the number of publications and citations and the QARA graph. Following that, the main themes related to the analyzed domain were investigated using the authors' keywords and their co-occurrence, identifying the driving research streams that contribute the most to the research development in the considered topic.

The findings indicate that there is still room for improvement and development in this field's literature. However, following COVID-19, such a topic resurfaced in the scientific community, sparking a significant publication trend. The second takeaway from the findings is that researchers are particularly interested in eight specific themes related to the investigated topic. A thorough review of current research reveals a lack of comprehensive analyses on integrating LF with order release methods. Furthermore, existing models fail to account for human factors like learning curves and worker heterogeneity. These findings emphasize the importance of studies examining the complexities of human dynamics in WLC systems.

We examine fundamental concepts such as flexibility, work environment and conditions, human resource and personnel management, technical and organizational issues, employee well-being and satisfaction, workforce planning and development, leadership and cultural impact, and legal and policy frameworks in LF and WLC. Our research focuses on LF, workforce adaptation to changing tasks, and operational flexibility. Flexible work arrangements include remote work and adaptable job roles. Human resources and personnel management prioritize employee training, reskilling, and upskilling. Advanced scheduling algorithms and real-time data analytics enhance labor allocation and shift management. Technical and organizational issues are explored to determine how WLC systems can integrate technologies to improve efficiency and responsiveness. Flexible work arrangements and fewer repetitive tasks improve employee morale and satisfaction, resulting in more meaningful work. Workforce planning and development uses predictive analytics and flexible staffing models to form and reconfigure teams based on future skill needs quickly. Supportive leadership and a culture of continuous improvement and adaptability are required. Labor laws and industry regulations have an impact on flexible work arrangements. Our findings help to advance the integration of LF and WLC. Data analytics and flexible working arrangements improve decision-making, operational efficiency, and employee empowerment. Strategic workforce planning allows organizations to adapt to market and technological changes. These practical applications can help manufacturers improve operational efficiency, workforce management, and competitiveness by leveraging our review's findings.

The findings distinguish the most productive and influential countries, journals, and authors, implying that both are worth studying. The study's limitations stem from the manual screening of the dataset used to create the bibliometric review. This was necessary because the terms "load leveling" and "load balancing" and their synonyms or abbreviations had different semantic meanings, resulting in articles unrelated to the analyzed topic. Instead, this work aims to outline the characteristics of the current body of knowledge on the subject, which will allow researchers to identify existing gaps in the literature as well as future research challenges and opportunities. The subsequent post-bibliometric analysis will include a thorough review of the bibliographic literature. As a result, the following bibliometric review aims to synthesize, critique, and discuss the existing research on LF within WLC. The literature review will also address the identified research gap by providing insights and recommendations for future research in this critical study area. The study's detailed, multi-step process aims to rigorously explore and map the landscape of LF within WLC, thereby filling a critical research gap in the existing literature.

Future research must develop innovative approaches to employee work distribution and empirical studies to validate these new methodologies in various manufacturing settings. Some significant developments will determine the future of LF and WLC. Advances in Industry 4.0 technologies, such as Artificial Intelligence (AI) and the Internet of Things (IoT), will be combined to improve decision-making efficiency through predictive workforce modeling and scheduling automation. A better understanding of human dynamics in workforce management will be required, particularly learning curves and worker heterogeneity. Innovative work distribution methods will emerge to accommodate flexible staffing and changing shift patterns. The value of empirical studies will increase as theoretical models are tested in various real-world settings to determine their practical implications. Cross-disciplinary frameworks will combine insights from multiple fields, resulting in a more comprehensive approach to LF and WLC. Organizational cultures will shift to prioritize flexibility, adaptability, and employee well-being, focusing on leadership styles that value workforce needs. Strategies will evolve to meet changing cultural, legal, and social contexts as the workforce becomes more universal and diverse. Finally, sustainability and ethical considerations will be integrated to ensure that workforce management practices are both socially responsible and environmentally sustainable. These advancements will boost operational efficiency, employee satisfaction, and organizational adaptability to changing work environments.

Acknowledgments. This research did not receive a specific grant from any financial institution in the public, commercial, or not-for-profit sectors.

Disclosure of Interests. The author(s) reported no potential conflict of interest.

References

1. Cappelli, P.: A Market-Driven Approach to Retaining Talent (2000). https://hbr.org/2000/01/a-market-driven-approach-to-retaining-talent
2. Costa, F., Portioli-Staudacher, A.: Labor flexibility integration in workload control in Industry 4.0 era. Oper Manag Res. **14**, 420–433 (2021). https://doi.org/10.1007/s12063-021-00210-2
3. Stevenson, M., Hendry, L.C., Kingsman, B.G.: A review of production planning and control: the applicability of key concepts to the make-to-order industry. Int. J. Prod. Res. **43**, 869–898 (2005). https://doi.org/10.1080/0020754042000298520
4. Ryalat, M., ElMoaqet, H., AlFaouri, M.: Design of a smart factory based on cyber-physical systems and Internet of Things towards industry 4.0. Appl. Sci. **13**, 2156 (2023). https://doi.org/10.3390/app13042156

5. Nagy, M., Lăzăroiu, G., Valaskova, K.: Machine intelligence and autonomous robotic technologies in the corporate context of smes: deep learning and virtual simulation algorithms, cyber-physical production networks, and industry 4.0-based manufacturing systems. Appl. Sci. **13**, 1681 (2023). https://doi.org/10.3390/app13031681
6. Torres-Hernández, M.A., Escobedo-Barajas, M.H., Guerrero-Osuna, H.A., Ibarra-Pérez, T., Solís-Sánchez, L.O., Martínez-Blanco, M. del R.: Performance analysis of embedded multilayer perceptron artificial neural networks on smart cyber-physical systems for IoT environments. Sensors (Basel). **23**, 6935 (2023). https://doi.org/10.3390/s23156935
7. Jaikumar, R.: Japanese flexible manufacturing systems: Impact on the United States. Jpn. World Econ. **1**, 113–143 (1989). https://doi.org/10.1016/0922-1425(89)90006-6
8. Treleven, M.: A review of the dual resource constrained system research. IIE Trans. **21**, 279–287 (1989). https://doi.org/10.1080/07408178908966233
9. Thürer, M., Stevenson, M., Silva, C.: Three decades of workload control research: a systematic review of the literature. Int. J. Prod. Res. **49**, 6905–6935 (2011). https://doi.org/10.1080/00207543.2010.519000
10. Thürer, M., Stevenson, M., Silva, C., Land, M.J., Fredendall, L.D., Melnyk, S.A.: Lean control for make-to-order companies: integrating customer enquiry management and order release. Prod. Oper. Manag. **23**, 463–476 (2014). https://doi.org/10.1111/poms.12058
11. Stevenson, M., Hendry, L.C.: Aggregate load-oriented workload control: a review and a reclassification of a key approach. Int. J. Prod. Econ. **104**, 676–693 (2006). https://doi.org/10.1016/j.ijpe.2005.05.022
12. Soepenberg, G.D., Land, M.J., Gaalman, G.J.C.: Adapting workload control for job shops with high routing complexity. Int. J. Prod. Econ. **140**, 681–690 (2012). https://doi.org/10.1016/j.ijpe.2012.03.018
13. Chen, I., Calantone, R., Chung, C.-H.: The marketing-manufacturing interface and manufacturing flexibility. Omega **20**, 431–443 (1992). https://doi.org/10.1016/0305-0483(92)90018-3
14. Tsourveloudis, N.C., Phillis, Y.A.: Manufacturing flexibility measurement: a fuzzy logic framework. IEEE Trans. Robot. Autom. **14**, 513–524 (1998). https://doi.org/10.1109/70.704212
15. Singh, T.P.: Role of manpower flexibility in lean manufacturing. In: Sushil, Stohr, E.A. (eds.) The Flexible Enterprise. Flexible Systems Management, pp. 309–319 (2014). Springer, New Delhi. https://doi.org/10.1007/978-81-322-1560-8_18
16. Costa, F., Kundu, K., Rossini, M., Portioli-Staudacher, A.: Comparative study of bottleneck-based release models and load-based ones in a hybrid MTO-MTS flow shop: an assessment by simulation. Oper. Manag. Res. **16**, 33–48 (2023). https://doi.org/10.1007/s12063-022-00276-6
17. Bertolini, M., Romagnoli, G., Zammori, F.: Simulation of two hybrid production planning and control systems: A comparative analysis. In: 2015 International Conference on Industrial Engineering and Systems Management (IESM), pp. 388–397 (2015)
18. Ferraro, S., Leoni, L., Cantini, A., De Carlo, F.: Trends and recommendations for enhancing maturity models in supply chain management and logistics. Appl. Sci. (Switzerland). **13**, (2023). https://doi.org/10.3390/app13179724
19. Strozzi, F., Colicchia, C., Creazza, A., Noè, C.: Literature review on the 'Smart Factory' concept using bibliometric tools. Int. J. Prod. Res. **55**, 6572–6591 (2017). https://doi.org/10.1080/00207543.2017.1326643
20. Cantini, A., Ferraro, S., Leoni, L., Tucci, M.: Inventory centralization and decentralization in spare parts supply chain configuration: a bibliometric review. In: Presented at the Proceedings of the Summer School Francesco Turco (2022)

21. Callon, M., Courtial, J.P., Laville, F.: Co-word analysis as a tool for describing the network of interactions between basic and technological research: the case of polymer chemistry. Scientometrics **22**, 155–205 (1991). https://doi.org/10.1007/BF02019280
22. Brookes, B.C.: Sources of information on specific subjects by S.C. Bradford. J. Inf. Sci. **10**, 173–175 (1985). https://doi.org/10.1177/016555158501000406
23. Cobo, M.J., López-Herrera, A.G., Herrera-Viedma, E., Herrera, F.: An approach for detecting, quantifying, and visualizing the evolution of a research field: a practical application to the fuzzy sets theory field. J. Informet. **5**, 146–166 (2011). https://doi.org/10.1016/j.joi.2010.10.002

Integrating Ontology with Cobot Execution for Human-Robot Collaborative Assembly Using Heterogenous Cobots

Yee Yeng Liau and Kwangyeol Ryu(✉)

Department of Industrial Engineering, Pusan National University, Busan 46241, Republic of Korea
kyryu@pusan.ac.kr

Abstract. The manufacturing industry has heavily relied on robotic automation since the third industrial revolution. However, traditional robots struggle to adapt to the variability of modern production demands. Collaborative robots (cobots) offer a solution by providing flexible automation tailored to personalized manufacturing requirements. Human-robot collaboration (HRC) systems, particularly beneficial for small-scale manufacturing, integrate human capabilities with technological advancements. Cobots enhance productivity, offer ergonomic benefits, and facilitate automation transformation through connectivity and data analytics. Effective communication between resources is crucial for enabling the execution of shared subtasks and ensuring the safety and efficiency of collaborative assembly. This study proposes integrating ontology with cobot execution programs to enhance collaborative assembly operations. The ontology encompasses knowledge related to the products to be assembled, the HRC environment (including resources, tools, and regions), and the relations between these entities and monitoring data. Additionally, establishing a connection between cobot controllers enables the seamless exchange of commands for coordinated execution. By enabling communication between resources, collaborative assembly tasks can be executed either sequentially or simultaneously, monitored for progress, and adjusted as necessary without manual intervention. The proposed approach is applied in the human-robot collaborative mold assembly for cobot execution. The integration of ontology simplifies cobot programming, modifications according to order changes, and facilitates cobot execution control based on monitoring results. This study contributes to streamline cobot execution and decision-making processes in manufacturing environments by incorporating ontology-based knowledge into cobot execution programs.

Keywords: Human-robot Collaboration Systems · Mold Assembly · Ontology · Decision-making

Published by Springer Nature Switzerland AG 2024
M. Thürer et al. (Eds.): APMS 2024, IFIP AICT 729, pp. 251–265, 2024.
https://doi.org/10.1007/978-3-031-65894-5_18

1 Introduction

With advancements in sensor technology and artificial intelligence, collaborative robots, commonly referred to as cobots, are revolutionizing industrial automation by enabling safer, more interactive, and adaptive HRC in manufacturing environments. Unlike traditional industrial robots, cobots are specifically designed to operate in close proximity to humans without the need for physical safety barriers. This paradigm shift enables direct collaboration between humans and robots, facilitating cooperative task performance. Cobots excel in tasks such as assembly, pick-and-place operations, quality control, and material handling, where they complement human skills and capabilities [1, 2]. By working collaboratively with humans, these systems not only enhance productivity but also contribute to greater flexibility and safety in manufacturing environments [3].

Recently, HRC systems have experienced rapid growth due to the development of advanced deep learning algorithms in machine vision and prediction applications. These deep learning algorithms empower robots to perceive and understand the environment within the HRC workspaces, ensuring smooth collaborative operations execution and human safety [4–7]. However, the management of knowledge within the HRC assembly environment remains a challenge. This includes handling the information exchange between humans and robots, understanding the functions, usage, and relationships of various models involved. Efficient management of this knowledge is crucial for establishing standardized procedures that facilitate cobot execution, such as cobot programming. This encompasses determining the necessity of specific models, collecting relevant data or information, making informed decisions, and directing these decisions for execution.

In this study, we present an approach to utilize the ontology developed for enabling subtasks execution in a human-robot collaborative mold assembly that involved collaboration between a human worker and two heterogeneous cobots. The control of cobot execution is centralized on a main computer, which connects with various components including cobot simulation software, monitoring devices, including Modbus and webcam, and OwlReady for accessing and updating the ontology. Integrating ontology with cobot execution programming aims to streamline the programming process for production personnel by adjusting cobot operations based on decisions obtained from the ontology, minimizing human intervention in planning cobot operations prior to assembly. This integration allows workers to request decisions from the ontology to maintain consistency in decision-making according to specific circumstances. The overarching goal is to organize knowledge acquired from experience and utilize knowledge reasoning for efficient programming, enabling new personnel to quickly adapt to collaborative task execution. The proposed approach aims to answer the research question: How can an ontology-based approach improve knowledge management, simplify information exchange, and facilitate programming cobots for execution in HRC mold assembly environments?

The paper is structured as follows: Sect. 2 reviews existing literature on communication and the application of ontology in HRC systems. Section 3 provides a brief description of the developed ontology, its connection with other functional models and communication within the HRC system. Section 4 details the integration of the ontology with the cobot program for monitoring and execution. Finally, Sect. 5 concludes the study and discusses future work.

2 Literature Review

2.1 Communication in Human-Robot Collaborative Systems

Communication and interaction between humans and robots within an HRC system are essential for effective collaboration during tasks. Communication in the HRC systems can be categorized into human-to-robot and robot-to-human communication [8]. Communication serves various purposes, including understanding and predicting human intentions [9, 10], controlling robot tasks [11], facilitating human-assisted and collaborative operations [12], and monitoring human and robot motions. Communication between humans and robots can be established through speech or voice commands [13], vision-based methods such as gesture recognition [14], and semantic [15] approaches. Verbal communication in human-robot teams was proposed for enabling a robot to optimize decision-making between physical actions and verbal commands [16]. Two types of utterances are explored: verbal commands prompting human action and state-conveying actions providing internal robot information. Hand gestures serve as a natural and effective means of communication between humans and robots in industrial environments due to their natural expression and detectability from human pose [14]. A communication paradigm introduced in a study involves conveying instructions to humans and robots through haptic interfaces and screens, with agents informing the scheduling algorithm of their status [17].

Another key purpose of communication is to monitor human safety and facilitate task execution through human-robot interaction. Monitoring of manual assembly tasks includes hand detection, human action recognition, and progress monitoring based on object detection [18, 19]. A vision monitoring system for HRC assembly operations, including workspace and posture monitoring [20], was proposed to detect errors during assembly, assess assembly quality, and monitor human hands. An approach was proposed to monitor manual task execution for HRC implementation [21], involving object detection, human hand tracking, and verification of correct parts picked by humans according to predefined assembly sequences. During HRC assembly, the robot must adapt to human actions, and prediction models were developed to enable the robot to anticipate human actions [22], transition between manual and robotic tasks [23], and collaborate with human workers [24]. An HRC assembly system utilized machine vision to locate part positions, support robot movement calculations, and create a safe work envelope for a robotic arm within the collaborative area [7].

These existing studies focused on developing recognition models to monitor human actions for task transition between humans and robots but ignored the transition between robots. Additionally, the knowledge management related to the utilization of these models in HRC operations and communication among other operation planning stages, such as assignment, is not presented. This study examines communication between humans and cobots across two distinct stages: planning and execution. During the planning stage, communication occurs between external task assignment and part allocation modules to respond to the production orders, whereas in the execution stage, communication involves interactions between humans and cobots facilitated by vision-based monitoring, as well as communication between the two cobots.

2.2 Application of Ontology in HRC Systems

Ontology refers to a formal representation of knowledge within a specific domain, typically organized in a structured format comprising concepts, entities, relationships, and axioms [25]. Ontology plays a vital role in human-robot collaboration systems by providing a structured representation of knowledge that facilitates communication, decision-making, and task execution between humans and robots. Ontologies have been used for in robotic systems [26, 27]. However, there are some missing matters on ontology for robotic to be applied in the HRC domains, including the description of human capabilities, interactions and coordination between humans and robots [28]. Researchers have explored the use of ontologies to represent domain-specific knowledge in HRC systems. These ontologies capture concepts, relationships, and constraints relevant to the collaborative tasks performed by humans and robots. Ontologies are utilized to model context information such as environment conditions, human preferences, and task constraints [29]. This enables robots to adapt their behavior and decision-making processes according to the current context, enhancing overall performance and user satisfaction [30]. Ontologies facilitate knowledge sharing and learning between human and robotic agents in collaborative settings. By encoding domain knowledge in a structured format, ontologies enable robots to acquire new knowledge from human demonstrations, observations, and interactions, leading to improved task performance and adaptation capabilities. Additionally, combination of ontology with vision dataset enables understanding of human intentions in manipulation tasks by intelligent robots to perform activities [31].

In the smart manufacturing domain, the SOSA (Sensor, Observation, Sample, and Actuator) ontology, a lightweight ontology designed to describe Internet of Things (IoT) platforms, sensors, observations, actuators, and their relationships [32]. SOSA provides a comprehensive framework for describing sensor data and related concepts in the IoT domain. Additionally, the Smart Manufacturing Ontology (SMO) was developed to define the concepts, relationships, and properties relevant to smart manufacturing systems [33]. This ontology provides a standardized way of representing and exchanging information about machines, equipment, materials, products, processes, and workflows involved in smart manufacturing. It allows for the seamless exchange of data and knowledge between different stakeholders, such as manufacturers, suppliers, and customers, thereby facilitating more efficient and collaborative manufacturing processes. However, this study did not use vocabularies widely used such as SMO, SOSA, etc. because their use may not sufficiently describe the exact meaning of the problem and situations in the mold assembly considered in this study. When generalizing our research to other smart manufacturing domains or applications, for example, we need to building a kind of rules for mapping our ontologies to existing ones in order to provide general knowledge or common understanding.

The missing aspect in these studies include is the absence of monitoring and execution stages, which are crucial during HRC operations. Although representation of positioning using ontology has been proposed, the presence of humans has not been taken into account. Furthermore, most studies have not considered the ontology connecting humans and robots during industrial assembly operations, especially the phase of controlling and executing subsequent robot tasks based on human progress to reduce assembly time and avoid collisions. Further research is needed to explore the use of ontologies tailored to specific application domains and collaborative scenarios for monitoring and execution, and to facilitate the development of more intelligent and adaptive robotic systems. In this study, ontologies were used to retrieve information needed to generate programs for cobot execution based on the operational context.

3 Facilitating HRC Mold Assembly: Ontology Construction and Resource Communication

3.1 Ontology of HRC Mold Assembly

The details of the ontology development are included in our previous study, and this study focuses on integrating the ontology developed to support cobot execution and programming [34]. Figure 1 illustrates the ontology developed for HRC mold assembly operation in our previous study. This ontology is developed using Protégé, an open source platform for ontology development [35]. The ontology comprises classes categorized into two contexts: assembly planning and operational.

The assembly planning context (highlighted with a yellow background) includes ten classes: mold, mold parts (grey colored bounding box), part characteristics, assembly tasks, subtask types and characteristics (green colored bounding box), resources (red colored bounding box), tools (brown colored bounding box), and regions (purple colored bounding box). The mold class consists of core and cavity subclasses, representing the core and cavity subassemblies of a mold, which contain mold parts with characteristics such as part type and shape. Assembly operations are decomposed into assembly tasks based on main parts, further broken down into feed subtasks and assembly subtasks. Feed subtasks involve moving a part from its initial position to its final assembly position, while assembly subtasks involve assembling a part at the final assembly position. These subtasks are categorized based on characteristics such as the action to be performed by resources, tolerance, and the type of part handled. The resources class includes subclasses representing each individual resource - human worker and two heterogeneous cobots - as well as a subclass representing collaboration between both cobots. Subtasks can be assigned to individual resources or the collaboration of both cobots. Tools are used by resources during assembly operations and are categorized into subclasses such as manual tools, fingers, and vacuum grippers used by the cobots. Additionally, four resource regions are defined based on the reachable distance of both cobots, comprising positions for locating parts, tool regions for tool placement, assembly regions, and tool.

The classes within the operational context (highlighted with a yellow background) pertain to monitoring and decision-making during the execution of HRC mold assembly operations. Monitoring-related classes include monitoring devices, monitoring data (yellow colored bounding box), assembly task states, resources, and tool states (pink colored bounding box), and cobot moving position (mint colored bounding box). Monitoring devices are utilized to collect monitoring data. Camera and cobots' Modbus are devices used to collect human action and cobots' tool center point (TCP). Assembly task, resource, and tool states are determined based on monitoring data, guiding decisions within the ontology regarding execution steps for resources.

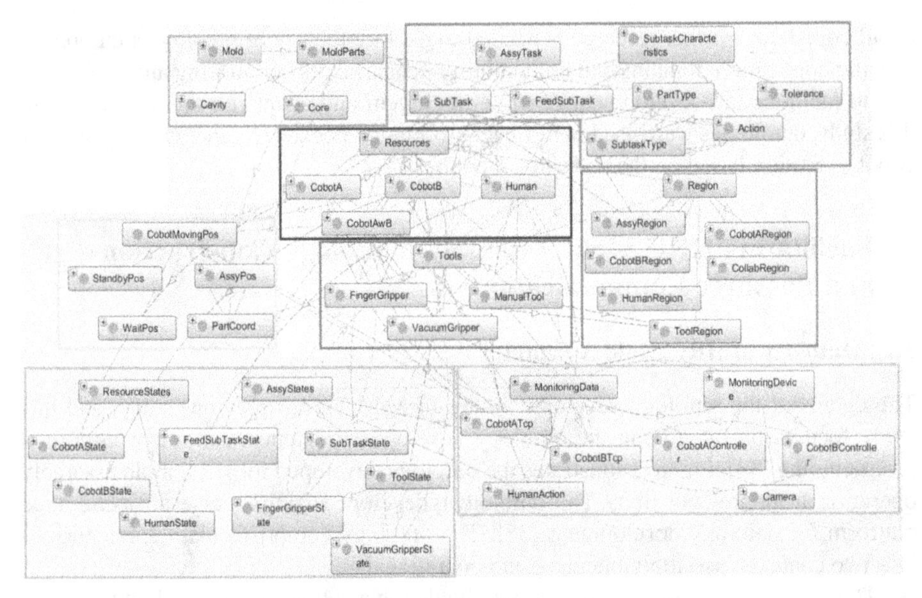

Fig. 1. Ontology for HRC mold assembly operation.

3.2 Establishing Communication Between Human and Cobots

In the proposed HRC mold assembly operation, effective communication between humans and cobots is essential for enabling cobot execution without direct human control and ensuring human safety within the shared HRC workspace. In this study, communication between humans and cobots occurs during the execution of subsequent subtasks by the cobot following the completion of the current manual subtask. While the specifics of the recognition model for human actions are not detailed in this study, the data flow of communication from humans to robots is elucidated in Sect. 4.2. The results from human action recognition are input to the ontology to update the subtask and human state. Simultaneously, we employed the PyModbus package [36] to encode the states as values and transmit them to the cobots. The values were set to 10, 20, or 30, representing “Start,” “In-progress,” and “End” states, respectively. An illustrative example of PyModbus configuration at the human action recognition processing computer and Cobot A is provided in Fig. 2. Initially, the IP address of Cobot A was defined at the processing computer. The processing computer acted as the client or master, while Cobot A functioned as the server or slave. Subsequently, the processing computer was connected to Cobot A using the client.connect() function. The value of the state was then written to address 130 based on the recognition outcome. Cobot A was programmed to wait for a value of 30, indicating the “End” state, before moving towards the final position using the get_modbus_slave(130) command, as the value was written to address 130.

At Processing Computer (Modbus client or master):

```
from pymodbus.client import ModbusTcpClient
#Define IP address of Cobot A
M1013_ip = '192.168.137.30'
client = ModbusTcpClient(M1013_ip)
client.connect()
#Write to value 'v' M1013 Modbus address 130
#Value 'v' indicates the state of subtask based on the recognition result
client.write_register(130, v)
```

At Cobot A (Modbus server or slave):

```
#read value written by computer at address 130
a = 0
while True:
    a = get_modbus_slave(130)
    if a == 30:
      break
movel(final_pos, vel=30, acc=30)
```

Fig. 2. PyModbus setting for human-robot communication.

3.3 Establishing Communication Between Cobots

We utilize the inherent Modbus TCP feature of the cobots to establish connectivity between them within our HRC system. The Modbus data model comprises four fundamental data types: discrete inputs, coils, input registers, and holding registers. Leveraging the built-in Modbus TCP, which supports both master and slave functions, we utilize holding registers to facilitate data transmission from Cobot A to Cobot B and vice versa (refer to Fig. 3). This enables reciprocal communication between the cobots, allowing them to issue instructions to each other during assembly operations. Communication between the cobots occurs in two main scenarios. Firstly, during the execution of subtasks assigned to the collaboration of both cobots, such as lifting and relocating a heavy mold plate from its initial position to the final assembly position. In this context, the sequence in which both cobots approach the part is critical to prevent collisions. In our setup, Cobot A is programmed to move to a predefined position on the part, while Cobot B remains stationary, awaiting a signal from Cobot A (as depicted in Fig. 3a). Upon completion of its motion, Cobot A sends a signal to Cobot B, prompting Cobot B to move to another predefined position on the same part (Fig. 3b). Subsequently, Cobot B signals Cobot A to jointly lift the part (Fig. 3c) and relocate it to the final assembly position (Fig. 3d). The second scenario arises when Cobot A executes the current subtask, and the subsequent subtask is assigned to Cobot B (Sect. 4.2). In such instances, communication becomes necessary to relay the position of the Tool Center Point (TCP) between both cobots via Modbus. In this study, we employ Modbus addresses to store the TCP position values for each cobot and share them with the other cobot. As shown in Fig. 4, the distance between TCPs of both cobots are calculated, and the execution of subsequent subtask by another cobot is decided based on the safety distance and the state of cobot partner.

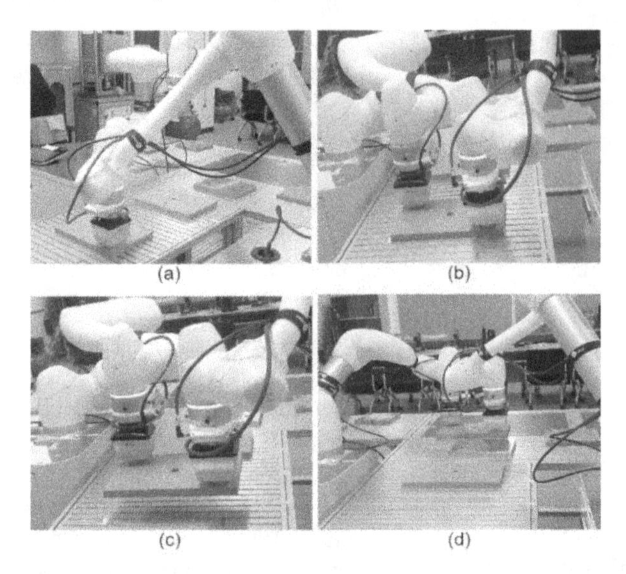

Fig. 3. Modbus communication for both cobots collaboration (Scenario 1).

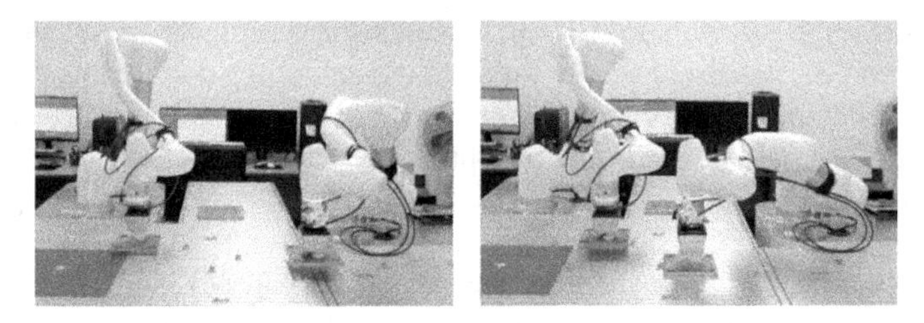

Fig. 4. Position sharing between cobots (Scenario 2).

4 Integration of Ontology with Cobot Execution

4.1 Integration of Ontology into HRC Assembly Operation

In this study, the ontology plays roles in decision making at planning and execution stages. Figure 5 illustrates the integration of the developed ontology with functional modes designed for task input, execution, and monitoring stages. During the task input stage, data related to the mold to be assembled and available resources are input into the ontology. Mold-related data includes information about mold parts and dimensions such as width, length, thickness, and weight, as well as assembly tasks and subtasks required for the assembly operation. Data concerning resources and tools encompass available resources, end-effectors used, and technical specifications of cobots and end-effectors, including payload and reachable distance. Additionally, regions and part positions are adjusted based on the reachable distance of the cobots. The instances and relevant data are updated in the ontology using OwlReady 2.0, a Python library for ontology-oriented programming [37]. Task assignment involves defining the subtask sequence, subtask-to-resource assignment, and determining the parts to be handled by the assigned resources.

Once all information is generated during task planning, it is input into the ontology. Utilizing defined rules and reasoning, the ontology allocates parts to available part positions within the assigned resource's region. Furthermore, it determines the tool to be used by the assigned resource to execute the subtask based on the part type if assigned to a cobot or based on the action if assigned to a human. Along with the assignment results, the positions of parts to be picked and placed, and the tools used by resources are input into the main execution program written in Python. The main execution program is linked to the ontology via OwlReady and connected to the cobot program via PyModbus. Manual subtasks are directed to the user interface, which is under development. For subtasks executed by cobots, the main execution program triggers the execution of corresponding subprograms based on the located part positions. Each cobot's subprogram is pre-programmed and includes commands for cobot movement from designated coordinates of a part position to an assembly position, as well as tool controlling commands, velocity, and acceleration settings.

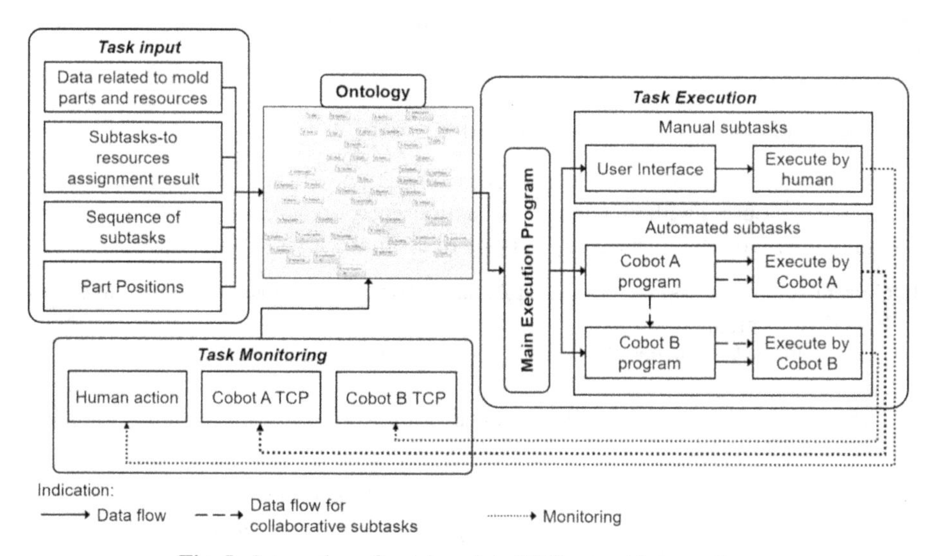

Fig. 5. Integration of ontology into HRC assembly operation.

During the execution phase, the ontology continues to play a pivotal role in guiding decision-making processes by facilitating various essential functions. Primarily, it oversees the selection of subtasks for execution, establishing connections such as the "nextsubtask" relation and utilizing information on assembly states to determine the current status for executing subsequent subtasks based on monitoring data. We set the cobot moving speed according to the moving condition. The moving condition of a cobot is also determined by the cobot TCP. If the next TCP is coordinates of a part position, then it means the cobot move without part, and the cobot speed is set to "Fast." If the next TCP is the assembly position, then it means the cobot is moving with parts, and the cobot speed is set to "Slow." As the HRC collaborative assembly operation progresses, monitoring of subtask and resource states is conducted. In this research, a camera captures input for human action recognition, while cobot controllers are equipped with Modbus functionality to retrieve and transmit Tool Center Points (TCPs). These TCPs are used to identify the movement conditions of cobots and determine their speed, which, in this study, is not directly integrated into the cobot program but is instead conveyed to a user interface to alert human workers. However, the specifics of the user interface are not addressed in this study.

Additionally, ontology assists in determining the z-coordinate where a part to be placed based on thickness of parts assembled in previous subtasks. This method eliminates the manual input of z coordinates for parts to be placed at each assembly subtask. Furthermore, ontology enables real-time decision-making by orchestrating the initiation of the next subtask in response to human actions and the Tool Center Point (TCP) of the ongoing subtask, while also directing cobot execution, pausing, or resuming tasks as necessary. Lastly, ontology aids in enhancing worker safety by obtaining robot speed insights based on cobot movement conditions, thereby alerting workers to potential hazards and ensuring a secure working environment. Through these functionalities, ontology serves as a critical decision support system, optimizing the efficiency and safety of HRC in assembly operations.

4.2 Execution of HRC Assembly Operation

Communication between resources is crucial to enable smooth and continuous execution by different resources without human interruption There are three cases during the HRC assembly operation that required monitoring of assembly and resource states:

1. Current subtask is assigned to Cobot A, and subsequent subtask is assigned to Cobot B, and vice versa.
2. Current subtask is assigned to Cobot A or B, and subsequent subtask is assigned to Human.
3. Current subtask is assigned to human, and subsequent subtask is assigned to Cobot A or B.

Figure 6 shows the ontology of cobot execution based on the state of the subtask. Relationships and instances (boxes with diagonal symbols) of classes related to cobot execution during the HRC collaborative assembly. Decisions made are executed by human, Cobots A and B. Since this study focuses on integration of ontology into cobot execution, only the decision made for cobots are included. The decision on actions to be executed by cobots includes execution, stopping, resuming, and standing by. For instance, "CobotAExecute" signifies that Cobot A is carrying out the assigned subtask, "CobotAStop" denotes the stoppage of Cobot A's activity, "CobotAStandby" indicates that Cobot A is waiting and standing by at the designated position during execution, and "CobotAResume" implies that Cobot A resumes its activity after being on standby. In this study, subtasks are categorized into "FeedSubtask" and "SubTask," where "FeedSubtask" refers to moving parts from the initial position to assembly posi-tion, and "SubTask" refers to assembling parts at the assembly position. The states of subtasks are represented by instances such as "AssemStart," "AssemInProgress," "AssemPending," and "AssemCompleted," which respectively signify the start, ongoing execution, pending status, and completion of the subtask.

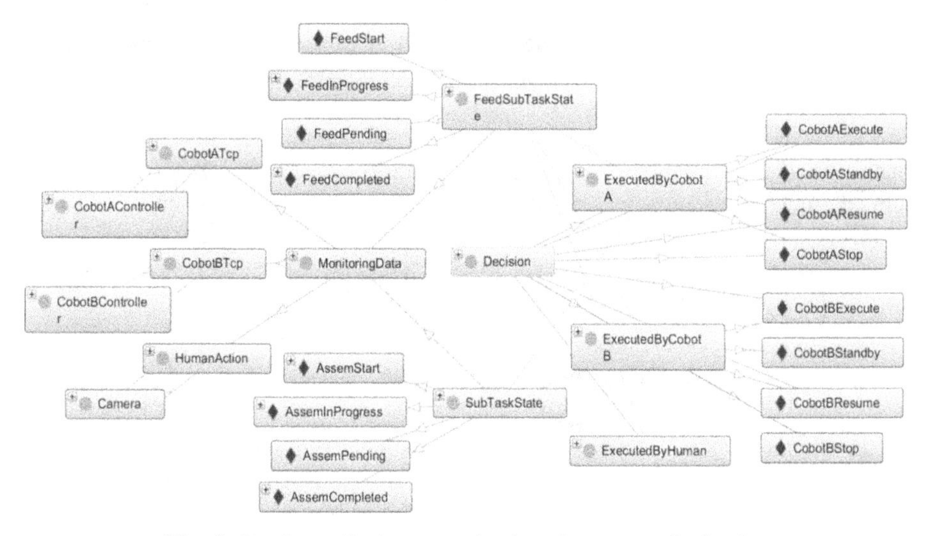

Fig. 6. Ontology of cobot execution based on states of subtasks.

Using the execution of Cases 1 and 2 as an example, the subsequent "FeedSubtask" is designated for execution by "CobotB" or "Worker" while the current "FeedSubtask" is being carried out by "CobotA." When the subtask is assigned to "CobotA," it initiates the execution process, prompting the cobot's state to switch to "CobotABusy." "CobotA" proceeds to execute the assigned subtask by lifting the relevant part, often identified as feedsubtask x. Meanwhile, the main program retrieves the subsequent TCP of the cobot ("CobotATCP") to ascertain the assembly state of feedsubtask x. If the next TCP corresponds to a part's location or assembly position, indicating that the task is in progress, the assembly state is labeled "FeedInProgress." At this point, if the subsequent subtask is designated for another cobot, "CobotB," the latter cobot commences its subsequent tasks by lifting the part and relocating it to a predefined coordinate located outside the assembly position. Conversely, if the next TCP corresponds to the cobot's standby position, it means that the current subtask has been sucecessfully completed, and the feed subtask state switches to "FeedComplete." Subsequently, a decision of "CobotBResume" is sent to "CobotB," and "CobotB" resumes its subsequent subtask and enters the assembly position to continue the collaborative operation. However, if neither condition is met, indicating that the cobot's task is not yet complete, a decision of "CobotBStandby" is made and "CobotB" remains idle at the waiting coordinates until the state of the current subtask switches to "FeedComplete," ensuring synchronization of tasks within the assembly process.

By taking another example of the execution of Case 3, the subsequent feed subtask is designated for execution by a cobot, "CobotA" while the current assembly subtask is being carried out by a human worker. The state of the manual progress is determined, as "AssemStart," "AssemInProgress," and "AssemCompleted," based on the recognized actions and then transmitted to the Modbus signal address via PyModbus. Meanwhile, "CobotA" assigned to execute the subsequent subtask remains stationary at the predefined standby position, awaiting the receipt of the "AssemInProgress" state as a signal to commence the task. Upon receiving the state "AssemInProgress," the cobot initiates the subsequent subtask based on the decision of "CobotAExecute," such as picking up plate A and proceeding towards the final assembly position. Subsequently, "CobotA" waits for the decision of "CobotAResume" to enter the assembly region when the state is changed to "AssemCompleted." This strategic approach ensures that the cobot refrains from entering the assembly region while the human worker is currently engaged a subtask. Once the human worker vacates the assembly region, signaling the completion of the task, the "AssemCompleted" state is transmitted to the cobot, allowing the cobot to proceed into the assembly region.

5 Conclusion and Future Works

The development of an ontology is vital for organizing and managing the knowledge required for HRC mold assembly operations, covering various aspects such as molds, tasks, resources, tools, and regions. This study addresses the research question by demonstrating how the integration of ontology can improve knowledge management, simplify information exchange and facilitate programming cobots for execution in HRC mold assembly operations. Our proposed framework facilitates the integration of this ontology with the execution program, enabling seamless communication between resources and functional models such as assignment and monitoring. During planning, linking the ontology with task assignment allows efficiently receiving subtask-resource assignments for part allocation and tool selection, the results of which are fed to the execution program connected to the cobot simulation program. Throughout the execution of HRC collaborative assembly operations, the ontology serves to provide crucial subtask sequence information and part data to the execution program, while receiving monitoring results to guide subsequent steps and provide feedback. This study contributes to a practical approach to support the production workforce, including new personnel with no experience in accessing and utilizing the ontology developed for programming cobot execution according to mold orders, to ensure in decision-making and enabling rapid response to execution reconfiguration.

The concept of "smart" has been introduced in manufacturing operations to recognize the context during operations and react accordingly. The ontology plays an important role in smart manufacturing operations by providing information about when, where and how data should be collected and analyzed, thereby facilitating HRC. In this study, we have described a conceptual framework for integrating ontologies into HRC practices, and we are currently working on simulation and evaluation of the framework to create a smart HRC mold assembly system. To expand this study to encompass not only mold assembly but also the HRC mold production system, we will explore the application of the SOSA and SMO ontologies, which are specifically designed for smart manufacturing and IoT implementation. Additionally, future efforts will focus on designing a user interface connected to the ontology to visually represent HRC assembly progress for human workers, thus enhancing overall efficiency and understanding in the manufacturing process.

Acknowledgments. This work was supported by the National Research Foundation of Korea (NRF) grant funded by the Korea government (Ministry of Science and ICT) (No.

2021R1A2C2009984). This work was supported by project for Smart Manufacturing Innovation R&D funded by Korea Ministry of SMEs and Startups (No. RS-2022–00140411).

Disclosure of Interests. The authors have no competing interests to declare that are relevant to the content of this article.

References

1. Wang, L., et al.: Symbiotic human-robot collaborative assembly. CIRP Ann. **68**(2), 701–726 (2019)
2. Aaltonen, I., Salmi, T.: Experiences and expectations of collaborative robots in industry and academia: barriers and development needs. Procedia Manuf. **38**, 1151–1158 (2019)
3. Nicora, M.L., Ambrosetti, R., Wiens, G.J., Fassi, I.: Human–robot collaboration in smart manufacturing: robot reactive behavior intelligence. J. Manuf. Sci. Eng. **143**(3), 031009 (2021)
4. Lou, P., Li, J., Zeng, Y., Chen, B., Zhang, X.: Real-time monitoring for manual operations with machine vision in smart manufacturing. J. Manuf. Syst. **65**, 709–719 (2022)
5. Li, X., Wang, J., Xu, F., Song, J.: Improvement of Yolov3 algorithm in workpiece detection. In: 9th Annual International Conference on CYBER Technology in Automation. Control, and Intelligent Systems (CYBER), pp. 1063–1068. IEEE, Suzhou, China (2019)
6. Wang, K.J., Rizqi, D.A., Nguyen, H.P.: Skill transfer support model based on deep learning. J. Intell. Manuf. **32**, 1129–1146 (2021)
7. Dang, A.T., Hsu, Q.C., Jhou, Y.S.: Development of human–robot cooperation for assembly using image processing techniques. Int. J. Adv. Manuf. Technol. **120**(5–6), 3135–3154 (2022)
8. Salehzadeh, R., Gong, J., Jalili, N.: Purposeful communication in human-robot collaboration: A review of modern approaches in manufacturing. IEEE Access **10**, 129344–129361 (2022)
9. Karami, A., Jeanpierre, L., Mouaddib, A.: Human-robot collaboration for a shared mission. In: 5th ACM/IEEE International Conference on Human-Robot Interaction (HRI), pp. 155–156. IEEE, Osaka, Japan (2010)
10. Chang, M.L., Guitierrez, R.A, Khante, P., Short, E.S., Thomaz, A.L.: Effects of integrated intent recognition and communication on human-robot collaboration. In: 2018 IEEE/RSJ International Conference on Intelligent Robots and Systems (IROS), pp. 3381–3386. IEEE Madrid, Spain (2018)
11. Nuzzi, C., et al.: MEGURU: a gesture-based robot program builder for meta-collaborative workstations. Robot. Comput.-Integr. Manuf. **68**, 102085 (2021)
12. Mutlu, B., Huang, C.: Coordination mechanisms in human-robot collaboration. In: Proceedings of the Workshop on Collaborative Manipulation. In: 8th ACM/IEEE International Conference on Human-Robot Interaction, pp. 1–6. IEEE (2013)
13. Plappert, M., Mandery, C., Asfour, T.: Learning a bidirectional mapping between human whole-body motion and natural language using deep recurrent neural networks. Robot. Auton. Syst. **109**, 13–26 (2018)
14. Ziaeefard, M., Bergevin, R.: Semantic human activity recognition: a literature review. Pattern Recogn. **48**(8), 2329–2345 (2015)
15. David, J., Coatanéa, E., Lobov, A.: Deploying OWL ontologies for semantic mediation of mixed-reality interactions for human–robot collaborative assembly. J. Manuf. Syst. **70**, 359–381 (2023)
16. Nikolaidis, S., Kwon, M., Forlizzi, J., Srinivasa, S.: Planning with verbal communication for human-robot collaboration. ACM Trans. Hum.-Robot Interact. (THRI) **7**(3), 1–21 (2018)

17. Maderna, R., Pozzi, M., Zanchettin, A.M., Rocco, P., Prattichizzo, D.: Flexible scheduling and tactile communication for human–robot collaboration. Robot. Comput. Integr. Manuf. **73**, 102233 (2022)
18. Kaczmarek, S., Hogreve, S., Tracht, K.: Progress monitoring and gesture control in manual assembly systems using 3D-image sensors. Procedia CIRP **37**, 1–6 (2015)
19. Oyekan, J., Fischer, A., Hutabarat, W., Turner, C., Tiwari, A.: Utilizing low cost RGB-D cameras to track the real time progress of a manual assembly sequence. Assem. Autom. **40**(6), 925–939 (2020)
20. Kozamernik, N., Zaletelj, J., Košir, A., Suligoj, F., Bracun, D.: Visual quality and safety monitoring system for human-robot cooperation. Int. J. Adv. Manuf. Technol. **128**(1–2), 685–701 (2023)
21. Andrianakos, G., Dimitropoulos, N., Michalos, G., Makris, S.: An approach for monitoring the execution of human based assembly operations using machine learning. Procedia CIRP **86**, 198–203 (2019)
22. Zhang, Z., Peng, G., Wang, W., Chen, Y., Jia, Y., Liu, S.: Prediction-based human-robot collaboration in assembly tasks using a learning from demonstration model. Sensors **22**(11), 4279 (2022)
23. Moutinho, D., Rocha, L.F., Costa, C.M., Teixeira, L.F., Veiga, G.: Deep learning-based human action recognition to leverage context awareness in collaborative assembly. Robot. Comput. Integr. Manuf. **80**, 102449 (2023)
24. Zanchettin, A.M., Casalino, A., Piroddi, L., Rocco, P.: Prediction of human activity patterns for human–robot collaborative assembly tasks. IEEE Trans. Industr. Inf. **15**(7), 3934–3942 (2018)
25. Uschold, M., Gruninger, M.: Ontologies: Principles, methods and applications. Knowl. Eng. Rev. **11**(2), 93–136 (1996)
26. Schlenoff, C., et al.: An IEEE standard ontology for robotics and automation. In: 2012 IEEE/RSJ International Conference on Intelligent Robots and Systems, pp. 1337–1342. IEEE, Vilamoura-Algarve, Portugal (2012)
27. Carbonera, J.L., et al.: Defining positioning in a core ontology for robotics. In: 2013 IEEE/RSJ International Conference on Intelligent Robots and Systems, pp. 1867–1872. IEEE, Tokyo, Japan (2013)
28. Umbrico, A., Orlandini, A., Cesta, A.: An ontology for human-robot collaboration. Procedia CIRP **93**, 1097–1102 (2020)
29. Ding, Y., Xu, W., Liu, Z., Zhou, Z., Pham, D.T.: Robotic task-oriented knowledge graph for human-robot collaboration in disassembly. Procedia CIRP **83**, 105–110 (2019)
30. Conti, C.J., Varde, A.S., Wang, W.: Human-robot collaboration with commonsense reasoning in smart manufacturing contexts. IEEE Trans. Autom. Sci. Eng. **19**(3), 1784–1797 (2022)
31. Jiang, C., Dehghan, M., Jagersand, M.: Understanding contexts inside robot and human manipulation tasks through vision-language model and ontology system in video streams. In: 2020 IEEE/RSJ International Conference on Intelligent Robots and Systems (IROS), pp. 8366–8372. IEEE, Las Vegas, NV, USA (2020)
32. Janowicz, K., Haller, A., Cox, S.J., Le Phuoc, D., Lefrançois, M.: SOSA: a lightweight ontology for sensors, observations, samples, and actuators. J. Web Semant. **56**, 1–10 (2019)
33. Yahya, M., Breslin, J.G., Ali, M.I.: Semantic web and knowledge graphs for industry 4.0. Appl. Sci. **11**(11), 5110 (2021)
34. Liau, Y.Y., Ryu, K.: Conceptual ontology-based context representation for human and two heterogeneous cobots collaborative mold assembly. In: Silva, F.J.G., Ferreira, L.P., Sá, J.C., Pereira, M.T., Pinto, C.M.A. (eds.) Flexible Automation and Intelligent Manufacturing: Establishing Bridges for More Sustainable Manufacturing Systems. FAIM 2023. Lecture Notes in Mechanical Engineering. Springer, Cham (2024). https://doi.org/10.1007/978-3-031-38165-2_62

35. Musen, M.A.: The protégé project: a look back and a look forward. AI Matters **1**(4), 4–12 (2015)
36. PyModbus. GitHub. https://github.com/pymodbus-dev/pymodbus. Accessed 31 Mar 2024
37. Lamy, J.B.: Owlready: ontology-oriented programming in Python with automatic classification and high level constructs for biomedical ontologies. Artif. Intell. Med. **80**, 11–28 (2017)

A Study on Production Scheduling Methods for Ready-Made Meal Industries

Hinari Hamada[1](✉), Nobutada Fujii[1], Shunsuke Watanabe[1], Takehide Soh[1], Ruriko Watanabe[2], Kohei Nakayama[3], Yuji Mishima[3], and Kazuo Yoshinaga[3]

[1] Graduate School of System Informatics, Kobe University, 1-1 Rokkodai-cho, Nada, Kobe, Hyogo 657-8501, Japan
231x055x@stu.kobe-u.ac.jp, nfujii@phoenix.kobe-u.ac.jp, s_watanabe@penguin.kobe-u.ac.jp, soh@lion.kobe-u.ac.jp

[2] Faculty of Science and Engineering, Waseda University, 3-4-1 Okubo, Shinjuku-ku, Tokyo 169-8555, Japan
r.watanabe@aoni.waseda.jp

[3] Rockfield Ltd., 2280 Shimonobu, Iwata, Shizuoka 438-0112, Japan
{k-nakayama,y-mishima,k-yoshinaga}@rockfield.co.jp

Abstract. In ready-made meal industries, which are the focus of this study, maintaining freshness is just as important as meeting delivery deadlines due to the handling of fresh ingredients. To prevent deterioration in freshness, it is necessary to reduce the time from the start of processing to shipping. In this paper, a production scheduling method that simultaneously considers maintaining freshness and due date adherence, is proposed and verified the effectiveness of the proposed method through computational experiments.

Keywords: Ready-made meal industries · Job shop scheduling · Optimization

1 Introduction

In ready-made meal industries, low labor productivity is a major challenge due to difficulties in mechanizing operations and high turnover rates among employees [1, 2]. Furthermore, despite the fact that many processes are required to produce a single product in ready-made meal industries, as economies have grown and societies have become more affluent, people have begun to demand more personalized products that cater to their individual preferences. Consequently, food preferences have diversified, leading to a demand for a wider variety of products. This, in turn, has led to the need for multi-product manufacturing, resulting in the increasing complexity of production systems [3]. However, at the factory, production scheduling currently relies on human experience and intuition, highlighting the need for production scheduling methods that do not depend on manual efforts. Ready-made meal industries have various characteristics that make production scheduling difficult. Firstly, there are many manual operations, and the processing time of these operations varies depending on the number and skill level

Published by Springer Nature Switzerland AG 2024
M. Thürer et al. (Eds.): APMS 2024, IFIP AICT 729, pp. 266–277, 2024.
https://doi.org/10.1007/978-3-031-65894-5_19

of workers [4]. Secondly, it is challenging to determine the production volume because the sales of prepared meals fluctuate due to various factors [5]. Thirdly, since freshness deteriorates over time, it is difficult to keep products in inventory, and short delivery deadlines are set [6]. This makes it crucial to shorten the lead time from processing to shipping, not only to meet delivery deadlines but also to maintain freshness.

The purpose of this study is to propose a production scheduling method that takes into account the characteristics of ready-made meal industries. As a first step, this paper reports on the optimization of production sequencing considering both freshness and due date adherence based on real data.

2 Target Factory

This chapter describes the factory targeted in this study. Figure 1 illustrates the process flow of how a single food product is made in the target factory. Each ingredient constituting the product is transported to the corresponding machine and processed according to a predetermined processing sequence. In some cases, there may be more than one machine that can be processed. Processed ingredients are combined at a certain point, and then undergo treatments such as sterilization and packaging to finally become the food product. Therefore, this problem can be considered as a flexible job shop problem with assembly processes.

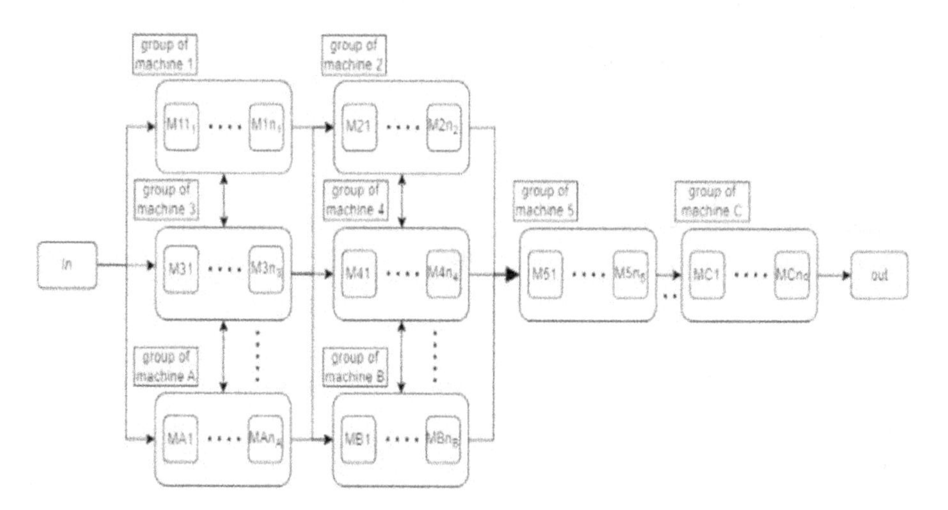

Fig. 1. Production process model of target factory

3 Production Planning Considering Due Date Adherence and Freshness

A production scheduling method that considers both due date adherence and freshness maintenance is proposed. In Sect. 3.1, the notations used in the formulation of the proposed method is defined, and in Sect. 3.2, an overview of the proposed method and its specific formulation are provided.

3.1 Notation

The notations used in proposed method are shown as follows:

Parameters:

- i : item number ($i = 1, \ldots, I$)
- k : operation number of each item ($k = 1, \ldots, k_i$)
- m : machine group number ($m = 1, \ldots, M_{ik}$)
- M_{ik} : machine group that can process operation k of item i
- l : machine number ($l = 1, \ldots, L_m$)
- d_{ik} : due date of operation k of item i
- P_{ik} : processing time of the operation k of item i
- B_k : Set of preceding operations of operation k
- α, β : Parameter to adjust the tolerance for tardiness

Variables:

- s_{ikml} : start time of operation k of item i at machine l in machine group m
- c_{ikml} : completion time of operation k of item i at machine l in machine group m
- x_{ikml} : $\begin{cases} 1: \textit{if operation k of item i is processed at machine l in machine group m} \\ 0: \textit{otherwise} \end{cases}$
- $y_{iki'k'ml}$:

$$\begin{cases} 1: \textit{if operation k of item i precedes the operation k' of item i' at machine l} \\ 0: \textit{otherwise} \end{cases}$$

In the proposed method, the decision variables are s_{ikml}, x_{ikml} and $y_{iki'k'ml}$.

3.2 Overview of the Proposed Method and Its Specific Formulation

The formulation of the flexible job shop scheduling problem is shown below. The proposed method aims to maintain freshness and achieve due date adherence, which are important factors in ready-made meal industries. By minimizing total due date deviations in the objective function, the production lead time is reduced, and the freshness of food ingredients is maintained. By using parameters, the allocation of operations to positions that result in tardiness is prevented, taking into account due date adherence.

$$\min. \sum_i \sum_k T_{ik} \tag{1}$$

$$\text{where } T_{ik} = \sum_m \sum_l g|c_{ikml} - d_{ik}|\{\forall i, \forall k, m \in M_{ik}, l \in L_m\} \tag{2}$$

$$g = \begin{cases} \alpha(c_{ikml} - d_{ik}) > 0 \\ \beta(c_{ikml} - d_{ik}) < 0 \end{cases} \tag{3}$$

$$c_{ikml} = s_{ikml} + x_{ikml} * p_{ik}\{\forall i, \forall k, m \in M_{ik}, l \in L_m\} \tag{4}$$

$$d_{ib} = d_{ik} - p_{ik}\{\forall i, \forall k, b \in B_k\} \tag{5}$$

$$\text{sub. to } s_{ikml} \geq 0\{\forall i, \forall k, m \in M_{ik}, l \in L_m\} \tag{6}$$

$$\sum_l x_{ikml} = 1\{\forall i, \forall k, m \in M_{ik}, l \in L_m\} \tag{7}$$

$$\begin{gathered} s_{ikml} \geq s_{ibk'l'} + p_{ib} \\ \left\{\forall i, \forall k, \forall b, b \in B_m, m \in M_{ik}, m' \in M_{ib}, l \in L_m, l' \in L_{m'}\right\} \end{gathered} \tag{8}$$

$$\begin{gathered} s_{i'k'ml} \geq y_{iki'k'ml} * c_{ikml} \\ \left\{\forall i, i', \forall k, k', \forall b, b \in B_m, m \in M_{ik}, l \in L_m\right\} \end{gathered} \tag{9}$$

Equation (1) represents that the objective function is to minimize the total due date deviations. Equation (2) is the definition of due date deviation, which is expressed by the absolute value of the difference between the completion time of the operation k of item i on machine l in machine group m and the given due date for each operation. Equation (3) is the coefficient for the tardiness tolerance. By giving larger weights to due date tardiness, allocations that lead to tardiness are suppressed. Equation (4) is the definition of the completion time of the operation k of item i on machine l in machine group m. Equation (5) is the definition of the due date for the operation k of item i. This equation ensures that the due date for each operation is given as the latest completion time within the final operation meeting the due date. Equations (6) to (9) are constraint equations. Equation (6) represents that the processing start time is non-negative. Equation (7) indicates that the operation k of item i is processed only once by a single machine. Equation (8) represents the precedence constraint between operations within item i, stating that operation k cannot start until all its preceding operations k are completed. Equation (9) represents the precedence relationship constraint between operations processed on the same machine. If the operation k of item i precedes the operation k' of item i', the start time of the operation k of item i must be earlier than the completion time of the operation k' of item i'.

By minimizing total due date deviations, the proposed method considers maintaining freshness. Additionally, by adjusting the weights of the parameters α and β, the method prevents operation allocations that result in tardiness, thereby achieving due date adherence.

4 Computational Experiment 1: Sensitivity Analysis

In the proposed method, freshness is considered by minimizing due date deviation, but there are two types of due date deviation: tardiness and earliness. To achieve due date compliance, a parameter β is set to determine the tolerance for tardiness. An analysis is conducted on how changes in this value actually affect the due date deviation and tardiness.

4.1 Experimental Conditions

The experiments are performed with the following conditions. [a, b] means that it is an integer constant between a and b. (a, b) means that it is a real number constant between a and b.

The due dates and processing time required for the operation are set based on real data. As an initial stage of this study, to simplify the problem, solving the flexible job shop problem as a job shop problem. In doing so, it is assumed that each machine group consists of only one machine, and that single machine has the capacity equivalent to the actual number of machines in the group.

- The number of items (I) : 18
- The number of operations for each item (k): [3,44]
- The number of machines (M) : 25
- Processing time for each process (P_{ik}) : $(0.01, 4569.5)$
- The due date of the last operation for each item: 10:00 $a.m.\ or$ 11:00 $a.m.$
- parameters that determines the coefficient for the due date deviation tolerance:$\alpha = 1, \beta = 1, 10, 100$

4.2 Experimental Result

Table 1 summarizes the total values of tardiness in production plans obtained for each value of the parameter β, while Table 2 summarizes the total due date deviations.

Tables 1 and 2 show that increasing the weight of parameter β reduces the value of tardiness, and when $\beta = 100$, the value of tardiness become zero. In contrast to the decrease the value of tardiness, the value of due date deviation increased, indicating that there is a trade-off relationship between the two.

Table 2 shows that when changing from $\beta = 10$ to $\beta = 100$, the due date deviation increased by 1300.1 min in order to reduce the 72.9 min of tardiness to zero. This is thought to be a result of the fact that the start time of the entire process had to be moved earlier in order to reduce tardiness to zero, resulting in the occurrence and accumulation of due date deviations in many processes.

Table 1. The impact on tardiness by changing parameter β

		$\alpha = 1, \beta = 1$	$\alpha = 1, \beta = 10$	$\alpha = 1, \beta = 100$
Item	Required operations	Sum of tardiness	Sum of tardiness	Sum of tardiness
1	35	75.9	26.4	0.0
2	19	93.8	0.0	0.0
3	16	26.6	9.5	0.0
4	8	0.0	0.0	0.0
5	17	2.8	0.0	0.0
6	9	72.2	0.0	0.0
7	11	124.0	00	0.0
8	5	0.0	0.0	0.0
9	5	22.0	0.0	0.0
10	5	2.9	0.0	0.0
11	19	131.3	0.0	0.0
12	21	0.0	0.0	0.0
13	5	0.0	0.0	0.0
14	3	0.0	0.0	0.0
15	19	27.9	0.0	0.0
16	44	1.5	5.0	0.0
17	11	77.5	0.0	0.0
18	15	49.6	32.0	0.0
The total due date deviation for all items		**708.0**	**73.0**	**0.0**

Figure 2 shows Gantt charts of the production plans obtained from experiments with different values of β, which is a parameter that determines the tolerance for the tardiness. In these Gantt charts, the vertical axis represents machines, and the horizontal axis represents time. The blue vertical lines indicate two types of due dates. Below the charts, a legend shows the correspondence between each item and its color.

As can be seen from the Gantt chart, when $\beta = 1$, tardiness occurs, and when $\beta = 100$, tardiness does not occur. This is because, although it is important to allocate as many processes as possible near the due date in order to minimize due date deviation discrepancies, when $\alpha = 1$ and $\beta = 1$, the earliness and tardiness are the same value, so some processes are actively allocated to positions where they will be tardiness, thereby making the overall process allocation near the due date.

Table 2. The impact on due date deviation by changing parameter β

		$\alpha = 1, \beta = 1$	$\alpha = 1, \beta = 10$	$\alpha = 1, \beta = 100$
Item	Required operations	Sum of due date deviation	Sum of due date deviation	Sum of due date deviation
1	35	240.4	255.9	180.1
2	19	99.4	90.4	47.1
3	16	26.6	9.8	125
4	8	48.4	143.1	290.1
5	17	28.1	8.9	142.7
6	9	72.2	125.8	222.9
7	11	124	235.5	262.8
8	5	2.1	46.5	60.2
9	5	24.1	13.5	93.2
10	5	2.9	160.5	207.3
11	19	202.2	299.6	560.6
12	21	13.5	15.5	95.3
13	5	60.4	290.7	337.7
14	3	21	52	83.2
15	19	75.2	420.5	436.3
16	44	98.6	147.4	142.4
17	11	130.1	368.8	488.8
18	15	49.6	32.4	241.1
The total due date deviation for all items		**1319.1**	**2716.8**	**4016.9**

5 Computational Experiment 2: Comparison with Dispatching Rule-Based Methods

In order to evaluate the performance of the proposed method, compared experiments were conducted using production schedules generated by two types of dispatching rules. One dispatching method uses the total processing time required to create an item as the rule criterion, while the other dispatching method uses the number of operations required to create an item as the rule criterion. The proposed method, generates a schedule by solving the optimization problem formulated in Chapter 3 using IBM ILOG CPLEX 22.10 [7].

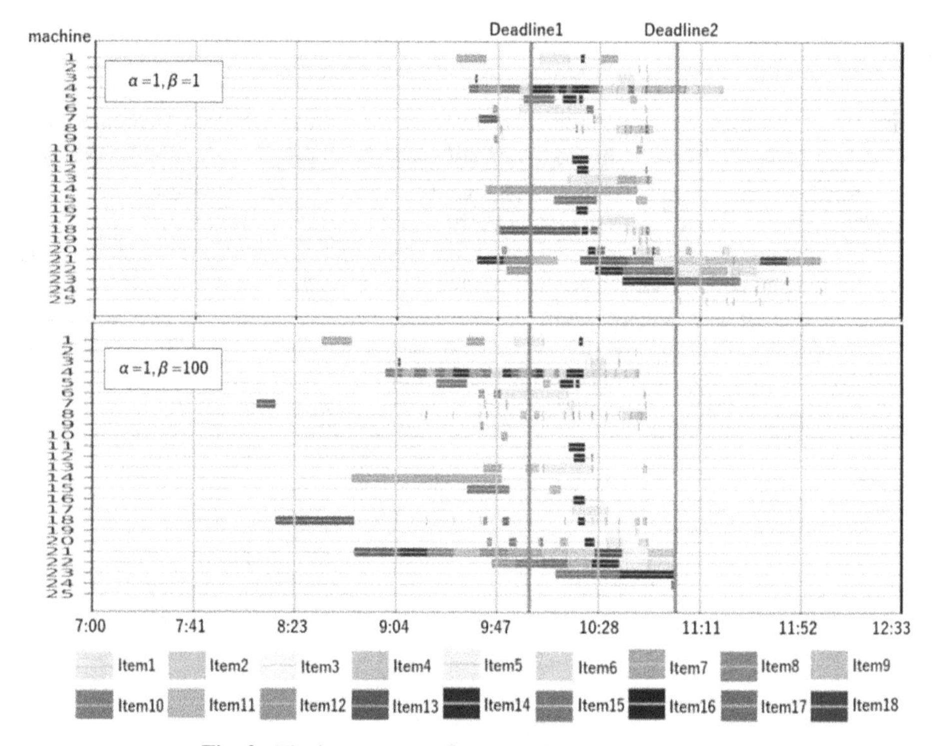

Fig. 2. The impact on tardiness by changing parameter β

5.1 Detailed Explanation of the Two Types of Dispatching Rules

The first dispatching method creates a schedule in the following steps.

STEP 1. Calculate the total processing time required to create each item.

STEP 2. Assign higher priority to items with smaller required processing times.

STEP 3. Allocate operations with the highest priority to minimize due date deviation to the extent that they do not result in tardiness.

The second dispatching method creates a schedule in the following steps.

STEP 1. Calculate the total processing time required to create each item.

STEP 2. Assign priority to items with a larger number of required operations.

STEP 3. Allocate the operations of higher priority items in a way that minimizes the due date deviation.

Hereafter, the production schedule generated by the first dispatching rule will be referred to as Compared Method 1, and the production schedule generated by the second dispatching rule will be referred to as Compared Method 2.

5.2 Experimental Conditions

The experimental conditions are the same as those in Chapter 4, "Sensitivity Analysis of Parameter" β with β set to 100.

5.3 Experimental Results

Table 3 shows the total due date deviation for each item and the sum of these deviations in the production schedules generated by each method. First, the results confirm that production planning by each method does not cause tardiness. From Table 3, it shows that the proposed method can create a production schedule with the smallest total due date deviation, confirming its effectiveness. Regarding the total due date deviation, the value of Compared Method 2 is smaller than that of Compared Method 1, indicating that under the current conditions, the number of operations is an important factor in planning that considers due date deviation minimization.

Table 4 shows the due date deviation per operation for each item in the production schedule generated by the proposed method. From these results, it became clear that in production plans that do not allow tardiness, the average due date deviation is small

Table 3. Comparison of due date deviation values between the proposed method and Compared methods

		Proposed	Compared Method 1	Compared Method 2
Item	Required operations	Sum of due date deviation	Sum of due date deviation	Sum of due date deviation
1	35	180.1	969.1	251.7
2	19	47.1	727.3	43.8
3	16	125.0	242.6	213.2
4	8	290.1	219.7	382.9
5	17	142.7	842.7	837.2
6	9	222.9	66.1	132.9
7	11	262.8	707.6	316.7
8	5	60.2	129.1	54.0
9	5	93.2	118.1	87.0
10	5	207.3	0.0	300.4
11	19	560.6	313.5	556.9
12	21	95.3	237	13.8
13	5	337.7	195.7	331.4
14	3	83.2	93.0	139.0
15	19	436.3	420.4	399.8
16	44	142.4	1410.8	98.7
17	11	488.8	22.3	713.6
18	15	241.1	358.1	661.8
The total due date deviation for all items		**4016.9**	**7073.2**	**5534.9**

for items with a large number of processes, and large for items with a small number of processes. This means that processes for items with many processes are preferentially allocated close to the due date, while processes for items with few processes are not allocated close to the due date. The following factors can be considered as the reasons for this trend. When a process is allocated to a position with leeway from the due date, the processes that need to be completed before that process will inevitably also be allocated to positions with leeway. If such an allocation is made for processes of items with a large number of processes, many processes will be allocated to positions with leeway from the due date, causing the due date deviation value to swell. Therefore, it is thought that the above trend appeared as a result of the computer avoiding such allocations in the process of minimizing the objective function value.

Table 4. Due date deviation per operation for each item in the proposed method's results

		Proposed method	
Item	Required operations	Sum of due date deviation	Average due date deviation per operation
1	35	180.1	5.1
2	19	47.1	2.5
3	16	125.0	7.8
4	8	290.1	36.3
5	17	142.7	8.4
6	9	222.9	24.8
7	11	262.8	23.9
8	5	60.2	12.0
9	5	93.2	18.6
10	5	207.3	41.5
11	19	560.6	29.5
12	21	95.3	4.5
13	5	337.7	67.5
14	3	83.2	27.7
15	19	436.3	23.0
16	44	142.4	3.2
17	11	488.8	44.4
18	15	241.1	16.1

6 Conclusion

In this paper, a job shop scheduling optimization method was proposed for minimizing the total due date deviation time, considering freshness, which is an important aspect of the ready-made meal industries. The effectiveness of proposed methods was confirmed by results of computational experiments; the proposed method was able to reduce the due date deviation and bring it to zero by adjusting the parameter β. At this time, it was confirmed that there is a trade-off relationship between the two, as the due date deviation increases along with the decrease the tardiness. Furthermore, when changing from $\beta = 10$ to $\beta = 100$, it can be seen that the due date deviation increases by 1300.1 min in order to reduce the tardiness of 72.9 min to 0. This is thought to be the result of the need to generally advance the start times of processes in order to make the tardiness zero, resulting in an increase and accumulation of due date deviations for many processes.

Computational experiments also confirmed that the proposed method could reduce the total due date deviation time without causing tardiness compared to production schedules based on dispatching rules. Furthermore, a characteristic of the proposed method was observed: items with a larger number of operations tend to have smaller average due date deviations, while items with fewer operations have larger average due date deviations. The fact that Compared Method 2 outperformed Compared Method 1 also suggests that under the current conditions, the number of operations is a highly important factor in production planning aimed at minimizing due date deviations.

On the other hand, there are two major points where the method needs improvement. The first is the issue of computation time. The proposed method solves the problem using the restricted method with a solver; due to the long computation time, it can only solve small-scale problems. As a result, it was unable to solve the production planning for an entire day in a real-world setting and could only solve problems targeting items with due dates before noon. In the future, investigate methods such as metaheuristics for large-scale experiments that consider factors such as the number of items and alternative machines is planned. The second point is the consideration of variability in processing times. One of the important characteristics of the prepared meal industry is the high proportion of manual operations and the resulting large fluctuations in processing times. The current production plan is scheduled to produce items just before their due dates to minimize the production lead time; with such scheduling, it may result in tardiness if the processing time becomes longer. Therefore, by working on methods that consider the uncertainty of processing times and designing buffer times to absorb variability, it is aimed to develop production plans that can be applied in real-world settings.

References

1. Ministry of Agriculture, Forestry and Fisheries: Basic Knowledge for Improving Food Industry Productivity (in Japanese). https://www.maff.go.jp/j/shokusan/sanki/soumu/attach/pdf/seisansei-5.pdf. Accessed 16 May 2023
2. Hironaka, Y.: The cause of low productivity of Japanese food manufacturing industry is low real utilization rate. J. Jpn. Product. Manag. Assoc. **26**(1) (2019). (in Japanese)
3. Jiang, Z., Yuan, S., Ma, J., Wang, Q.: The evolution of production scheduling from industry 3.0 through industry 4.0. Int. J. Product. Res. **60**(11), 3534–3554 (2022)

4. Ministry of Agriculture, Forestry and Fisheries: Vision for Overcoming Labor Shortage in the Food Manufacturing Industry (2019). (in Japanese)
5. Syntetos, A., Babai, Z., Boylan, J.E., Kolassa, S., Nikolopoulos, K.: Supply chain forecasting: theory, practice, their gap and the future. Eur. J. Oper. Res. **252**(1), 1–26 (2016)
6. Nose, T., Kuriyama, S.: A study on ordering and inventory policies for fresh foods considering uncertainties in logistics environment. J. Jpn. Ind. Manag. Assoc. **43**(4), 276–280 (1992). (in Japanese)
7. ILOG CPLEX. https://www.ibm.com/jp-ja/products/ilog-cplex-optimization-studio. Accessed 16 May 2023

Experimentation and Evaluation of the Usability of an AR-Driven Zero Defect Manufacturing Solution in a Real Life Complex Assembly Setting

João Soares[1(✉)], Jerome Martins[1], Emrah Arica[2], Robert Schmitt[3], Jochen Wacker[3], Christoph Rettig[3], Daryl Powell[2], and Manuel Oliveira[1]

[1] KIT-AR, London, United Kingdom
joao.soares@kit-ar.com
[2] SINTEF Manufacturing, Raufoss, Norway
[3] Dentsply-Sirona, Bensheim, Germany

Abstract. This paper reports the outcomes of a case study on evaluating the usability of AR-driven quality control solution conducted in real life production setting with highly complex, high value production characteristics. The AR solution has been evaluated by the production personnel from every organizational level, with main emphasis given to operators and process engineers. The results clearly indicate a significant usability of the solution in avoiding the quality defects in the assembly line driven by manual operations.

Keywords: Digitalization · Manufacturing · Industry 5.0 · Zero defect manufacturing · Quality Management · Human-centred Manufacturing

1 Introduction

The zero defect manufacturing (ZDM) paradigm aims to eliminate quality defects in manufacturing processes, through a range of strategies, including detection, repair, prevention, and prediction [1]. Detection and repair strategies address identifying defects, while prevention and prediction strategies anticipate and avoid defects before they occur.

Digitalization technologies offer numerous opportunities to advance manufacturing firms towards achievement of data-driven preventive and predictive ZDM strategies [2]. Initial steps involve utilizing data acquisition technologies (e.g., sensors) for continuous data collection from equipment, processes, and products. This enables advanced analysis through big data, AI-based, or simulation-based algorithms, providing a more precise understanding of the relationship between process/product deviations and defects. With this data-driven knowledge, process conditions can be closely monitored, and defects predicted.

Nevertheless, recent research findings indicate that human component remains crucial in the data-driven advancement of ZDM strategies, yet to receive adequate attention

M. Thürer et al. (Eds.): APMS 2024, IFIP AICT 729, pp. 278–289, 2024.
https://doi.org/10.1007/978-3-031-65894-5_20

and requires further investigation [1, 3]. Manufacturing quality is significantly influenced by the human-in-the-loop, especially with their knowledge and flexibility to obtain new insights and improvements in the manufacturing quality.

Human-technology collaboration is particularly crucial for closing the data-driven ZDM loops in human-driven quality inspection and control processes. This collaboration is essential for addressing the limitations of both humans and technology, as discussed in a recent case study by Arica et al. [4]. In this context, the term "augmented operator" or commonly known as "operator 4.0" [5] refers to the collaboration between humans and technology in production. It involves leveraging enabling technologies to enhance the skills and capabilities of operators, as well as to support their cognitive and physical tasks [6]. Augmented Reality (AR), as an emerging technology in production and quality control, can play a critical role in achieving a human-technology collaborative data-driven ZDM loop [4]. In addition to sensing and capturing data from complex processes via embedded sensors, it can assist operators in work performance by augmenting the work instructions, as well as automate the quality verification by combining process analytics and machine learning tools.

In this context, there is a growing interest in academia and industry in testing and evaluating the AR-driven quality control solutions in real life complex manufacturing settings. Nevertheless, there is a lack of literature on field-based studies that report on actual testing and evaluation in complex manufacturing settings, crucial for guiding future implementation projects. While some studies offer evaluation guidelines for AR implementation in industry [7], comprehensive evaluation results are lacking. This paper presents the results of a case study that aims to evaluate the usability of AR-driven quality control and assembly assisting solution in a manufacturer of complex dental instruments. The case study was conducted as part of an EU H2020 project, designated DAT4Zero, that focused on developing a digitally enhanced quality management system for involved case companies.

2 Methodology

2.1 Research Scope and Methodology

This research study has taken as part of testing and validation runs of an AR-driven quality control solution in a use case company involved in the EU H2020 DAT4.Zero project (https://dat4zero.eu/). As AR technology gains traction in industrial contexts, this research aims to thoroughly assess its usability and effectiveness within a complex real life manufacturing setting. This evaluation encompasses examining both user experience and performance enhancements facilitated by the solution.

The study utilizes a case study research methodology which enables an in-depth exploration of specific instances within their real-life context to comprehend complex phenomena and their multifaceted interactions [8]. It employs both quantitative and qualitative data collection and analysis techniques, outlined in Fig. 1, to provide a holistic understanding of the participants' interaction with the AR solution. This structured approach includes gathering demographic information before the experiment, assessing usability afterward, and gathering additional qualitative insights through semi-structured interviews. By adopting this multifaceted systematic approach, the study ensures a comprehensive assessment of the AR solution's impact on workers' perspectives. This integrated approach further enhances the validity and reliability of the findings by facilitating triangulation and cross-verification of information [8].

Prior to participation, all individuals involved in the study were asked to provide informed consent. This ethical prerequisite ensured that participants were fully aware of the study's objectives, potential risks, and the data handling procedures. The informed consent process adheres to established ethical guidelines and reflects the commitment to safeguarding the rights and well-being of the participants [9].

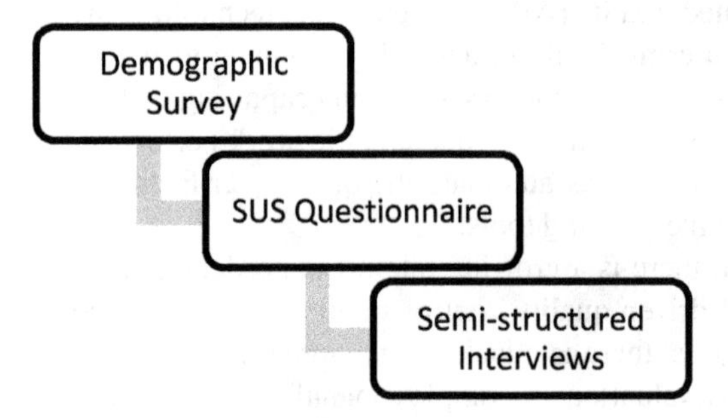

Fig. 1. Data collection and analysis techniques

2.2 Demographic Survey for Profiling of Study Participants

The demographic survey aimed to characterize the diverse participant profiles for cross-analysis and assessment of how their backgrounds and contextual factors may have influenced their interaction with the AR solution. The sample analysed consisted of 14 shopfloor workers with some experience in the particular workstation and assembly process that the study focused on. Before experimenting the AR solution, each participant filled a short survey composed of 7 demographic characterisation questions listed in Table 1. The data was collected using Microsoft Forms as an online survey tool. All the questions and answers were translated into German for participants that are more comfortable with it.

2.3 Experiment and SUS Questionnaire for Quantitative Evaluation of Usability

After the demographic questionnaire the participants underwent an onboarding process designed to familiarize them with the AR solution's interface and the Head-Mounted Display (HMD). This orientation was facilitated through a concise video presentation, enabling participants to gain a preliminary understanding of the technology, and fostering a conducive environment for subsequent tasks [10, 11].

Table 1. Survey questions and answer choices

Question	Answer choices
1.Your Gender?	Male/Female/Other
2.Your Age?	16–20/20–29/30–39/40–55/> 56 (in years)
3.For how much time are you in this company?	<1/1–2/3–5/5–10/10–20/> 20 (in years)
4.Years of experience at this workstation/production line	0 (none)/ <1/2–5/> 5 (in years)
5.How comfortable are you with digital technology (smartphone/ tablet/ computers)?	1 to 10 Scale (From 1 "I'm not comfortable" to 10 "I'm tech savvy")
6.What is your level of understanding concerning Augmented Reality?	None, I never heard of it./I've heard of it, but I don't know about it./I understand what it is./I'm know a lot about it;
7.Have you ever experienced/worked with Augment Reality?	No, never. /Yes but only with Smartphone/tablet./Yes, also with a Head Mount Device

The core of the study involved participants engaging with the AR solution to execute an assembly process. This hands-on experiment provided insights into the practical applicability of the AR solution in a real-world context, evaluating its effectiveness in guiding users through tasks and minimizing errors in the assembly process [12].

After completing the experiments, the participants went through the System Usability Scale (SUS) questionnaire to evaluate the usability of the AR solution quantitatively. SUS is a widely recognized and validated tool used to assess the subjective experience and usability of systems, products, or services [13]. With its simplicity and effectiveness, the SUS is particularly well-suited for assessing a user's overall satisfaction and perception of ease of use [14]. The scale's adaptability to various domains, including technology and software, makes it a valuable tool for comparing and benchmarking usability across different contexts [15, 16].

The SUS consists of the ten statements below, each rated on a 5-point Likert scale, ranging from Strongly Disagree to Strongly Agree. The statements cover aspects such as perceived complexity, ease of use, and overall confidence in using the system. Participants were asked to score each statement individually, and the cumulative scores were then normalized to derive a composite usability score.

1. I think that I would like to use this system frequently.
2. I found the system unnecessarily complex.
3. I thought the system was easy to use.
4. I think that I would need the support of a technical person to be able to use this system.
5. I found the various functions in this system were well integrated.
6. I thought there was too much inconsistency in this system.
7. I would imagine that most people would learn to use this system very quickly.

8. I found the system very cumbersome to use.
9. I felt very confident using the system.
10. I needed to learn a lot of things before I could get going with this system.

After participants completed the questionnaire, individual item scores were converted to a scale of 0 to 4 (Strongly Disagree = 0, Strongly Agree = 4). The odd-numbered statements (1, 3, 5, 7, 9) were scored by subtracting 1 from the participant's response, while the even-numbered statements (2, 4, 6, 8, 10) were scored by subtracting the participant's response from 5. The cumulative scores across all ten statements provided a composite usability score, with higher scores indicating a more positive perception of usability. The SUS questionnaire was also conducted using Microsoft Forms as an online survey tool and presented to the participants in German language, using as reference the translation validated by Brix et al. [17].

2.4 Semi-structured Interviews

To complement the quantitative data from the System Usability Scale, semi-structured interviews were conducted to obtain in-depth insights into the participants' subjective experiences and perceptions of the AR-driven quality control solution and its usability. For participants more comfortable in German, a staff member from the case study company was available to assist in the interview process. This combination of interviews allowed for a deeper exploration of participants' perspectives, capturing nuanced insights that may not be discernible through quantitative measures [18] as well as the emergence of unanticipated insights organically [19].

The questions were crafted to encourage participants to reflect on their interactions with the AR solution, focusing on aspects such as improvement suggestions and the likelihood of recommending the solution. To ensure a consistent yet exploratory approach, the interviews were designed with a set of two predetermined open-ended questions:

1. What would you improve in the solution? Why?
2. Would you recommend the solution to a colleague? Why?

Key annotations and short responses from the semi-structured interviews were collected and subjected to a thematic analysis to identify recurring patterns and emergent themes [20]. This process involved categorizing data into codes, searching for patterns, and developing overarching themes that encapsulated participants' feedback. By extracting key ideas, the analysis aimed to uncover nuanced perspectives on the AR solution's usability, providing qualitative depth to complement the quantitative SUS data. The combination of these techniques facilitated a holistic understanding of the AR solution's usability from both quantitative and qualitative perspectives.

3 Research Context

3.1 The EU H2020 DAT4.Zero Project

The DAT4.Zero project endeavours to pioneer a digitally enhanced quality management and ZDM system. It is built upon a structured data driven ZDM framework comprising data acquisition, management, analysis, and modelling/utilization layers. The data acquisition and management layers are tasked with the generation, aggregation, and organization of data stemming from various shop floor resources, processes, and products, facilitated by distributed multi-sensor networks and data integration technologies. Meanwhile, the data analysis and modelling layers are responsible for processing, simulating, and leveraging the acquired data to support the formulation of strategies, facilitate informed decision-making, and implement feedforward control mechanisms. These endeavours are aimed at advancing the realization of ZDM within smart factories and their associated ecosystems.

3.2 KIT-AR: The AR-Driven ZDM Solution

The KIT-AR solution is designed to digitally enhance shopfloor workers and assist them in complex assembly and quality management operations. The solution incorporates hardware and software technologies that enable the data driven ZDM layers defined above. A Head Mounted Display (HMD) and AR technology provides overlay of digital information over what is physically perceived. The HMD have additional sensors beyond cameras, namely depth camera, accelerometer, gyroscope, magnetometer, four cameras and a 5-channel audio microphone array. The use of sensor rich device supports the acquisition of numerous additional data streams related to the workflow that assist in establishing a comprehensive work context. Combined with a process mining tool that helps analysing the data collected from sensors and from other sources (e.g., Information and Communication Technologies- ICT), the operator is empowered to acquire seamlessly all data in the data-driven ZDM loop for analysis and traceability of the processes, as well as to detect and categorize defects, and to generate inspection reports. As such, the errors are minimized and the efficiency is enhanced by ensuring the worker follows a defined process, step by step.

3.3 The Case Company: Dentsply-Sirona

Dentsply Sirona is a global player in the dental industry with the focus to deliver complete and integrated solutions for dentists and dental technicians. One important production site is based in Bensheim, Germany, where important product segments are produced due to high competence. Its focus is the development and production of dental instruments, such as turbines and electrical highspeed handpiece. With a deep level of manufacturing competence, most of the small and complex metal parts, including high precision gears, are produced inhouse. Main challenges are the size which lies in the range of a few millimetres and the extremely strict tolerances, down to a few micrometres. One of the main quality goals is to reduce the operating sound created by dental rotary tools, as this is highly valued by the costumer and user. Besides the challenging manufacturing processes, the assembly processes are critical due to the high variety of different product variants and the fact that most process steps are done manually by an operator.

The AR-driven ZDM solution was purposely configured to assist in this process. The experimentation and evaluation of the solution was focused on a part of the assembly process to a high rotational speed dental contra angle. The goal was improving the quality of operator involvement in the assembly process by providing a visualization of the correct information and supporting highlights at the right time to make the operator effectively and safely follow the correct assembly workflow in this complex environment.

4 Evaluation Results

4.1 Results of the Demographic Survey – Participants' Profile

The results of the demographic survey indicate that the study predominantly comprised male participants, with only three females included. In terms of age distribution, the majority (64%) fell within the 20 to 39 age range, while a notable proportion (28.5%) were aged over 56. Similarly, the tenure of participants at the case study company was diverse, with 57% having been employed for less than two years, and 35.7% (five participants) having more than ten years of service.

Concerning experience with the assembly process targeted by the AR solution, participants were evenly distributed. Four workers had less than one year of experience, five had one to five years, and the remaining five had over five years of experience with the specific workstation assembly process.

Given the nascent state of AR solutions in industrial markets, it was anticipated that participants might have little to no experience with such technologies. This expectation could have influenced the perceived usability scores in the SUS questionnaires. To assess participants' comfort with digital technologies, the demographic questionnaire included inquiries about their familiarity. Results showed that only 35.7% of participants felt somewhat comfortable with technology, while 50% considered themselves passive users. None of the participants identified as tech-savvy, and two even expressed discomfort with digital technologies.

In questions focused specifically on participants' familiarity with AR, most indicated little to no understanding of the technology. Only 42.8% claimed to understand what AR is, although none considered themselves highly knowledgeable about it. Additionally, in terms of actual experience with AR, the majority of participants (78.5%) had never encountered it in any form. Only three participants reported having experienced AR, and only one of them had used an HMD.

4.2 Quantitative and Qualitative Results for Usability of the Solution

The findings of SUS suggest a predominantly positive user experience, with 71% of respondents indicating that they could effectively focus on the utility of the solution. These findings result in a SUS score surpassing the threshold of 68, which is used by some as indicative of satisfactory usability [15, 16]. Bangor et al. [21] also suggest a following grading scale for interpreting SUS scores:

- Grade A (85 or above): Excellent
- Grade B (70 to 84): Good
- Grade C (50 to 69): OK
- Grade D (below 50): Poor

Still, while Bangor et al. [21] don't specifically mention 68 as a cutoff for usability, the interpretation is commonly made that scores above 68 are indicative of acceptable usability. However, it's essential to keep in mind that usability is context-dependent, and what constitutes acceptable usability may vary based on the specific application and user needs.

The results from the interviews were significantly positive, with only four out of 14 participants showing lower scores on the SUS questionnaire. However, when the participants' SUS scores are compared with their overall acceptability and feedback of the AR solution obtained from the semi-structured interviews it was possible to check that those scores where somehow false negatives. These scores could arguably have been highly biased by scepticism of what is new and the fear of technology replacing experienced workers. When analysing the SUS score combined with the semi-structured interviews' results, in only one of the participants it was found considerable struggle with the use of the solution.

A significant portion of the interview documentation focused on participants' suggestions for improving the solution based on their perceived needs. While each participant had unique insights, some common themes emerged. Firstly, there was unanimous agreement on the importance and utility of the AR solution for new employees. It was recognized that these individuals often require extensive learning and support to reach competency quickly, making them the primary beneficiaries of the solution.

Interestingly, there was also discussion regarding the benefits of the solution for more experienced workers. Key points highlighted the necessity of providing tailored features depending on the user's expertise level. While all features would be available to senior and expert workers, they could selectively utilize them based on their requirements. Similarly, participants emphasized the importance of granularity in work instructions. They suggested the ability to adjust the level of detail within a step, particularly for experienced workers who may need clarification or assistance when encountering tasks they haven't performed in some time.

5 Discussion

This case study was planned and conducted with multiple research methods to gain a better understanding and evaluation of the contextual factors influencing the usability and performance of the AR-driven ZDM solution being implemented. As such, the case study was characterized by administering a demographic questionnaire to each participant before the experiment and the SUS evaluation.

From the analysis, an initial characteristic of the sample was the disparity in ages and in the self-evaluation of comfort with technology. The following cross table indicates a potential correlation between the age of the workers participating in the study and their responses regarding their comfort with technologies (Table 2).

This observation is much in line with the literature, as older individuals tend to have less experience with digital technologies and to create some resistance to the adoption of new technologies [22, 23].

Table 2. Cross table Age of participants VS Comfort with Digital Technology.

	Not comfortable	Passive	Comfortable	Tech Savvy	Total
20–29 Years	0	**2**	**3**	0	**5**
30–39 Years	0	**2**	**2**	0	**4**
40–55 Years	0	**1**	0	0	**1**
>56 Years	**2**	**2**	0	0	**4**
Total	**2**	**7**	**5**	0	**14**

Based solely on direct and objective analyses of the results, the AR Solution being experimented in the assembly workstation appears to be perceived as useful by the assembly workers participating in the study. Despite four participants receiving lower scores in the SUS questionnaire, indicating near-poor usability [21], the overall results are promising for the utilization of the AR solution to support shopfloor workers in the assembly process.

Nevertheless, it is important to note that despite the generally positive scores from the SUS questionnaire, a significant portion of participants demonstrated a low level of familiarity with digital technologies, particularly with AR. None of them had prior experience with the experimental onboarding of the AR solution or with the HMD used (i.e., Hololens 2), which has a reduced Field of View. This limitation was mentioned in some of the interviews as having a significant impact on their experience with the AR solution.

In addition, the experimentation of the AR solution faced technical challenges including integration with existing systems, optimizing performance for real-time use, and mitigating environmental factors. Collaborative efforts with local IT teams addressed the integrations, connection, and compatibility issues, while iterative testing, refined performance and calibration procedures enhanced accuracy of the AR solution to implement and evaluate.

While this study focused on a specific industrial setting, the principles and functionalities of the AR solution are inherently adaptable and scalable to various manufacturing environments. Key factors influencing scalability include the flexibility of the AR platform to accommodate different workflows and processes, the modularity of the solution to support customization based on specific industry requirements, and the scalability of infrastructure to handle increased user volumes and data processing demands.

It is also worth noting that each user had only one opportunity to experiment with the AR solution and the HMD. On average, users spent approximately 15 to 20 min completing the full work instruction set. One can question whether the outcome of usability assessments would remain consistent if participants would have used the solution/HMD for a longer time. While the SUS provides a snapshot of user perceptions after a single interaction, prolonged usage of the AR solution via head-mounted displays (HMDs) could introduce negative factors such as fatigue, discomfort, or cognitive load, or positive factors such as familiarization with the interface, experience with the interaction gestures or habituation with the system, potentially impacting usability scores. However, despite these limitations, the results were favourable for the AR solution as the user's experience was considered as initial part of the onboarding of the solution being implemented. Moreover, through analysis of the characterization questionnaire and annotations from individual interviews, two main discussion insights emerged: Firstly, contextual variables may have influenced the SUS score, and secondly, two outliers were identified for further study.

First one is focused on analysing the five participants that gave a lower score than the rest of the group regarding the AR solution. The primary correlation observed was between the duration of participants' tenure within the company and their scepticism toward the solution. Specifically, participants who scored lower on the SUS had over 10 years of experience. However, during the interviews, these were the participants who became more engaged in brainstorming ideas on how the solution could also benefit more senior roles, after further discussion and understanding the potential of the solution to aid junior roles. As such, using the interviews as complementary method to SUS, some of the participants gained a new perspective regarding the usability of the solution and contributed to discussions for how its utility could be increased. In addition, one highlights demographic results by those participants who 1) have higher seniority within the company; 2) are over 56 years old; 3) have over 5 years expertise within the workstation; and 4) have lower comfort with digital technology (below 6 points); it is also possible to observe a clear relation with those SUS scores below the threshold of 68 points.

The second major insight focused on analysing the two outliers identified in the SUS results. One outliner is a worker with less than 55 years old and less than 2 years of experience in the company. This worker had genuine difficulties with the AR solution, and in general with interacting with digital interfaces, being completely aware of it and mentioning that in the interview. The second outlier is a worker with over 56 years old, and over 10 years of experience within the company. This outlier reported a very low level of comfort with digital technology and gave a SUS score of 95 to the AR solution. The individual interview was critical in understanding this outlier. Thus, this worker had a very particular and interesting approach to the AR solution, since despite having trouble with digital technology the AR solution was perceived as an aid to support the worker memory in tasks not performed for a long time and also as key tool to become independent – as the worker has seen the opportunity to be fully independent - by relying on the AR assist solution and to not have to rely on other colleagues.

6 Conclusion

This paper has focused on evaluation of the usability of AR solution for supporting human-driven quality control of complex assembly processes in a real-life manufacturing setting. A case study was performed with a manufacturer of dental instruments with several techniques involved, including a demographic survey, SUS questionnaire, and semi structured interviews. Based on the analysis of the collected data, several key insights emerge regarding the perception and usability of the AR solution among assembly workers. Firstly, there is a noticeable link between the age of the workers and comfort with technology, that suggests older individuals may exhibit less familiarity and resistance to adopting new technologies. Further, there is a correlation between participants' tenure within the company and their scepticism toward the AR solution. The longer the tenure is the higher the scepticism. Despite of the scepticism, there is a generally positive perception of the AR solution with most participants viewing it as useful for their assembly tasks. This has also been observed on an outlier participant that had low comfort with technology, supporting its potential to enhance independence and task performance.

It is also important to consider the impact of limited exposure of participants to digital technologies and the AR solution itself on the results obtained from the study. Each participant had only one opportunity to use the AR solution, which may have impacted their familiarity and comfort level, and hence the results.

Further studies are recommended to take this aspect into account. Future studies should explore the usability dynamics over extended periods to assess how factors like user fatigue, adaptation influence and perceived usability, ultimately providing insights into optimizing user experiences for sustained usage in real-world industrial environments. Other possibility could even delve into the possibility for a longitudinal usability study to better understand the sustained impact of AR solutions on user experience and productivity in industrial settings as current usability studies based on one-time only SUS may only capture users' immediate perceptions of usability and may not fully capture long-term user experiences or changes in usability over time. Another future work is to extend the study to more case companies both in similar and different manufacturing contexts to extrapolate the results. The evaluation results obtained in this study should be considered as preliminary.

In conclusion, while there are challenges associated with introducing AR technology into the workplace, the analysis highlights the potential benefits and opportunities for improving usability and acceptance through further engagement and understanding of users' needs and perspectives.

Acknowledgments. This study was funded by EU H2020 project DAT4.Zero with grant number 958363.

References

1. Psarommatis, F., May, G., Dreyfus, P.-A., Kiritsis, D.: Zero defect manufacturing: state-of-the-art review, shortcomings and future directions in research. Int. J. Prod. Res. **58**, 1–17 (2020). https://doi.org/10.1080/00207543.2019.1605228
2. Powell, D., Magnanini, M.C., Colledani, M., Myklebust, O.: Advancing zero defect manufacturing: a state-of-the-art perspective and future research directions. Comput. Ind. **136**, 103596 (2022)
3. Wan, P.K., Leirmo, T.L.: Human-centric zero-defect manufacturing: state-of-the-art review, perspectives, and challenges. Comput. Ind. **144**, 103792 (2023)
4. Arica, E., Oliveira, M., Pedersen, T., Mannhardt, F., Myklebust, O.: Human in the data-driven zero defect manufacturing loop: case examples from manufacturing companies. In: Presented at the International Conference on Flexible Automation and Intelligent Manufacturing (2023)
5. Romero, D., Stahre, J., Taisch, M.: The Operator 4.0: Towards socially sustainable factories of the future (2020)
6. Arica, E., Oliveira, M.F., Powell, D.J.: Augmenting the production operators for continuous improvement. In: Proceedings of the 2022 IEEE International Conference on Industrial Engineering and Engineering Management. Kuala Lumpur (2022)

7. Dahl, T.L., Oliveira, M., Arica, E.: Evaluation of augmented reality in industry. In: Presented at the IFIP International Conference on Advances in Production Management Systems (2020)
8. Yin, R.K.: Case Study Research: Design and Methods. Sage Publications, Thousand Oaks, California (2009)
9. Emanuel, E.J., Wendler, D., Grady, C.: What makes clinical research ethical? JAMA **283**, 2701–2711 (2000)
10. De Pace, F., Manuri, F., Sanna, A.: Augmented reality in industry 4.0. Am. J. Comput. Sci. Inf. Technol. **6**, 17 (2018)
11. Bottani, E., Vignali, G.: Augmented reality technology in the manufacturing industry: a review of the last decade. Iise Trans. **51**, 284–310 (2019)
12. Stewart, K., Williams, M.: Researching online populations: the use of online focus groups for social research. Qual. Res. **5**, 395–416 (2005)
13. Brooke, J.: SUS-A quick and dirty usability scale. Usability Eval. Ind. **189**, 4–7 (1996)
14. Bangor, A., Kortum, P.T., Miller, J.T.: An empirical evaluation of the system usability scale. Intl. J. Hum. Comput. Interact. **24**, 574–594 (2008)
15. Lewis, J.R., Sauro, J.: The factor structure of the system usability scale. In: Presented at the Human Centered Design: First International Conference, HCD 2009, Held as Part of HCI International 2009, San Diego, CA, USA, July 19–24, 2009 Proceedings 1 (2009)
16. Saunders, M., Lewis, P., Thornhill, A.: Research methods for business students. Pearson education (2009)
17. Brix, T.J., Janssen, A., Storck, M., Varghese, J.: Comparison of German translations of the system usability scale–which to take? In: German Medical Data Sciences 2023–Science. Close to People, pp. 96–101. IOS Press (2023)
18. Kvale, S., Brinkmann, S.: Learning the Craft of Qualitative Research Interviewing. Sage Publications, Thousands Oaks (2009)
19. Gibbs, G.R.: Thematic coding and categorizing. Analyzing Qual. Data. **703**, (2007)
20. Braun, V., Clarke, V.: Using thematic analysis in psychology. Qual. Res. Psychol. **3**, 77–101 (2006)
21. Bangor, A., Kortum, P., Miller, J.: Determining what individual SUS scores mean: adding an adjective rating scale. J. Usability Stud. **4**, 114–123 (2009)
22. Czaja, S.J., et al.: Factors predicting the use of technology: findings from the center for research and education on aging and technology enhancement (CREATE). Psychol. Aging **21**, 333 (2006)
23. Morris, M.G., Venkatesh, V.: Age differences in technology adoption decisions: implications for a changing work force. Pers. Psychol. **53**, 375–403 (2000)

Enriching Scene-Graph Generation with Prior Knowledge from Work Instruction

Zoltán Jeskó[1], Tuan-Anh Tran[1], Gergely Halász[1], János Abonyi[2,3], and Tamás Ruppert[1,2](✉)

[1] Department of System Engineering, University of Pannonia, Egyetem str. 10, Veszprém, Hungary
ruppert@abonyilab.com
[2] HUN-REN-PE Complex System Monitoring Research Group, University of Pannonia, Egyetem str. 10, Veszprém, Hungary
[3] Department of Process Engineering, University of Pannonia, Egyetem str. 10, Veszprém, Hungary

Abstract. With the current focus on human resources in Industry 5.0, analysing the work movements of industrial operators is the important first step in optimising labour performance. Thanks to the popularity of camera sensors, vision-based Human Activity Recognition models have become useful engines for real-time monitoring tools, in which scene-graphs play an important role. Traditional scene-graph generation methods rely primarily on visual data for perception, neglecting a valuable source of process-oriented prior knowledge: the work instruction. Therefore, an extension of the scene-graph paradigm by integrating ground truth elaborated on elements from the work instruction is elaborated to complement and enhance the understanding of human activities in industrial environments, and improve the tracking capability with micro and repetitive movements. This conceptual paper discusses the basic design of this approach with potential applications in industrial environments, which is validated by a simulated use case of an electronic assembly process. Based on the proposed extension, the Human Activity Recognition model can be lightweight and robust. Further integration of multi-modal sensory inputs beyond visual cues, such as environmental and human-centric data, can enrich scene interpretation and provide a more comprehensive understanding of work behaviour, paving the way for more effective labour utilisation and improved productivity.

Keywords: Scene-graph · Human-centered · Activity recognition · Industry 5.0 · Operator 4.0

Published by Springer Nature Switzerland AG 2024
M. Thürer et al. (Eds.): APMS 2024, IFIP AICT 729, pp. 290–302, 2024.
https://doi.org/10.1007/978-3-031-65894-5_21

1 Introduction

With machines and automated robots still struggling to replace the cognitive and motor skills of human operators [20,27], the human presence in manufacturing systems is irreplaceable, thus urging the effective management of human resources [25]. To improve human performance, Human Activity Recognition (HAR) solutions have been developed for automated detection and assessment, most of which use inertial measurement units (IMUs) or accelerometers on the target operator [26]. This sensor-based approach has the inherent disadvantage of requiring the wearer to carry the device or smartphone, which is not comfortable for industrial operators during work performance [15]. With the rapid development of image processing techniques, vision-based HAR has become a trending and preferred solution [5] to capture natural, hands-free work gestures. However, a major challenge for a continuous vision-based monitoring application is predicting the interaction between people and objects with an appropriate level of accuracy at the right time [1]. From a productivity improvement perspective, considering that work movements and micro-movements in an industrial context can be as short as a quarter of a second [7], tracking human activities with high sensitivity to repetitive micro-movements is valuable for productivity and performance assessment purposes [22], and indirectly supports the design and optimisation of macro-work tasks [4,30] especially in labour-intensive industries such as electronics assembly.

To this end, object-graph representations have been used in several previous efforts, such as a hybrid network of scene and temporal graphs for complex activity detection [18], a spatio-temporal action graph network for multi-object interactions in near-collision events [13], or a scene context-aware graph using common-sense knowledge mapping on skeletal data [34]. Similarly, a set of actions can be automatically recognized with an image-based scene graph prediction model such as Action Genome [17]. Although promising results have been obtained, these studies were initially developed to recognize everyday activities in a general context, thus under-utilizing the specific characteristics of manufacturing processes, such as repetitiveness, or the abundance of available information, such as the work instruction for a given process. In industrial production, where processes are tightly controlled with predefined productivity and quality requirements, repetitive manual tasks are well defined within certain limits based on work instructions [2] and the heuristic behavior of operators. Although work instructions may vary within companies and industries without a general standard, they contain the prior knowledge of the process, including the interaction between work-pieces (which include parts and components), work tools, and their spatial or even temporal relationships.

Based on the advancement of Industry 4.0 (I4.0) technologies, the development of human-centered solutions has flourished to meet the needs of human operators during their work [29], as formulated in the core concept of Operator 4.0 (O4.0) [28]. In the context of Industry 5.0 (I5.0) [3], human centricity has received particular attention [9,24] urging a deep human integration into a resilient manufacturing system as the generation of Operator 5.0 (O5.0).

Motivated by these circumstances, this study proposes an extended scene graph principle, specifically elaborated for the industrial environment with repetitive micro movements, by integrating the information from the work instruction. The process-oriented prior knowledge in each work step formulates a knowledge graph as the backbone for recognizing the activity pattern. The detected scene can be updated with cyclical behaviors, improving accuracy while providing robustness against occlusion, image loss, etc. Based on this extension, the vision-based HAR solution can be lighter and more accurate, but still computationally efficient for real-time industrial monitoring applications. A concept of an ideal HAR solution based on the proposed extension is formulated, which suggests how the scope of a HAR model can be extended to include various process and environmental sensory inputs, as well as the physiological parameters of an operator during the work session. By avoiding poor quality work instructions, higher efficiency and job satisfaction can be achieved [12]. This study expands the scope of O4.0 technologies to enhance human performance, facilitating the transition to O5.0 [8].

The paper is structured as follows: Sect. 2 introduces the main components and structure of the proposed extension for scene graph generation from the work instruction. Section 3 presents a conceptual use case of applying the extension in recognizing and extracting meaningful activities from video recordings of a simulated workstation with repetitive movements, while Sect. 4 delivers remarks on the proposed concept, and suggestions for future research.

2 Formalisation of the Proposed Extended Scene Graph

A scene graph is a structural representation that describes the content of a scene in detail, including objects, their attributes, and the associated relationships between them. A scene graph generation model takes an image as an input and generates a visually-grounded scene graph [35], with objects visualized by a set of nodes belonging to different classes and having corresponding attributes. A set of edges connecting the object nodes encode the relationships between objects.

Casual techniques of scene graph generations can be depicted on the left side of Fig. 1. Various segmentation algorithms and object recognition models are deployed to extract the objects from the input image, which also assign labels to the objects. Semantic segmentation [11] treats the elements in the image as uncountable and amorphous. Consequently, similar objects (e.g., people) are grouped into a class label. It assigns all the pixels of the image to a class. In the case of instance segmentation [10], overlapping masking is performed per object in addition to labeling, and the overlaps are resolved using non-maximum suppression. Then the predicted segment is sorted by its confidence value, and segments with low scores are removed. The algorithm goes through the sorted confidence values for each object, checking whether the previous object contained a pixel from the current one. The segment is accepted after this iteration if enough of the segment remains, otherwise, it is deleted [19]. However, these efforts all suffer from problems such as occlusion (both temporarily and permanently), poor lighting conditions, etc. As an aftermath, unknown objects appear,

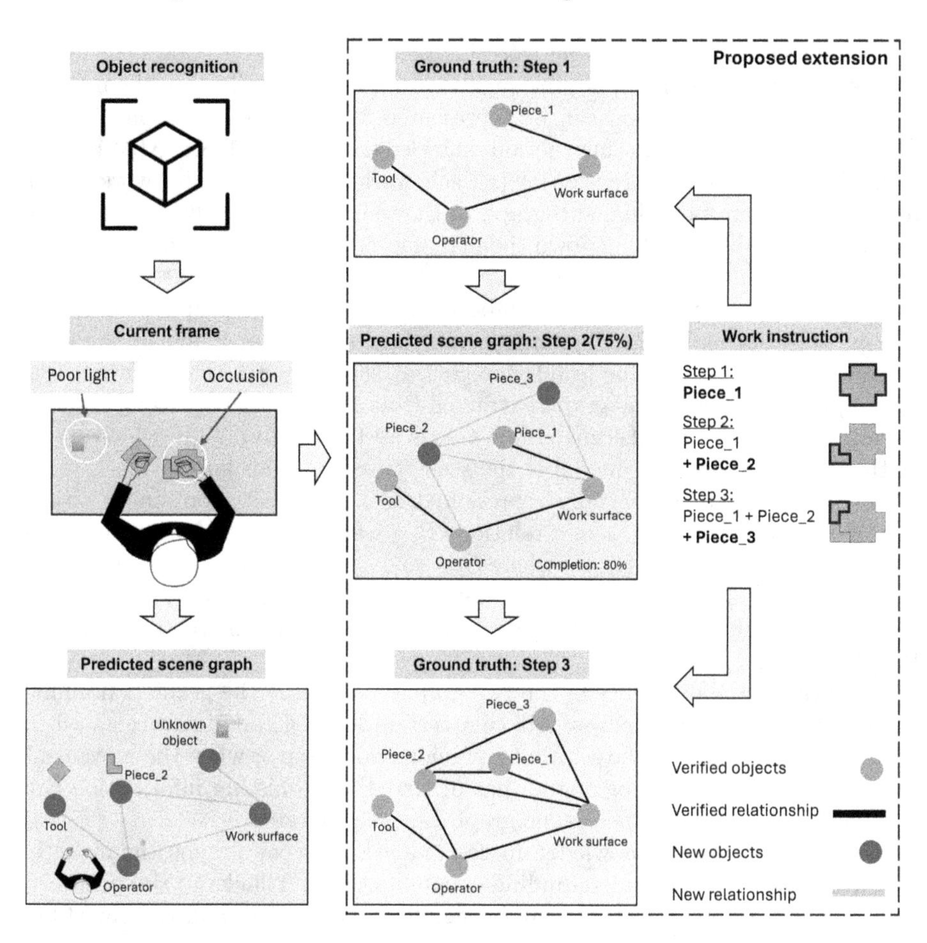

Fig. 1. The scene graph generated by the normal approach on the left, with missing objects in the predicted scene graph due to poor light and occlusion. The proposed extension is described in the right block, which utilizes the work instruction to generate the ground truth scene graphs for the previous and next steps. Thanks to these ground truths, the current frame is predicted as the second work step with 80% completed, with a 75% probability.

or some details are missed from the predicted graph. The more effort to cope with these problems, the more complicated the image processing techniques become.

Our proposed extended scene graph, can be described as an extension of a scene graph generation model, which utilizes the available data from the work instruction of the ongoing process. Assuming that a well-written work instruction contains sufficient information in each work step, including necessary work motions, used tool, and new work-piece that appears with the expected relationship with existing work-pieces from the previous step [2], the scene graph for each step can be elaborated to create the ground truth (or expected condition).

The information derived from work instructions into a G_W knowledge graph begins by structurally mapping the various process steps outlined in the instructions. The work instruction can be represented as a knowledge graph, like the AWI-KG (assembly work instruction knowledge graph) [21] or ARWI (augmented reality work instruction) [6]. Each work step within the workflow is represented as an individual sub-graph in the temporal knowledge graph of the whole process, thus breaking down the manufacturing workflow into manageable sub-processes.

The construction of the $G_{W,n}$ sub-processes knowledge graph of the n-th process step begins by the identification of the core entities found in the work instruction. The draft scene graph detected at the current frame is denoted as G_D, representing the scene graph extracted from the observation along with its uncertainty. The expected scene graph, G_E, is abstracted from G_W, which serves as the ideal or theoretical model of the scene based on prior knowledge.

The graphs can be formally represented as a given sets containing object classes C, attribute types A, and relations R, a scene or knowledge graph G can be defined as a tuple $G = (O, E)$ where $O = \{o_1, \ldots, o_n\}$ is the set of objects, and $E \subseteq O \times R \times O$ is the set of edges. All objects are denoted as $o_i = (c_i, A_i)$, where $c_i \in C$ represent the class of the object and $A_i \subseteq A$ represent the attributes of the object [32].

By forming the difference of the two graphs, we obtain the graph containing the uncertain nodes and edges. Above a certain level of confidence (e.g., when information has been passed on from a previous work step or when the movement of the hand of the operator or tool has occurred), the missing information can be injected, and the enriched scene graph can be generated.

Incorporating prior knowledge in the analysis of scene graphs significantly enhances holistic scene understanding capabilities [35]. Thanks to the proposed extension, the calculation effort can be significantly reduced, and the algorithm will become more robust against problems such as poor lighting, occlusion, etc. The objects in the scene graph only need to be updated when there is a change in the detected objects. For example, if the workpiece remains stationary and no work is being performed on it, there is no need to update the node and its edges. Establishing a new relationship between objects is necessary only when a new object appears on the scene or when the registered objects undergo some form of interaction.

By having a priori knowledge of the cycle time and process time of the process elements, unnecessary edges can be revised and gradually removed. After the specified time has elapsed, the task is considered completed. This allows for the removal of irrelevant objects and edges. Based on the currently detected ground-truth scene graph, these transitions can be used for anomaly detection or alarm management. For example, they can be used to signal to the worker or supervisor when the production cycle time is continuously increasing due to repetitive work load, thus predicting worker exhaustion. In such cases, the worker can be sent on a break or transferred to another workstation where they can perform different tasks. This change in working conditions may help the worker regain a state of flow, enabling them to continue working productively.

A demonstration is given in the next section, to describe the resultant advantages in the industrial work environment with micro and repetitive tasks.

3 Demonstration Study of the Extended Scene Graph

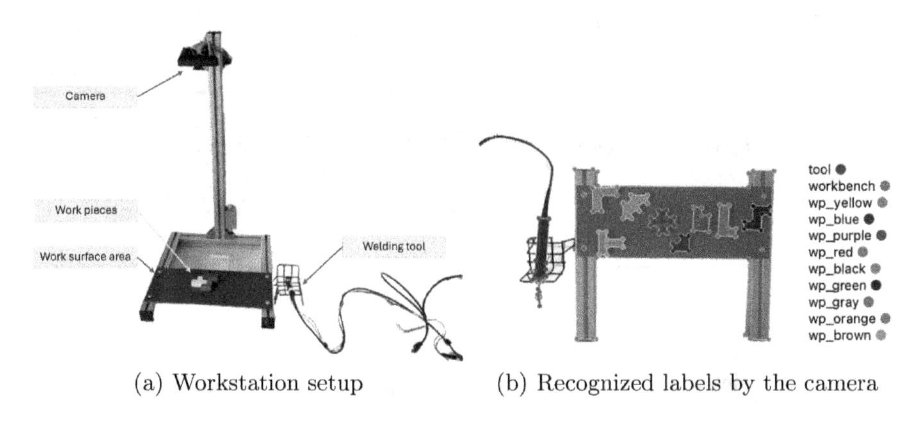

(a) Workstation setup (b) Recognized labels by the camera

Fig. 2. The constructed workstation with a camera mounted on top to capture the working activities conducted on the work surface area (a) and the work-pieces placed on the work surface area, with recognized labels by the camera (b).

In this demonstration, a simulated workbench similar to an industrial workstation is constructed as shown in Fig. 2(a). The production process focuses on assembling work-pieces by using a welding tool, similar to the manual process in electronics production. Figure 2(b) shows the work-pieces on the work surface area and the welding tool used in the assembly process, from the camera point-of-view. The objects are recognized by using a general segmentation model. During the study, an operator was required to weld the work-pieces together to complete the final product, following the work instructions provided in Table 1.

The work process is repeated to create a dataset for later data processing and improvement planning. The primary goal is to improve the processing of the output of the scene graphs by enabling the prediction of the actual process step based on the elements of the graph. Figure 3 is demonstrating the application of the proposed scene graph extension on the fifth step of the work process. The work instruction requires the operator to use both hands: holding the tool in one hand while fixing the work-pieces with the other hand during the welding process.

The "draft scene graph" (see in Fig. 3) constructed from the "segmentation map" of the "current frame" contains uncertainties, which result from blind spots on the work-pieces caused by the occlusion of the left hand and its shadowing

Table 1. The work instruction with five elementary steps and corresponding generated visual ground-truth scene graph. Firstly, the process began with placing the white part in the designated work area. Secondly, the brown part is placed and aligned with the white one, then they are joined by using the welding tool. The green part was then placed and aligned with the previously assembled block of white and brown parts.

Step	Figure	Description	Ground-truth scene graph
1		Placing a white part on the table	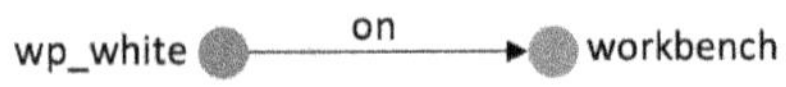
2		Fitting a brown work-piece to a white one	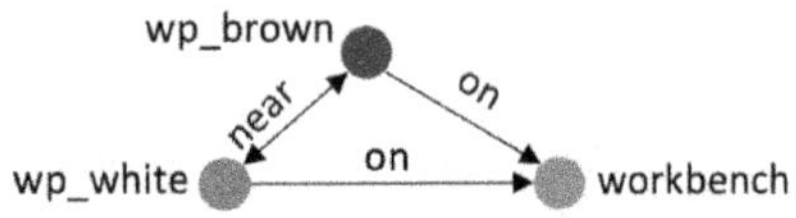
3		Welding brown and white work-pieces	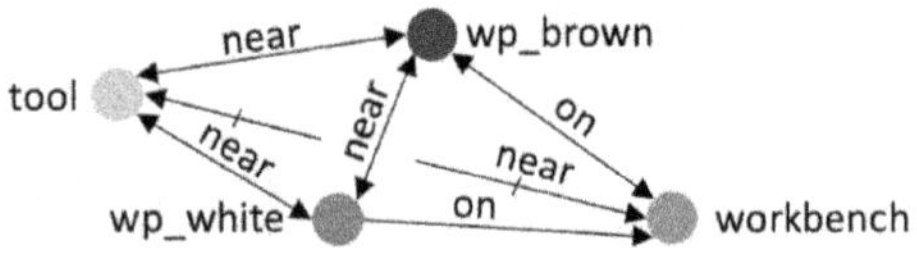
4		Fitting green work-piece to brown and white work-piece	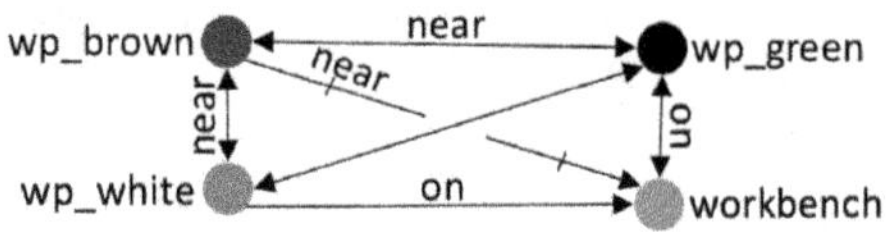
5		Fitting green work-piece to brown and white work-piece	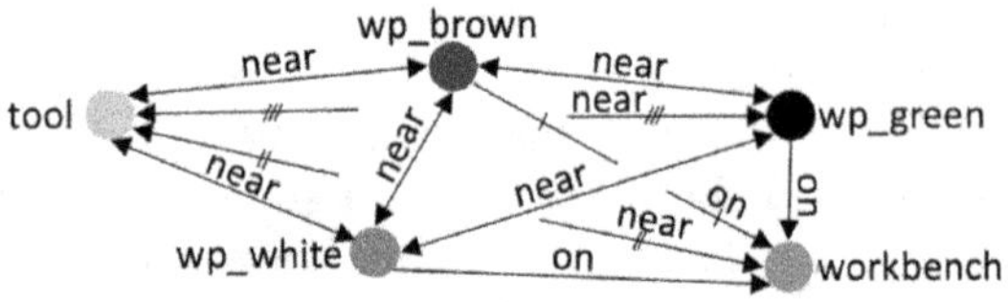

of the visible part. Even if the visible part were well-illuminated, detection with high precision would be challenging for most object segmentation models since less than 20 percent of the object is visible. The "draft scene graph" cannot predict which workflow will take place without prior information. All work steps where the tool is to be used can be considered as potential predictions, in this case, the third and the fifth steps. Using our proposed extension by including prior knowledge, an "enriched scene graph" can be generated with a high degree of predicted certainty, which accurately describes the scene. Based on the previous iterations, the current activity is expected to align with the analysis of the

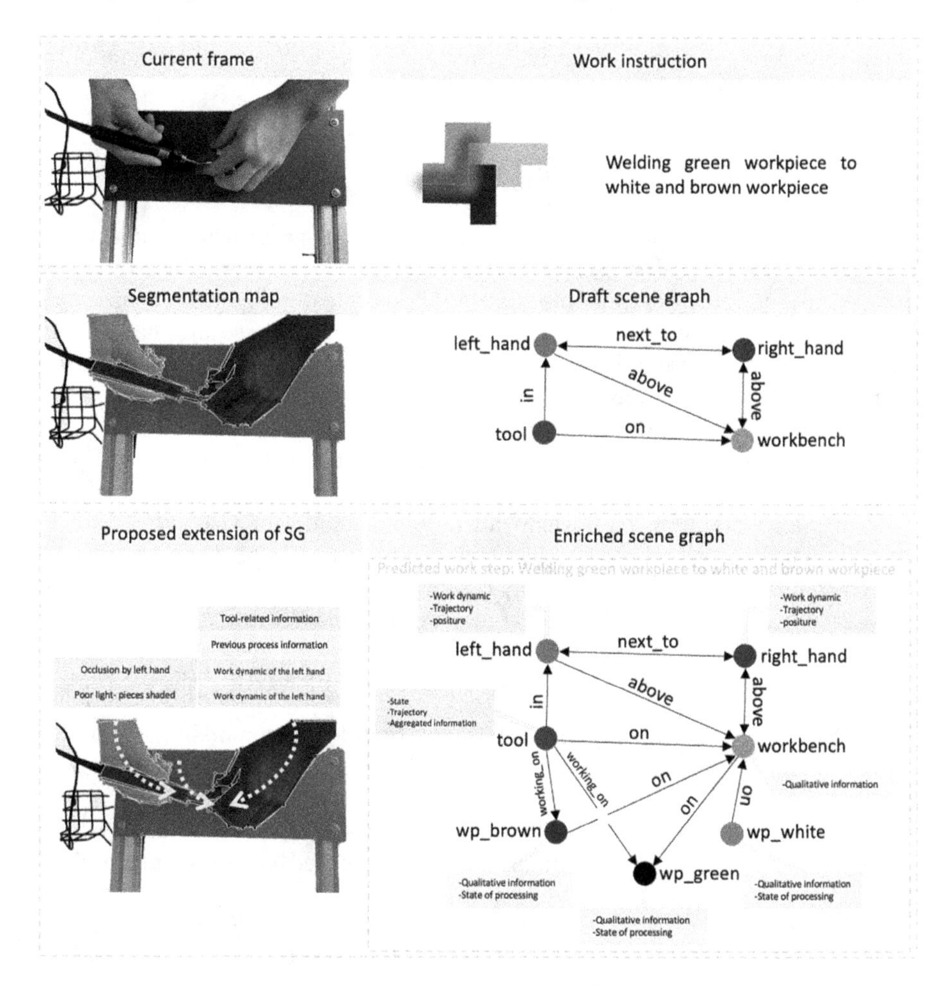

Fig. 3. Example of generating the enriched scene graph with prior knowledge from work instruction, from one frame recorded by the camera during the use case.

ontology derived from the work instruction and by following the manufacturing process step-by-step. Considering these constraints, the expected activity should be the fifth process step, welding the green work-piece to the white and brown work-pieces.

The partially detected graph can also be completed based on the knowledge graph stored as a sequence of ground-truth scene graphs and the information from the position and processing of the work-pieces in the previous steps. As the white, brown, and green work-pieces should be positioned under the hands according to the work instruction, and considering the dynamics of the left hand, it is highly possible that the work-pieces can be found under the hand of the operator. By storing temporary information generated and captured during the production process in the model, with a timestamp, in the nodes of the graphs as feature vectors, enables further improvement of the accuracy of the model by considering the features of each node.

4 Conclusion and Future Works

In this study, a conceptual method for enriching scene-graph generation in the industrial environment is suggested, based on the prior knowledge derived from the work instruction. A detailed use case study will be developed in future work to prove the applicability of the proposed method, with a setup that closely imitates the manual assembly processes commonly used in factories, where components are either semi-automated or completely assembled by human labor using tools. The main focus of the study will be to compare the results of the optical-only approach and the extended prior knowledge approach in different workflow scenarios with different uncertainties.

We are also planning to develop an ontology model of the work instruction. The ontology model can be segmented into four sets of domain ontology classes because the knowledge graph comprises multiple sub-ontologies, such as resource, process, product, and monitoring ontology. Time also appears as a domain-independent core ontology [14]. The following list summarizes the industry-specific name-spaces and the ontologies can be used for the concept:

- smo- Smart Manufacturing Ontology [33] – Ontology for modelling I4.0 production lines and smart factories based on the RAMI 4.0 digitalisation framework [31].
- bbd- Body-based Gestures - An Ontology for Reasoning on Body [23] provides a framework for modeling human body model-based gestures within their context of use, utilizing extensible gesture representation.
- sosa—Sensor, Observation, Sample, and Actuator ontology [16] - Ontological description for modeling the interaction between the entities involved in observing, actuating and sampling.

Besides the concept serving as a lightweight basis for a HAR model, this approach suggests a hint for individualizing the work instructions based on the detected activities, thus assigning an operator with an appropriate workload or temporal demand, considering the level of environmental comfort and the available cognitive capacity of the operator.

For that purpose, additional multi-modal sensory inputs with environmental data (e.g., intensity and direction of light condition, machine vibration, noise), and human-centric data (e.g., acceleration of hand or body, eye response such as pupil dilation) can be integrated as illustrated in Fig. 4. While the work instruction supports the ground truth scene graph and a draft scene graph can be achieved from the current frame, additional data provide useful information about the expected error and uncertainty of the current scene graph. To collect environmental data, a light sensor can provide the illumination intensity and direction, which generate an uncertainty score for the region belonging to a certain node. Shading and occlusion can be detected with the same principle,

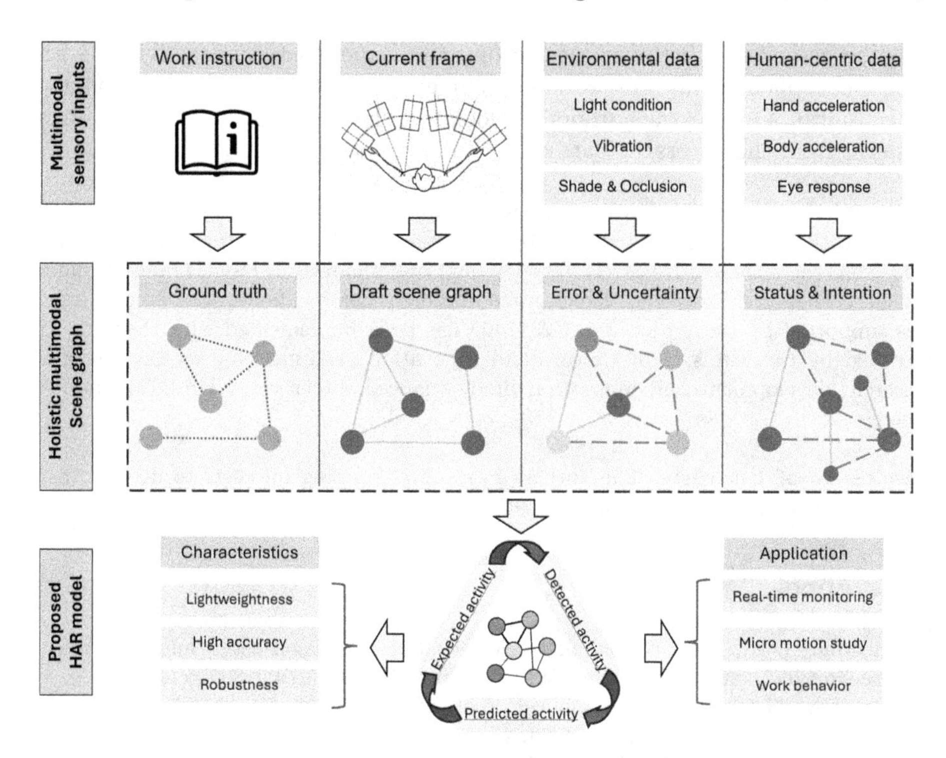

Fig. 4. The proposed video-based HAR with the extended scene graph.

or from another camera that captures the surroundings of the workstation. To avoid the cases when the operators forget about the work instructions, or do the wrong steps, their intention can be strengthened with additional human-centric data from wearable such as accelerometers, or eye movements. This data generates extra nodes or edges as the possible intention of the operator for the next step.

With this extended scene graph that contains the enriched ground truth, possible error and uncertainty, human status and intention, a HAR model can be elaborated as the core of a real-time monitoring solution, that serves the purpose of establishing the "expected activity", capturing the current "detected activity", and delivering the "predicted activity". The built-in characteristics are the lightweightness, high accuracy, and robustness. The solutions that are developed on this approach are applicable for micro-motion study, as the model not only can capture each body part of the operator such as hands, but also can keep track of each temporal relationship between objects. The results from the micro-motion study enable deeper consideration of the work behavior of the operator, based on subtle gestures that suggest hesitation, forgetting work steps, wrong and unergonomic gestures, etc. The abstractions from both sensory inputs and visual domains can generate patterns or principles of the work movement in close association with surrounding conditions and human factors, providing insights into the association between specific operator experiences and activities with the process phenomena, thus deriving more general patterns and subtle connections. In future works, the authors will look for relevant methods to extract such patterns from the generated dataset.

Acknowledgments. This work has been implemented by the TKP2021-NVA-10 project with the support provided by the Ministry of Culture and Innovation of Hungary from the National Research, Development and Innovation Fund, financed under the 2021 Thematic Excellence Programme funding scheme. The work of Zoltán Jeskó was supported by the project no. C2317469 has been implemented with the support provided by the Ministry of Culture and Innovation of Hungary from the national research, development and innovation fund, financed under the KDP-2023 funding scheme.

Disclosure of Interests. The authors have no competing interests to declare that are relevant to the content of this article.

References

1. Beddiar, D.R., Nini, B., Sabokrou, M., Hadid, A.: Vision-based human activity recognition: a survey. Multimedia Tools Appl. **79**(41), 30509–30555 (2020)
2. Bragança, S., Costa, E.: An application of the lean production tool standard work. Jurnal Teknologi (Sci. Eng.) **76**(1), 47–53 (2015)
3. Breque, M., De Nul, L., Petridis, A., for Research, D.G., Innovation: industry 5.0: towards a sustainable, human-centric and resilient European industry. Res. Innov. Policy (2021)
4. Cimini, C., Romero, D., Pinto, R., Cavalieri, S.: Task classification framework and job-task analysis method for understanding the impact of smart and digital technologies on the operators 4.0 job profiles. Sustainability **15**(5), 3899 (2023)
5. Dang, L.M., Min, K., Wang, H., Piran, M.J., Lee, C.H., Moon, H.: Sensor-based and vision-based human activity recognition: a comprehensive survey. Pattern Recogn. **108**, 107561 (2020)
6. Geng, J., et al.: A systematic design method of adaptive augmented reality work instruction for complex industrial operations. Comput. Ind. **119**, 103229 (2020)
7. Ghani, U., Hayat, M., Khalid, Q.S., Azam, K.: Productivity improvement through time and motion method. Int. J. Eng. Technol. **12**(2), 108–123 (2020)
8. Gladysz, B., Tran, T.A., Romero, D., van Erp, T., Abonyi, J., Ruppert, T.: Current development on the operator 4.0 and transition towards the operator 5.0: a systematic literature review in light of industry 5.0. J. Manuf. Syst. **70**, 160–185 (2023)
9. Grosse, E.H., Sgarbossa, F., Berlin, C., Neumann, W.P.: Human-centric production and logistics system design and management: transitioning from industry 4.0 to industry 5.0 (2023)
10. Hafiz, A.M., Bhat, G.M.: A survey on instance segmentation: state of the art. Int. J. Multimedia Inf. Retrieval **9**(3), 171–189 (2020)
11. Hao, S., Zhou, Y., Guo, Y.: A brief survey on semantic segmentation with deep learning. Neurocomputing **406**, 302–321 (2020)
12. Haug, A.: Work instruction quality in industrial management. Int. J. Ind. Ergon. **50**, 170–177 (2015)

13. Herzig, R., et al.: Spatio-temporal action graph networks. In: Proceedings of the IEEE/CVF International Conference on Computer Vision Workshops (2019)
14. Hobbs, J.R., Pan, F.: An ontology of time for the semantic web. ACM Trans. Asian Lang. Inf. Process. (TALIP) **3**(1), 66–85 (2004)
15. Hussain, Z., Sheng, M., Zhang, W.E.: Different approaches for human activity recognition: a survey. arXiv preprint arXiv:1906.05074 (2019)
16. Janowicz, K., Haller, A., Cox, S.J., Le Phuoc, D., Lefrançois, M.: Sosa: a lightweight ontology for sensors, observations, samples, and actuators. J. Web Semant. **56**, 1–10 (2019)
17. Ji, J., Krishna, R., Fei-Fei, L., Niebles, J.C.: Action genome: actions as compositions of spatio-temporal scene graphs. In: Proceedings of the IEEE/CVF Conference on Computer Vision and Pattern Recognition, pp. 10236–10247 (2020)
18. Khan, S., Teeti, I., Bradley, A., Elhoseiny, M., Cuzzolin, F.: A hybrid graph network for complex activity detection in video. In: Proceedings of the IEEE/CVF Winter Conference on Applications of Computer Vision, pp. 6762–6772 (2024)
19. Kirillov, A., He, K., Girshick, R., Rother, C., Dollár, P.: Panoptic segmentation. In: Proceedings of the IEEE/CVF Conference on Computer Vision and Pattern Recognition, pp. 9404–9413 (2019)
20. Lagorio, A., Cimini, C.: Towards 5.0 skills acquisition for students in industrial engineering: the role of learning factories. Procedia Comput. Sci. **232**, 317–326 (2024)
21. Li, W., Wang, J., Jiao, S., Liu, M.: Augmented assembly work instruction knowledge graph for adaptive presentation. In: Liu, X.-J., Nie, Z., Yu, J., Xie, F., Song, R. (eds.) ICIRA 2021, Part I. LNCS (LNAI), vol. 13013, pp. 793–803. Springer, Cham (2021). https://doi.org/10.1007/978-3-030-89095-7_75
22. Mani, N., Kisi, K.P., Rojas, E.M.: Estimating labor productivity frontier: a pilot study. In: Construction Research Congress 2014: Construction in a Global Network, pp. 807–816 (2014)
23. Ousmer, M., Vanderdonckt, J., Buraga, S.: An ontology for reasoning on body-based gestures. In: Proceedings of the ACM SIGCHI Symposium on Engineering Interactive Computing Systems, pp. 1–6 (2019)
24. Padovano, A., Cardamone, M., Woschank, M., Pacher, C.: Exploring human-centricity in industry 5.0: empirical insights from a social media discourse. Procedia Comput. Sci. **232**, 1859–1868 (2024)
25. Poláková, M., Suleimanová, J.H., Madzík, P., Copuš, L., Molnárová, I., Polednová, J.: Soft skills and their importance in the labour market under the conditions of industry 5.0. Heliyon **9**(8) (2023)
26. Reining, C., Niemann, F., Moya Rueda, F., Fink, G.A., ten Hompel, M.: Human activity recognition for production and logistics – a systematic literature review. Information **10**(8), 245 (2019)
27. Rikala, P., Braun, G., Jarvinen, M., Stahre, J., Hamalainen, R.: Understanding and measuring skill gaps in industry 4.0—a review. Technol. Forecast. Soc. Change **201**, 123206 (2024)
28. Romero, D., Stahre, J.: Towards the resilient operator 5.0: the future of work in smart resilient manufacturing systems. Procedia CIRP **104**, 1089–1094 (2021)
29. Ruppert, T., Jaskó, S., Holczinger, T., Abonyi, J.: Enabling technologies for operator 4.0: a survey. Appl. Sci. **8**(9), 1650 (2018)
30. Sopidis, G., Ahmad, A., Haslgruebler, M., Ferscha, A., Baresch, M.: Micro activities recognition and macro worksteps classification for industrial IoT processes. In: Proceedings of the 11th International Conference on the Internet of Things, pp. 185–188 (2021)

31. Spec, D.: 91345: 2016-04 reference architecture model industrie 4.0 (rami4. 0). Din **4**, 2016 (2016)
32. Xu, P., Chang, X., Guo, L., Huang, P.Y., Chen, X., Hauptmann, A.G.: A survey of scene graph: generation and application. IEEE Trans. Neural Netw. Learn. Syst. **1**, 1 (2020)
33. Yahya, M., Breslin, J.G., Ali, M.I.: Semantic web and knowledge graphs for industry 4.0. Appl. Sci. **11**(11), 5110 (2021)
34. Zhang, W.: Scene context-aware graph convolutional network for skeleton-based action recognition. IET Comput. Vision (2023)
35. Zhu, G., et al.: Scene graph generation: a comprehensive survey. arXiv preprint arXiv:2201.00443 (2022)

Designing Augmented Reality Assistance Systems for Operator 5.0 Solutions in Assembly

Chiara Cimini[1(✉)], Francesca Tria[1], Alexandra Lagorio[1], Tamas Ruppert[2], and Sandra Mattsson[3]

[1] Department of Management, Information and Production Engineering, University of Bergamo, Viale Marconi 5, BG, Dalmine, Italy
{chiara.cimini,alexandra.lagorio}@unibg.it, f.tria@studenti.unibg.it

[2] HUN-REN-PE Complex System Monitoring Research Group, University of Pannonia, Egyetem Street. 10, Veszprem, Hungary
ruppert@abonyilab.com

[3] RISE Research Institutes of Sweden, Argongatan 30, 43153 Mölndal, Sweden
sandra.mattsson@ri.se

Abstract. Industry 5.0 emphasises how technology may benefit humans and marks a move towards a socio-technical paradigm. This study looks at how Augmented Reality (AR) can be integrated into human-centered smart manufacturing systems to improve operator performance, especially when it comes to assembly and disassembly work. Relevant AR applications in manufacturing are found through a methodical assessment of the literature, emphasising the necessity of human-centered design methodologies. The paper then offers basic recommendations for integrating AR systems into manual workstations in an efficient manner with the goal of enhancing operator productivity and welfare. The background, motivation and methods are discussed. The main findings include specific considerations for supporting the AR design in assembly, discussing the relevance of targeting group of users, choicing the suitable devices according to the usability and developing effective instructions.

Keywords: Augmented Reality · Operator 4.0 · Operator 5.0 · Assembly

1 Introduction

Industry 5.0, as delineated by Xu et al. [1], marks a departure towards a socio-technical paradigm. Unlike the focus of Industry 4.0 on what humans can achieve with technology, Industry 5.0 demands a paradigm shift, primarily focusing on how technology can serve humanity [2]. While Industry 4.0's first aim was to enhance operational and business performance, Industry 5.0 addresses concerns of environmental and social sustainability, utilising the unique characteristics of humans, encouraging a smooth transition to intelligent manufacturing and logistics processes through a socio-technical approach [3]. Additionally, increasing recognition of the interconnectedness between industrial

Published by Springer Nature Switzerland AG 2024
M. Thürer et al. (Eds.): APMS 2024, IFIP AICT 729, pp. 303–317, 2024.
https://doi.org/10.1007/978-3-031-65894-5_22

development and societal impacts has led to a reevaluation of the effects of technological innovation on human labour [2], calling for methods and tools to predict the impacts of technologies on human factors, job roles, and skills. Integrating human expertise and autonomous machinery within manufacturing facilities could enhance process efficiency and leverage human cognitive capabilities and creativity alongside intelligent systems [4]. To achieve this goal, the research field of human-centered industrial engineering focuses on designing tailored production and logistics solutions while considering human factors and workforce characteristics [5]. The human-centered design theory advocates for an approach that emphasises usability, multidisciplinary skills, user-centered design, and user involvement throughout the design and development process [6] aiming not only to improve effectiveness and efficiency but also to promote human well-being and satisfaction, thus positively impacting health and safety. Recently, various technological solutions aimed at augmenting human capabilities have emerged, leading to the concept of Operator 4.0/5.0 to describe the new generation of workers equipped with enabling technologies [7].

Among the technologies expected to enhance human capabilities in human-centered smart manufacturing systems, Augmented Reality (AR) stands out. AR, classified as an "immersive technology" alongside virtual reality (VR) and mixed reality (MR) [8], enhances human-environment interaction without replacing reality. It enriches user experiences by overlaying virtual elements in the real world. Devices supporting AR, including smart glasses, smartphones, tablets, and projectors, offer various modalities, each with its strengths and weaknesses, requiring careful evaluation for effective implementation [9]. AR finds numerous applications in manufacturing, notably in guiding assembly operators' manual tasks [10]. However, the design of AR guidance procedures from a human-centered perspective is still underexplored. Given the importance of operators in manual assembly tasks, especially for complex, multi-variant products requiring cognitive processes [11], designing solutions to aid operators, known as "cognitive automation," from a human-centered viewpoint is crucial for effectively augmenting human capabilities [12]. Indeed, according to Mattsson et al.[13], the cognitive processes for performing assembly should be intuitive and information and content signals are useful to support operators in that cognitive processes only if they are properly structured and designed.

Aiming to filling this gap, thus approaching the lens of human factors, this article grounds on a comprehensive systematic literature review to discuss the adoption of AR to support operators during assembly and disassembly tasks in production systems. After presenting the most relevant applications that can be found in literature, this research aims to provide insight and preliminary guidelines to design and implement effectively AR system in assembly and disassembly workstations that require manual activities performed by operators in order to increase their capabilities and well-being.

The remainder of the paper is organised as follows. Section 2 provides the background and motivation of this study. Section 3 presents the methodological flow that guided the literature review, while in Sect. 4 the main results are presented. Starting from them, Sect. 5 discusses the main steps that AR design process should follow. In Sect. 6, conclusions and future research are depicted.

2 Augmented Reality for Operator 4.0

Since the first introduction of Industry 4.0, it appeared the need for constant competitive production, achievable through the implementation and integration of technologies, increasingly demanding human-machine interaction and allowing more flexible manufacturing systems that more promptly answer to the desire of production efficiency related with customisation [14]. Among several technologies, Augmented Reality has become a prominent opportunity for industrial companies, thanks to its potential to improve the human-machine interaction, enhancing the productivity and the efficiency of the manufacturing field [15].

Indeed, AR allows operator to see the real environment and meanwhile be immersed in an extended reality environment thanks to projection, head-worn or hand-worn devices, which can display several types of information, ranging from simple data to multimedia and holograms [16].

For this reason, since the first conceptualisation of the Operator 4.0 paradigm [17], the *augmented operator* typology has been defined in relation to those applications that support real-time the manual operations performed by humans, constituting a digital assistance system. Also Segura et al. [18] appoints AR among the most relevant visual computing technologies, identifying a list of supported tasks, representative of the work carried out by operators in manufacturing environments, i.e. assembly, maintenance, quality control, training, inventory, and machine operation.

According to previous studies, introducing Augmented Reality in the production brings numerous advantages [19]. In particular, during production activities, by unifying the physical workspace with virtual space, AR enables workers to access real-time instructions and data without relying on less modern and less accurate paper manuals, it also facilitates remote assistance and guidance without the need of a physical real-time support, thus minimising the problem-solving time. These aspects enhance efficiency and accuracy, reducing errors and speeding up tasks.

Despite the positive sides of integrating AR in the production systems, previous studies highlight that companies can face challenges in combining this technology with traditional manual production [20]. Integrating the already existing systems with new advanced technologies requires, at the beginning, more highly skilled operators and software experts to create a seamless workflow and process. This aspect implies more expensive investments, in terms of time and labour, at the first stage of the integration. Moreover, developing the correct procedure for giving AR instructions to the operators requires specific tests and experiments since the performance and, therefore the expected increase in productivity depends on the efficiency of giving the appropriate instructions.

The possibility of including a customisation of the specific instructions according to the different operator level and experience represent an important factor for the company, and can be applied for training newer operators in the first time approaching an assembly or for more experienced operators approaching more complex assembly.

Studies have been done also for investigating the effects of AR on human factors, mostly related to the change in the cognitive load caused by the use of AR in the production. The mental effort of the workers performing the task with AR systems must be evaluated carefully. Aiming to shedding lights on the most relevant aspects in successful AR applications for Operator 5.0, this paper will be mostly focused about how AR can be applied in assembly and disassembly operations within companies, supporting the training for new operators and providing assistance for complex and more advanced assembly processes. Compared to previous studies in the field, such as the comprehensive surveys by Wang et al. [21, 22], the novelty of this research concern the specific focus on supporting the Operator 5.0 applications. Indeed, even if previous studies already addressed the use of AR in manufacturing, discussing mostly the technical aspects related to visual reepresentation and interaction, they aimed at analysing the current state of scientific maturity and industrial application of the technology, lacking the perspective of how AR can provide increased capabilities to human working in factories, both augmenting sensing, decreased learning time and increased understanding.

3 Methodology

A Systematic Literature Review (SLR) approach has been chosen for the purpose of this research. Indeed, this method is deemed as an efficient and replicable way to identify what has been already published on a distinctive topic, enabling the construction of well-structured overviews about established knowledge and the identification of actual gaps [23].

This methodology consists in a review of a collection of existing papers concerning the topic of interest, made by searching according to specific keywords and then evaluating and selecting the most appropriate papers for the research. The steps followed to select the relevant and useful papers for the SLR are described in this section and are based on the work of Kitchenham and Brereton [24].

3.1 Step 1: Inclusion/Exclusion Criteria

At first, the research topic of interest has been defined. For this paper, the literature review focuses on the application of Augmented Reality for the guidance of workers in assembly or disassembly operations, investigating the applicability in the industrial context, the challenges in the implementation and the effects of AR systems on human factors. Some keywords have therefore been identified for the research, including *Augmented Reality* (also in the abbreviated form of *AR*), *assembly* or *disassembly*, *manual operations*, *workers*, *learning*, *instructions*, *training*. Taking into account the described focus keywords, the search proceeded following these research steps to elaborate a final proper and comprehensive query with a proper combination of the chosen keywords. Scopus has been chosen as the most suitable database to retrieve literature.

The final query used for the SLR is: (TITLE-ABS-KEY ("augmented reality" OR ar) AND TITLE-ABS-KEY (assembly OR disassembly) AND TITLE-ABS-KEY (instruction*) AND TITLE-ABS-KEY (manual)) AND PUBYEAR > 2013 AND PUBYEAR < 2025 AND (LIMIT-TO (SUBJAREA, "engi")) AND (LIMIT-TO (LANGUAGE, "english")).

Further inclusion criteria have been defined as reported in Table 1. After that, the corpus of papers included 54 articles.

Table 1. Inclusion criteria

Inclusion Criteria	Description
Language	English
Document types	Articles and Conference papers
Field	Engineering
Time Interval	Since 2013

3.2 Step 2: Selection Based on Title and Abstract

After defining the corpus of relevant literature works, a paper database has been created. From this point, papers have been reviewed starting from the title and the abstract evaluating if the title matched the research topic, and if the abstract revealed interesting focus points to be later investigated. Also, a first summary classifications of the found papers has been made, considering the area of the application of the AR system, the article type, if case studies/applications/laboratory tests were presented and, in this case, the devices used and the effects of the implementation.

3.3 Step 3: Selection Based on Full Text and Snowballing

To ensure that the SLR covered all the most relevant previous research in the field, we further adopted a backward snowballing approach to leverage the reference list for uncovering potential new papers for inclusion. We thoroughly examined titles and abstracts then made informed decisions on whether to incorporate them into the final sample. Additionally, we employed forward snowballing to identify additional papers by analysing those citing the initial set of 54 papers. The process of reviewing papers in this method mirrored that of the backward approach [25] and finally led to adding 2 new papers to the corpus.

4 Results

The literature analysis gave insights in five main topics, that are discussed in the following sections, namely AR applications, devices, ergonomic factors, benefits and target groups.

4.1 AR Applications

The main topics emerging from the literature review are about the potential uses of AR in production field, in complex assembly operations and in the use of AR for training new operators. In addition to that, some studies focus on the comparison between assembly guidance technologies and between different AR devices, testing different modes of visualisation of instructions, for example comparing the effects on conveying assembly instruction with paper-based manual, video-display and AR devices. Moreover, the effect of using AR devices in production field has been evaluated testing different level of experience of the workers and for workers with disabilities and assessing the cognitive workload of workers and their performance of the process guided by AR technology [26].

A remarkable number of papers investigate case studies and application examples of AR, proposing tests and experiments on target groups to measure and evaluate the effects of the new solution proposed. Application examples of implementation are found in assembly of wooden trusses [27], disassembly of vehicle power batteries [28], even in the assembly of architectural elements [29], in the automotive industry [30], electronics field [31] and the automation of smart assembly lines [32].

It finally emerged that AR can be used for training new operators or giving assistance and support in complex assembly operations to more skilled workers, while only a few articles discuss disassembly operations.

4.2 Devices

The analysis of the AR devices described in the literature case studies was performed. The most common spatial AR device found in the review is the Microsoft HoloLens, in its newer or older versions. This consists of smart glasses, also named HMD (Head Mounted Display) that shows the real environment with the addition of virtual instructions and figures.

Several systems exploit AR projection solutions: the AR projects instructions directly on the workstation and sense the objects with visual imaging system.

Other AR systems are camera-based systems and they can also be combined with projectors [33]. RGB-D camera, like Kinect 2, sense the position and the posture of the worker in a marker-based or marker-less environment. Among these typology, it is relevant to mention the devices that can assess the hand movements and track their location through the use of wearable sensors like haptic gloves [34].

Among the software adopted to create virtual AR instruction, it is possible to find Unity 3D and Blender, normally used for 3D games [35]. With these software it is possible to create the virtual environments to be projected in the AR devices. In the HMD, for example, virtual instructions created in the 3D software are displayed and showed to the operator through AR glasses. Also several applicatons and systems are used, like Dynamics 365 Guides, used to create virtual instructions and environments and Unifeye Engineer, a marker-based tracking system that recognise the workstation environment.

A common first step in developing an AR system is the creation of a library that contains frequent used models, like arrows, signals, geometries that are later used in the 3D software [36]. HoloLens devices exploit the HoloToolKit software library for the same purpose. AR instructions are then created in a STEP file from the CAD software, then exported together with the models from the libraries in a FBX objects to be read by Unity and Blender [35]. Developers need programming skills to create the code that allows the display of the created AR environment in the real world, evaluating the overlay of objects, their positions, size, traslation and rotation movements [36].

4.3 Ergonomic Factors and Usability

Almost all the articles concerning new AR systems' proposal describe prototypes and testing activities based on the choice of devices and software, in order to test their applicability and benefits. Ergonomic and human factors are the focus points in developing effective AR systems, since evaluating the effectiveness of an AR solution depends precisely on these factors: assembly systems are still designed for workers therefore they should be taylored for human handling. To evaluate AR devices in specific applications, some researchers proposed the net promoter score (NPS), used to indicate how much some product are more recommended to others. Its range is from -100 to 100. As an example, in [37], the AR device HoloLens got an NPS of 11, 5 points below the industry average of 16. SUS score (System Usability Scale) and NASA-TLX can be employed to measure the usability of AR devices as well, even if, in some applications, it has been highlighted that, compared to traditional paper-based instruction manual, AR has still a lower score [37].

Experiments were conducted also for evaluating if AR could have impact on vision and on the fatigue of operators using HoloLens smart glasses. In the study of Drouot et al. [38], no significant effects were found as concern the optometric study after 30 min of use. About subjective parameters, AR led to a more blurred vision and slight headache after the use, suggesting a discrepancy between objective and subjective parameters, but no nausea or dizziness were perceived from the participants, since the real environment is always visible.

In order to increasingly improve the AR usability, ergonomics is an important factor to be improved in the devices, and this is more relevant for wearable devices. They are becoming more lightweight and less obtrusive in the recent years [37]. Eswaran et al. [10] affirm that user-acceptance is determined by ergonomics factors like the weight, the field of view, the light sensitivity, the time of usage, the discomfort in vision, especially for HMD. Indeed, they still have some disadvantages: the weight, the low resolution in some cases, the delay in the movements, the fatigue and dizziness created due to long usage time. However, AR systems do not only involve HMD: projector and camera-based motion capture systems represent a solution to the issues mentioned before, even if they are able to provide a less immersive environment. [36].

4.4 Benefits of Adopting AR

Currently, the majority of assembly and disassembly tasks is still performed by means of manual operations, which are guided by standard operating procedures (SOPs) presented in paper manuals or video presentations provided to the workers [26]. However, these traditional methods present some challenges concerning the product quality, the limited automation, especially for products that come in a lot of varieties and customisable choices, the efficiency and, in some cases, the safety of the process, as it can happen in disassembly operations of critical products, such as decommissioned power batteries. AR comes in help in maintaining a competitive pace in assembly or disassembly of products that require high degree of variability and customisation [28].

Moreover, the most advanced industrial environments are rich of an increasing amount of data and data collection, that increase the burden of information acquisition, manipulation and analysis for the operators. AR systems, combined with machine-learning algorithms, is expected to simplify this cognitive burden [39].

To assess actual benefits, the performance of AR systems havs been tested in the majority of articles, focusing mostly on the evaluation of the task completion time, the errors during the operations, the product quality, the cognitive workload of the worker, the need of technical external support, the learning curve during the task, the usability of the systems, and the fatigue load. In particular, in the majority of the studies, the effects have been measured comparing AR system with video-based or paper-based manuals.

Li et al. [28] proved that the efficiency of implementing AR in production increases, also for disassembly and also for non-technical workers that achieved a more proficient performance in disassembly as compared to utilising paper-based disassembly instructions, especially for complex products. In the work of Dorloh et al. [26], the number of assembly errors decreases by about the 50% using AR instructions as compared to paper-based manuals for assembly process but the time to complete the task with the AR system, assessed by using a stopwatch, resulted to be higher than performing the task following a printed manual. This time was also higher for female than for male participants. AR systems provided, therefore, an increase in the completion time but also an increase in the quality of the final assembled product. Also other studies discuss that, using AR, the completion time increases since the user needs more time to comprehend the assembly instruction provided with the augmented reality system and understand the correct positioning and orientation of the component to assembly [29]. Nevertheless, different applications showed that using AR devices can reduce the cycle time of assembly. According to Hoover et al. [40], using HoloLens device is 29% faster than following instructions on a video-based manual. The reason can be attributed to the quicker availability of AR instruction directly into the real environment. Concerning the need of requested support, also in this case AR systems could require more external supports from technicians than video display or printed instructions [26].

Focusing more about using AR for training operators, a study confirmed that the learning rate increase by 22% using an AR marker-less motion capture system. In addition to that, it showed a reduction up to -51% in the time needed to complete the first assembly cycle as compared as the paper-based instruction manual [41]. About the cognitive load and others more subjective effects evaluated in people with disabilities, it has been found that AR is a good technology to cognitively support operators because it can reduce the stress level and the perceived complexity [42]. In this way AR can be used to support the operator to focus on specific things.

4.5 Target Groups in Case Studies

In the analysed literature works, the target groups selected for the experiments range from 12 to 60 individuals, aged from 18 to 54 years old people, covering all the workers' ages. Partitions of the test group sometimes concern workers with different levels of experience, divided into junior/non-technical workers and high skilled operators [28], or participants who had a technical background in engineering field and a slight familiarity with AR technology [40, 43], or even students without any experience in assembly or disassembly operations [26, 30], or among other stakeholders from the company which is performing the test [27]. Moreover, studies have been done mostly on people without visual diseases testing with Landolt C chart and color book to examine the visual acuity and color deficiency of the participants [26]. For some studies, color-blind people, people with visual acuity lower that 10/10 and with ocular pathology were not admitted [28, 38]. Another study focused on considering the hand dexterity (right handed or left handed) of the participants performing, for example, the Purdue Pegboard Test (PPT) and the Minnesota Dexterity Test (MDT) [26]. The study focused mostly on right-handed participants, since they represented the majority of the target group. One study was finally focused on evaluating the effects of AR among people with cognitive and motor disabilities [42].

Table 2. Summary of the literature user studies

Ref	Topic	Domain	Measured effects	Total participants (female)
[28]	Disassembly efficiency	Vehicle batteries disassembly	Instruction design and efficiency, product quality	40 workers with different level of experience
[43]	Occlusion problems	Machining maintenance	Completion time, accuracy, and cognitive load	42 (6) within the university with knowledge in engineering and AR
[26]	Comparison between instruction methods focused on performance	Computer assembly and disassembly	Efficiency, help-seeking behaviour, product quality, usability	21 (10) healthy students without any experience in assembly/disassembly
[27]	Interaction design	Wooden trusses assembly	Completion time, visibility, product quality	12 different stakeholders (designer, managers, developers, researchers) + 8 (1) professional assemblers
[42]	Comparison between instruction methods focused on cognitive load	Appliances assembly	Learning curve, quality, stress, help-seeking behaviour, perceived complexity, physical effort and frustration	44 (20) people with special needs, cognitive or motor disabilities

Table 2 presents some of the most recent literature case studies analysed. From a general evaluation, presented studies concern comparisons between devices, the cognitive load analysis and tests about process efficiency. A critical observation is that most participants are male individuals selected from academic environment. Only few studies investigate on target groups of practitioners (i.e. assembly workers), also with competences in AR field. This can be explained by the relative ease of recruiting researchers and students compared to workers and, sometimes, testing prototypes with people who are new to the area and are uninfluenced by prior knowledge of AR and assembly can be advantageous. However, this approach may limit the potential of studies and claim for further testbeds of the technology in the industrial context.

5 Discussion

5.1 Designing AR for Assembly

Analysing comprehensively the literature, it emerged that the principal enablers of successful development and implementation of AR systems supporting assembly operations are: flexibility, also through augmented intelligence, user-centered design and worker feedback, visualisation and efficient creation of digital assembly instructions. Flexibility is an important characteristic that lets AR adapt to changes in the product variety or in the sequence of operations. Integrating algorithms of artificial intelligence can make AR systems more flexible and promptly adaptable in providing new instructions according to the performed task in the real environment. Feedback from workers and user-centered design approaches are recognised as crucial to a successful implementation and integration of the AR system into the existing production. To achieve this, in the design of AR applications, real-time data acquisition about the operations performed by the workers is essential, to ensure a continuous alignment and efficient workflow, while ensuring the monitoring of the worker's state of wellbeing. In particular, before developing targeted instructions and customized applications, the target group of users should be identified. This is relevant also for the choice of the most suitale devices to adopt. Indeed, for specific AR devices, potential issues due to uncomfortable wearing for prolonged periods of time could appear, affecting the successful implementation for different groups of users.

Concerning the specific development of instructions for assembly, visualisation of the information and instructions is essential for the efficiency of performing the tasks. Since AR instructions are visual information guiding the operator in all the assembly or disassembly steps overlapping with the physical environment, in designing AR assistance solutions, the main issues are related to the generation and the presentation of AR instructions [21]. Traditionally, according to the different devices and methodologies used in industrial settings, there are multiple ways to convey instructions to the operators for the assembly and disassembly processes. They can be presented in paper-based manuals that show descriptive texts, graphs and images explaining the operations, or they can be presented with videos on laptop computers or tablets, with the possibility to click and stop the explanation of the task. These traditional ways can be more time-consuming and with a lower cognitive efficiency of the workers. To overcome these issues, AR instructions are capable of highly stimulate operators enhancing the attention, the concentration and therefore the performance efficiency. On the other hand, the quality of the perceived information depends on the way in which the information is conveyed and also on how the instruction are presented. It is important to tailor the instruction methods to the different users, according to the level of experience and possible disabilities.

It is crucial that the operator understands precisely the task to be performed, therefore a precise guidance should not have different interpretations or uncertainties from the worker's point of view. The instruction should precisely describe 'what' to do, in which position and at which time during the assembly process. These are the key information to be given to avoid mistakes in the operations. Moreover, to facilitate the interaction between workers and the AR system, the execution should be intuitive, without the need of external support. In reaching this, it is necessary to provide responsive feedback to the worker when the task is successfully completed. Another this that is needed to perform good quality is that the operator understand 'why' something is performed which can be added afterwards.

Further, AR software and systems present a semi-immersive virtual-real environment. In order to create a more immersive and realistic spatial rendering, it is important to provide a geometric consistency with the virtual and real worlds, in accordance with the choice of AR devices. Geometric factors like perspective and positioning in the real environment help the users to be more involved in the assembly tasks, improving the performance efficiency. Lightning consistency is also a key factor to implement a virtual world as similar as possible to the real one, since it leads to a better integration between AR and the surroundings, enhancing user's experience. Moreover, in some cases, also, instructions in the virtual environments could reduce the field of view of the real assembly environment due to obstruction problems [40]. For this reason, occlusion problems for blind areas should be one of the main aspects to explore, since these lead to generate a loss of information.

To further support the design of efficient AR solutions for assembly, from the literature review, the main issues that can be encountered and need to be properly evaluated emerge. In particular, accurate alignment between the virtual description and the position of the physical assembly part should be always provided [40]. In addition to that, AR devices need to be set with responsive and fast recognition of the assembly step in the real environment, in order to avoid longer assembly time and users' frustration [29].

5.2 Future Development Paths

Case studies that emerged from the literature review suggested new developments and innovative implementations of future AR systems as well as solutions to overcome previously mentioned limitations.

RGB camera based methods, integrated with AI algorithms, such as deep neural networks, can be the starting point for solving the occlusion problems, since the AI software can recreate the loss of information in the real environment into a digital image reconstruction [34].

Since another limitation for the application of spatial projector AR systems is that they could be not suitable for color-blind people, future developments can focus on implementing tailored instruction schemes that do not require colored instructions. Same consideration applies for left-handed people, since most of tests are conducted on the majority of right-handed people [26]. Instructions should be tailored for both and customisable for both right and left handed workers.

An interesting development involves integrating AR systems with AI based on machine learning algorithms, to real-time detect the working environment, assessing the precision of the task and detect the real-time efficiency of the manual assembly [39]. Based on this principle, AI can be also used to instantly and automatically generate assembly instructions adjusting them to the user's needs [33]. AI could be useful also for complex assembly in terms of a lot of variety in the customisation of the products. The product variance needs to meet the stakeholders' requirements and be directly and efficiently translated into manufacture realisation. Developing a system for augmented reality performing the different product varieties with AI generated instructions means exploiting a high degree of adaptability and combinations to properly perform the complex assembly.

6 Conclusion

The objective of this article was to conduct a literature review investigating how innovative AR systems can be integrated into the production area, focusing on the evaluation of the potential benefits and effects of this implementation to support manual assembly work. Implementing AR technology certainly brings benefits to industrial companies because AR seems able to makes production more efficient, reduces errors during assembly, and speeds up the learning rate during operators training. However, in some cases, AR leads to longer cycle times due to a more difficult initial understanding of the system functions. Further research involve e.g. using AI to develop instructions and to study how the applications in AR could be developed to support color-blindness.

This research has limitations since it has been focused more on analysing assembly and disassembly operations rather than incorporating other areas such as maintenance or quality control, even if they present different balance in reasoning and intuition during the task execution. Literature suggests that maintenance and repair operations could be fruitfully supported by AR, so further studies could embed the relevant literature and knowledge in this fields to guide the design of AR applications for Operator 5.0.

Another limitation lies in the number of articles analysed, which is limited to a total of around 50 articles, and the proposed query did not perform well in incorporating more articles regarding disassembly, since a significant majority of papers focused only on assembly.

Future implementations and researches can focus on expanding this literature review to the use of AR systems in other industrial areas, also delving deeper tests and practical verifications of the different KPIs found.

Acknowledgments. The research has been funded by European Union – "NextGenerationEU", within the program "MUR-Fondo Promozione e Sviluppo - DM 737/2021" withing the project named "HUMARWISE".

Disclosure of Interests. The authors have no competing interests to declare that are relevant to the content of this article.

References

1. Xu, X., Lu, Y., Vogel-Heuser, B., Wang, L.: Industry 4.0 and Industry 5.0—Inception, conception and perception. J. Manuf. Syst. **61**, 530–535 (2021)
2. European Commission. Directorate General for Research and Innovation: Industry 5.0: towards a sustainable, human centric and resilient European industry. Publications Office, LU (2021)
3. Sony, M., Naik, S.: Industry 4.0 integration with socio-technical systems theory: a systematic review and proposed theoretical model. Technol. Soc. **61**, 101248 (2020)
4. Nahavandi, S.: Industry 5.0—A Human-Centric Solution. Sustainability. 11, 4371 (2019)
5. Sgarbossa, F., Grosse, E.H., Neumann, W.P., Battini, D., Glock, C.H.: Human factors in production and logistics systems of the future. Annu. Rev. Control. **49**, 295–305 (2020)
6. International Organization for Standardization: ISO 9241–210:2019. Ergonomics of human-system interaction — Part 210: Human-centred design for interactive systems. https://www.iso.org/cms/render/live/en/sites/isoorg/contents/data/standard/07/75/77520.html
7. Gladysz, B., Tran, T., Romero, D., van Erp, T., Abonyi, J., Ruppert, T.: Current development on the Operator 4.0 and transition towards the Operator 5.0: a systematic literature review in light of Industry 5.0. J. Manuf. Syst. **70**, 160–185 (2023)
8. Suh, A., Prophet, J.: The state of immersive technology research: a literature analysis. Comput. Hum. Behav. **86**, 77–90 (2018)
9. Elia, V., Gnoni, M.G., Lanzilotto, A.: Evaluating the application of augmented reality devices in manufacturing from a process point of view: an AHP based model. Expert Syst. Appl. **63**, 187–197 (2016)
10. Eswaran, M., Gulivindala, A.K., Inkulu, A.K., Raju Bahubalendruni, M.V.A.: Augmented reality-based guidance in product assembly and maintenance/repair perspective: a state of the art review on challenges and opportunities. Expert Syst. Appl. **213**, 118983 (2023)
11. Li, D., Mattsson, S., Salunkhe, O., Fast-Berglund, Å., Skoogh, A., Broberg, J.: Effects of information content in work instructions for operator performance. Procedia Manuf. **25**, 628–635 (2018)
12. Mattsson, S., Fast-Berglund, Å., Li, D., Thorvald, P.: Forming a cognitive automation strategy for operator 4.0 in complex assembly. Comput. Ind. Eng. **139**, 105360 (2020)
13. Mattsson, S., Fast-Berglund, Å.: How to support intuition in complex assembly? Procedia CIRP. **50**, 624–628 (2016)
14. Cimini, C., Boffelli, A., Lagorio, A., Kalchschmidt, M., Pinto, R.: How do industry 4.0 technologies influence organisational change? An empirical analysis of Italian SMEs. JMTM. **32**, 695–721 (2021)
15. Aquino, S., Rapaccini, M., Adrodegari, F., Pezzotta, G.: Augmented reality for industrial services provision: the factors influencing a successful adoption in manufacturing companies. J. Manuf. Technol. Manage. ahead-of-print (2023)
16. Lagorio, A., Pasquale, V.D., Cimini, C., Miranda, S., Pinto, R.: Augmented Reality in Logistics 4.0: implications for the human work. IFAC-PapersOnLine. **55**, 329–334 (2022)
17. Romero, D., et al.: Towards an Operator 4.0 Typology: a human-centric perspective on the fourth industrial revolution technologies. In: Proceedings of the International Conference on Computers and Industrial Engineering (CIE46), pp. 1–11 (2016)
18. Segura, Á., et al.: Visual computing technologies to support the Operator 4.0. Comput. Ind. Eng. **139**, 105550 (2020)
19. Mourtzis, D., Zogopoulos, V., Xanthi, F.: Augmented reality application to support the assembly of highly customized products and to adapt to production re-scheduling. Int. J. Adv. Manuf. Technol. **105**, 3899–3910 (2019)

20. Syberfeldt, A., Holm, M., Danielsson, O., Wang, L., Brewster, R.L.: Support systems on the industrial shop-floors of the future – operators' perspective on augmented reality. Procedia CIRP. **44**, 108–113 (2016)
21. Wang, Z., et al.: A comprehensive review of augmented reality-based instruction in manual assembly, training and repair. Rob. Comput. Integr. Manuf. **78**, (2022)
22. Wang, P., et al.: AR/MR remote collaboration on physical tasks: a review. Rob. Comput. Integr. Manuf. **72**, 102071 (2021)
23. Lagorio, A., Cimini, C., Gaiardelli, P.: Reshaping the concepts of job enrichment and job enlargement: the impacts of lean and industry 4.0. In: Dolgui, A., Bernard, A., Lemoine, D., von Cieminski, G., and Romero, D. (eds.) Advances in Production Management Systems. Artificial Intelligence for Sustainable and Resilient Production Systems, pp. 721–729. Springer International Publishing, Cham (2021). https://doi.org/10.1007/978-3-030-85874-2_79
24. Kitchenham, B., Brereton, P.: A systematic review of systematic review process research in software engineering. Inf. Softw. Technol. **55**, 2049–2075 (2013)
25. Wohlin, C.: Guidelines for snowballing in systematic literature studies and a replication in software engineering. In: In 8th International Conference on Evaluation and Assessment in Software Engineering (EASE 2014, pp. 321–330. ACM (2014)
26. Dorloh, H., Li, K.-W., Khaday, S.: Presenting job instructions using an augmented reality device, a printed manual, and a video display for assembly and disassembly tasks: what are the differences? Appl. Sci. **13**, 2186 (2023)
27. Tobisková, N., Malmsköld, L., Pederson, T.: Head-mounted augmented reality support for assemblers of wooden trusses. Presented Procedia CIRP (2023)
28. Li, J., Liu, B., Duan, L., Bao, J.: An augmented reality-assisted disassembly approach for end-of-life vehicle power batteries. Mach. **11**, 1041 (2023)
29. Chu, C.-H., Liao, C.-J., Lin, S.-C.: comparing augmented reality-assisted assembly functions-a case study on dougong structure. Appl. Sci. (Switzerland). **10** (2020)
30. Lampen, E., Teuber, J., Gaisbauer, F., Bär, T., Pfeiffer, T., Wachsmuth, S.: Combining simulation and augmented reality methods for enhanced worker assistance in manual assembly. Procedia CIRP. **81**, 588–593 (2019)
31. Becerra, E.J., Hovanski, Y., Tenny, J., Peterson, R.: Assessing Enterprise Level, Augmented Reality Solutions for Electronics Manufacturing, Presented at the SAE Technical Papers (2023)
32. Horejsi, P., Novikov, K., Simon, M.: A smart factory in a smart city: virtual and augmented reality in a smart assembly line. IEEE Access. **8**, 94330–94340 (2020)
33. Thamm, S., et al.: Concept for an augmented intelligence-based quality assurance of assembly tasks in global value networks. Procedia CIRP. **97**, 423–428 (2021)
34. Fang, W., Hong, J.: Bare-hand gesture occlusion-aware interactive augmented reality assembly. J. Manuf. Syst. **65**, 169–179 (2022)
35. Neb, A., Strieg, F.: Generation of AR-enhanced Assembly Instructions based on Assembly Features. Presented at the Procedia CIRP (2018)
36. Fiorentino, M., Uva, A.E., Gattullo, M., Debernardis, S., Monno, G.: Augmented reality on large screen for interactive maintenance instructions. Comput. Ind. **65**, 270–278 (2014)
37. Wang, C.-H., Tsai, N.-H., Lu, J.-M., Wang, M.-J.J.: Usability evaluation of an instructional application based on Google glass for mobile phone disassembly tasks. Appl. Ergon. **77**, 58–69 (2019)
38. Drouot, M., Le Bigot, N., Bolloc'h, J., Bricard, E., de Bougrenet, J.-L., Nourrit, V.: The visual impact of augmented reality during an assembly task. Displays. **66**, 101987 (2021)
39. Li, W., Xu, A., Wei, M., Zuo, W., Li, R.: Deep learning-based augmented reality work instruction assistance system for complex manual assembly. J. Manuf. Syst. **73**, 307–319 (2024)

40. Hoover, M., Miller, J., Gilbert, S., Winer, E.: Measuring the performance impact of using the microsoft HoloLens 1 to provide guided assembly work instructions. J. Comput. Inf. Sci. Eng. **20**, (2020)
41. Pilati, F., Faccio, M., Gamberi, M., Regattieri, A.: Learning manual assembly through real-time motion capture for operator training with augmented reality. Presented Procedia Manuf. (2020)
42. Vanneste, P., Huang, Y., Park, J.Y., Cornillie, F., Decloedt, B., Van den Noortgate, W.: Cognitive support for assembly operations by means of augmented reality: an exploratory study. Int. J. Hum. Comput. Stud. **143**, (2020)
43. Laviola, E., Gattullo, M., Evangelista, A., Fiorentino, M., Uva, A.E.: In-situ or side-by-side? A user study on augmented reality maintenance instructions in blind areas. Comput. Ind. **144**, 103795 (2023)

Inclusive Work Systems Design: Applying Technology to Accommodate Individual Worker's Needs

An Examination of the Limited Adoption of Personalized Work Instructions in Assembly to Accommodate Individual Worker's Needs

Jos A. C. Bokhorst, Sabine Waschull(✉), and Christos Emmanouilidis

University of Groningen, PO Box 800, 9700 AV Groningen, The Netherlands
{j.a.c.bokhorst,s.waschull}@rug.nl

Abstract. Creating a good fit between an individual's information needs and the information provided in work instructions by personalizing its content and form facilitates allocating a more diverse group of operators to a particular assembly task, stimulating the labor force participation rate and inclusiveness. While ongoing technological development should make the personalization of work instructions easier, adoption in practice is still limited. Analyzing nine companies through a multiple case study approach, this paper investigates the challenges organizations face when creating, maintaining, and using personalized work instructions in practice. Overall, organizations struggle to adopt personalization due to its demanding nature, involving complex and time-consuming maintenance to keep personalized, and thus a large number of instructions, up-to-date. Personalization found in the cases mainly involves the use of a personal mentor to adapt the information detail to the characteristics of the worker, as well as allowing employees to conduct minor edits or to create their own instructions. Despite its possibilities, technology hardly played a role in creating and communicating personalized work instructions. To facilitate a more inclusive work design in practice, creating, implementing, and maintaining work instructions must become more accessible and manageable.

Keywords: Work Instructions · Personalization · Assembly

1 Introduction

Work instructions provide information to operators on how to perform their tasks. With increasing customization and more product variants assembled in smaller volumes, good-quality instructions are becoming even more important [1, 2]. While instructions were traditionally paper-based, the ongoing digitalization and further advancements in smart working technologies [3] have led to many alternative ways to provide information. These technological developments open up opportunities to personalize instructions, which may facilitate a more diverse group of people to participate in the labor market by supporting them in executing tasks at the desired performance level [e.g., 4, 5]. Yet, technology-enabled personalization is still rarely observed in practice.

M. Thürer et al. (Eds.): APMS 2024, IFIP AICT 729, pp. 321–335, 2024.
https://doi.org/10.1007/978-3-031-65894-5_23

When designing work instructions, decisions must be made concerning the (information) content [6] and the (communication) form [7, 8]. The content distinguishes the scope of the instruction (i.e., product, task, activity), the level of standardization, and the level of detail. The form distinguishes the communication channel (digital vs. paper) and the form type (text vs. images). Traditional work instructions usually take the form of a standardized paper-based document describing sequential processing steps in words. More advanced work instructions are digital, ranging from digital documents with written instructions to Augmented Reality projected digital objects showing the orientation of the next part to be assembled.

Adapting the content or form of work instructions to a user's needs can be regarded as personalization of work instructions. For example, in terms of content, precisely the right amount of support can be provided to an operator by choosing a fitting level of instructional detail [9], or instructions can be provided with graphical or audio support (form) for illiterate operators [8]. This personalization may facilitate the design of inclusive work [10], providing job opportunities for a diverse group of people, such as those with physical or intellectual disabilities or belonging to other minority groups. Facilitating inclusive work not only benefits the people who currently have difficulties fully participating in the labor market but also solves some of the shortages of skilled labor that employers face [11].

While much research has focused on the technical designs of advanced worker assistance systems, comparing their performance in controlled experimental settings [e.g., 1, 9, 12, 13], the creation of inclusive work through personalized work instructions lags behind in practice. Hence, this paper explores the challenges of personalization in practice through a multiple case study.

2 Background

2.1 Interpretative Research Framework

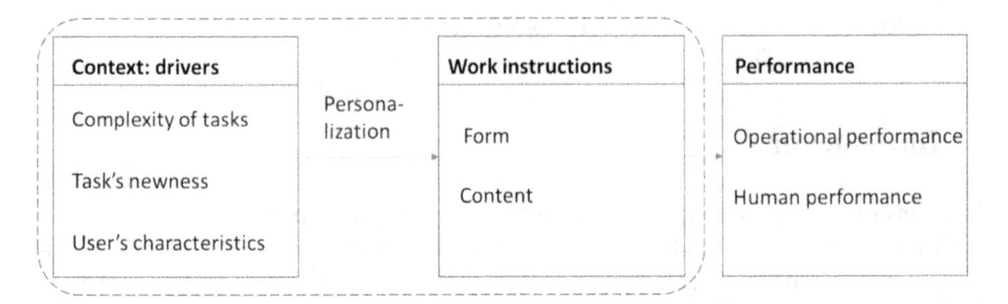

Fig. 1. Interpretative model linking the concepts

Figure 1 shows an interpretive model that provides an overview of the concepts relevant to this research. Section 2.2 reviews the literature on work instructions and performance, while Sect. 2.3 delves into the drivers of personalization. The empirical analysis of the cases in Sect. 4 uses the same concepts while investigating the challenges of personalization, focusing on the concepts within the dashed line.

2.2 Work Instructions and Performance

Work instructions must be created before they can be used and they must be maintained over time. These activities involve significant time and costs [7], which also depend on the chosen form and content. There are different enabling digitization technologies that assist workers by delivering (i.e., form) the information content in innovative ways. In addition, digitization may facilitate the flexible adaptation of instruction content. Mark et al. [14] provided an overview of worker assistance systems in manufacturing, dividing them into three categories as put forward by Romero et al. [15]: sensorial, cognitive, and physical worker assistance systems. Provision of work instructions through, for example, augmented reality, tablets, in-situ projection, computer-assisted instructions, monitors, and head-mounted displays generally falls within the cognitive worker assistance systems category. A benefit of digital systems compared to paper-based instructions is that the information is easily accessible for users and those who make or update the instructions.

Usually, the instructions' users (e.g., operators, assemblers) differ from the instructions' creators and maintainers (e.g., manufacturing engineer, process engineer, quality engineer). Since the users are the assembly experts, they are capable of providing feedback on the instruction quality. Therefore, it is important to have an efficient feedback loop between the user and the creator.

The complexity of the creation process also depends on how advanced the work instruction system is. For example, for AR-based instructions, not only the digital information content and form need to be designed, but also the tracking of actions, registrations, and identification of objects requires attention [16]. Because of this dependency on knowledge of the assembly operation in the real environment, it is even more important that the instructions' user is involved in the creation process [16].

The literature distinguishes operational and human performance impacts of applying work instructions. The operational performance measures commonly used to assess work instructions are the time to accomplish a task and the quality of the output [12, 17]. Human performance effects in the literature often relate to operator cognitive workload, commonly measured using the NASA Task Load Index [18, 19]. Other aspects considered are the usability of the work instruction system and user experience [7].

Numerous studies have experimentally explored the performance effects of applying different work instructions regarding content and form in generic manufacturing or assembly settings, using either (simplified) industrial tasks or abstract tasks using non-industrial products (e.g., LEGO products). Related to content, Haug [6] presented a framework indicating which types of information quality are relevant in work instructions to obtain higher performance. However, most studies focused on comparing the performance effects of different forms of work instruction, while few studies addressed both [e.g., 17]. An important finding related to form is that work instructions with images result in better operational and human performance than text-based instructions [e.g., 20, 21]. Lemathe and Rößler [12] showed that digitally animated, interactive work instructions perform better with respect to quality and time than paper-based instructions. Bosch et al. [22] showed that projected work instructions lead to higher productivity and quality compared to electronic instructions on screen, while the cognitive load of the operators was significantly decreased.

2.3 Drivers for Choosing Content and Form

While the content and form of work instructions individually are shown to impact performance, the overall effect also depends on the application context, such as the task's newness, complexity, and assemblers' personal characteristics. Therefore, the content and form of work instructions must be designed to fit the context in which they are used, which can be regarded as the personalization of the instructions to accommodate the user's needs and promote inclusive work [23].

Task's Newness: Learning Versus Operational Phase

An important driver to adapt work instructions is the newness of the task for a worker. To better analyze the support of work tasks with cognitive automation, Mattsson et al. [24] distinguished three modes operators move through. In the learning mode, the operators learn new tasks or technologies when a new product is introduced. In the operational mode, the operators perform the assembly tasks required for routine day-to-day work. Finally, operators perform non-routine tasks such as handling disturbances or machine failures in the disruptive mode. Similarly, Palmqvist et al. [25] distinguish between (1) educational instructions to support learning when the operator or job is new and (2) simplified instructions when the operators are experienced. In the learning phase, extensive videos or face-to-face instructions could be used [17], supporting the 'reasoning' needs of the operators, while in the operational phase, the 'intuition' needs should be supported [24] with simplified instructions using reminders and relevant and accurate information [24, 25]. The more advanced digital work instruction systems, using digitally animated and interactive work instructions, are particularly useful when workers have to learn new or quickly changing tasks [12].

Complexity of Assembly Tasks

Over the years, the complexity of many manufacturing and assembly processes has increased due to product evolution and greater product diversity [26]. Several authors have noted a relationship between the assembly complexity context and the potential of different work instruction systems. For example, Wiedenmaier et al. [27] noted that AR guidance has more potential for more complex tasks than for easy ones. They also showed that assembly time was similar for simple tasks when using a paper manual or AR-based instructions. Similarly, Uva et al. [28] concluded that AR is "more effective for difficult tasks than for simple ones." Radkowski et al. [29] showed that the type of visual feature used in an AR instruction for a particular assembly operation must correspond to its relative difficulty level. In line with this, Mattsson et al. [30] claim that complexity can be managed/reduced by presenting information in a simplified way, with different work instruction systems having different capabilities to do so.

User's Characteristics

In theory, personalized instructions can vary from person to person, but they can also change over time for the same person. When the instruction system is context-aware and, for example, the status of the assemblers or feedback on their performance is included, personalization can even be real-time. Funk et al. [31] provide an example where the level of feedback (beginner, advanced, or expert mode) is adjusted based on the errors detected, or the frequency and intensity of the feedback are changed based on the worker's stress level and vital parameters.

Several papers have indicated that beginners need more detailed instructions than experienced workers [4, 32, 33]. Also, Kolbeinsson et al. [20] showed that beginners preferred step-by-step instructions with all steps included and in the correct order, while experienced workers could handle more types of instructions and more complex assembly tasks and needed little support for standard assemblies. When unnecessary information is provided, this may lead to frustration. In contrast, withholding important (detailed) information when there is a need may lead to assembly errors [34].

Another literature stream focuses on the personalization of work instructions for users with different types of disabilities, such as visual, hearing, motor, and cognitive impairments [4, 5, 8, 10, 35, 36]. Depending on the type of impairment, the user interface can be adapted to specific user needs, facilitating a better person-job fit.

Heinz-Jakobs et al. [10] showed that people with cognitive disabilities generally see potential in guidance through projection-based assistance systems. Similarly, Jost et al. [5] designed an adaptive AR-based instruction system for operators with cognitive impairments, with user tests showing strong learning rates and successful completions of new tasks. Also, in-situ projection allowed cognitively impaired workers to assemble more complex products faster and with fewer errors [35]. Vanneste et al. [36] included the cognitive skills of operators and their level of experience with the task and showed that AR instructions lead to fewer errors than oral or paper-based instructions. In contrast to oral or paper-based instructions, with AR instructions less experienced operators performed equally well as experienced operators.

Peltokorpi et al. [8] included a more extensive range of forms for work instructions (i.e., paper-based, animations, projection, and adaptive projection) to study how they accommodate the needs of illiterate, psychosocially disabled, and cognitively disabled workers, and how they impact industrial assembly performance. They concluded that incorporating more instructional elements and cognitive assistance through projection enhances initial learning cycles but may also lead operators to become more dependent on these guidance systems. They also found that tailoring the complexity of instructions based on the operator's experience promotes independence. However, this approach may not suit individuals with psychosocial issues who find reduced assistance uncomfortable. Furthermore, their results showed that highly intuitive and adaptive instructions are ineffective for cognitively disabled operators due to their unpredictable performance and significant need for mentor support. This shows the importance of individualized, unbiased instructional design, recognizing the diverse responses to cognitive assistance across and within disability groups, to effectively meet each operator's unique needs.

3 Multiple Case Study

To explore the challenges faced in creating, maintaining, and using personalized work instructions in practice, we conducted a multiple case study involving nine Dutch organizations involved in the manufacturing or assembly of a wide array of different products. This allowed us to find commonalities and differences across cases [37], which led to a better understanding of the challenges of personalization in different contexts.

Table 1. Overview and description of cases.

Case	FTE	Function of Interviewee(s)	Case Description
A	3500	Senior Production Engineer	Sheltered workplace assembling an extensive range of products for a large range of customers
B	225	Production Manager	A company focused on the assembly of elevators
C	350	Production Manager, R&D Process Engineer & Assistant Unit Foreman	Production company offering a wide range of tools and parts for metal bending operations
D	250	Production Manager & Head Business Office	Sheltered workplace performing an extensive range of assembly services
E	700	Manufacturing Engineering Manager	Part of a large international organization responsible for assembling a specific range of (port) lift structures
F	120	Manager Operations	Production site selling semi-finished electro-technical products
G	180	Team leader Master Data & Team leader Process Engineering	Production of air suspension systems with a large variety of products offered
H	450	Production Manager	A company offering the design and production of, mostly, custom-made truck trailers
I	95	Production Manager	Sheltered workplace with a large variety of products. Mainly focusing on metalworking and electrical installation

An important case selection criteria included the use of manufacturing instructions, which is typical for high-variety/low-volume production environments. Moreover, three sheltered organizations were included in the research (cases A, D, I). Table 1 provides an overview and a short description of the cases. An interview protocol was used for the semi-structured interviews with in total 28 questions on the following topics: (1) Generation phase of instructions, (2) Work instructions in the workplace, (3) Content of work instructions, (4) Form of work instructions, (5) Personalized work instructions, (6) Performance of work instructions, and (7) Challenges. The interviews were transcribed, summarized, and analyzed.

4 Results

Section 4.1 first describes the work instructions of the cases, how they are created and maintained and what content and form they take. Section 4.2 then describes whether personalization is used and what drivers are. Section 4.3 highlights the challenges for personalization faced by the cases. Finally, Section 4.4 provides a comparative analysis.

4.1 Work Instructions

Table 2 provides an overview of the creation and maintenance of work instructions of the cases. Instructions are either created and maintained by staff in the production department (B, F, H, I) or in the engineering domain (A, C, E, G). In case D, the work instructions are provided by the client. The majority of cases that create work instructions in the production department directly involve end-users during the creation process (F, H, I). The majority of cases where work instructions are created by the engineering disciplines (A, C, G) indirectly involve end-users during the usage phase of instructions through varying feedback channels (e.g., feedback written on the instruction, consultation with the supervisor, etc.). The cases use various but simple tools for creating and maintaining instructions, ranging from MS Word applications (A,C,I), to self-developed tools (E,F,H) to somewhat more sophisticated Virtual Knowledge Systems (B,G).

Table 2. Overview of cases related to the creation and maintenance of work instructions

Case	Creator function	Involvement end-user	Tool	Complexity of creation
A	Product engineer	Indirectly	MS Word / 3d software	Low (requires knowledge over process)
B	Teamlead	Indirectly	VKS Software	Low
C	Product engineer (creation) & Teamlead (maintenance)	Indirectly	MS Word / Solidworks	Low
D	Client	Indirectly	-	-
E	Manufacturing engineer	Directly	Self-developed tool; photoshop	High
F	Teamlead	Directly	Self-developed tool	Low
G	Process & product engineer	Indirectly	VKS software	High
H	Production department employee	Directly	Self-developed tool	Low
I	Teamlead	Directly	MS word	Product dependent

Cases (A, B, C, F, H) indicate that creating instructions is not complex, mostly referring to the limited amount of time it takes to create an instruction. For case D, poor information provision by their clients makes generating work instructions complex. Case E considers creating work instructions quite complex, as attempts to outsource these activities to India failed due to a lack of knowledge about the assembly process and organizational specifics. Case G stated that creating instructions is not inherently complex, but processing all feedback into the instructions to maintain them takes a significant amount of time. Finally, case I indicates that creating instructions is more complex for products with a higher complexity.

Table 3 provides an overview of the information content (i.e., scope, level of standardization, and level of detail), and the form (i.e., communication channel and form type) for the different cases. Furthermore, the feedback channels are presented, and the level of autonomy in following the instructions, ranging from having to follow the instructions step-by-step (low autonomy) to using instructions as a reference and tailoring work to their preferences (high autonomy).

Table 3. Overview of cases related to content and form

Case	Scope	Level of standardization	Level of detail	Communication channel	Form type	Feedback channels	Level of autonomy
A	Product	High	High - task level	Paper	Images	Creator	Low
B	Product	High	High - task level	Digital	Images	VKS system	Low
C	Product	High	Low - task level	Digital	Text	Creator	High
D	Product	Low	Varies - depends on customer	Paper	Text	Supervisor	Low
E	Product	High	High but also varies - depends on task complexity	Paper	Text	Feedback meetings	High
F	Product	High	Low but also depends on team	Digital	Images	Employee	High
G	Product	Varies	High - activities	Paper	Images/text	Supervisor	Low
H	Product	High	High - but also varies on module	Digital	Images/text	Employee or supervisor	High
I	Product	Low	Varies per product	Paper	Image	Not present	High

For all cases, instructions are created per product as opposed to task or activities, and for the majority of cases (A, B, C, E, F, G, H), a highly standardized format for presenting information in instructions is adopted (i.e., the information elements can be found in the same places within the format). The work instructions contain, at least, the activities and associated tasks. Other information elements found in the work instructions include assembly drawings, important remarks, and parts lists. The parts lists have been intentionally removed from the work instructions by some cases (A, G) because, in practice, a parts list may change without the need to modify the work instruction.

The level of detail of information presented in the work instructions varies across case organizations and within cases. The majority of cases (A, B, E, G, H) deploy very detailed instructions (i.e., high level of detail), and a few (C, F) use more generic instructions (i.e. low level of detail). Multiple cases indicated variations in the level of detail deriving from different drivers, including the complexity of the production task (E), the type of product (H, I), and the type of team working on the product (F).

In four cases (B, C, F, H) the work instructions are displayed through computer screens at the employee's workstation. Different systems are used to get the instructions on the screen, which are either developed internally or acquired from a third party. Some systems have a manual connection or even integration capability with the ERP system. The other five cases (A, D, E, G, I) present the work instruction on paper, available in a binder at a central location, at workstations, or linked to work orders.

The work instructions are primarily conveyed through a combination of images/illustrations and text. For five cases (A, B, F, G, I), images/illustrations predominate over text. These cases utilize various types of images, including photos of components or sub-/final assemblies, engineering drawings, or 3D objects.

The majority of the cases (all except C) are currently or have in the past conducted pilots for the application of modern technologies to communicate work instructions, with one case (I) having a full implementation of a Smart Beamer. Generally, the pilots have involved visual systems, such as augmented reality combined with 3D objects and exploded views, guidance through beamer projection, and virtual reality. Additionally, one case (G) has a pilot running with order picking based on voice instructions.

Except for one case (I), a clear multidirectional flow of information occurs between instruction users and creators, although there are significant differences in which feedback channels are used. In some cases (F, H), the employees themselves can directly make adjustments in the work instruction itself, while for the other cases (A, B, C, D, E, G), the employee must report to their immediate supervisor or the creator of the work instruction. For some cases (A, E) the flow of information and involvement of other individuals is standardized in a procedure. Several cases (D, E) distinguish a phase where an initial run of new products is introduced with (draft) work instructions. In this phase, end users are heavily involved and have many opportunities for feedback.

Five cases (C, E, F, H, I) indicate that employees are highly autonomous in following the work instructions, although this sometimes leads to errors. The other cases (A, B, D, G) expect employees to follow instructions step-by-step (low autonomy), although deviations occur, with workers changing the order of steps or using the instruction as a reference for repetitive tasks. The two cases utilizing less detailed instructions (C, F) also reported a high level of autonomy in following the work instructions, while for the five cases with very detailed instructions, three reported a low level.

4.2 Personalization and Drivers

Table 4 provides an overview of whether personalization is used in the different cases, how it impacts work instructions (i.e., which aspects of content or form), what the drivers are or could be (in brackets), and what the perceived usability is of personalization based on user's characteristics. In several cases, personalization was found (D, E, F, G, H, I). In two cases (D, I) personalization is achieved through a supporting mentor/work supervisor who provides personal guidance to employees with a distance to the labor market. In two other cases (F, G) employees themselves are able to personalize their instructions: employees can execute small changes in the instructions (F), or create their own personal instructions (H). In one case (E), two 'types' of instructions are generated, relating to the task's newness. For the learning phase, a complete and detailed 'sequence of events' list is created, while in the operational phase, the level of detail in the instruction depends on

Table 4. Overview of cases related to personalization

Case	Presence of personalized instructions	Impact of personalization on content and/or form	Drivers for personalization	Perceived usability of personalization based on user's characteristics
A	No			Low due to high perceived workload
B	No		(Experience)	Low, perceived as challenging and time-consuming
C	No		(Complaints/feedback)	High, especially in tight labor market
D	Yes (mentor)	Level of detail, Form	Skills	Low due to high variety
E	Yes	Form, level of detail	(Achieved level of certificate), (language), product complexity, task's newness	Low due to high costs of maintaining instructions
F	Yes	Content: small changes and comments	Experience	
G	Yes (1 person)	Level of detail, Form (exploded views, photos, videos)	Experience, technical skills, age, complexity and amount of activities	Useful for future
H	Yes		Language	Not suitable due to large variation of products
I	Yes (mentor)	Level of detail	Work memory, physical strain, independence, culture, experience, age, language	Very high

the complexity of the assembly tasks. Finally, case G provides personalized instructions for a specific employee with a distance to the labor market. Other cases did not have personalization (A, B, C) but saw potential (C) or mentioned the required time as a reason for not applying it (A, B).

All cases where some form of personalization was found report to personalize the content of instructions, primarily the level of detail. In addition, several cases (D, E, G) refer to the form of instructions as an important way to personalize instructions, e.g. through providing extra videos and extended photos (G).

As drivers for personalization, user's characteristics were mentioned, such as the employees' experience (B, C, F, G). In addition, technical skills (C, G), age (related to digital skills) (G, I), language (E, H, I), and achieved level of certificate for process execution (E) were mentioned. Two cases mentioned the complexity of the product (E, G) and the number of required assembly activities (G) as drivers for personalization. The task's newness was mentioned in case E. Additionally, a third case (C) mentioned that complaints and feedback from a learning management system could also be a relevant driver for personalization.

4.3 Challenges

The maintenance of the work instructions is most frequently mentioned as the biggest challenge. In almost all cases, the maintenance of personalized work instructions is perceived as complex. Making specific adjustments to personalize instructions or implementing changes across all instructions through an appropriate modular structure requires significant time and resources. Thus, organizations find it unfavorable to implement personalized work instructions from a work instruction management perspective, as this personalization puts further pressure on the time and resources required to maintain work instructions.

Additionally, some cases (B, G, H) mention the challenge of establishing an effective process for creating the instructions and ensuring their ownership.

4.4 Comparative Analysis

Personalization of work instructions is seen as a very challenging, if not impossible, task, especially from the perspective of maintaining a large number of work instructions.

There are no apparent differences between regular companies and sheltered workshops. However, the two companies that achieved personalization through a supporting mentor are sheltered workshops. A possible explanation for this could be that the specific needs of these employees can likely only be identified through close contact with the employee. Furthermore, two of the sheltered workshops mention specific employee characteristics, such as skills and work memory, as being relevant for personalization. Case D further argues that personalization allows sheltered workshops to have their employees carry out more complex work.

All cases that create and personalize their instructions based on user's characteristics do so in the production department with close user interaction, as opposed to cases where the engineering disciplines are responsible and no personalization takes place, or personalization is based on task's newness and assembly task complexity (E). In cases where instructions are the responsibility of the production department, workers are heavily involved in the creation phase of the instructions, compared to cases where instructions are created and maintained by the engineering domain. This may suggest that worker needs can be better addressed when there is a closer link between the creator and the assembly worker.

Several cases (B, E, F, G, I) view employee experience as a reason for personalization. Case B mentions that while work instructions are evidently necessary since skipping them leads to errors, their employees do go through a learning curve. Initially, they find the instructions very helpful, but after a year, some may find the instructions too detailed. This presents an excellent opportunity for personalization in terms of the level of detail. The cases that applied personalization mainly focused on the content (level of detail), while some cases considered personalization in form additionally.

In the cases, advanced technology was hardly used to create or communicate work instructions. However, many cases have had pilot programs that received positive feedback but raised practical questions about the maintenance, necessary expertise, and time investment these technologies require. The initial creation of an instruction is not seen as an issue in most cases. However, nearly all cases find maintaining the instructions very complex and time-consuming, especially when making minor to major adjustments. The two cases that implemented a VKS reported that changes and comments can be communicated and implemented quickly, resulting in more up-to-date instructions. Interestingly, this has not led to more personalization in these cases yet.

5 Discussion

We found that many organizations do not use personalized work instructions yet or see limited usability due to its challenges related to maintainability. In their current way of working, instructions are created per product, with little possibility of reusing instructions between products. A technical solution for this issue was demonstrated in Waschull et al. [38], where a standardized and modular library of instructions was developed, allowing the reusability of instructions on different levels. Advanced technologies to more easily assist workers in a personalized way are increasingly available. However, the cases show that their implementation in practice falls behind. This aligns with the observations of Letmathe and Rößler [12], who attributed it to organizational inertia. However, most cases of our dataset have had pilot programs, meaning that they have tried new technologies but were faced with significant time investments and costs for maintaining instructions. Technological solutions based on context-awareness could possibly be used to lower the adoption threshold. Such solutions need to follow sound design principles for designing context-aware systems [39] and are increasingly considered as key technology enabler particularly relevant to designing, implementing, and operating inclusive work support system [40]. The drivers for personalization mentioned in the literature were apparent in the cases. The user's characteristic 'experience' was most frequently mentioned as a driver for personalization, while product complexity also played a role. The task's newness was referred to as a driver in only one case.

6 Conclusion

Despite the relevance of personalization for creating an inclusive work design and the expected performance benefits described in the literature, organizations currently struggle to adopt personalized work instructions. The complex and time-consuming maintenance required to keep personalized instructions, and thus a large number of instructions, up-to-date is mentioned as a central challenge. Creating, implementing, and maintaining work instructions must be accessible and easy; otherwise, it becomes time-consuming and costly.

While advanced digital technology can support the personalization of work instructions, it was hardly used to create or communicate instructions. Therefore, future research should focus on the adoption of such technologies in practice or on the development of new technologies or context-awareness features to facilitate the maintenance of personalized instructions.

Acknowledgements. The research presented in this paper is embedded in the RAAK.MKB17.017 project on Flexible work instructions, led by the HAN University of Applied Sciences. The authors acknowledge the collaboration with HAN staff and all company representatives who participated in the case interviews.

References

1. Papetti, A., Ciccarelli, M., Palpacelli, M.C., Germani, M.: How to provide work instructions to reduce the workers' physical and mental workload. Procedia CIRP **120**, 1167–1172 (2023)
2. Funk, M., Kosch, T., Schmidt, A.: Interactive worker assistance: comparing the effects of in-situ projection, head-mounted displays, tablet, and paper instructions. In: Proceedings of the 2016 ACM International Joint Conference on Pervasive and Ubiquitous Computing, pp. 934–939. ACM (2016)
3. Frank, A.G., Dalenogare, L.S., Ayala, N.F.: Industry 4.0 technologies: implementation patterns in manufacturing companies. Int. J. Prod. Econ. **210**, 15–26 (2019)
4. Stöhr, M., Schneider, M., Henkel, C.: Adaptive work instructions for people with disabilities in the context of human robot collaboration. In: IEEE 16th International Conference on Industrial Informatics (INDIN), pp. 301–308. IEEE (2018)
5. Jost, M., Luxenburger, A., Knoch, S., Alexandersson, J.: PARTAS: a personalizable augmented reality based task adaption system for workers with cognitive disabilities. In: Proceedings of the 15th International Conference on Pervasive Technologies Related to Assistive Environments, pp. 159–168. ACM (2022)
6. Haug, A.: Work instruction quality in industrial management. Int. J. Ind. Ergon. **50**, 170–177 (2015)
7. Pimminger, S., Neumayr, T., Panholzer, L., Augstein, M., Kurschl, W.: Reflections on work instructions of assembly tasks. In: 2020 IEEE International Conference on Human-Machine Systems (ICHMS), pp. 1–4. IEEE (2020)
8. Peltokorpi, J., Hoedt, S., Colman, T., Rutten, K., Aghezzaf, E., Cottyn, J.: Manual assembly learning, disability, and instructions: an industrial experiment. Int. J. Prod. Res. **61**(22), 7903–7921 (2023)
9. Vanneste, P., et al.: Towards tailored cognitive support in augmented reality assembly work instructions. J. Comput. Assist. Learn. **40**(2), 797–811 (2024)
10. Heinz-Jakobs, M., Große-Coosmann, A., Röcker, C.: Promoting inclusive work with digital assistance systems: experiences of cognitively disabled workers with in-situ assembly support. In: 2022 IEEE Global Humanitarian Technology Conference (GHTC), pp. 377–384. IEEE (2022)
11. Mark, B.G., Hofmayer, S., Rauch, E., Matt, D.T.: Inclusion of workers with disabilities in production 4.0: legal foundations in europe and potentials through worker assistance systems. Sustainability **11**(21), 5978 (2019)
12. Letmathe, P., Rößler, M.: Should firms use digital work instructions?—Individual learning in an agile manufacturing setting. J. Oper. Manag. **68**(1), 94–109 (2022)

13. Li, D., Mattsson, S., Fast-Berglund, Å., Åkerman, M.: Testing operator support tools for a global production strategy. Procedia CIRP **44**, 120–125 (2016)
14. Mark, B.G., Rauch, E., Matt, D.T.: Worker assistance systems in manufacturing: a review of the state of the art and future directions. J. Manuf. Syst. **59**, 228–250 (2021)
15. Romero, D., Bernus, P., Noran, O., Stahre, J., Fast-Berglund, Å.: The operator 4.0: human cyber-physical systems & adaptive automation towards human-automation symbiosis work systems. In: Advances in Production Management Systems. Initiatives for a Sustainable World: IFIP WG 5.7 International Conference, pp. 677–686. Springer International Publishing, Iguassu Falls, Brazil (2016)
16. Geng, J., et al.: A systematic design method of adaptive augmented reality work instruction for complex industrial operations. Comput. Ind. **119**, 103229 (2020)
17. Li, D., Mattsson, S., Salunkhe, O., Fast-Berglund, Å., Skoogh, A., Broberg, J.: Effects of information content in work instructions for operator performance. Procedia Manufacturing **25**, 628–635 (2018)
18. Buchner, J., Buntins, K., Kerres, M.: A systematic map of research characteristics in studies on augmented reality and cognitive load. Compu. Edu. Open **2**, 100036 (2021)
19. Hoover, M., Miller, J., Gilbert, S., Winer, E.: Measuring the performance impact of using the Microsoft HoloLens 1 to provide guided assembly work instructions. J. Comput. Inf. Sci. Eng. **20**(6), 061001 (2020)
20. Kolbeinsson, A., Fogelberg, E., Thorvald, P.: Information display preferences for assembly instructions in 6 industrial settings. In: 14th International Conference on Applied Human Factors and Ergonomics (AHFE) and the Affiliated Conferences, pp.152–161. AHFE International Open Access (2023)
21. Mattsson, S., Fast-Berglund, Å., Li, D.: Evaluation of guidelines for assembly instructions. IFAC-PapersOnLine **49**(12), 209–214 (2016)
22. Bosch, T., Könemann, R., De Cock, H., Van Rhijn, G.: The effects of projected versus display instructions on productivity, quality and workload in a simulated assembly task. In: Proceedings of the 10th International Conference on PErvasive Technologies Related to Assistive Environments, pp. 412–415 (2017)
23. Schlund, S., Kostolani, D.: Towards Designing Adaptive and Personalized Work Systems in Manufacturing. Digitization of the work environment for sustainable production, 81–96 (2022)
24. Mattsson, S., Fast-Berglund, Å. Li, D., Thorvald, P.: Forming a cognitive automation strategy for operator 4.0 in complex assembly. Comput. Ind. Eng. **139**, 105360 (2020)
25. Palmqvist, A., Vikingsson, E., Li, D., Fast-Berglund, Å., Lund, N.: Concepts for digitalisation of assembly instructions for short takt times. Procedia CIRP **97**, 154–159 (2021)
26. Samy, S.N., ElMaraghy, H.: A model for measuring products assembly complexity. Int. J. Comput. Integr. Manuf. **23**(11), 1015–1027 (2010)
27. Wiedenmaier, S., Oehme, O., Schmidt, L., Luczak, H.: Augmented reality (AR) for assembly processes design and experimental evaluation. Int. J. Hum. Comput. Interac. **16**(3), 497–514 (2003)
28. Uva, A.E., Gattullo, M., Manghisi, V.M., Spagnulo, D., Cascella, G.L., Fiorentino, M.: Evaluating the effectiveness of spatial augmented reality in smart manufacturing: a solution for manual working stations. Int. J. Adv. Manuf. Technol. **94**, 509–521 (2018)
29. Radkowski, R., Herrema, J., Oliver, J.: Augmented reality-based manual assembly support with visual features for different degrees of difficulty. Int. J. Hum. Comput. Int. **31**(5), 337–349 (2015)
30. Mattsson, S., Li, D., Fast-Berglund, Å.: Application of design principles for assembly instructions – evaluation of practitioner use. Procedia CIRP **76**, 42–47 (2018)

31. Funk, M., Dingler, T., Cooper, J., Schmidt, A.: Stop helping me - I'm bored! Why assembly assistance needs to be adaptive. In: Adjunct Proceedings of the 2015 ACM International Joint Conference on Pervasive and Ubiquitous Computing and Proceedings of the 2015 ACM International Symposium on Wearable Computers, pp. 1269–1273. ACM Press (2015)
32. Mourtzis, D., Xanthi, F., Zogopoulos, V.: An Adaptive framework for augmented reality instructions considering workforce skill. Procedia CIRP **81**, 363–368 (2019)
33. Stockinger, C., Stuke, F., Subtil, I.: User-centered development of a worker guidance system for a flexible production line. Hum. Factors Ergon. Manuf. Serv. Ind. **31**(5), 532–545 (2021)
34. Wolfartsberger, J., Heiml, M., Schwartz, G., Egger, S.: Multi-modal visualization of working instructions for assembly operations. Int. J. Mech. Ind. Aerosp. Sci. **13**(2), 107–112 (2019)
35. Funk, M., Mayer, S., Schmidt, A.: Using in-situ projection to support cognitively impaired workers at the workplace. In: Proceedings of the 17th International ACM SIGACCESS Conference on Computers & Accessibility, pp. 185–192. ACM Press (2015)
36. Vanneste, P., Huang, Y., Park, J.Y., Cornillie, F., Decloedt, B., Van den Noortgate, W.: Cognitive support for assembly operations by means of augmented reality: an exploratory study. Int. J. Hum. Comput. Stud. **143**, 102480 (2020)
37. Eisenhardt, K.M., Graebner, M.E.: Theory building from cases: opportunities and challenges. Acad. Manag. J. **50**(1), 25–32 (2007)
38. Waschull, S., Bokhorst, J.A.C., Wortmann, J.C., Molleman, E.: The redesign of blue- and white-collar work triggered by digitalization: collar matters. Comput. Ind. Eng. **165**, 107910 (2022)
39. Cena, F., Console, L., Matassa, A., Torre, I.: Principles to design smart physical objects as adaptive recommenders. IEEE Access **5**, 23532–23549 (2017)
40. Philips, G.R., Huang, M., Bodine, C.: Helping or Hindering: inclusive design of automated task prompting for workers with cognitive disabilities. ACM Trans. Accessible Comput. **16**(4), 1–23 (2024)

Skills and Information Needed for Operator 5.0 in Emergency Production

Sandra Mattsson[1(✉)] and Martin Kurdve[1,2]

[1] RISE Research Institutes of Sweden, Argongatan 30, 431 53 Mölndal, Sweden
sandra.mattsson@ri.se

[2] Department of Industrial and Materials Science, Chalmers University of Technology, Hörsalsvägen 7A, 412 96 Gothenburg, Sweden

Abstract. This paper explores what skills and information are needed to meet the challenges of emergency production. In the near future, Operator 5.0 will operate within Industry 5.0, a sector focused on fostering innovation for all stakeholders, including the environment. One of the core pillars of Operator 5.0 is resilience which means being able to manage emergencies and uncertainties. Achieving this poses a challenge, as the industry struggles to acquire the necessary skills recommended. A framework for skills and information needed for Operator 5.0 to perform emergency production was suggested and used in a case study for face mask production. The results are as follows: 1) skills needed for emergency production are the cognitive/physical ability to perform a task and basic overall digital skills, and 2) the information needs are standards, instructions, and training materials. To create information the following demands on the system were suggested: Universal design, minimize unexpected events, productivity and product quality and safety. The framework could be used with existing contingency planning and preparatory emergency production to plan for better management of emergencies in Sweden or Europe.

Keywords: Skill · Instructions · Resilience · Operator 5.0 · Emergency production

1 Introduction

Humans have been identified as key enablers for managing Industry 5.0 (I5.0) as well as when managing crisis [1, 2]. *I5.0* is a vision presented in 2023 by the European commission stating that industry should go beyond the production of goods and services for profit and focus on all stakeholders e.g. environment, operator health and society [1]. The vision places humans in the center in terms of managing new technology while also solving challenges of sustainability and resilience. A big challenge that was managed was the pandemic, in which humans were seen as an invaluable part [2]. During the pandemic many countries were lacking protection products and after the pandemic new strategic work was identified crucial to support nations in producing their own

Published by Springer Nature Switzerland AG 2024
M. Thürer et al. (Eds.): APMS 2024, IFIP AICT 729, pp. 336–349, 2024.
https://doi.org/10.1007/978-3-031-65894-5_24

safety products [3]. The following issues were seen: 1) there was a common shortage of products and materials for e.g. test kits and personal protection equipment [3], and 2) emergency preparation was seen as something urgently in need. *Emergency preparation* is defined as "improving the speed, volume, and quality of the emergency response" [4, Emergency preparedness]. In preparing for emergencies, production companies should form contingency plans to manage unexpected events as well as identify how emergency production could be deployed.

The focus of this paper is therefore to describe how operators working in I5.0 could work with emergency production. The biggest challenges are handling the lack of competence (described in Sect. 1.1) and finding a process for how emergency production could be done (described in Sect. 1.2).

1.1 Skill Gap for Operator 5.0

From a societal perspective this study is relevant due to a lack of competence in industry [5, 6]. I5.0 is defined as "going beyond producing goods and services for profit" [1; p. 13]. This means that industry must take responsibility for innovation for all stakeholders, including not only customers and investors but also the society and the environment, while maintaining the strive for the opportunities of technology advances presented through Industry 4.0 (I4.0) [1].

Specifically, there is a lack of digital culture and skills in their organization [5] required to manage technologies such as e.g. Industrial Internet of Things, Cybersecurity, Additive Manufacturing, Augmented Reality, Machine-learning and Big Data. Skills needed were according to Hernandez-de-Menendez et al. [7] creativity, entrepreneurial thinking, problem-solving, conflict solving, decision making, analytical skills, research skills as well as efficiency orientation.

The operator working in the technologically driven industry previously described as I4.0 is called Operator 4.0. *Operator 4.0* was defined as: "A smart and skilled operator who performs not only cooperative work with robots but also work aided by machines as and when needed by means of human cyber-physical systems…" [8, section 2]. Moving into meeting I5.0 the Operator 5.0 will have to meet the challenges of I4.0 while also being resilient. *Operator 5.0* is defined as a smart and skilled operator that is empowered by information and technology and uses human creativity, ingenuity, and innovation [9] (supported by [10–12]). In addition, Operator 5.0 is an individual that is job seeking and has digital skills [13], is predicted to collaborate or communicate with new technology in a symbiotic way [9, 14–17] and can react to unexpected or difficult situations [9]. The skills required for Operator 5.0 are a "combination of technical skills, as well as soft skills such as communication, critical thinking, problem-solving, and emotional intelligence" [18, p. 1601].

The labor shortage of industrial personnel is challenging the definitions of Operator 4.0 and 5.0 since requiring the skills needed will not be possible. In Sweden, demographic changes forecasted by Statistics Sweden [19] suggest that over the next 50 years, the population aged 18–64 will decrease by 53%, accounting for 57% of the current population within that age range. In addition, the World Economic Forum reports that metal, machinery, and related trades workers have the second most common labor shortages by occupations in 2022 in Europe (the biggest labor shortage was building and related trades) [24]. Exemplifying the urgent need, 1000 people will be employed in the new battery factory in north of Sweden, Borlänge 2024, and in mid Sweden 3000 new people will be employed in the automotive industry (Tuve 2025).

There is therefore an urgency to understand who might be able to perform the work and what training or education is needed. Some authors have stressed working with inclusive technology systems and how to support impaired workers e.g. Zambiasi et al. [12], that suggested how softbot tutors could be used to train hearing-impaired and color-blind workers. Allemang-Trivalle et al. [14] suggested that robots could be used to decrease fatigue for operators and Mattsson et al. [21], described how workplaces should be designed to fit all workers not just workers with normal functionality.

It is likely that the operator needed i.e. Operator 5.0 will need to learn and adapt without having prior knowledge in the production area. Looking at the digital skill level, Eurostat presented that in Sweden (in general) over 60% had basic overall digital skills in 2021 (ages 16–74, [22]). In the EU this number was 54% in general. Basic overall digital skill included aspects of data literacy, communication and collaboration, digital content creation, safety, and problem-solving skill. Having a basic overall digital skill could mean that the person has created a profile on social media, can send and use an email, handles their password in a safe way e.g. did not choose "1234" as their password and can solve problems such as manage connectivity issues. Understanding what minimum skill and information that is needed to run I4.0 and I5.0 needs to be explored.

1.2 Operator 5.0 for Emergency Production of Face Masks

Emergency production can refer to an event that is not normal which requires production processes that are not continuous [23–25]. *Emergency* is defined as "an exceptional event that exceeds the capacity of normal resources and organizations to cope with it" [23, section 1.1.1 The nature of disaster]. An example of emergency production is "a production line that is run in case of emergency and is usually not in continuous use" [21, p. 4 section 2.1].

As stated in the introduction, emergency preparation is something that is crucially needed. Due to the pandemic several issues were raised connected to the lack of protection products and legal issues arouse including a) Fraudulent and counterfeit products, b) Producers ramping up production: manufacturing errors, c) Novel producers and novel production methods and d) Newly developed products [3]. Here, operator skills and knowledge were highlighted as crucial. As an example, in a small study by Kurdve and Arvidsson [2] on industrial resilience in Sweden during Covid, the common enabler to adapt to the crisis was that personnel production resources could be utilized and counteract disturbances (as connected to crisis). One indication of the study was that emergencies should be trained for and that several emergency preparation plannings should be set in motion. Creating training for emergency production is one step in the direction of creating emergency preparedness according to the International Organization for Migration (IOM). IOM suggests the following approach for emergency preparation: 1) perform risk analysis, 2) take minimum preparedness actions (not risk specific and 3) perform contingency planning [4].

At Research Institutes of Sweden AB, RISE, a Face Mask Machine TESTEX APL 80 was installed as an emergency production resource to be utilized if a similar crisis as

COVID would give a national shortage of face masks, see Fig. 1. The production process consists of two serial automated operations, i) folding joining and cutting masks and ii) connecting ear-strings to the mask. One operator operates both operations, changes input of raw material, and takes care of the produced masks.

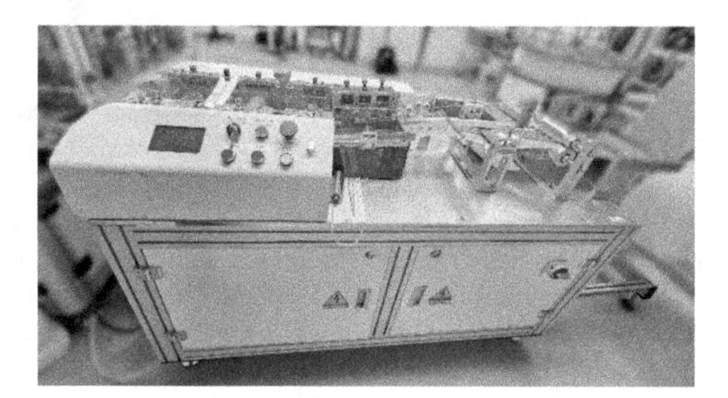

Fig. 1. Face mask machine at RISE.

As according to the suggested emergency preparation, a vulnerability analysis for the emergency production was performed (Kurdve et al. 2024) and four issues were identified:

1. raw materials supply
2. spare part supply
3. availability of trained operators, and to
4. keep operations knowledge updated between crises.

It was suggested raw material supply could be switched to dual sourcing within the EU (managing risk 1) and that spare parts were deemed to be possible to manage with the support workshop available that includes both subtractive and additive manufacturing capability (managing risk 2). To ensure emergency production the availability of operators with skill is needed (risk 3) as well as to understand what skills and information are needed to keep operations knowledge between crisis (risk 4).

1.3 Purpose and Delimitation

This article addresses the identified research gap of exploring what skill and information that is needed for emergency production for Operator 5.0. The authors define **Operator 5.0 as having basic overall digital skills and basic understanding of production systems whilst being able to manage unexpected events.** This is done to state a minimum requirement for Operator 5.0. While some of these abilities may already be present in operators, others will need to be developed through training or education.

The purpose of the paper is to describe a framework that can better help to prepare Swedish production companies and facilities for emergency. This way lack of competency can be planned for. The target of the research are production companies, and the objectives are to:

1. suggest a framework for what skills and information Operator 5.0 needs for emergency production,
2. identify skills and information needed to manage the face mask machine, and
3. implement 2) in practice and draw conclusions on how 1) can be used in practice to better prepare for emergency productions.

1.4 Method

To meet the above objectives, an exploratory approach is adopted. An explorative approach is particularly suitable considering the novelty of the research in this field and the inherent challenge associated with data collection [26]. To reach the first objective, deductive research is used to combine theory and previous case study results to form a basis of the phenomena (forming the framework for skills and information). *Deductive research* is used to draw conclusions from logical statements e.g. draw conclusions from theory [27]. Then *inductive research* is used to draw conclusions from empirical findings [27] through a case study so that the framework can be strengthened, and general conclusions could be indicated (objective 2). Inductive and deductive research is used to reach objective 3, i.e. combining results from objective 1 and 2 to discuss how and if knowledge can be generalized so that in practice companies can better prepare for emergency production.

By combining theory, previous studies and empirical data, the results are strengthened through triangulation. *Triangulation* means that the results can be validated but also that the understanding can be increased [28]. There is strength in combining the two research approaches however since the topic is complex to understand there is also risk and uncertainty in its findings e.g. due to that it is a single case, and that used theory could provide a bias for the findings [29, 30].

2 Theory

Skills can be defined as "the ability to bring about some end result with maximum certainty and minimum outlay of energy, or time, or of time and energy." [31, p. 136]. Gaining a skill is attained through using previous knowledge and experience and by repeating a task [32]. Having a certain skill can be divided into several aspects ranging from e.g. personal expertise, cognitive abilities and social skills connected to a job [33]. However, having skills for tasks at work can be separated from the skills needed in the organization to complete a task. It can therefore be difficult to define skills since cognitive processes of a specific task do not represent the complete work performed.

In the case the task for operator was to produce safe and efficient production of face masks. Except for running the machine with a minimum outlay of resources this could include e.g. collaborating with a co-worker, changing the material for the machine and calling for help when difficult problems occur.

2.1 Advancing from Operator 4.0 to Operator 5.0 in Emergency Production

The transition from Operator 4.0 to 5.0 requires careful planning, as the challenges associated with I4.0 have not yet been thoroughly studied or investigated in the literature

[34]. To make a good transition, the operator needs must be included in the design of the solutions [35] and more human factors aspects must be included in the implementation of technology [34]. As an example, Yitmen et al. [36] state that the worker knowledge and autonomous technology should be combined to enhance production to I5.0. This would require human factors such as cooperation and trust (Ibid.). In terms of being able to handle unexpected events, social skills and how to be able to manage organizational changes should be needed i.e. how to ensure quality as well as being able to work in an organization that changes [33]. This could be handled through training and information, which should be included in the framework.

Considering the authors' definition of Operator 5.0, the person should already have basic overall digital skills and knowledge of the production system whilst being able to manage unexpected events. What should be considered is what 'basic knowledge in production' means as well as what is needed to be able to manage unexpected events. Basic knowledge in production could be connected to having a holistic understanding of the production system or having a system perspective which is needed when forming a resilient and sustainable production system [37]. This means that the operator should be able to understand why parts of a machine or system break down to be able to problem solve it or to at least understand when an expert is needed.

2.2 Information Design for Emergency Production

Information is commonly referred to as facts about something and could in the context of emergency production be facts about the machine and routines that could support already trained behavior. When presenting information the operators cognitive processes should be investigated [38]. This means studying what task the operator is performing in the activity or operation needed [39]. There are two types of cognitive processes that are relevant when supporting the operator: intuition and reasoning [40]. In an operational state the operator uses *intuition*, which is automatic, fast, and effortless. This process is based on logic and correlates with general intelligence [40]. *Reasoning*, which is stated to be the contrast cognitive process, requires effort, and is used for decision-making and is slow and based on analysis and rules [41].

Emergency production instructions need to support intuition so that they can be followed easily. Intuition should be supported by training or education [42, 43]. The training is for learning how to read the instruction and not just how a specific product is built [39], to perfect skill and set goals that they can reach [42] and to perform work in the most efficient way fitting that person's natural abilities [43]. The instructions could be used as a guide and not as a manual (which would support reasoning).

To support operators with information the method Design of information presentation (DFIP) could be used which has six steps [44]:

1. select station and divide the operation into subtasks
2. map cognitive processes
3. analysis of the operator's perception of the environment
4. analysis of data regarding mental limitations
5. analyzing data based on individual differences and preferences (flow)
6. analyzing data regarding the location of information carriers and information.

The method could be used to support operators so that information is easy to understand.

To design for people with disabilities a company Husmuttern AB in Sweden developed a concept for universal design for people not currently employed. The concept was used during the pandemic for emergency production in the assembly of 450 000 protective aprons. According to the studies made the following aspects are key in the training for production emergency: see, try and assembly independently without help [21]. Husmuttern's concept is easy to use (the assembly patterns are simple and intuitive), it is flexible, it requires low physical strength and are error tolerant and is designed to fit independent of the individuals' size, posture, or mobility and can be connected to the seven principles for universal design [45]. In this sense the concept accommodated people with a wide range of preferences by being equitable and flexible, the assembly patterns are simple and intuitive, they are perceptible due to that the animated video used requires low effort to replay and therefore also error tolerant, they are designed to fit independent of size, posture, or mobility (size and space and low physical effort).

The Husmuttern concept will be used to set demands for skill and information in the framework.

3 Results

A framework for skills and information needed for Operator 5.0 in emergency production is presented in Sect. 3.1 (objective 1). The skills and information needed to run the face mask machine is presented in Sect. 3.2 (objective 2).

3.1 Framework for Skills and Instructions for Operator 5.0 in Emergency Production

In Fig. 2 a framework for Operator 5.0 skill and information needs is presented. The framework can be seen as an operator wearing a big hat. The two first rows are the hat, then the circle is the face, and the contents is the body of the operator. The hat consists of skills derived from the article and demands coming from Sect. 2. The model should be read from left to right to indicate order of activities when using the model. The concept is described below.

The two skills that are needed are 1) the cognitive and physical ability to perform a task and 2) having basic overall digital skills. These two skills are relevant since the defined task affects the cognitive process [38, 39]. When the task has been analyzed e.g. to assemble a product with help from a digital screen, it is possible to design the needed information together with the operators i.e. set a standard, create instructions, and develop a training according to production demands. The basic overall digital skills are needed if a system is used. This can be a system for scanning materials, reading the instructions, or logging errors. If such a system does not exist, then this skill is not needed. Often such skills are needed e.g. for writing emails, writing error reports, or reading instructions on an Ipad.

When the skill has been identified the company should carefully make choices according to the suggested demands (on the system): a) Universal design, b) Minimizing

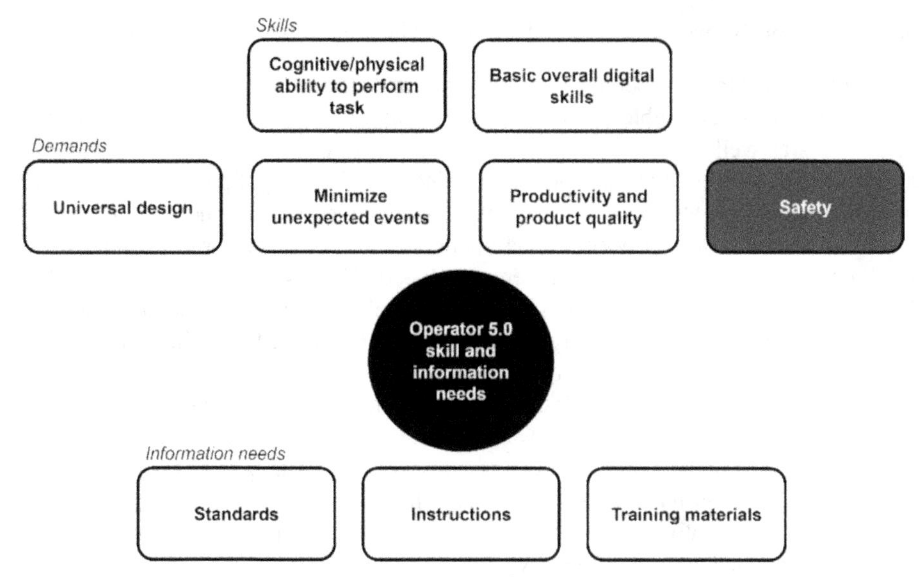

Fig. 2. Framework for operator skills and information needs in emergency production in the form of an operator wearing a hat. The hat contains skills and demands which should be considered before creating the information needs, which is the body of the operator. The circle is the face of the operator.

Unexpected events, c) Productivity and product quality and d) Safety. The first demand is *Universal design*. This should be applied according to the Husmuttern concept so that it is e.g. easy to use (the assembly patterns are simple and intuitive), it is flexible, it requires low physical strength and are error tolerant, they are designed to fit independent of the individuals' size, posture, or mobility. The second demand is to identify *unexpected events* to *minimize* them. This should be used when making Instructions and Training materials so that the operator can train and manage unexpected events i.e. to make them expected and simple to follow. An example of an unexpected event is identifying the uncommon errors when using the face mask machine. The third demand is to design the task so that *Productivity and product quality* is reached. The fourth demand *Safety* is always important and includes not only safety when performing the task but safety in being at the production facility etcetera.

The third row concerns the information needs i.e. i) Setting a Standard, ii) Creating Instructions and iii) Creating Training materials. These could be performed in that order. When the task has been defined and it has been tested and iterated using the demands, a standard for how it should be performed in the best possible way can be set. Then an instruction can be created and based on the standard training material can also be created. Creating information content can be difficult and should be considered e.g. by using the six steps for information presentation [44].

3.2 Skills and Information Needed for Face Mask Emergency Production

The skills needed for the operator to produce safe and efficient production of face masks was identified as being able to mimic behavior, and either listen to and see, or read instructions as well as to have a general knowledge in production. In addition, basic digital literacy is needed to manage the machine (see Fig. 3 right picture). In terms of information, the operator needed to rely on standards, instructions, and training.

To assure the availability of trained operators (risk 3 from the vulnerability analysis), operator training instructions were developed. The instruction manual, together with interviewing assigned operators and making trials were used to make Standardized Operations Procedures (SOP's) for the operator work tasks, see Fig. 3. In addition, one-point-lessons were made for operator tasks i.e. start/stop sequence or for common errors. Both the SOP's and the one-point lessons followed the DFIP method e.g. identifying the task and then assuring that intuition was supported through simple and easy to follow instructions based on operator need. It was estimated that training of new operators can be finalized within one day, probably less if visualized instructions are developed further.

To keep the operator skill (risk 4), it was estimated that the process needed to be started and operated at least once per year.

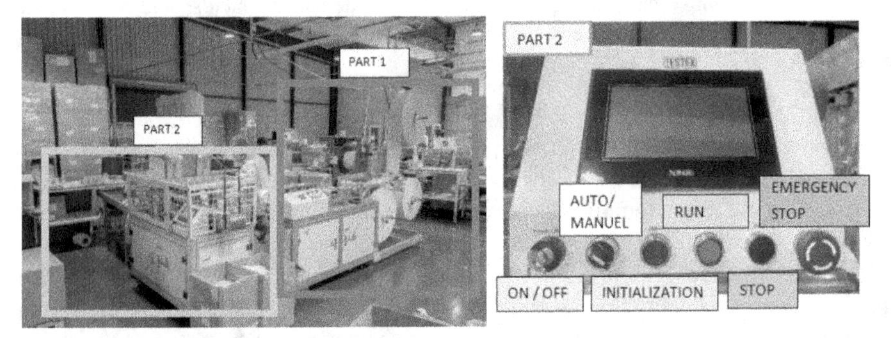

Fig. 3. Instructions for the face mask machine. To the left the machines parts are presented and to the right is a photo of the start-up and control panels.

To ensure emergency preparation, it was stated that raw materials may be further developed in the future. Therefore, also a plan for yearly training needed to involve sourcing of raw materials, maintenance of the equipment, and quality testing of the final product. Also, it should be considered what additional operator skills might be needed that can help the company to become resilient or sustainable.

4 Discussion

The results of the first two objectives are discussed in Sect. 4.1, and the case study results, and the third objective is discussed in Sect. 4.2. To make a good transition between Operator 4.0 to Operator 5.0 it was stated that careful planning is needed [34] and it is important to include operators in the design of the solutions [35]. This was achieved in the face mask case since careful consideration was placed into making a vulnerability analysis and by defining the task e.g. through setting a standard, developing instructions, and setting training routines. Operators were included in the continuous process development of these processes.

4.1 Relevance of the Suggested Framework and Case Study Results

The framework was based on theoretical concepts and tested in the case study which increases its validity. Due to the fact that labor will be lacking it is reasonable that only two skills were stated as needed. Having a cognitive and physical ability to perform the task is relevant due to that the task controls the operators' cognitive processes which is connected to what stimuli that is needed when interacting with technology or to suggest what type of instruction or information that is suitable [38]. In addition, having physical ability to perform the task is a prerequisite. In the production of face masks, the training was said to be completed in one day however the upkeep of the skill needed to be repeated at least once per year. It is reasonable that general digital skills are needed to enable I5.0 and since over half of Europe reaches that [22] this is a reasonable minimum skill level. In general, basic overall digital skills were needed and for the face mask machine it concerns interacting with the screen (Fig. 3, right). Although it was not deemed as necessary in the face mask example, this is suggested as a needed skill since many, if not all, production systems include some kind of digital tool. Especially for mobile emergency production it may be preferred to store digital instructions to make training material available before the physical equipment is in place. And in general, some instructions could be presented on screens as was done in the production of protective aprons in the concept developed by Husmuttern. In the definition of Operator 5.0 a relevant skill was being able to handle unexpected events, which is useful if the production is not acting as it should. This, however, requires more of the operator in terms of social and or emotional intelligence and it was suggested to train for managing unexpected or unlikely events, e.g. a fire as an unlikely event, or also new unknown events can be trained. The event should be made explicit so that the operator can learn how to handle such events. This is the same in normal production i.e. that fire drills are trained for. Before they are trained a standard is developed and instructions are made that are easy to follow so that everyone knows what to do.

The skills in the framework should be tested and explored further to include:

- Basic understanding of the production system e.g. requirements on the product
- Holistic view of the production system [37]
- Understanding or skill in resilience or sustainability [1]
- Handling unexpected events that is not trained for e.g. emotional intelligence and stress management.

To adapt the working environment so that instructions are easy to follow and can be used by anyone universal design can be used e.g. as suggested by [45]. This could enable more people can be able to work in I5.0. Burgstahler described seven principles may be perceived as not so concrete when implementing universal design, then and the adaptation performed by Husmuttern can be copied as a first step. Since they developed and used parts of the principles of universal design for emergency production of a high number of protective aprons, their ideas and ways of working could successfully be included to help companies be more inclusive and to create a good working environment (ensuring I5.0) in emergency production. In addition, universal design places the individual in focus which is useful to perfect skill and set individual goals so that work can be efficient for that individual [42, 43]. Having a universal design approach could enable operators that e.g. are not fluent in the language to be hired and that less errors are made by the operators. The framework therefore supports inclusion and individual adaptation of the workplace due to Universal design. Having a Universal design approach means that all the work and the workplace is designed to support a wide range of work preferences e.g. in terms of size, posture and mobility and it is evaluated to make instructions easy to understand and to follow [45].

The other demands in the framework are useful since they are a normal part of industrial work i.e. productivity, quality, and safety. In the case of emergency production, it was seen as especially important to ensure safety due to the fact that the operator might be working in a non-normal working environment.

4.2 Practical Contribution

The vulnerability analysis provided a good start on how to work with the preparations for emergency production of the face masks. The analysis showed risks connected to skills and information, e.g. ensuring the availability of trained personnel as well as keeping operators trained. By using the framework, it was possible to identify which skills are needed and to discuss how information should be developed to ensure found risks. Due to the fact that skills are difficult to define and measure [33] the suggested skills could be discussed together with a company to define and structure the emergency preparation. For the face mask machine, the task included to be able to run the machine and ensure product quality which requires that the operator has a basic understanding of how production and machines work. In this sense it could also require that the operator understands material handling holistically [37] and that the combination of raw materials affect the product. The operator must also be able to follow the requirements of the production e.g. must be able to understand an order. This can be assessed e.g. through testing. The developed instructions should be trained which is one step in the direction of creating emergency preparedness [4].

5 Conclusions

The skills and information suggested for Operator 4.0 do not apply for Operator 5.0 in emergency production. Instead, to supply Operator 5.0 with the needed skill and information, a framework with nine factors was suggested: cognitive/physical ability to perform task, basic overall digital skills, universal design, minimize unexpected events, productivity and product quality, safety, standards, instructions, and training materials. The results from the case study show that the factors are relevant, and companies can draw conclusions from it when they start working with emergency production. Being able to perform emergency production is a way to improve the company's resilience and sustainability i.e. to become I5.0. The framework could be used with existing contingency planning and preparatory emergency production to plan for better management of emergencies in Sweden or Europe.

Acknowledgments. This paper was written as part of a Swedish Civil Contingencies Agency funded project (MSB 2022–09211) and the Vinnova funded project RESPIRE (RESPIRE: Rethinking the management of unexpected events for resilient and sustainable production, 2021–03685.) and DIGITALIS (DIGITAL work InStructions for cognitive work (DIGITALIS), 2022–01280.). It is connected to sustainable production research in XPRES. The support is gratefully acknowledged.

Disclosure of Interests. The authors have performed research together with Husmuttern AB since 2016. The company is currently involved in the RESPIRE and the DIGITALIS project.

References

1. European Commission Homepage: Directorate-General for Research and Innovation, Breque, M., De Nul, L., Petridis, A., Industry 5.0 – Towards a sustainable, human-centric and resilient European industry. Publications Office of the European Union. https://data.europa.eu/doi/10.2777/308407. Accessed 03 Apr 2024
2. Kurdve, M., Arvidsson, A.: Adaptive industrial production systems to respond to societal demand at time of crisis. In: 29th International Annual EurOMA Conference, 3–6 July 2022. University of Sussex Business School Berlin, Germany (2022)
3. Fairgrieve, D., Feldschreiber, P., Howells, G., Pilgerstorfer, M.: Products in a pandemic: liability for medical products and the fight against COVID-19. Eur. J. Risk Regul. **11**(3), 565–603 (2020)
4. UN International Organization for Migration Homepage: Guidance Note Preparedness for Emergency Response in IOM. https://emergencymanual.iom.int/emergency-preparedness. Accessed 18 Apr 2024
5. Simic, M., Nedelko, Z.: Development of competence model for industry 4.0: a theoretical approach. In: Economic and Social Development: Book of Proceedings, pp. 1288–1298 (2019)
6. World Economic Forum Homepage, The Future of Jobs Report 2018. http://www3.weforum.org/docs/WEF_Future_of_Jobs_2018.pdf. Accessed 03 Apr 2024
7. Hernandez-de-Menendez, M., Morales-Menendez, R., Escobar, C.A., et al.: Competencies for industry 4.0. Int. J. Interact. Des. Manuf. **14**, 1511–1524 (2020)
8. Romero, D., Wuest, T., Stahre, J., Gorecky, D.: Social factory architecture: social networking services and production scenarios through the social Internet of Things, services and people for the social operator 4.0. In: Lödding, H., Riedel, R., Thoben, K.D., von Cieminski, G., Kiritsis, D. (eds.) APMS 2017. IFIPAICT, vol. 513, pp. 265–273. Springer, Cham (2017). https://doi.org/10.1007/978-3-319-66923-6_31
9. Romero, D., Stahre, J.: Towards the resilient operator 5.0: the future of work in smart resilient manufacturing systems. Procedia CIRP **104**, 1089–1094 (2021). ISSN 2212-8271
10. Simeone, A., Grant, R., Ye, W., et al.: A human-cyber-physical system for operator 5.0 smart risk assessment. Int. J. Adv. Manuf. Technol. **129**, 2763–2782 (2023)
11. Bechinie, C., Zafari, S., Kroeninger, L., Puthenkalam, J., Tscheligi, M.: Toward human-centered intelligent assistance system in manufacturing: challenges and potentials for operator 5.0. Procedia Comput. Sci. **232**, 1584–1596 (2024). ISSN 1877-0509

12. Zambiasi, L.P., Rabelo, R.J., Zambiasi, S.P., Lizot, R.: Supporting resilient operator 5.0: an augmented softbot approach. In: Kim, D.Y., von Cieminski, G., Romero, D. (eds.) APMS 2022. IFIPAICT, vol. 664, pp. 494–502. Springer, Cham (2022). https://doi.org/10.1007/978-3-031-16411-8_57
13. Mourtzis, D., Angelopoulos, J., Panopoulos, N.: Operator 5.0: a survey on enabling technologies and a framework for digital manufacturing based on extended reality. J. Mach. Eng. **22**(1), 43–69 (2022)
14. Allemang-Trivalle, A., et al.: Modeling fatigue in manual and robot-assisted work for operator 5.0. IISE Trans. Occup. Ergon. Hum. Factors, 1–13 (2024)
15. Alves, J., Lima, T.M., Gaspar, P.D.: Is industry 5.0 a human-centred approach? A systematic review. Processes **11**, 193 (2023)
16. Golovianko, A., Terziyan, V., Branytskyi, Vladyslav, V., Malyk, D.: Industry 4.0 vs. industry 5.0: co-existence, transition, or a hybrid. Procedia Comput. Sci. **217**, 102–113. (2023). ISSN 1877-0509
17. Leng, J., et al.: Industry 5.0: prospect and retrospect. J. Manuf. Syst. **65**, 279–295 (2022)
18. Hattinger, M., Stylidis, K.: Transforming quality 4.0 towards resilient operator 5.0 needs. Procedia CIRP **120**, 1600–1605 (2023). ISSN 2212-8271
19. Statistics Sweden Homepage: The Future Population of Sweden 2019–2070. Sveriges framtida befolkning 2019–2070. (in Swedish). https://www.scb.se/contentassets/24496c5905454373b2910229c29001ec/be0401_2019i70_sm_be18sm1901.pdf. Accessed 01 Mar 2023
20. World Economic Forum Homepage: The Future of Jobs Report 2023. https://www.weforum.org/publications/the-future-of-jobs-report-2023/ . Accessed 17 Apr 2024
21. Mattsson, S., Kurdve, M., Almström, P., et al.: Framework for universal design of digital support and workplace design in industry. Int. J. Manuf. Res. **18**(4), 392–414 (2023)
22. Eurostat Homepage: How many citizens had basic digital skills in 2021? – Product Eurostat News – Eurostat. https://ec.europa.eu/eurostat/web/products-eurostat-news/-/ddn-20220330-1. Accessed 17 Apr 2024
23. Alexander, D.: Principles of Emergency Planning and Management. Dunedin Academic Press Ltd: e-book (2014). https://books.google.se/books?id=c1xwDwAAQBAJ&hl=sv&source=gbs_navlinks_s. Accessed 03 Apr 2024
24. Al-Dahash, H., Thayaparan, M., Kulatunga, U.: Understanding the terminologies: disaster, crisis and emergency. Presented at Association of Researchers in Construction Management (ARCOM), Manchester, UK, September 2016
25. Kurdve, M., Mattsson, S., Thuresson, U., Stenlund, P., Ström, M., Raaholt, B.: Continuity analysis method for manufacturing: case study of emergency production of masks. In: Proceedings of NOFOMA (2024)
26. Van de Ven, A.H.: Engaged Scholarship. Oxford University Press, Oxford (2006)
27. Thurén, T.: Vetenskapsteori för nybörjare. Prinfo/Team Offset & Media, Malmö (2002)
28. Olsen, W.K., Haralambos, M., Holborn, M. (ed.): Triangulation in social research: qualitative and quantitative methods can really be mixed. In: Developments in Sociology. Causeway Press Ltd. (2004)
29. Flynn, B., Sakakibara, S., Schroeder, R.G., Bates, K.A., Flynn, E.J.: Empirical research methods in operations management. J. Oper. Manag. **9**(2), 250–284 (1990)
30. Eisenhardt, K.M.: Building theories from case study research. Acad. Manag. Rev. **14**(4), 532–50 (1989)
31. Guthrie, E.R.: The psychology of learning (Rev. ed.). Harper (1952). https://gwern.net/doc/cat/psychology/1960-guthrie-thepsychologyoflearning.pdf. Accessed 17 Apr 2024
32. Koivo, A.J., Repperger, D.W.: Skill evaluation of human operators. In: 1997 IEEE International Conference on Systems, Man, and Cybernetics. Computational Cybernetics and Simulation, vol. 3, pp. 2103–2108. IEEE, October 1997

33. Grugulis I., Stoyanova, D.: Skill and performance. Br. J. Ind. Relat. **49**(**3**), 515–536 (2011)
34. Gladysz, B., Tran, T., Romero, D., van Erp, T., Abonyi, J., Ruppert, T.: Current development on the operator 4.0 and transition towards the operator 5.0: a systematic literature review in light of industry 5.0. J. Manuf. Syst. **70**, 160–185 (2023)
35. Kaasinen, E., et al.: Empowering and engaging industrial workers with operator 4.0 solutions. Comput. Ind. Eng. **139**, 105678 (2020)
36. Yitmen, I., Almusaed, A., Alizadehsalehi, S.: Investigating the causal relationships among enablers of the construction 5.0 paradigm: integration of operator 5.0 and society 5.0 with human-centricity, sustainability, and resilience. Sustainability **15**, 9105 (2023)
37. Säfsten, K., Harlin, U., Johansen, K., Larsson, L., Vult von Steyern, C., Öhrwall Rönnbäck, A.: Resilient and sustainable production systems: towards a research agenda. Int. J. Manuf. Res. **18** (4), 343–365 (2023)
38. Jamieson, G.A., Vicente, K.J.: Ecological interface design for petrochemical applications: supporting operator adaptation, continuous learning, and distributed, collaborative work. Comput. Chem. Eng. **25**(7–8), 1055–1074 (2001)
39. Mattsson, S., Fast-Berglund, Å.: How to support intuition in complex assembly? Procedia CIRP **50**, 624–628 (2016)
40. Tsujii, T., Watanabe, S.: Neural correlates of dual-task effect on belief-bias syllogistic reasoning: a near-infrared spectroscopy study. Brain Res. **1287**, 118–125 (2009)
41. Evans, J.St.B.T.: In two minds: dual-process accounts of reasoning. Trends Cognit. Sci. **7**(10), 454–459 (2003)
42. Csikszentmihaly, M.: Flow the Psychology of Optimal Experience. Harper Perennial Modern Classics, New York (1990)
43. Taylor, F.W.: The principles of scientific management (2005)
44. Mattsson, S., Li, D., Fast-Berglund, Å.: Application of design principles for assembly instructions – evaluation of practitioner use. Procedia CIRP **76**, 42–47 (2018)
45. Burgstahler, S.: Creating Inclusive Learning Opportunities in Higher Education: A Universal Design Toolkit, p. 47–48. Harvard Education Press, Boston (2020)

Augmenting the One-Worker-Multiple-Machines System: A Softbot Approach to Support the Operator 5.0

Ricardo J. Rabelo[1(✉)], Lara P. Zambiasi[1,2,4], Saulo P. Zambiasi[3], Mina Foosherian[4], Stefan Wellsandt[4], David Romero[5], and Karl Hribernik[4]

[1] UFSC – Federal University of Santa Catarina, Florianópolis, Brazil
ricardo.rabelo@ufsc.br
[2] IFSC – Federal Institute of Santa Catarina, Chapecó, Brazil
[3] UNISUL – University of the South of Santa Catarina, Florianópolis, Brazil
[4] BIBA at the University of Bremen, Hochschulring 20, 28359 Bremen, Germany
[5] Tecnológico de Monterrey, Mexico City, Mexico

Abstract. Industry 4.0/5.0 workplaces are characterized by humans surrounded by massive digitalization, huge data generation, and data-driven management. However, this brings more complexity to the operators as they are exposed to vast amounts of data to reason about as well as to many situations of overwhelming cognitive load, leading them to potentially less assertive and stressful decision-making. This becomes challenging when operators should manage two or more machines simultaneously as in a 'One-Worker-Multiple-Machines' (OWMM) working environment, including critical processes and equipment. This paper proposes a softbot approach to address these issues devising an OWMM smart cockpit environment where an intelligent softbot supports an operator in several production situations. A software prototype was developed to show the potential and benefits of the softbot approach in OWMM environments.

Keywords: Softbots · Intelligent Digital Assistants · Operator 5.0 · Industry 5.0 · One-Worker-Multiple-Machines (OWMM)

1 Introduction

Industry 5.0 envisions a production landscape where humans and intelligent machines collaborate synergically. This collaboration is fueled by cutting-edge technologies like AI, IoT, Robotics, and Immersive Reality, involving Digital Twins, Augmented (AR), Virtual (VR), and Mixed Realities (MR) [1].

Published by Springer Nature Switzerland AG 2024
M. Thürer et al. (Eds.): APMS 2024, IFIP AICT 729, pp. 350–366, 2024.
https://doi.org/10.1007/978-3-031-65894-5_25

In line with the concepts of *Operator 4.0* [2] and *Operator 5.0* [3], the emerging concept of the "Cognitive Operator" represents knowledge workers and their cognitive interactions with systems, enabling a more symbiotic execution of tasks. This facilitates the development of systems that exhibit deeper perception, awareness, and understanding by humans [4]. *Augmented technologies* provide opportunities to develop a human-centric vision of manufacturing to benefit both businesses and their employees [5].

Industry 4.0 and *Industry 5.0* fundamentally rely on massive industry digitalization, (big) data generation, and data-driven management [6]. However, such data are typically spread over disparate and isolated silos as well as embedded over several legacy systems and industrial equipment. They are implemented by adopting different technologies, formats, semantics, user interfaces, and (cyber)security schemas [7]. This makes searching, accessing, understanding, and using the required data within given operational contexts very challenging for decision-makers at all company levels [7, 8].

Moreover, despite the benefits of *data-driven manufacturing environments,* they can bring more complexity to operators. It has been observed that operators are more and more exposed to massive amounts of data about several companies' machines (mostly in real-time), processes, technical norms, etc., leading to very stressful and cognitively overloaded situations. This can result in lower productivity and less comprehensive or wrong analyzes and decision-making due to the many checks, analyzes, supervision actions, etc. that need to be more often and rapidly performed by operators via different systems and dashboards [9–11]. Tedious, repetitive but relevant tasks can also be refrained from being correctly done [5]. Yet to consider that machines' features (logical or physical) as well as operation protocols and internal norms change from time to time, demanding a permanent 'mental updating' by operators, exposing them to many situations of error, forgetfulness, and unnoticed events. This can become even worse when dealing with critical equipment with processes that demand very careful and close monitoring and quality control, including those that have some degree of operation, material (inbound logistics), or data (control) interdependence [10, 11].

One of the most important goals of *Industry 4.0* is to improve *operational efficiency* [12]. In this direction, many companies are used to applying the *One-Worker-Multiple-Machines (OWMM)* manufacturing concept on their shop floor. OWMM generally means making a single operator manage and operate two or more machines "simultaneously" in a production system. Although an old concept, OWMM is still considered a good industrial practice [13, 14] and it can also be a feasible alternative to cope with the increasing workforce shortage [15]. On the other hand, all those issues can get more cognitively overwhelming to be handled by *OWMM operators* given they should manage several machines at the same time with the required quality and attention during their entire working shifts.

This paper proposes a *softbot approach* [16] to help address these issues. A *softbot* can be defined as "a virtual system deployed in a given computing environment that automates and helps humans in the execution of tasks by combining capabilities of conversation-like interactions, system intelligence, autonomy, proactivity, and process automation" [17]. Its potential and applications in smart manufacturing and Industry 4.0 are many [16, 18]. In its vision for a future-ready workforce, WEF foresees that workers in industrial settings will have visual guides and intelligent digital assistants, making them more efficient and improving work outputs [19].

The envisaged goal in this research is to equip *OWMM operators* with a so-called *smart cockpit environment,* through which they can better and more symbiotically manage and operate several machines simultaneously. By decreasing overwhelmed cognitive situations, it is expected to enhance operators' productivity, operations compliance and safety, decision-making support, workplace satisfaction, and operators' mental health [16, 20].

After conducting a systematic literature review, no work was found proposing or implementing a *softbot-assisted OWMM smart cockpit environment* for shop floor operators. Under the *Action-Research* method, this ongoing research aims to investigate how such an environment can help *OWMM operators* in their cognitively demanding work in Industry 4.0/5.0 shop floors.

This paper is organized as follows. Section 1 has pointed out the tackled problem, its motivation, and the intended goal. Related works are presented in Sect. 2. Section 3 describes the proposed *softbot-assisted OWMM smart cockpit environment* and its architecture. Section 4 presents the implemented environment. Section 5 provides some discussion and assessment. Finally, Sect. 6 provides conclusions and further work.

2 Related Works

Supporting operators with digital management and/or operating environments is not new. In modern industries, *OWMM operators* already use some applications (although often proprietary) and devices (e.g., tablets and smartphones) provided by different manufacturers to handle machines.

Solutions in the market and the ones proposed/developed as academic prototypes have offered different types of support. They include integrated cockpits for enterprise management [21]; remote supervision with on-line dashboards (coping with more than one single machine, like advanced supervisory/MES systems [22]); local supervision (walking through the shop floor and jumping from one machine to another [23]) with different levels of immersion (e.g., AR/VR/MR and industrial metaverse) [24, 25]; interaction with intelligent digital assistants and chatbots [26, 27] as well as with multiple collaborative softbots [10]; real-time visualization and data analyzes, as digital twins [28] and 'cognitive' digital twins with actuation on the physical machine [29]; intelligent reasoning (as in AI-based systems) [30, 31] and adaptiveness [32]; assisted training [33]; and support for OWMM scheduling and workload balance [34].

Despite the very important contributions brought by such authors about *individual* issues that could be somehow applied to the envisaged environment, no research works have been found in the searched literature *combining* interaction and reasoning capabilities, remote and local work, immersion, supervision and actuation, and intelligent digital assistants to support *OWMM operators* in handling multiple machines simultaneously into one *integrated* and *softbot-assisted* environment. Other more comprehensive papers and literature reviews on *softbots* (and "equivalent" terms, such as intelligent digital assistants and chatbots), as [16, 35, 36]), have not addressed OWMM scenarios either.

3 Proposed Softbot Environment for OWMM Operators

3.1 OWMM Smart Cockpit Environment Basic Rationale

As an *Action-Research* work, this started by identifying the needs of *Operators 5.0* for being assisted by *intelligent softbots* in *OWMM scenarios* to decrease their cognitive overload and to improve their general productivity. A set of requirements and derived design principles (see below) were defined, an implementation architecture was further conceived, and a *POC software prototype* was implemented and generally evaluated. These artifacts were refined in cycles. This paper represents the first consolidated result of this process. *Testing scenarios* were selected for pilot implementation through meetings with a team from a food industry company, combined with some cases from the analyzed literature review. The team's expertise also helped define the 3D model and choose relevant components and functional software aspects, as well as to better understand user requirements and to collect feedback for further developments.

The design principles were defined as general guidelines used to drive the cockpit's system architecture. They were a result of a rational compilation and adaptation of key foundations for industrial softbots: advanced (industrial) *softbots* requirements [16, 37], *human-centric systems* [20, 38, 39], and *levels of human interaction* [40].

i. *Cognitive:* such an environment should provide operators with an intelligent digital assistant they can interact with during their activities using user-friendly and adaptive communication ways. This means a more symbiotic and intuitive cockpit to better exploit operators' cognition capabilities as well as to offer them richer user experience and higher work satisfaction. This assistance should enable proactive problem detection, insights, and suggestions for more effective decision-making, also considering operational safety, correctness, and efficiency. It should facilitate the understanding and execution of complex tasks, also automating repetitive and some manual tasks. Different problem-solvers and external services can be used to support the softbot's intelligence. This assistance can also offer personal feedback to operators;
ii. *Digital:* all data about and from the machines/equipment under the operator's control and supervision should be digital;
iii. *Seamless Interoperability:* the softbot acts as an integrated hub for OWMM operators to monitor and control all machines under their responsibility using different devices (multi-tenancy) and communication protocols. This includes hiding the intrinsic heterogeneity of machines, PLCs, legacy wrappers and systems, etc. An OWMM smart cockpit environment should support real-time tracking and updates on machine status. It should hide the many differences between the diverse enterprise information systems a softbot can have access to when performing its tasks and attending to operators' requests;
iv. *Comprehensiveness:* to deliver a broad view of both the physical space and the machines' processes, ensuring operators have a full grasp of production status and any other machines' related issues. This view can involve both individual and independent machines, and groups of inter-dependent machines;
v. *Evolving:* such an environment can learn from machines' operations and operators' actions, and adapt or optimize its knowledge and actions accordingly, as well as that machines' properties and their operating protocols and technical norms can change throughout their lifecycles;
vi. *Governance:* such an environment should grant its access not only to the authorized operators but also to limit/control their actions according to predefined action rights (e.g., to access information, to act on the machine, etc.). This also refers to the system's autonomy set-up;
vii. *Virtual:* all the operators of such an environment are provided with an integrated virtual and web-based (operating system independent) ambient through which they can receive data from machines and other enterprise information systems, visualize the machines in proper dashboards and 3D GUI, interact with the cockpit and machines in different ways and using different technologies and devices, make analyzes and simulations, make decisions, and can act on a proper physical machine(s). This can be done either remotely or locally close to the real machines on the shop floor. The environment can be deployed either in a local server or in the cloud;
viii. *Immersive:* such an environment should allow the possibility of operators to navigate and walk through the machines and surrounding shop floor virtually immersed;
ix. *Scalable:* an OWMM smart cockpit environment should allow the entrance or leaving of machines and cockpit functionalities;
x. *Security:* such an environment should be protected against cyber-attacks and from unauthorized accesses and users;
xi. *Auditing:* such an environment should allow the generation of automatic reports about operators' actions when piloting the cockpit for different purposes, such as management control, identification of training needs, formal requests from the machines' manufacturers regarding machines' insurance, etc.;
xii. *Open:* using open ICT standards as much as possible as well as having the possibility to access/invoke software services offered (under different modes) by ecosystems of external providers;
xiii. *Service-oriented:* the architectural style of designing, implementing, integrating, and making functionalities available.

When seen as a whole, these principles make the proposed environment a *smart system*. It can do and help *OWMM operators* in doing tasks in clever, automatic, and proactive ways based on acquired data, digitally communicating and seamlessly interoperating with other systems and humans. In general, it can sense the ambient (i.e., operators, company's systems, and shop floor equipment), reason about it to make decisions of diverse complexities, and act on it when necessary and adaptively according to the operators' profile, shop floor status, and company's governance model.

Besides this, by being smart, it also wanted to exploit the *Operators 5.0 autonomy*. This means that an operator, depending on the current need and possibilities, can have the flexibility to either use the *smart cockpit* remotely or locally; either immersive (i.e., with AR, VR, or MR) or in a desktop-like interactive environment; either handling one single machine or several ones simultaneously; either only supervising the machines or also acting on them; and requesting virtual assistance just when needed. This kind of weighting is not an automatic process but rather a decision taken by the own operators based on their experience, knowledge, autonomy level, and company norms.

3.2 OWMM Smart Cockpit Environment Architecture

In general, a derived *softbot* presents three types of behavior modes, which in turn can be used in a combined way when coding each *softbot's action:* (i) *reactive* – when the *softbot* acts in response to direct user requests via chatting (e.g., to ask about more detailed information from a given machine); (ii) *planned* – when the *softbot* acts in response to predefined scheduled tasks (of different types and complexities), bringing their results to users after their execution (e.g., to generate consolidated performance reports weekly); and (iii) *pro-active* – when the *softbot* performs predefined tasks autonomously on behalf of users or company, bringing their results to users if needed (e.g., to continuously checking machines operation and promptly take measures to solve some problem, including sending warnings and alarms to operators). A *softbot* can also learn from the actions taken, make predictions, and recommend actions.

Considering the *OWMM smart cockpit environment's* design principles as well as a reference architecture for industrial softbots [33], Fig. 1 shows its general implementation architecture, developed and used as the basis for its prototype.

Users

- *OWMM Operator:* it is the one for this assisted environment that is developed for – the 'end-user'.
- *OWMM (smart) Cockpit Configurator:* it is the user who does all the required configurations in the *OWMM smart cockpit environment* according to the company's rules, operators, and machines to consider, etc., and that can have access to the other cockpit's modules (as security, knowledge base, etc.) for maintenance purposes.

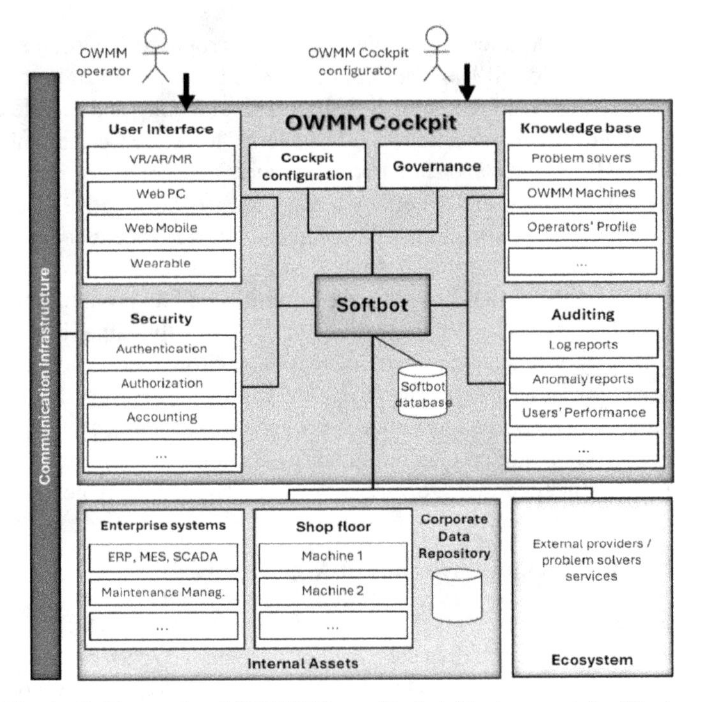

Fig. 1. Softbot-assisted OWMM Smart Cockpit Environment Architecture

OWMM Cockpit

- *User Interface:* this is the basic interface through which the *OWMM operator* can interact with the (smart) cockpit. This architectural component generally incorporates the features *i, v, viii,* and *xii* (see Sect. 3.1). The (smart) cockpit can be used with different devices and technologies. The operator can log in via single sign-on access, accessing the (smart) cockpit wherever (s)he is.
- *Softbot:* this is the core architectural (smart) cockpit component. It also has a database for its internal purposes. It generally incorporates the features *ii, iii, iv, vii, xii, xiii,* and *xiv*.
- *Cockpit Configuration*: this component supports the configuration of several aspects related to users', systems', and machines' configurations. It mainly refers to the features *ii, iii,* and *x*.
- *Knowledge Base:* it stores information (historical data, profiles, etc.) about operators, machines, and surrounding shop floor, internal problem-solvers (based on AI, Operations Research, *ad-hoc* algorithms, etc.), simulator(s), etc. In general, it refers to the features *iv* and *vii*.
- *Security:* it essentially handles cybersecurity issues. It comprises feature *ix*.
- *Governance:* it defines and controls the operators' rights. It refers to feature *x*.
- *Auditing:* this component involves the storage of different types of data for further analysis by control sectors. It incorporates feature *xi*.

Internal Assets

- *Enterprise (information) systems:* this component involves other data sources to support analysis and decision-making. It refers to the features *vi, vii, xii, xiii,* and *xiv.*
- *Shop floor:* this component encompasses all the machines, equipment, connected IoT devices, etc., the OWMM operator has to deal with, and real-time data that are collected; it refers to the features *ii, iii, vi, xii,* and *xiii.*

Ecosystem

- This component refers to auxiliary software that is provided by external services (usually "as-a-service") as a complement to the softbot's knowledge base. It generally refers to the features *vi, vii, xiii,* and *xiv.*

Communication Infrastructure

- It supports communication, integration, and interoperation between all the (smart) cockpit components as well as with the industrial machines, the other internal enterprise information systems, and the external ecosystem. It more directly involves the features *vi, xiii,* and *xiv.*

4 Implementation Aspects and Scenarios

The *POC software prototype* has been developed as a VR 3D mock-up model. The equipment's parts were previously designed in a CAD system (*Solid Works* software) and provided by the company. After being converted to a proper 3D format, colors, and textures were added to the parts using the *Blender* software. The entire environment was implemented within the *Godot Engine* framework/tool. All animations and parts' choreographies, buttons' design, and coding (in *C# scripts*) were implemented using *Blender* and *Godot* tools. The immersive part uses the *Quest 2* glass from Meta.

The *softbot* itself has been developed using the *ARISA NEST*[1] cloud-based platform, which allows the derivation of instances of both single and groups of *service-oriented softbots. ARISA* supports different communication protocols in its API, being *Telegram* used as the internal messaging mechanism used between a *derived softbot* and the *ARISA's kernel* (deployed in the cloud). All dialogues between a softbot and users were inspired by the *AIML*[2] concept, where keywords are defined and mapped into contexts.

The implemented POC has three main areas from the operator's viewpoint: equipment dashboards; digitalized real-time images from the equipment and surrounding production area (like a digital twin); and the interaction space.

[1] ARISA NEST tool for softbot derivations – https://arisa.com.br/ [in Portuguese].

[2] https://web.archive.org/web/20070715113602/http://www.alicebot.org/press_releases/2001/aiml10.html.

4.1 Scenarios' Background

Application scenarios differ in various ways, including where the *OWMM operator* uses the system (at home, in the office, on the shop floor), and how the interaction occurs (e.g., immersive or not). Operators' actions may also differ in complexity and degree of assistance when supervising (i.e., checking data and events virtually) or (physically) actuating on the machines.

Regarding this paper's goal, hypothetical scenarios have been devised as a basis to illustrate the potential of the proposed approach in terms of how *softbot-assisted OWMM smart cockpit environments* can help *OWMM operators* in their cognitively demanding work in Industry 4.0/5.0 shop floors. These scenarios were inspired by the WEF's vision of a future smart shop floor in 2030 [19], adapted to a real industrial case of a food company from the South of Brazil, and implemented in a simulated way.

In general, this company receives and processes raw meat (including some with small bones inside), transforms and further packs them in proper cardboard boxes as final products to customers. There is one operator in one sector who handles three pieces of equipment: two machines and one manipulator robot. These machines work either independently or connected to each other to form a small manufacturing cell, depending on the different types of production orders. When working as a manufacturing cell (to increase production efficiency) following sanitation protocols, the robot takes predefined portions of smashed and grinded meat that are delivered by *Machine 1* to feed *Machine 2. Machine 1 (deboning)* does a critical operation, grinding the meat in the very right granularity. This specific operation is done by a special blade part, which should be cleaned, sanitized, and checked for its sharpness up to four times a shift.

This company adopts the *OWMM strategy* in some parts of its shop floor. Operators do their activities essentially manually and on the equipment. They interact with the different equipment via their PLCs' HMI, also having some managerial support via some basic dashboards provided by the company's MES.

The company's engineer stated that operators are only trained basically when deeper changes in the equipment are implemented (in their software, in some of their devices, or in some of their internal parts for maintenance purposes). Other technical training is generally assumed that operators have when they are hired, whilst sanitation or general protocols are mostly taught just once. Operators sometimes do not perform the operations correctly, they forget or jump some operations or protocols' steps overestimate their experience, do not read technical manuals at all, do not always follow the recommended best practices, and it is very difficult to monitor all the operators' activities closely and daily, among other problems. These are quite common situations in many industries, especially in SMEs [43, 44].

In the implemented POC, the following scenarios' assumptions have been made:

- These machines are cyber-physical systems, capable of sending and receiving data from them and about what they are doing;
- The company has a corporate data repository (local database or cloud) containing machines' data and operation history, technical manuals and norms, best practices, working and sanitation protocols, etc., as well as some enterprise systems (ERP, MES & SCADA). Some "problem solvers" can be accessed "as-a-service" from the open/external ecosystem;

- The (smart) cockpit works 24 × 7, it is always ready to help the operator, and does several actions automatically (as planned) or proactively (considering the defined cockpit's governance), no matter the operator in the shift.

4.2 Scenario 1 – At the Office

Mrs. Lara is the operator of the current work shift (see Fig. 2b). She supervises the manufacturing cell under her responsibility "remotely", from the factory's office, using the *OWMM smart cockpit* in her smartphone (see Fig. 2a) or desktop computer.

Through the *OWMM smart cockpit*, Mrs. Lara can follow the involved production orders and equipment in real-time via some dashboards; monitor the equipment's OEE (*Overall Equipment Effectiveness*); turn machines' on/off; have access to enterprise or shop floor data and reports; and interact with the *softbot* (see Fig. 2). The equipment's OEE is calculated by an external service accessed as-a-service. Other services could also be used, e.g., to proactively predict future bottlenecks, maintenance activities, etc.

During the supervision, the *softbot-assisted OWMM smart cockpit* suddenly warns Mrs. Lara about a drop in *Machine 2's OEE* (see Fig. 2c). This would have been realized after the *softbot* had checked the current value against the reference values stored in the company's database as Mrs. Lara could not remember it by heart among so many other data to bear in mind. Mrs. Lara checks *Machine 2's dashboard,* which presents production details in real-time as well as the equipment and surrounding production area's digital images (see Fig. 2a).

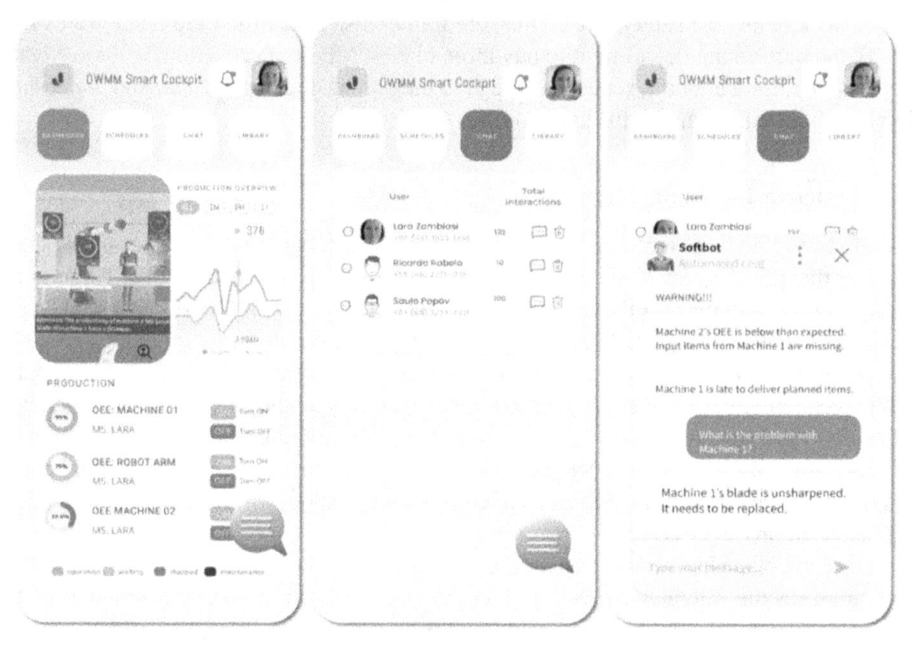

Fig. 2. Softbot-assisted OWMM Smart Cockpit Environment at the Office: (a), (b) and (c) from Left to Right

Uncertain about the effective cause of the problem, Mrs. Lara asks the *softbot* about it. After reasoning about the question and communicating with the MES enterprise system, the *softbot* identified that *Machine 2* had actually stopped. Checking the sensors' data from the three pieces of equipment, the *softbot* also identified that *Machine 1's blade* got unsharpened (see Fig. 2a to 2c), causing a quality problem. Thus, *Machine 1* (and the *Robot*) cannot feed *Machine 2* as planned. This means that the cockpit provides Mrs. Lara with a more comprehensive view of the manufacturing cell she needs to simultaneously supervise.

Mrs. Lara asks, via the *softbot,* about which actions might be taken to solve it. The *softbot,* in turn, internally formulates a question to *ChatGPT*, which brings a list of actions back; one of them refers to replacing the part.

Given this is a serious issue, Mrs. Lara decides to turn *Machine 1* off (pressing 'OFF' in the smart cockpit's GUI), releasing the *Robot* and *Machine 2* for doing other eventual production orders that are not dependent on *Machine 1.* This is possible because the *softbot* has checked the governance model and saw that Mrs. Lara's user is allowed to do this.

There would be a set of actions that Mrs. Lara would have to do by herself related to this type of problem. In this case, the *softbot* helps her in doing so. For example, in order to minimize machine downtime, the *softbot* asks Lara if she can replace the blade right now (see Fig. 3).

This (previously designed business process) involves checking the blade in the enterprise's inventory system as well as requesting it to the warehouse management department to take it to *Machine 1.* The *softbot* observes that this action will make the blade part's stock below the safety level. Thus, the *softbot* also asks Mrs. Lara if she wants to notify the purchasing department to buy more blades. Mrs. Lara confirms it, the *softbot* notifies Mrs. Lara about it, sends a message to the purchasing department, and creates a log file about this event for further auditing.

4.3 Scenario 2 – On the Shop Floor

Mrs. Lara should go to the shop floor to replace the blade on *Machine 1.* Once there, she realizes that this is a new blade version, having different types of blade fitting. Although similar to the previous part's version, Mrs. Lara prefers to be assisted because it is a very costly and critical part; this involves some steps that should be executed in a proper sequence; it is hard to handle it as it is a bit heavy and sharpy; she is not sure about which version this part refers to so as to replace it correctly; she recognizes the urgency of swiftly resolving this issue to get the machine up and running again, thus minimizing its downtime. She then wears a VR glass and is assisted by the *softbot* on how to uninstall and reinstall the blade. This guidance includes warnings about wrong actions during this replacement process (see Fig. 4).

Once the blade has been replaced, she needs to restart the whole manufacturing cell. Still wearing the VR glass, she can make a final check on the other equipment and surrounding environment to see if everything is all right. In this case, the *softbot* warns her that the manufacturing cell is not ready and lists the issues (see Fig. 5a). She can also use the *softbot* to intermediate a chat with other operators to have access to other enterprises' information systems data, etc.

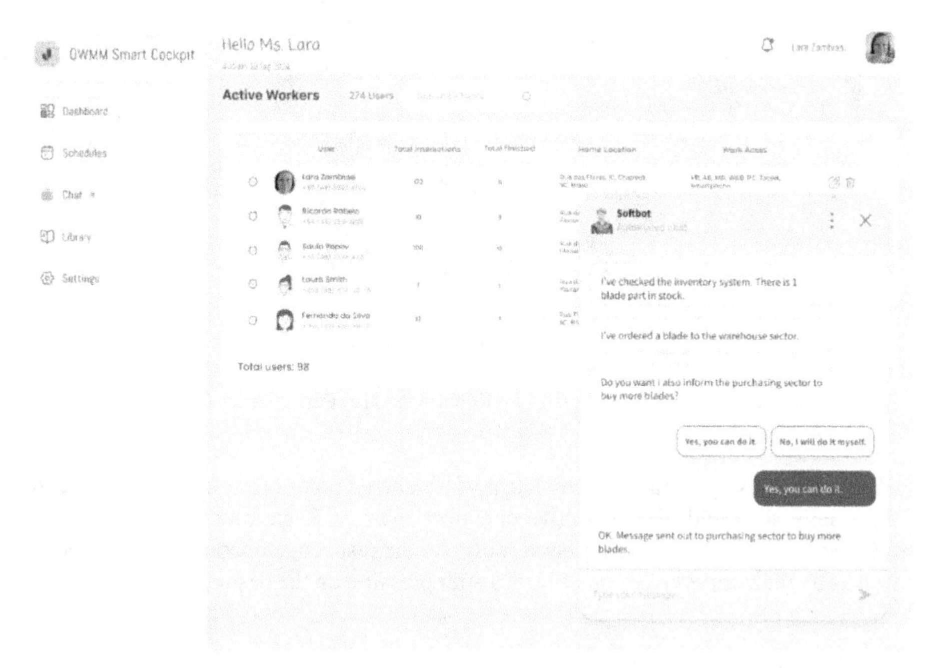

Fig. 3. Softbot Suggesting and Doing Supporting Actions

Fig. 4. Softbot Assisting the Operator in Part Replacement

After having finished the entire operation, the *softbot* automatically updates the log files (see Fig. 5b). It can also look for a suitable on-line or immersive course on the related issues, including a time suggestion based on Mrs. Lara's agenda.

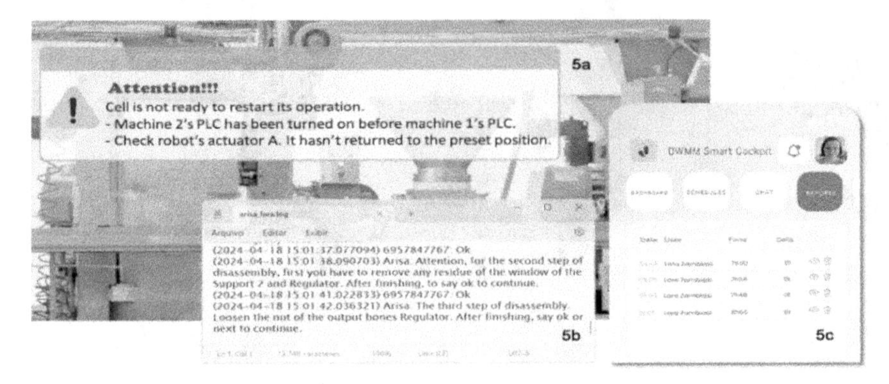

Fig. 5. (a) Using AR/VR, (b) Log Reports, and (c) Performance Reports

The *softbot* also generates a complete performance report of the machines at the end of each daily shift with the respective operators' names. Regarding transparency and governance, this report can be accessed (only) by the respective operators when needed as well as by their supervisors (see Fig. 5c) after pressing on the desired day on the GUI.

4.4 Scenario 3 – At Home

On a remote workday, Mrs. Lara is at home (or even on business travel), monitoring the production cell through the *smart cockpit.* She can do all the activities she could do while working at the company's office (see Scenario 1).

Mrs. Lara can virtually walk through the factory to supervise in real-time the manufacturing cell under her responsibility (see Fig. 6), as in Scenario 2. Thanks to the *smart cockpit's capabilities* (chatting and immersion), Mrs. Lara can also make remote meetings with her supervisor as well as talk with the duty worker who is supervising other machines to help her in some issues she cannot handle at all being remote.

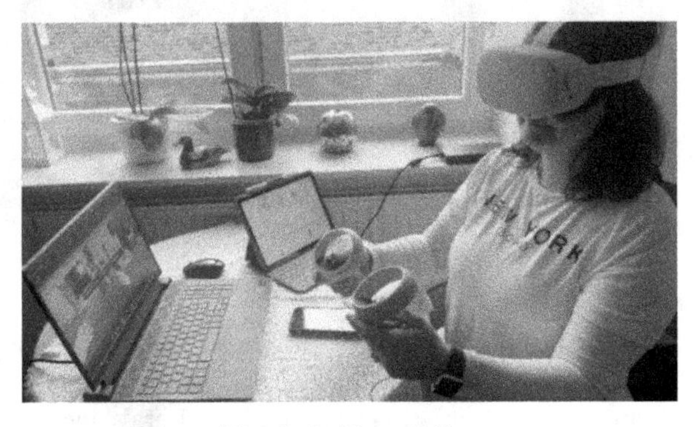

Fig. 6. In Home Office

5 Preliminary Assessment

The implemented POC could implement several usual situations related to the future Operators 4.0–5.0 and how a *softbot-assisted* environment can help them in diverse shop floor operations.

After seeing the POC and being inquired about it, the company's engineer has also highlighted the great potential of the proposed approach. On the other hand, he has highlighted some obstacles the adoption of a technology like that could bring in terms of costs, organization culture (e.g., operators' acceptance, and new process definitions), and systems/technologies integration.

Despite such potential benefits, this POC has also allowed us to realize that a full-fledged implementation of an environment like this can be very complex at several levels. It involves many issues, as: the usage and integration of many (legacy) systems, ICT and industrial protocols, cybersecurity and data protection, different problem solvers, the development of more sophisticated inference reasoning mechanisms, user interaction types (i.e., UI/UX, wearables, AR/VR/MR glasses technological evolution) and learning strategies, new models for operators' training, among others.

Being a kind of information system, the quality of information to be handled by the softbot is totally dependent on how equipped the involved machines are in terms of sensors, network connectivity and interoperability, data exchange capabilities and reliability, information/knowledge representation, and its management. Poor, wrong, or outdated data can bias some analyzes and hence hazard decisions' effectiveness [41].

The access and usage of operators' data performance should be carefully handled to respect legal privacy. There is also the risk of decision-makers blindly ceding responsibility to the softbot; simply ignoring its recommendations; or interrupting its actions inadvertently [22, 42]. The boundaries separating humans and softbots' duties, including the definition of the right level of softbots' automation/ autonomy, are not clear at all also because they can vary from company to company [20].

On the other hand, thanks to the evolving and scalable properties of the architecture, those issues can be more easily introduced as well as new features can be supported (such as machine learning, to work based on each operator's profile and experience, to predict undesired production situations, etc.).

6 Final Considerations and Next Steps

This paper has presented a *softbot approach* to augment operators in *OWMM environments* towards the "human-centric smart manufacturing [38]" paradigm.

A software prototype has been developed as a POC regarding the devised design principles to conceive a *softbot-assisted OWMM smart cockpit environment* to support an *OWMM operator* in its cognitively demanding work in an Industry 4.0/5.0 shop floor.

The implemented scenarios have allowed us to envision the good potentials of the proposed environment in terms of helping OWMM operators in their tasks in many ways and levels and, at the same time, to decrease their cognitive overwhelming. In doing so, there is a good range of opportunities to increase OWMM operators' efficiency, productivity, decision-making confidence, safety, and mental health.

Considering the current stage of this research as well as the aspects that arose from the initial assessment, the main next steps of this research work include: (i) a deeper understanding of using generative and explainable AI to operators for better decision-making; (ii) the evaluation of the effect of the increasing levels of machines' automation and embedded local intelligence/autonomy/self-supervision in the actions to be taken respecting the functional boundaries of machines, enterprise systems (as MES and digital twins), the operator, and the softbot; (iii) the identification of proper performance indicators to measure OWMM operators in this kind of new virtual assisted environment; and (iv) the deployment of real pilots from the developed POC to more properly test and assess the proposed approach.

Acknowledgments. This work has been partially funded by (i) CAPES, The Brazilian Agency for Higher Education, project PrInt "Automation 4.0" and "Finance Code 001", and (ii) the European Union's Horizon Europe project WASABI "White-label shop floor digital intelligent assistance and human-AI collaboration in manufacturing" (GA 101092176). The authors thank Mr. Bruno T. Guedes, the Hightech company's engineer (www.hightech.ind.br), for having provided the information that mainly served for the scenarios' conception of this work.

References

1. Mourtzis, D., Angelopoulos, J., Panopulos, N.: The future of the human–machine interface (HMI) in society 5.0. Future Internet **15**(5), 162 (2023)
2. Romero, D., Stahre, J., Taisch, M.: The operator 4.0: towards socially sustainable factories of the future. Comput. Ind. Eng. **139**, 106128 (2020)
3. Romero, D., Stahre, J.: Towards the resilient operator 5.0: the future of work in smart resilient manufacturing systems. Procedia CIRP **104** 1089–1094 (2021)
4. Thorvald, P., Fast-Berglund, Å., Romero, D.: The Cognitive Operator 4.0. Advances in Transdisciplinary Engineering, pp. 3–8. IOS Press (2021)
5. WEF: World Economic Forum Whitepaper: Augmented Workforce: Empowering People, Transforming Manufacturing (2022)
6. Breque, M., De Nul, L., Petridis, A.: Industry 5.0 – towards a sustainable, human-centric and resilient European industry. European Commission (2021)
7. Romero, D., Vernadat, F.: Enterprise information systems state of the art: past, present and future trends. Comput. Ind. **79**, 3–13 (2016)
8. Spasojevic, I., Havzi, S., Stefanovic, D., et al.: Research trends and topics in IJIEM from 2010 to 2020: a statistical history. Int. J. Ind. Eng. Manag. **12**, 228–242 (2021)
9. Mcdermott, A.: Information Overload Is Crushing You. Here are 11 Secrets That Will Help. Workzone (2017). https://www.workzone.com/blog/information-overload/
10. Rabelo, R.J., Zambiasi, S.P., Romero, D.: Collaborative softbots: enhancing operational excellence in systems of cyber-physical systems. In: Camarinha-Matos, L.M., Afsarmanesh, H., Antonelli, D. (eds.) PRO-VE 2019. IFIPAICT, vol. 568, pp. 55–68. Springer, Cham (2019). https://doi.org/10.1007/978-3-030-28464-0_6
11. Park, E., Jung, Y., Kim, I., Lee, U.: Charlie and the semi-automated factory: data-driven operator behavior and performance modeling for human-machine collaborative systems. In: Conference on Human Factors in Computing Systems, pp. 1–16 (2023)
12. Kusiak, A.: Smart manufacturing. Int. J. Prod. Res. **56**, 508–517 (2018)
13. Petroni, A.: Critical factors for MRP implementation in small and medium firms. Int. J. Oper. Prod. Manag. **22**, 329–348 (2002)

14. Benkalai, I., Rebaine, D., Baptiste, P.: Assigning operators in a flow shop environment. In: 6th International Conference on Information Systems, Logistics and Supply Chain (2016)
15. BCG – Boston Consulting Group Whitepaper: The Global Workforce Crisis (2014)
16. Rabelo, R.J., Zambiasi, S.P., Romero, D.: Softbots 4.0: supporting cyber-physical social systems in smart production management. Int. J. Ind. Eng. Manag. **14**, 63–94 (2023)
17. Kim, J.H.: Ubiquitous robot. In: Reusch, B. (eds.) Computational Intelligence, Theory and Applications. Advances in Soft Computing, vol. 33, pp. 451–459. Springer, Heidelberg (2005). https://doi.org/10.1007/3-540-31182-3_41
18. Wellsandt, S., et al.: Fostering human-AI collaboration with digital intelligent assistance in manufacturing SMEs. In: Alfnes, E., Romsdal, A., Strandhagen, J.O., von Cieminski, G., Romero, D. (eds.) APMS 2023. IFIPAICT, vol. 689, pp. 649–661. Springer, Cham (2023). https://doi.org/10.1007/978-3-031-43662-8_46
19. WEF: World Economic Forum Whitepaper: Navigating the Industrial Metaverse: A Blueprint for Future Innovations (2024)
20. Lu, Y., Zheng, H., Xia, W., Xu, X.: Outlook on human-centric manufacturing towards industry 5.0. J. Manuf. Syst. **62**, 612–627 (2022)
21. Loss, L., Rabelo, R.J., Luz, D., Pereira-Klen, A., Klen, E.R.: Using data mining for virtual enterprise management. In: Camarinha-Matos, L.M. (eds.) BASYS 2004. IFIPIFIP, vol. 159, pp. 443–450. Springer, Boston (2005). https://doi.org/10.1007/0-387-22829-2_47
22. Alberdi, E., Povyakalo, A., Ayton, P.: Why are people's decisions sometimes worse with computer support? In: 28th International Conference of Computer Safety, Reliability, and Security, pp. 18–31 (2009)
23. Freire, S., et al.: Lessons learned from designing and evaluating CLAICA: a continuously learning AI cognitive assistant. In: 28th International Conference on Intelligent User Interfaces, pp. 553–568 (2023)
24. Zambiasi, L.P., Rabelo, R.J., Zambiasi, S.P., Lizot, R.: Supporting resilient operator 5.0: an augmented softbot approach. In: Kim, D.Y., von Cieminski, G., Romero, D. (eds.) APMS 2022. IFIPAICT, vol. 664, pp. 494–502. Springer, Cham (2022). https://doi.org/10.1007/978-3-031-16411-8_57
25. NSFLOW. https://nsflow.com/industries/augmented-reality-in-manufacturing-industry
26. Longo, F., Nicoletti, L., Padovano, A.: Smart operators in industry 4.0: a human-centered approach to enhance operators' capabilities and competencies within the new smart factory context. Comput. Ind. Eng. **113**, 144–159 (2017)
27. Rabelo, R.J., Romero, D., Zambiasi, S.P.: Softbots supporting the operator 4.0 at smart factory environments. In: Moon, I., Lee, G., Park, J., Kiritsis, D., von Cieminski, G. (eds.) APMS 2018.IFIPAICT, vol. 536, pp. 456–464. Springer, Cham (2018). https://doi.org/10.1007/978-3-319-99707-0_57
28. Sterling, S., et al.: Cognitive twin: a cognitive approach to personalized assistants. In: AAAI Spring Symposium Combining Machine Learning with Knowledge Engineering (2020)
29. Rabelo, R.J., Romero, D., Zambiasi, S.P., Magalhães, L.C.: When softbots meet digital twins: towards supporting the cognitive operator 4.0. In: Dolgui, A., Bernard, A., Lemoine, D., von Cieminski, G., Romero, D. (eds.) APMS 2021. IFIPAICT, vol. 634, pp. 37–47. Springer, Cham (2021). https://doi.org/10.1007/978-3-030-85914-5_5
30. Zajec, P., Rožanec, J.M., Novalija, I., Fortuna, B., Mladenić, D., Kenda, K.: Towards active learning based smart assistant for manufacturing. In: Dolgui, A., Bernard, A., Lemoine, D., von Cieminski, G., Romero, D. (eds.) APMS 2021. IFIPAICT, vol. 633, pp. 295–302. Springer, Cham (2021). https://doi.org/10.1007/978-3-030-85910-7_31
31. Bousdekis, A., et al.: Human-AI collaboration in quality control with augmented manufacturing analytics. In: Dolgui, A., Bernard, A., Lemoine, D., von Cieminski, G., Romero, D. (eds.) APMS 2021. IFIPAICT, vol. 633, pp. 303–310. Springer, Cham (2021). https://doi.org/10.1007/978-3-030-85910-7_32

32. Yigitbas, E., Sauer, S.: Self-adaptive digital assistance systems for work 4.0. Digit. Transform., 475–496 (2023)
33. Zambiasi, L.P., Rabelo, R.J., Zambiasi, S.P., Romero, D.: Metaverse-based softbot tutors for inclusive industrial workplaces: supporting impaired operators 5.0. In: Alfnes, E., Romsdal, A., Strandhagen, J.O., von Cieminski, G., Romero, D. (eds.) APMS 2023. IFIPAICT, vol. 689, pp. 662–677. Springer, Cham (2023). https://doi.org/10.1007/978-3-031-43662-8_47
34. Zhang, W., Gu, H.: Job-shop scheduling problems considering similar learning effect in one-worker and multiple-machine patterns. China Mech. Eng. **34**, 1701–1709 (2023)
35. Zheng, T., Grosse, E., Morana, S., Glock, C.: A review of digital assistants in production and logistics: applications, benefits, and challenges. Int. J. Prod. Res. (2024)
36. Pereira, R., Lima, C., Pinto, T., Reis, A.: Virtual assistants in industry 4.0: a systematic literature review. Electronics **12**(19), 4096 (2023)
37. Wellsandt, S., Hribernik, K., Thoben, K.D.: Anatomy of a digital assistant. In: Dolgui, A., Bernard, A., Lemoine, D., von Cieminski, G., Romero, D. (eds.) APMS 2021. IFIPAICT, vol. 633, pp. 321–330. Springer, Cham (2021). https://doi.org/10.1007/978-3-030-85910-7_34
38. Zhang, C., et al.: Towards new-generation human-centric smart manufacturing in industry 5.0: a systematic review. Adv. Eng. Inf. **57**, 102121 (2023)
39. Bechinie, C., et al.: Toward human-centered intelligent assistance system in manufacturing: challenges and potentials for operator 5.0. In: 5th International Conference on Industry 4.0 and Smart Manufacturing, pp. 1584–1596 (2024)
40. Parasuraman, R., Sheridan, T., Wickens, C.: A model for types and levels of human interaction with automation. IEEE Trans. Syst. Man Cybern. Part A Syst. Hum. Part A Syst. Hum. **30**(3), 286–297 (2000)
41. Parasuramam, R., Manzey, D.: Complacency and bias in human use of automation: an attentional integration. J. Hum. Fact. Ergon. Soc. **52**(3), 381–410 (2010)
42. Stieglitz, S., et al.: Collaborating with virtual assistants in organizations: analyzing social loafing tendencies and responsibility attribution. Inf. Syst. Front. **24**, 745–770 (2022)
43. Guastello, S.J.: Human Factors Engineering and Ergonomics: A Systems Approach. CRC Press, Boca Raton (2023)
44. Kernan Freire, S., et al.: Knowledge sharing in manufacturing using LLM-powered tools: user study and model benchmarking. Front. Artif. Intell. **7**, 1293084 (2024)

Quantitative Models for Workforce Management in a Large Service Operation

Siddhanth Shetty and Vittaldas Prabhu(✉)

Pennsylvania State University, University Park, PA 16802, USA
{sxs7100,vittal.prabhu}@psu.edu

Abstract. This paper develops quantitative models for optimizing employee scheduling in large labor-intense service operations that face complexities such high employee turnover, contractual service-level agreements (SLA), non-billable overtime cost, and rising minimum wages. The workforce analytics part of the model uses payroll data to characterize employee features such as consistency, overtime affinity, and retention. The business analytics part of the model ensures that adequate employees are assigned to a shift to comply with SLA while minimizing anticipated overtime cost based on the prevailing overtime salary. A Mixed Integer-Linear Program is formulated with the objective of balancing total overtime cost reduction, employee preferences, and employee features. Additionally, employees can be clustered to identify distinct patterns based on their features of consistency, overtime, and retention. The proposed approach has been applied at a North American service provider with over 4,000 employees across more than 100 sites. K-means clustering based on employee features identified four distinct clusters. Deeper analytics using mixed-effects analysis can show the contribution to profitability from reliable employees, which are limited in availability. Boosted tree importance scoring can establish the influence of moderately reliable employees on overtime cost. Furthermore, decision tree model highlighted that tactical hiring and scheduling must account for collective workforce variability rather than individual attributes in isolation. A partial dependence plot helps visualize the relationship between employee reliability and overtime, thereby helping to characterize the impact of workforce mix on cost. Key impact from this work is that the company management is working to improve its recruitment, retention, and scheduling policies to better align business needs and human capital.

Keywords: Workforce Analytics · Scheduling · Workforce Management

1 Introduction

In today's economy, the service sector accounts for about 80% of GDP consisting of industries such as retail, construction, healthcare, leisure, hospitality, restaurant, transportation, and call centers. In several labor-intense service industries it is difficult to attract and retain good employees even by offering competitive wages above the minimum permitted by regulations. This becomes even more challenging when employees

Published by Springer Nature Switzerland AG 2024
M. Thürer et al. (Eds.): APMS 2024, IFIP AICT 729, pp. 367–381, 2024.
https://doi.org/10.1007/978-3-031-65894-5_26

have to work in varying shifts or during weekends and holidays. Shift assignment needs are industry specific; for example, nurses may be assigned to specific shifts, while call centers might implement overlapping shifts to accommodate fluctuating demand over a short period of time [1, 2]. With labor costs constituting a significant portion of operational expenses, organizations must develop innovative strategies to attract, retain, and efficiently utilize their workforce [3].

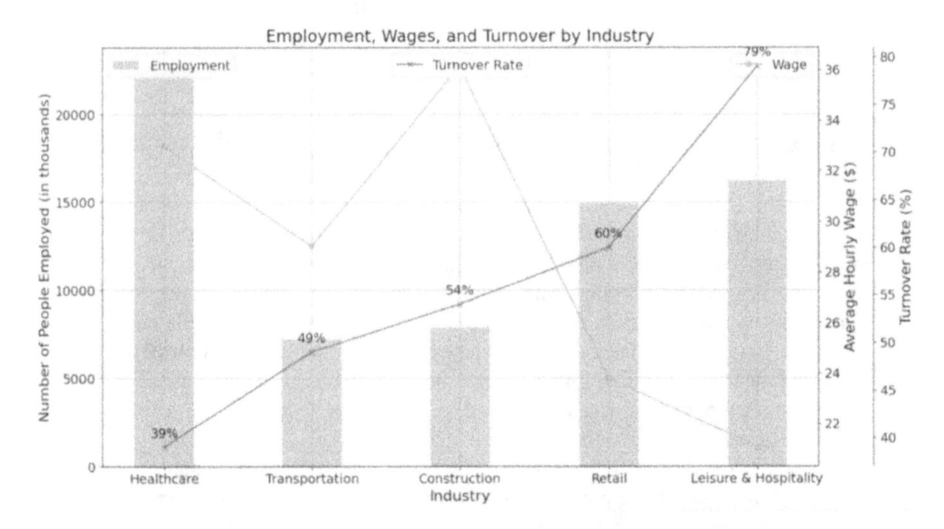

Fig. 1. Employment, Wages, and Turnover by Industry [4].

Figure 1 shows employment, wages, and employee turnover in prominent service industries such as healthcare, transportation, construction, retail, and leisure & hospitality based on data available from the U.S. Breau of Labor Statistics [4]. Employee turnover, which can be as high as 40 to 70%, is a critical issue in many service industries, leading to increased recruitment and training costs, reduced productivity, and diminished service quality [5]. In the healthcare sector, nurse turnover has been associated with adverse patient outcomes, increased mortality rates, and higher healthcare costs [6]. Similarly, in the retail and hospitality industries, high employee turnover can result in poor customer service, decreased sales, and damage to brand reputation [7]. Understanding the factors that influence employee retention and turnover is essential for developing targeted interventions and policies to mitigate these challenges.

Previous research has identified several factors that contribute to employee turnover, such as job satisfaction, organizational commitment, work-life balance, and compensation [8, 9]. For instance, a study by Lu et al. (2019) found that work-life balance and job satisfaction were significant predictors of nurse retention in Chinese hospitals [10]. Similarly, a meta-analysis by Griffeth et al. (2000) revealed that job satisfaction, organizational commitment, and perceived job alternatives were the strongest predictors of turnover intentions across various industries [11].

While these studies provide valuable insights into the determinants of employee turnover, there is a gap in research that specifically addresses the role of employee reliability and its impact on operational efficiency and cost reduction. Employee reliability, which includes factors such as consistency, punctuality, and dependability, is a critical aspect of workforce management that has received limited attention in the literature [12]. Understanding how employee reliability influences scheduling practices, overtime costs, and overall operational performance is essential for developing data-driven strategies to optimize the workforce.

Managing human capital in service enterprises presents a complex challenge, requiring innovative strategies to address these diverse needs effectively. Hamilton and Sodeman (2019) demonstrate the use of big data analytics in quantifying employee performance by analyzing diverse data sources, including social media and IoT. They highlight the ethical necessity of addressing privacy concerns and ensuring compliance with regulations such as GDPR, emphasizing transparency and consent in HR practices [13]. Manoharan, Muralidharan, and Deshmukh (2009) use Data Envelopment Analysis (DEA) for employee efficiency evaluation in manufacturing, analyzing factors like job knowledge and customer relations to set improvement benchmarks without predefined evaluation weights [14]. In addressing the complexities of workforce scheduling, particularly with part-time staff, this report builds upon existing literature to explore innovative approaches for operational efficiency. Mohan (2023) presents a model tailored for part-time employees, emphasizing the importance of aligning shift assignments with individual preferences and availability, a concept that resonates with the segmentation approach in our study [8]. Previous studies have explored various aspects of data-driven workforce management, such as employee preferences (Topaloglu & Ozkarahan, 2004) [15], skills (Firat & Hurkens, 2012) [16], and patient satisfaction (Maghsoodi et al., 2021) [17]. This study distinguishes itself by focusing on employee reliability as a key driver of performance. Specifically, in this paper we develop a metric for employee reliability that can be efficiently quantified using payroll data and effectively track dynamic changes in employee reliability.

The company under study is a rapidly growing contractor providing labor-intensive services across multiple sectors within the county, highlighting employees as a crucial asset. With numerous sites spread across North America, managing a substantial workforce, characterized by a notable employee underutilization, poses significant operational challenges. As a premium service provider committed to tight service level agreements (SLAs) centered around staffing positions, the firm faces the critical task of minimizing non-billable overtime costs that occur when covering staff shortfalls. This scenario highlights the importance of integrating workforce analytics, business analytics and advanced scheduling models that accommodate both employee preferences and performance metrics.

This paper explores the optimization of employee scheduling and operational efficiency. Through the analysis of extensive data, the research employs quantitative models like clustering and Mixed Integer Linear Programming (MILP) to improve scheduling practices. A key finding is the importance of integrating advanced scheduling models with employee preferences and performance data to optimize work shifts, thereby enhancing operational efficiency. It proposes innovative analytical approaches for scalable and adaptable scheduling solutions that consider the variability of employee reliability and preferences. By segmenting employees into reliability-based clusters and analyzing site-level variations, the study offers insights into reducing overtime through better workforce management.

2 Proposed Model

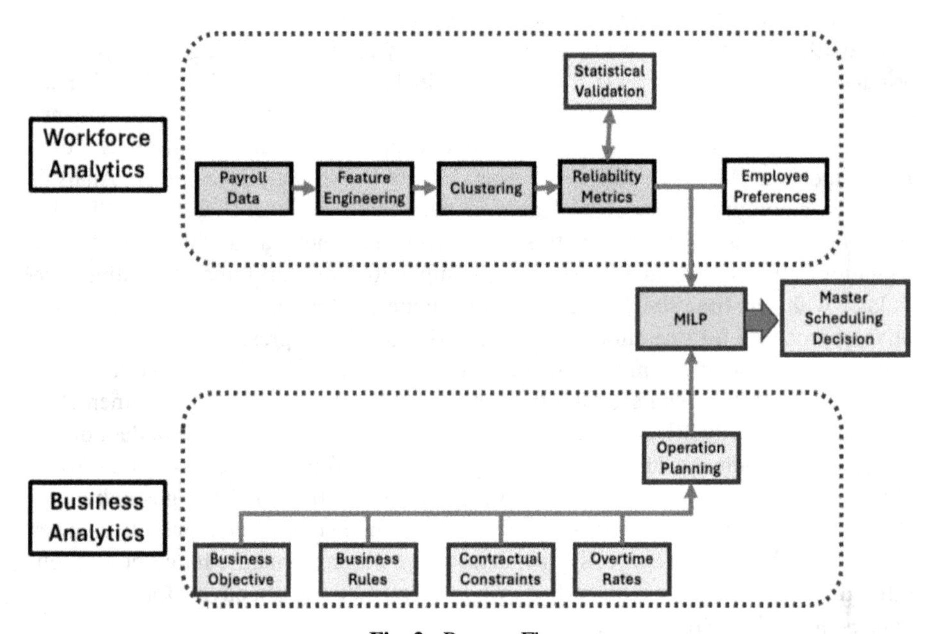

Fig. 2. Process Flow

As illustrated in Fig. 2, the proposed model has two major aspects: (a) workforce analytics driven by payroll data and employee preferences; and (b) business analytics which is driven by business objectives, constraints, and operational requirements. Workforce analytics clusters employees based on their performance after which the clusters are used to define an employee reliability metric (*PerfScores*). By incorporating employee shift preferences ($Pref_{ijt}$), we develop an objective function that uses these metrics to find an optimal schedule. This approach aims to balance the twin goals of accommodating employee preferences and achieving the business objective of reducing overtime cost.

2.1 Model Formulation

The main assumption in the model proposed in this work is that key data used for workforce analytics and business analytics are available. Specifically, these include payroll data, employee preferences, and business constraints.

Notation

Sets and Indices

- I: Set of employees, indexed by i.
- J: Set of shifts, indexed by j.
- T: Set of time periods, indexed by t.

Parameters

- L_j: Length of shift j in hours.
- H: Required hours per time period.
- $OvertimeRate_t$: Overtime rate for time period t.
- $PerfScores_i$: Performance score of employee i.
- $Pref_{ijt}$: Preference of employee i for shift j at time t.
- A_{ijt}: Availability of employee i for shift j at time t.
- $MaxHours_i$: Maximum allowable working hours for employee i during the scheduling period.

Decision Variables

- x_{ijt}: Binary variable indicating if employee i is scheduled for shift j at time t.
- s_{it}: Total hours worked by employee i at time t.
- o_{it}: Overtime hours for employee i at time t.

Objective Function

The objective is to minimize the total overtime costs, subtracting the weighted sum of employee preferences and performance scores:

$$\min \sum_{i \in I} \sum_{t \in T} OvertimeRate_t \cdot o_{it} - \sum_{i \in I} \sum_{j \in J} \sum_{t \in T} PerfScores_i \cdot Pref_{ijt} \cdot x_{ijt} \tag{1}$$

Constraints

Staffing Requirements

Ensure staffing meets or exceeds required hours.

$$\sum\nolimits_{i \in I} L_j \cdot x_{ijt} \geq H, \forall j \in J, t \in T \tag{2}$$

Employee Availability

Employees can only be scheduled for available shifts.

$$x_{ijt} \leq A_{ijt}, \forall i \in I, j \in J, t \in T \tag{3}$$

Overtime Calculation

Total and overtime hours are calculated per employee and time period.

$$s_{it} = \sum_{j \in J} L_j \cdot x_{ijt}, \forall i \in I, t \in T \tag{4}$$

$$o_{it} \geq s_{it} - MaxHours_i, o_{it} \geq 0, \forall i \in I, t \in T \quad (5)$$

One Shift at a Time
Employees are restricted to one shift per time period.

$$\sum_{j \in J} x_{ijt} \leq 1, \forall i \in I, t \in T \quad (6)$$

The approach employs statistical insights to evaluate the influence of employee performance on the business profitability goals and with Mixed Integer Linear Programming (MILP) to optimize employee scheduling across multiple facilities, ensuring efficient allocation and transfer of staff based on operational needs and employee reliability. By analyzing work records and categorizing employees, the model identifies optimal schedules that minimize overtime costs while maintaining high operational performance. This approach allows for dynamic scheduling, enabling organizations to adapt to unforeseen situations to include a resilient strategy. This model is applicable across various organizations to enable tailoring their workforce management strategies based on data-driven insights. It suggests actionable intelligence for optimizing hiring practices, scheduling policies, and workforce composition to reduce overtime and enhance operational efficiency. This methodology can be adapted to various large service enterprises to improve operational outcomes through better workforce management. The complexity of the model can be expressed as:

$$\text{Complexity} = O(|I| \text{ x } |J|) = O(|I| \text{ x } (|H| \text{ x } |L|). \quad (7)$$

Here I, J, and H represent the number of employees, shifts, and hours, respectively. The total hours for the site are fixed, the shift length is fixed at 8 h the employees can work a variable number of shifts, subject to availability and other constraints (Table 1).

Table 1. Model Complexity

Number of employees	Number of Decision variables	Number of Constrains	Complexity
9	423	477	O (405)
15	1,155	1,245	O (1,125)
27	3,699	3,861	O (3,645)

Figure 3 shows, as the number of employees increases, the complexity of the model grows significantly. The number of decision variables and constraints exhibits a polynomial growth rate with respect to the number of employees, shifts, and time periods. While this complexity analysis provides us a basis for how the compute time would scale, the actual computational time will depend on a variety of factors such as ability of the solver software to exploit multi-core architectures, memory hierarchy of the run-time environment, and the time to load and save data,

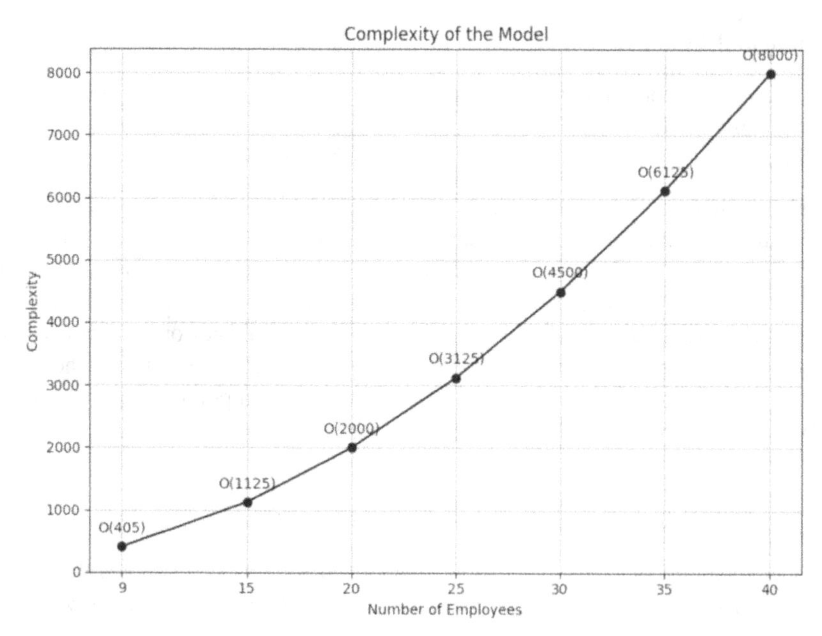

Fig. 3. Model Complexity

3 Case of a Large Service Provider

The challenges of managing a large, distributed workforce in the service industry—balancing profitability, employee satisfaction, and effective employee scheduling—are multifaceted. This research aims to categorize employees by reliability to optimize shift staffing and assess the impact of workforce composition on scheduling efficacy and profitability. By categorizing over 4,300 employees across 125 locations according to their reliability, the study aims to optimize shift staffing, thereby minimizing overtime costs and enhancing operational efficiency. Utilizing two years of data, statistical modeling was used to investigate the impact of workforce reliability on overtime expenses. The MILP model could be run in about 10 s for one site with 9 employees and 8-h shifts. Therefore, from Fig. 3, we can except a 40-employee site to take about 20 times longer, which is about 10 min. In this case however the clustering computations took about 10 min on a standard desktop PC, making it the dominant aspect of computing time.

The findings indicate a significant correlation between the composition of employee reliability clusters and overtime costs, offering valuable insights for the improvement of scheduling practices. The methodology involves categorizing and grouping employees based on their reliability before analyzing their influence on overtime rates. This approach, detailed in the following sections, highlights the study's efforts to optimize employee scheduling within a large, distributed service industry workforce.

3.1 Data and Methods

The raw dataset encompassed two years of aggregated employee payroll data. Domain expertise guided feature engineering to construct three metrics encapsulating employee reliability:

Consistency Score: This score evaluates the regularity of an employee's working hours. It's calculated by analyzing the mean and standard deviation of hours worked, considering factors like day of the week and quarter. A lower standard deviation, indicating more consistent work hours, results in a higher score. He formula is designed to measure consistency within a range of 0 to 1, where a score of 1 indicates optimal consistency, and a score approaching 0 indicates a greater degree of variability in work hours. It's important to note that this formula specifically excludes temporary employees due to their typically higher variance in work hours.

$$Consistency\_Score_{i,d,q} = 1 - \frac{\sigma_{i,d,q}}{\mu_{i,d,q}}$$

where i denotes the Employee, d denotes the day of the week, q denotes the quarter, $\sigma_{i,d,q}$ & and $\mu_{i,d,q}$ is the standard deviation and mean of hours worked by $employee_i$ on day d during quarter q.

Overtime Affinity: This metric assesses an employee's tendency to work overtime. It's determined by counting the number of times an employee's working hours exceed a typical 8-h workday. This count reflects their inclination to work beyond regular hours, which might indicate workload management, work ethic, or operational demands.

$$Overtime\_Affinity_i = \sum_{n=1}^{N_i} 1\left(\text{Total\_Hours}_{i,n} > 8\right)$$

where i represents the Employee, N_i is the total number of records for employee i,1 is an indicator function that returns 1 if the condition (Total hours greater than 8) is true, 0 otherwise.

Retention Score: This score represents an employee's tenure with the company. By calculating the number of days between an employee's first and most recent 'Counter Date', and then normalizing this number against the longest tenure in the dataset, a score between 0 and 1 is derived. A higher score indicates longer tenure, suggesting better employee retention, while a lower score suggests shorter tenure.

$$\text{Retention}_{\text{Score}\,i} = \frac{\text{Tenure}_i}{\max(\text{Tenure})}$$

Through feature engineering, three key metrics were established: Consistency Score, Overtime Affinity, and Retention Score. These metrics were aggregated across all quarters for each employee to calculate their mean and standard deviation, forming a basis for clustering analysis.

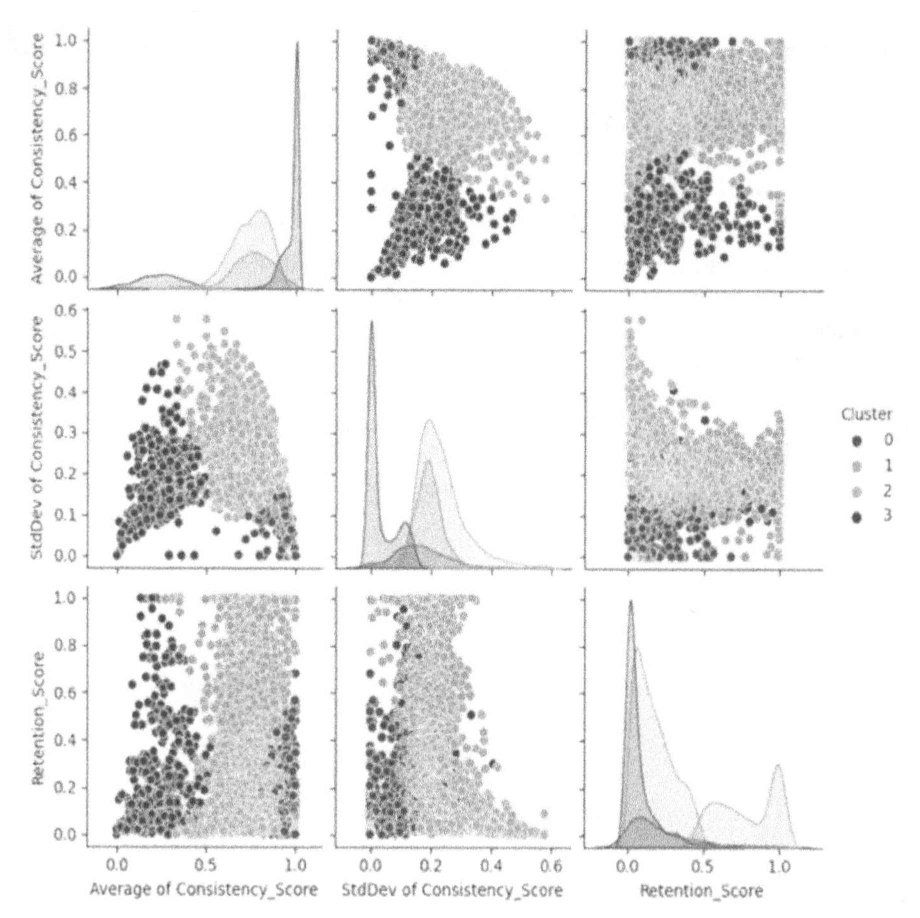

Fig. 4. Clustered Data Scatter Plots with Density Distribution.

Table 2. Clustered metrics

Cluster	Average of Consistency_Score	StdDev of Consistency_Score	Retention_Score
0 (Cluster-1)	0.976074	0.033610	0.791177
1 (Cluster-2)	0.759139	0.186283	0.783053
2(Cluster-3)	0.739454	0.234578	0.179326
3 (Cluster-4)	0.239243	0.163989	0.240207

Using the K-Means algorithm, as demonstrated in Fig. 4 employees were segmented into four clusters based on their performance metrics, identified through the Elbow Method as the optimal number of groups. These clusters were characterized by their distinct patterns in work regularity, overtime, and retention Table 2 summarizes the detailed performance metrics for each cluster:

- Cluster 1 (Highly Reliable): High consistency and retention scores.
- Cluster 2 (Moderately Reliable): Moderate scores across all metrics.
- Cluster 3 (Less Reliable): Lower retention scores.
- Cluster 4 (Least Reliable): Lowest consistency and retention scores.

The analysis further categorized Clusters 1 and 2 as more dependable, while Clusters 3 and 4 were considered less reliable, necessitating closer managerial supervision.

To understand the impact of employee cluster distribution on site performance, a mixed-effects linear model was applied, revealing a negative correlation between the proportion of Highly Reliable employees (Cluster 1) and overtime percentages, indicating their role in enhancing workforce efficiency. Additionally, a RandomForestRegressor analysis explored how different employee clusters contribute to overtime, highlighting the significant influence of Moderately Reliable employees (Cluster 2) on operational efficiency as shown in Fig. 5.

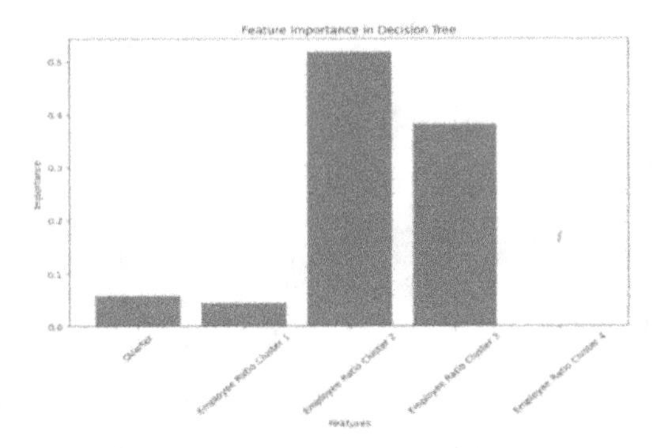

Fig. 5. Feature Importance Plot

4 Results and Discussion

When analyzing the impact of employee reliability on business objectives (NBOT), classifiers and regression techniques serve different purposes. Classifiers categorize employees into discrete groups, such as labeling them as 'reliable' or 'unreliable' and so on. However, this approach is inadequate for studying quantitative impacts on business outcomes. Instead, regression techniques, which predict continuous values, are more appropriate. For instance, using a random forest regressor allows us to model how varying levels of employee reliability affect business performance metrics. Notably, random forests can function as both classifiers and regressors, providing flexibility but highlighting the importance of choosing the right tool for the right analysis. Figure. 6 illustrates that notable negative trend exists between the 'Employee Ratio Cluster 1' and 'Overtime Percentage', indicating that a higher presence of Cluster 1 employees correlates with reduced overtime.

'Employee Ratio Cluster 2' stands out as the most impactful feature in overtime prediction, with Cluster 3 also being significant.

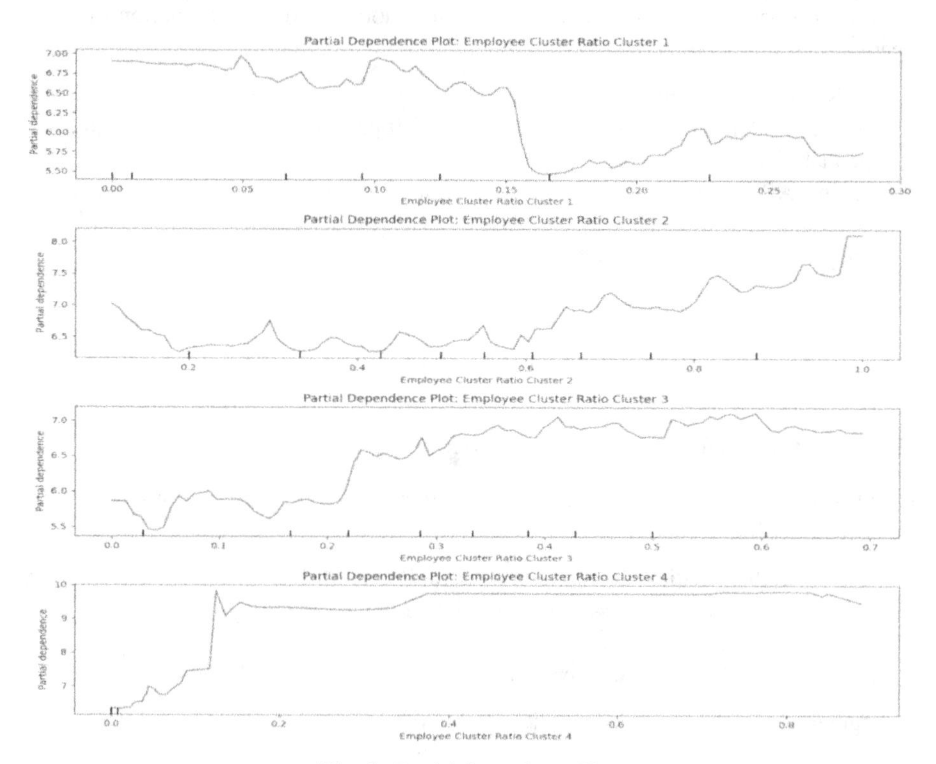

Fig. 6. Partial dependence Plot

- The analysis for 'Employee Ratio Cluster 1' reveals a predominantly negative trend, implying that an increased proportion of these employees is likely linked to enhanced scheduling efficiency.
- For 'Employee Ratio Cluster 2', the data depicts a sharp initial decline in overtime with small proportional increases, suggesting an optimal threshold for this cluster that minimizes overtime before the benefit levels off.
- The trend for 'Employee Ratio Cluster 3' is positively sloped, suggesting that a larger share of these employees may contribute to higher overtime due to factors like job requirements or scheduling challenges.

The findings provide a perspective on how employee composition influences overtime, with implications for optimizing workforce management and operational efficiency. The partial dependence plots offer a visual representation of these dynamics, which will be elaborated upon in the following sections.

4.1 Quantification Through Feature Engineering

Identifying employees who demonstrate consistent participation is crucial for building reliable operations [9]. Derived metrics must look beyond participation rates to encapsulate multifaceted reliability:

- Consistency Score evaluates regularity in weekly work hours using the coefficient of variation. It identifies employees with stable shift patterns imperative for scheduling dependability.
- Overtime Affinity determines historical likelihood of overtime based on the frequency of extra hours. Employees with high affinity can provide crucial workforce flexibility during unexpected shortfalls.
- Retention Score calculates employee tenure. Staff with longer tenures possess operational knowledge which is vital for both efficiency and service quality.

Thus, engineered features provide a 360-degree reliability measure.

4.2 Segmentation for Enhanced Decision Making

By clustering employees based on engineered metrics, groups with varying reliability and scheduling impact emerged:

- Highly Reliable: Critical for efficiency but limited in number.
- Moderately Reliable: Possess flexibility imperative for adaptation.
- Less Reliable: Prone to unpredictability requiring oversight.
- Least Reliable: Necessitate active management to minimize volatility.

Segmenting enables tailored policies catered to balance trade-offs like flexibility and consistency while optimizing operational efficiency.

4.3 Statistical Modeling for Actionable Intelligence

While descriptive analytics classified employees, predictive modeling generated specific decision-support:

- Mixed-effects analysis revealed that compared to highly reliable employees, other clusters inflate overtime, directly eroding profits.
- Boosted tree importance scoring identified moderately reliable employees as most influential in predicting overtime.
- Decision tree model highlighted that tactical hiring and scheduling must account for collective workforce variability rather than individual attributes in isolation.

4.4 Multidimensional Analytics for Competitive Advantage

Integrating descriptive categorization, predictive modeling, and decision optimization offers a multifaceted, analytics-driven approach crucial for sustaining long-term competitiveness. To achieve this, refining hiring practices based on reliability levels determined through modeling is essential. This approach ensures that the workforce is composed of reliable individuals, thereby minimizing the likelihood of overtime caused by less dependable additions. Simulating various workforce mix scenarios helps identify optimal employee ratios, ensuring an efficient and balanced workforce. Additionally, predictive analytics enables accurate forecasting of overtime costs, allowing companies to bid competitively for future contracts. Overall, these analytics techniques enhance workforce management and position companies for sustained success in a competitive landscape.

4.5 Data-Driven Scheduling Strategy

The new scheduling approach was introduced at a site requiring approximately 168 h of staffing, equating to the workload of about 5 full-time employees. This site was chosen due to its high levels of non-billed overtime, indicating inefficiencies in shift allocation and scheduling practices. The main aspect of this approach is the quantification of employee site preferences and the integration of their performance ratings into the scheduling process. This method aims to prioritize high-performing employees by aligning their work schedules with their preferences, thus ensuring critical shifts are consistently covered. The scheduling model is designed to be flexible, allowing adjustments according to business needs. Once a master schedule is established using this methodology, it can be fine-tuned on a weekly basis to reflect changes in employee location preferences and other operational requirements.

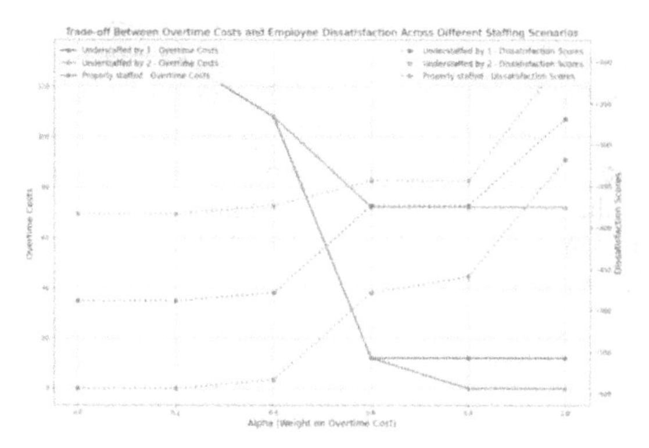

Fig. 7. Tradeoff Between Overtime costs and Employee Satisfaction

Figure 7 depicts as alpha increases, the emphasis on reducing overtime costs results in a decrease in preference scores and a corresponding increase in dissatisfaction scores. This indicates that while overtime costs are controlled, employee dis-satisfaction increases as their preferred shifts are less frequently assigned.

5 Conclusion

The study offers compelling evidence that employee reliability is a central pillar in designing scheduling framework. Through segmentation of employees into distinct reliability clusters, it has put forth the varied contributions of different groups towards operational efficiency, especially in the context of overtime management. This understanding, further enforced by predictive modeling techniques, presents the organizations with actionable insights. These insights facilitate more strategic scheduling decisions, anticipating the outcomes of various scenarios with greater accuracy, thus paving the way for a more informed decision-making process that aligns workforce with operational goals.

In response to these findings, the study suggests a strategic modification in workforce management, suggesting the adoption of data-driven decisions across hiring and scheduling practices. By enhancing hiring processes to include metrics on employee reliability, such as prior organization performance indicators including punctuality, job tenure, and reasons for leaving, organizations can identify candidates who are likely to be dependable. This approach ensures that new hires have a track record of reliability. The recruitment team should actively monitor performance against these metrics and intervene as needed when deviations occur. The objective is to maximize employee retention by ensuring employee satisfaction, fostering a stable and productive work environment. Moreover, the recommendation to develop a scheduling strategy around employee reliability highlights the importance of aligning the variety within the workforce with the operational demands of the business, thereby minimizing unproductive labor hours. These recommendations conclude on a holistic approach that integrates analytics, modeling, and industry know-how to transform workforce data into actionable intelligence. This not only aims to enhance current operational efficiency but also to provide a framework for navigating the complexities of the industry, ensuring that workforce strategies remain responsive and resilient in the face of evolving challenges. A future research direction would be extending the MILP model using techniques from multi-criteria decision making (MCDM) and Goal Programming if the application warrants more nuanced tradeoffs that may arise because of regulations or new policies [18, 19]. Additionally, given the practical challenges in getting companies in the service sector to adopt decision-making based on quantitative models, interactive MCDM approaches may also be fruitful area of research [20].

References

1. Kletzander, L., Musliu, N.: Solving the general employee scheduling problem. Comput. Oper. Res. **104794** (2019). https://doi.org/10.1016/j.cor.2019.104794
2. Ernst, A.T., Jiang, H., Krishnamoorthy, M., Sier, D.: Staff scheduling and rostering: a review of applications, methods, and models. Eur. J. Oper. Res. **153**(1), 3–27 (2004)
3. Defraeye, M., Van Nieuwenhuyse, I.: Staffing and scheduling under nonstationary demand for service: a literature review. Omega **58**, 4–25 (2016)
4. https://www.bls.gov/cew/
5. Van den Bergh, J., Beliën, J., De Bruecker, P., Demeulemeester, E., De Boeck, L.: Personnel scheduling: a literature review. Eur. J. Oper. Res. (2013). https://doi.org/10.1016/j.ejor.2012.11.029
6. Hamilton, R.H., Sodeman, W.A.: The questions we ask: opportunities and challenges for using big data analytics to strategically manage human capital resources. University of Mississippi, School of Business, University, MS 38677, USA, Clark University, Worcester, MA 01610, USA (2020)

7. Manoharan, T.R., Muralidharan, C., Deshmukh, S.G.: Employee performance appraisal using data envelopment analysis: a case study. Res. Pract. Hum. Resour. Manag. **17**(1), 92–111 (2009)
8. Mohan, S.: Scheduling part-time personnel with availability restrictions and preferences to maximize employee satisfaction. School of Global Management and Leadership, Arizona State University (2023)
9. Deceuninck, M., Fiems, D., De Vuyst, S.: Outpatient scheduling with unpunctual patients and no-shows. Department of Industrial Systems Engineering and Product Design, Ghent University; Department of Telecommunication and Information Processing, Ghent University (2018)
10. Li, Y., Chen, J., Cai, X.: An integrated staff-sizing approach considering feasibility of scheduling decision. Ann. Oper. Res. **155**, 361–390 (2007). https://doi.org/10.1007/s10479-007-0215-z
11. Farrell, K.: Understanding participant reliability: an examination of future intent, attendance, and generalizability in the context of workplace training research. Dissertation Abstracts International Section A, vol. 78 (2017)
12. Lu, Y., et al.: The relationship between job satisfaction, work stress, work–family conflict, and turnover intention among physicians in Guangdong, China: a cross-sectional study. BMJ Open **9**(5), e026141 (2019)
13. Griffeth, R.W., Hom, P.W., Gaertner, S.: A meta-analysis of antecedents and correlates of employee turnover: update, moderator tests, and research implications for the next millennium. J. Manag. **26**(3), 463–488 (2000)
14. Sturman, M.C., Trevor, C.O.: The implications of linking the dynamic performance and turnover literatures. J. Appl. Psychol. **86**(4), 684–696 (2001)
15. Topaloglu, S., Ozkarahan, I.: An implicit goal programming model for the tour scheduling problem considering the employee work preferences. Ann. Oper. Res. **128**(1), 135–158 (2004). https://doi.org/10.1023/B:ANOR.0000019103.83912.3a
16. Firat, M., Hurkens, C.A.: An improved MIP-based approach for a multi-skill workforce scheduling problem. J. Sched. **15**(3), 363–380 (2012)
17. Maghsoodi, A.I., Khalilzadeh, M., Brauers, W.K., Spronk, J.: GAPSO optimization for scheduling nursing personnel based on risk management and patient satisfaction. J. Adv. Nurs. **77**(4), 1933–1945 (2021)
18. Ruiz-Torres, A.J., Ablanedo-Rosas, J.H., Mukhopadhyay, S., Paletta, G.: Scheduling workers: a multi-criteria model considering their satisfaction. Comput. Ind. Eng. **128**, 747–754 (2019)
19. Cuevas, R., Ferrer, J.C., Klapp, M., Muñoz, J.C.: A mixed integer programming approach to multi-skilled workforce scheduling. J. Sched. **19**, 91–106 (2016)
20. Lauer, J., Jacobs, L.W., Brusco, M.J., Bechtold, S.E.: An interactive, optimization-based decision support system for scheduling part-time, computer lab attendants. Omega **22**(6), 613–626 (1994)

Evolving Workforce Skills and Competencies for Industry 5.0

A State-of-the-Art Review and Framework for Human-Centric Automation in Industry 5.0

Mohammed Yaqot[1(✉)], Brenno Menezes[1], Abdulfatah Mohammed[2], and Kim Moloney[3]

[1] College of Science and Engineering, Hamad Bin Khalifa University, Doha 34110, Qatar
moyaqot@hbku.edu.qa
[2] College of Islamic Studies, Hamad Bin Khalifa University, Doha 34110, Qatar
[3] College of Public Policy, Hamad Bin Khalifa University, Doha 34110, Qatar

Abstract. As we go beyond the concept of Industry 4.0 (I4.0), the point where industry and human participation meet becomes a matter of philosophical debate. What is the rationale for developing a substitute for human effort? The expected substitution of several work positions sparks debate on society's ability to adjust to this technology-driven change, necessitating a comprehensive investigation. The primary research question in this work addresses how Industry 5.0 (I5.0) can enhance human-machine collaboration to improve operational efficiency and worker satisfaction. The methodology employed in this work involves a systematic review of relevant literature, focusing on two particular aspects of human-centric manufacturing: the symbiotic relationship between humans and automation and the identification of the capabilities that automation can enhance in worker skills amidst the myriad challenges of digital governance. To address these challenges—such as job displacement, skill gaps, and ethical concerns in automation—the Enhanced Human-Automation Symbiosis (EHAS) framework is proposed. This framework aims to balance technological advancements with human needs by enhancing our understanding of the complex dynamics between human capacities and automation. In this paper, we present a comprehensive state-of-the-art review and outline our proposed framework, discussing future research directions, developmental prospects, and practical implementation issues.

Keywords: Industry 5.0 Operator 5.0 · Human-automation · Digital transformation · Enhanced human-automation symbiosis

M. Thürer et al. (Eds.): APMS 2024, IFIP AICT 729, pp. 385–400, 2024.
https://doi.org/10.1007/978-3-031-65894-5_27

1 Introduction

The journey to Industry 5.0 (I5.0) begins with digitization, converting analog information and assets into digital formats. In manufacturing, it involves converting analog machinery, product designs, and operational data into digital forms [52]. This step enables efficient data storage and transmission, laying the groundwork for integrating digital technologies into industrial ecosystems. Second, digitalization builds upon digitization by systematically employing digital technologies to enhance operations. It involves utilizing digital data, computing power, and connectivity to streamline processes and improve decision-making. By deploying sensors, IoT devices, and data analytics tools, digitalization enables real-time data collection and analysis in manufacturing systems and supply chains [32], facilitating data-driven decision-making based on process modeling and optimization. Third, digital transformation (DT) represents a holistic organizational shift beyond mere technology adoption. It involves reimagining business models and operations to fully leverage digital technologies for improved performance and broader impact [18]. DT integrates digital technologies into all aspects of an organization, fostering agility, innovation, and adaptability to meet evolving market demands and customer expectations [13]. Finally, the evolution of the industrial environment through digitization, digitalization, and digital transformation gives rise to Industry 5.0. This new paradigm aims to establish a human-centric and sustainable industry by leveraging advancements in information and communication technologies, modeling and solving algorithms, high-performance computation, and mechatronics. I5.0 signifies the integration of physical and digital dimensions with an emphasis on service provision and smart manufacturing centered on the human. Unlike Industry 4.0 (I4.0), which prioritizes automation and data-driven decision-making, I5.0 emphasizes human intelligence, intuition, and creativity in smart manufacturing. Human-machine interactions in I5.0 are collaborative, aiming for greater efficiency, customization, and long-term viability [28]. Figure 1 summarizes the sequential development of manufacturing and service paradigms, identifying significant shifts from digitization to Industry 5.0. The figure is created based on a synthesis of the literature review.

The fusion of human ingenuity and technology shapes socio-technical transitions, redefining agility, technology, and sustainability [34,39,48]. Intelligent manufacturing and service systems (IMSS) play a crucial role by integrating automation, big data analytics, and artificial intelligence. IMSS enable real-time decision-making, system optimization, and adaptability in a rapidly changing environment [50]. However, the escalating conflict between advanced technology and human involvement in IMSS presents a critical challenge. While technology advancements enhance efficiency, they raise concerns about job displacement and skills gaps [57]. Ethical issues regarding data privacy and algorithmic bias also arise [11]. This tension highlights the need for thoughtful action. The pace of technology evolution outstrips human adaptation, necessitating robust training and re-skilling programs to bridge the gap and empower individuals, known as Operator 5.0 (O5.0), in the digital era.

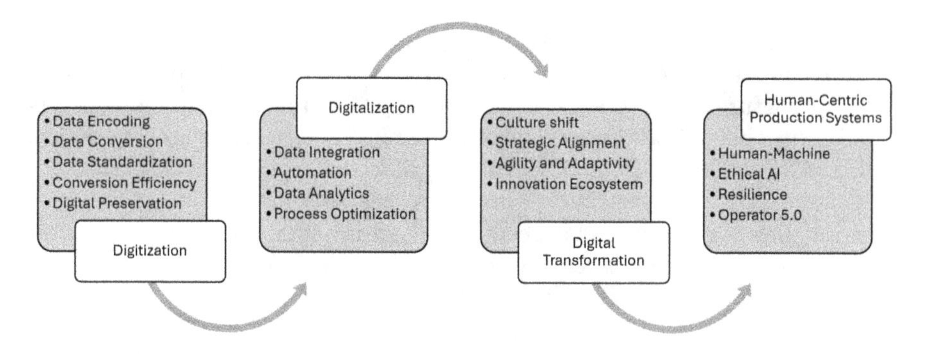

Fig. 1. Evolution of Smart Manufacturing Paradigms: From Digitization to Human-Centeric Systems

This study explores the dynamic symbiosis between humans and automation, examining possible conflicts between technology and humanity from a socio-technical standpoint. We propose an Enhanced Human-Automation Symbiosis (EHAS) conceptual framework.

2 Main Industry 5.0 Dimensions

To navigate the central discourse, it is critical to analyze key elements of I5.0 linked to the aims of this work and establish a clear trajectory for our discussion. This segment provides a concise overview of intelligent manufacturing, service systems, and human-automation symbiosis, setting the stage for subsequent investigation. The dimensions were identified through a comprehensive literature review, focusing on studies highlighting human-centric manufacturing, sustainability, and technological integration. Each dimension was chosen based on its relevance to enhancing human-machine collaboration and overall system efficiency.

2.1 Intelligent Manufacturing and Service Systems

Based on the literature, the concept of IMSS, propelled by I4.0, has emerged recently to meet the need for greater agility and adaptability in industrial operations [53]. Substantial progress in integration, interoperability, and interconnectivity is on the horizon as a result of the convergence of technological enablers, such as big data analytics, high-performance computing, and auxiliary robotics, within the context of industrial digital transformation [6]. Industrial Internet of Things (IoT) and cyber-physical systems (CPSs) are fundamental concepts for the development of IMSS [56]. The present perspective encourages proactive and integrated responses throughout the manufacturing life cycle via the virtualization of physical equipment that are often seen as separate in conventional factory floors.

Some studies conducted on the subject [4,20,60], demonstrate that IMSS typically places great importance on the creation of sophisticated process automation infrastructures, data processing, with a strong focus on self-automated learning and corrective actions such as machine learning, deep learning, reinforcement learning, transfer learning, etc., which are primary reasons for its characterization as "intelligent". Nevertheless, this study has also revealed a deficiency in comprehending the conceptual aspects of human involvement in the production process. Human roles are often limited to solely supervisory capacities or, in other instances, lack a clearly defined role altogether [14].

2.2 Human-Automation Symbiosis

The term "Human Cyber-Physical Systems" (HCPS) refers to a networked arrangement of physical and digital components, all of which are subject to the operation, management, and control of humans [61]. This dimension focuses on the evolving relationship between humans and technology, highlighting stages of human participation, from direct control to supervisory roles. The identification of this dimension is based on literature that discusses the enhancement of human capabilities through advanced automation and AI. Based on the literature, we have classified the development of this reciprocal relationship into discrete phases. Fundamentally, the emphasis continues to be on the ever-evolving interplay between technology and humans, which transpires through observable stages of human participation. The phases that have been delineated symbolize the progression of the Human Automation Symbiosis.

Emergence of Automation and Control Systems: During the initial stage, human control was prevalent in CPSs, with human operators actively overseeing system operations as "human-in-the-loop of CPSs". This period marked the beginning of a mutually beneficial relationship, where humans entrusted technology with tasks in which it excelled, leading to a robust co-extension [8]. Technology acted as an extension of human capabilities, enhancing skills and capacities, laying the foundation for subsequent stages of interaction and evolution.

Progression of Digital Computing: The transition to digital interfaces necessitates enhanced operations and control strategies capabilities, particularly with integrating big data analytics. Incorporating high-performance computing into diverse domains, such as manufacturing and service systems, fosters sophisticated human-computer interactions. This interplay represents a crucial juncture in the ongoing co-evolution of the relationship between humans and technology. Both entities engage in continuous adaptation and immediate human-computer interaction, signifying a dynamic process of co-action [8].

Maturity of Human-CPSs: The emergence of CPSs a decade ago marked a significant phase in integrating digital technologies into physical processes. Human involvement shifted to an "on-the-loop" role in CPSs, assuming supervisory responsibilities rather than direct real-time control. This phase fused information and communication technologies with physical operations, emphasizing co-action [26]. Examples include healthcare systems with remote patient monitoring, smart grids for energy management, etc. As technology advanced, human engagement in system control and decision-making increased, with big data analytics requiring enhanced automation capabilities.

State-of-the-art IoT and AI: The integration of evolving IoT and AI applications refines the concept of humans operating "on-the-loop" while introducing "out-of-the-loop" elements in CPSs. "Out-of-the-loop" implies systems functioning autonomously or with minimal human intervention. Advanced AI-enabled systems handle large-scale, complex tasks autonomously, with humans involved in higher-level oversight and decision-making. Examples include smart city applications, where human expertise is pivotal but augmented by AI-driven insights. This phase emphasizes co-dependency, as humans rely on technology for tasks once linked to daily life, highlighting the mutually dependent nature of this relationship [15].

Futuristic Human-Centric CPSs: The evolution of HCPSs is shifting towards a more human-centric approach, emphasizing user-friendly interfaces, responsible AI-human collaboration, and ethical considerations to ensure the well-being, safety, and inclusivity within HCPSs [2,22]. In parallel, IMSS is being strategically explored with a focus on the co-evolution of humans and technology. IMSS emphasizes ethical and responsible development [21], highlighting co-extension, co-action, and co-dependency as enduring elements of this intricate relationship. Building on the I5.0 paradigm, which harmonizes technology-driven advancements with human-centeredness, sustainability, and resilience, the concept of Operator 5.0 (O5.0) emerges. The O5.0 paradigm illustrates eight scenarios, from super-strength operators using exoskeletons to analytical operators leveraging big data analytics for decision-making [44,63]. These enhancements, already being implemented and expected to further materialize, signify a profound transformation in human-technology interaction.

3 Methodology

This section examines the current research patterns using specific keywords to address the following questions, forming our proposed framework's foundation.

1. What is the evolutionary trajectory of human-automation symbiosis?
2. What are the primary capabilities that automation can enhance for humans?
3. What are the challenges faced in transitioning towards human-centric manufacturing?

The Scopus database was utilized to search for and select appropriate articles. Key dimensions of I5.0 were identified through a systematic review of articles focusing on human-centric manufacturing, intelligent systems, and technological advancements. Criteria for selecting these dimensions included their impact on enhancing human-machine collaboration, supporting sustainable practices, and integrating advanced technologies. A range of keywords was systematically employed, tested, and evaluated to carry out the review phase. The most effective and pertinent keywords included "human-centric manufacturing", "Intelligent Manufacturing and Service Systems", "human cyber-physical systems", and

"Operator 5.0" which were applied as filters in the topic search (title, abstract, keywords). A total of 249 documents were initially identified and subsequently subjected to several refinement stages. Following rigorous inclusion (e.g., range 2012–2024, journal articles, English) and exclusion (else) criteria, 89 publications were selected. Then, in step 3, a first filtration was employed by screening documents and selecting only the most relevant contributions, which resulted in 75 documents. These documents went through full-text reading to leave only the publications aligned with the research questions and objectives addressed in this study. The final dataset ended up with 52 documents.

4 Discussion

The most relevant literature pertaining to the topic is categorized into two primary sub-themes: empowering human capabilities and regulatory control through governance. While much of the literature addresses various objectives, there is considerable causal relationship and overlap between these sub-themes. Nevertheless, they are categorized separately for analytical clarity purposes.

4.1 Operator 5.0 Essential Competencies

O5.0 recognizes the significance of corporate social responsibility beyond job creation and promoting economies [45], a concept known as Society 5.0. The goal is to provide sustainable wealth while ensuring industrial systems align with human capabilities and emphasize operator well-being in evolving IMSS. Adopting a human-centric approach ensures that operators remain at the core of the complex industrial transformation [16].

Skill development is vital for expansion and efficiency in the manufacturing and service sectors. Operators' proficiency plays a significant role in fostering sustainable economic development and augmenting output [25] [38]. However, each industrial revolution amplifies gaps in knowledge and capabilities [7]. Addressing various challenges is critical to narrowing this gap and improving operator skills. Human operators in O5.0 often encounter barriers in professional expertise, cognitive capabilities, and access to up-to-date information [3].

In Industry 5.0, artificial intelligence and humans interact through stages of co-extension, co-evolution, co-action, and co-dependency. Many scholars advocate for human-AI collaboration to leverage AI's multifaceted abilities, including communication, leadership, collaborative work, creativity, and interpersonal skills [37]. AI frees people from monotonous tasks, allowing them to focus on creative and value-added activities. Flexible work solutions enable individuals to pursue professional development and achieve higher job satisfaction. Additionally, suitable technology enhances the efficiency of elderly workers both at work and in their daily activities [23].

4.2 Operator's Challenges

Deploying human-centric solutions in Industry 5.0 is complex and interconnected, encompassing technological, organizational, and cultural dimensions [9]. Addressing these challenges requires an integrated, comprehensive approach considering the dynamics between humans and AI agents. The shift towards a human-centered industrial paradigm offers new growth potential for organizations, but only if they embrace innovation, collaboration, and adaptability in their strategies [43]. However, transitioning to a human-centric paradigm in O5.0 poses several challenges that require careful planning. A primary concern is efficiently integrating diverse agents like humans, AI, IoT, and robots [23]. Each agent brings unique capabilities to the manufacturing ecosystem, and harmonizing their interactions is crucial for optimal outcomes. Ensuring seamless collaboration between humans and AI is essential, involving not only technical integration but also addressing cognitive and organizational aspects to facilitate effective teamwork and decision-making [62].

Another significant challenge is developing methods for human-in-the-loop optimization [49]. While automation and AI technologies offer efficiency improvements, human oversight is indispensable for ensuring safety, reliability, and adaptability in dynamic environments. Therefore, designing systems that leverage human expertise while utilizing AI's computational power and predictive capabilities is crucial [1]. In parallel, creating frameworks that empower humans while maintaining control is essential. This involves balancing autonomy and oversight, where humans retain authority over critical decisions while delegating routine tasks to automated systems. Such frameworks must be adaptable to evolving requirements and preferences, enabling continuous optimization of human-technology interactions [43].

Furthermore, adaptable frameworks are needed to enhance collaboration within industrial plants. Traditional hierarchical structures can hinder information exchange and expertise, impeding innovation and problem-solving [40]. Embracing decentralized, collaborative approaches can foster knowledge-sharing and collective decision-making, enhancing overall operational efficiency and agility. Moreover, addressing the limitations of I4.0 is crucial in the transition to I5.0 [45]. While I4.0 focused on technological advancements, it overlooked human-centric considerations [27]. Thus, existing systems and practices must be reevaluated to align with human-centric values, emphasizing worker well-being and empowerment. The shift towards a value-driven, human-centric approach underscores the complexity of the transition [35]. It requires technological innovation and cultural and organizational changes to foster a supportive environment for human development [5]. By prioritizing human well-being alongside operational efficiency, organizations can achieve sustainable growth and resilience in evolving market dynamics.

Adopting human-focused approaches in O5.0 faces complex challenges related to digital ethics, citizenship, privacy, and human rights [55]. With growing integration of innovations into manufacturing systems, prioritizing ethical conduct and privacy rights is essential [33]. This requires developing robust digital governance frameworks that protect human rights, data privacy, and digital citizenship. Ensuring harmonious integration of technological advancements with ethical principles and regulatory requirements is crucial for establishing trust, accountability, and transparency in I5.0 ecosystems. This, in turn, facilitates inclusive and sustainable digital transformations.

5 Proposed Enhanced Human-Automation Symbiosis (EHAS) Framework

The Enhanced Human-Automation Symbiosis (EHAS) framework we offer provides a thorough method for comprehending the complex interaction between human capacities and automation, as illustrated in Fig. 2. The EHAS Framework classifies these improvements into four primary areas, providing a systematic perspective to examine the changing dynamics of human-automation collaboration. This paradigm explores how automation empowers individuals in physical & sensory, cognitive & emotional, communications & collaboration, and digital ethics and rights. The proposed EHAS framework provides a basis for analyzing the complexities of each category and studying how these improvements contribute interchangeably to human-automation partnerships.

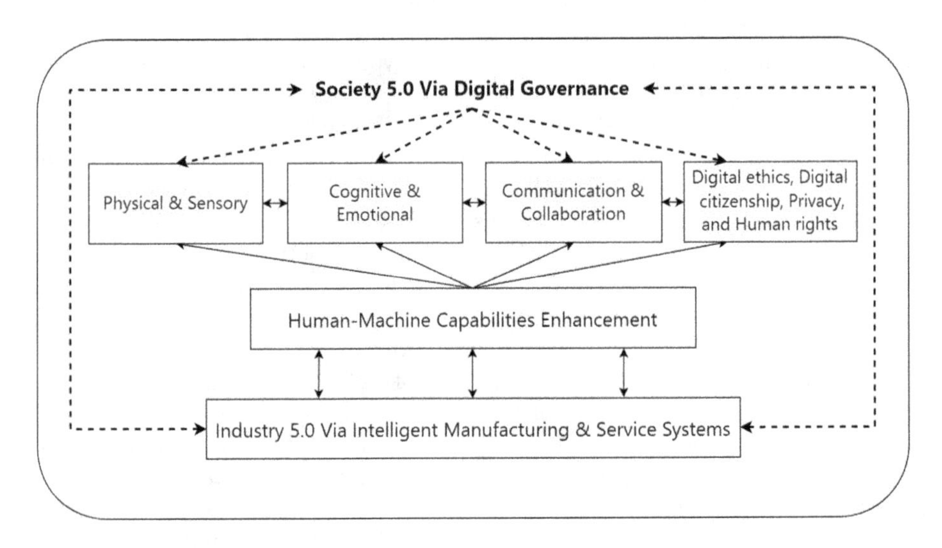

Fig. 2. Proposed Enhanced Human-Automation Symbiosis (EHAS) Framework.

Physical and Sensory Enhancement: Enhancing physical and sensory capabilities through automation technology is referred to as physical and sensory enhancement. This includes tools and technologies that augment sensory perception, physical strength, and endurance, optimizing performance across diverse functions. Examples include powered apparatus and wearable devices with sensors, cameras, and data processing tools that augment human senses [10]. However, ethical concerns must be addressed, as the widespread use of devices and sensors could raise data privacy issues. These technologies must be deployed responsibly and securely, with regulatory compliance to ensure adherence to ethical guidelines and safety standards. Updated regulations must be established and enforced by industrial organizations to ensure the responsible application of sensory and physical enhancements [24].

Cognitive and Emotional Enhancement: Cognitive and emotional enhancement can be broken down into three parts. First, enhanced cognitive capabilities involve using automation and technology to augment mental faculties and cognitive skills through AI, machine learning, and data analytics [58,59]. Second, enhanced emotional and social intelligence pertains to developing the capacity to identify, comprehend, regulate, and proficiently employ emotions, as well as navigating interpersonal relationships. In high-stress sectors like agriculture and mining, automation technology can help individuals better understand and control their emotions and those of others [31]. By shedding light on social dynamics, communication patterns, and relationship-building, enhanced emotional intelligence boosts social intelligence, making work environments more appealing and productive. Privacy and ethical concerns must be addressed during the implementation of these technologies to uphold individual boundaries and foster positive social exchanges. Third, enhanced adaptability and learning are fundamental elements of EHAS, enabling individuals to thrive in ever-changing work environments impacted by technological progress. Organizations must allocate resources to training infrastructure and digital education materials, considering each employee's unique learning requirements. Incorporating automation should align with the organization's talent development and knowledge management strategies, governed by ethical codes and regulatory oversight [42].

Communication and Collaboration Enhancement: Effective communication and cooperation are essential for the beneficial interaction between individuals and automated technology. Automation systems enhance information flow and dissemination, promoting speed and efficiency in industries like financial trading and healthcare. However, they also raise concerns about data security and privacy. According to Jaiswal [19], this concept includes elements like real-time data sharing, simplified messaging platforms, and data visualization tools. Automated systems provide quick data transmission for prompt decision-making in finance and enable seamless communication between patients and healthcare providers through telemedicine. Increased collaboration facilitated by automation strengthens teamwork across geographical distances [17]. Collaboration tools, project management software, and cloud-based platforms streamline collaborative work and information exchange. In research and development, automation helps scientists collaborate on experiments, assess data, and share results globally. In manufacturing, collaborative robots (cobots) work with human operators to achieve optimal output through high coordination. These factors highlight the crucial role of technology in promoting efficient communication and cooperation, enhancing productivity, creativity, and collaborative decision-making [30]. To fully utilize these improvements, organizations need to incorporate appropriate communication and collaboration tools and foster a culture of teamwork and information-sharing protocols that optimize automation technologies while adhering to ethical standards and human rights.

Digital Ethics, Digital Citizenship, Privacy, and Human Rights Enhancement: This element can be segmented into two sections, the first covers digital ethics and citizenship, while the second covers privacy and human rights.

Digital ethics and Digital Citizenship: Digital ethics is comparable to non-digital ethics in many respects. "Digital ethics" refers to the "normative principles for action and interaction in digital environments" [29]. At its core is the recognition that ethical conduct is necessary for a just society. Understanding this behavior as inherent to digital governance is more accurate than assuming it is exclusive to people, labor, industry, or public officials. Similar to medical ethics, digital ethics can establish ethical protocols, involve ethical committees, and uphold individual autonomy [51]. Ethical codes can manifest in various forms: choice (opt-in for privacy), corporate (socially aware algorithm design), or legal-regulatory (EU GDPR). Ethics committees are now more common than before. Most hospitals have ethics committees, and institutional review boards for research involving human subjects are far more prevalent now than decades ago. While discussions about AI Ethics Boards in the private sector are still in their infancy (e.g., Schuett et al. [47]), the circumstances surrounding Sam Altman's dismissal and reinstatement as CEO of OpenAI in November 2023 highlight the need for societal safety concerns related to AI to take precedence over financial gain.

Digital citizenship goes beyond one's legal status in a nation. Definitions of digital citizenship vary among scholars. It may refer to a basic understanding of technology or encompass an emancipatory concept emphasizing social justice and societal improvement [12]. Despite these disparities, a unifying element remains: ensuring individuals can actively participate, influence digital governance, and redefine the limits of digital ethics, citizenship, and governance as needed. This will be achieved through training, education, legislation, and public policy. Whether these limits are redefined by analog, augmented, automated governance, or a mix, digital citizenship should extend to society as a whole. I5.0 must consistently evaluate its social ramifications to meet modern society's demands.

Privacy and Human Rights: The ethical ramifications extend beyond job displacement, incorporating concerns about data privacy, security, biases in AI algorithms, and the lack of transparency in automated decision-making [54].

The rapid advancement of technology compared to human capabilities reveals significant gaps in ethical frameworks that need urgent attention. Developing solid principles is essential to balance technology and human values in smart systems. It is also crucial to guarantee digital governance, regulatory compliance, safety regulations, and strict automation standards. Ensuring a harmonious coexistence between humans and technology, considering ethical and compliance factors, is vital for fostering sustainable and ethical development of smart production systems. Moreover, the lack of clearly defined standards for the ethical use of AI raises concerns about responsibility and transparency [46], compromising the technology's dependability in preserving human values. Recognizing issues arising from new technology's impact on human rights, legal frameworks, and conventions highlights areas needing thoughtful examination. AI decision-making systems might threaten human rights due to potential biases and abuses [41]. The clash between cultural values and technical advancement necessitates integrating diverse cultural perspectives into technological development. Additionally, current legal frameworks are inadequate for regulating and overseeing the rapidly evolving technology landscape.

To effectively navigate the constantly evolving technological landscape while protecting fundamental human values and rights, it is crucial to foster interdisciplinary cooperation among professionals in technology, ethics, policymaking, and sociology [36]. Failing to adopt a global perspective often neglects the diverse economic, cultural, and social contexts in which technological progress is implemented. Addressing areas of low understanding and encouraging a comprehensive approach requires including many viewpoints and promoting inclusive discourse. Incorporating diverse perspectives into decision-making in governance and using different frameworks helps prioritize ethical issues in technology creation and implementation. The ethical dimension in our proposed framework emphasizes a co-evolutionary approach, recognizing the interdependence between technology and society. Incorporating ethical concerns into development acknowledges the ever-changing nature of societal values and the need for technical advancements. A proactive approach enables the thoughtful examination and resolution of ethical issues that emerge with technological and societal progress. This method helps develop a morally upright link between humanity and technology.

6 Remarks and Conclusion

In this work, we attempted to review and analyze the complex relationship between humanity and technology as it pertains to evolving IMSS. The EHAS framework presented here, derived from synthesizing the literature, offers a systematic approach to understanding this complex interaction. This paradigm outlines four crucial stages, starting with the introduction of automation to the ascent of sophisticated IoT and AI, which mirrors the progression of partnership between humanity and automation. Exploring the many aspects of human-automation enhancement, which include physical and sensory advantages as well as cognitive, emotional, and social upgrades, highlighted the important role of automation in empowering humans in different areas. In addition, the EHAS framework highlighted the importance of ethical issues such as digital ethics, digital citizenship, privacy, and human rights in relation to automation-driven advancements.

The increasing tension between sophisticated technology and human interaction in IMSS poses a dilemma that requires novel solutions. The existing ethical frameworks are lacking in addressing important issues such as job displacement, skills shortages, ethical consequences, and the unequal rate at which technology is advancing compared to human capabilities. To fill these gaps, it is necessary to have strict digital governance, compliance, and a seamless integration of technology with ethical and human-centered methods. As we navigate through this ever-changing environment, our fundamental goal is to promote a mutually beneficial connection between technology and people. It is crucial to achieve a harmonious balance between technological progress and ethical principles and human values in order to ensure the sustained development of IMSS in the Industry 5.0 age. Further research should explore the ethical consequences of the fast-paced development of artificial intelligence and automation. This research should specifically concentrate on narrowing the gap in skills, improving digital literacy, and promoting a mutually beneficial connection between technology, ethics, and human values. In addition, performing comparative case studies across sectors and examining the larger socio-economic ramifications may provide a more thorough picture of the changing environment of human-automation cooperation, specifically within each business.

References

1. Alimam, H., Mazzuto, G., Tozzi, N., Ciarapica, F.E., Bevilacqua, M.: The resurrection of digital triplet: a cognitive pillar of human-machine integration at the dawn of industry 5.0. J. King Saud Univ. Comput. Inf. Sci. 101846 (2023)
2. Alves, J., Lima, T.M., Gaspar, P.D.: Is industry 5.0 a human-centred approach? A systematic review. Processes **11**(1), 193 (2023)
3. Angulo, C., Chacón, A., Ponsa, P.: Towards a cognitive assistant supporting human operators in the artificial intelligence of things. Internet Things **21**, 100673 (2023)
4. Atieh, A.M., Cooke, K.O., Osiyevskyy, O.: The role of intelligent manufacturing systems in the implementation of industry 4.0 by small and medium enterprises in developing countries. Eng. Rep. **5**(3), e12578 (2023)
5. Banholzer, V.M.: From „industry 4.0 "to „society 5.0 "and „industry 5.0 ": value-and mission-oriented policies. Technol. Soc. Innovations–Aspects Syst. Trans. IKOM WP **3**(2), 2022 (2022)
6. Berardi, D., Callegati, F., Giovine, A., Melis, A., Prandini, M., Rinieri, L.: When operation technology meets information technology: challenges and opportunities. Future Internet **15**(3), 95 (2023)

7. Bousdekis, A., Apostolou, D., Mentzas, G.: A human cyber physical system framework for operator 4.0–artificial intelligence symbiosis. Manuf. lett. **25**, 10–15 (2020)
8. Brangier, É., Hammes-Adelé, S.: Beyond the technology acceptance model: elements to validate the human-technology symbiosis model. In: Ergonomics and Health Aspects of Work with Computers: International Conference, EHAWC 2011, Held as Part of HCI International 2011, Orlando, FL, USA, July 9-14, 2011. Proceedings, pp. 13–21. Springer (2011)
9. Cillo, V., Gregori, G.L., Daniele, L.M., Caputo, F., Bitbol-Saba, N.: Rethinking companies' culture through knowledge management lens during industry 5.0 transition. J. Knowl. Manage. **26**(10), 2485–2498 (2022)
10. Cristina, O.P., Jorge, P.B., Eva, R.L., Mario, A.O.: From wearable to insideable: is ethical judgment key to the acceptance of human capacity-enhancing intelligent technologies? Comput. Hum. Behav. **114**, 106559 (2021)
11. Dhirani, L.L., Mukhtiar, N., Chowdhry, B.S., Newe, T.: Ethical dilemmas and privacy issues in emerging technologies: a review. Sensors **23**(3), 1151 (2023)
12. Fernández-Prados, J.S., Lozano-Díaz, A., Ainz-Galende, A.: Measuring digital citizenship: a comparative analysis. In: Informatics. vol. 8, pp. 18. MDPI (2021)
13. Fernandez-Vidal, J., Perotti, F.A., Gonzalez, R., Gasco, J.: Managing digital transformation: the view from the top. J. Bus. Res. **152**, 29–41 (2022)
14. Gervasi, R., Barravecchia, F., Mastrogiacomo, L., Franceschini, F.: Applications of affective computing in human-robot interaction: State-of-art and challenges for manufacturing. Proc. Inst. Mech. Eng. Part B J. Eng. Manuf. **237**(6–7), 815–832 (2023)
15. Gil, M., Albert, M., Fons, J., Pelechano, V.: Designing human-in-the-loop autonomous cyber-physical systems. Int. J. Hum. Comput. Stud. **130**, 21–39 (2019)
16. Granata, I., Faccio, M., Boschetti, G.: Industry 5.0: prioritizing human comfort and productivity through collaborative robots and dynamic task allocation. Procedia Comput. Sci. **232**, 2137–2146 (2024)
17. Hjorth, S., Chrysostomou, D.: Human-robot collaboration in industrial environments: a literature review on non-destructive disassembly. Robot. Comput. Integr. Manuf. **73**, 102208 (2022)
18. Imran, F., Shahzad, K., Butt, A., Kantola, J.: Digital transformation of industrial organizations: toward an integrated framework. J. Chang. Manag. **21**(4), 451–479 (2021)
19. Jaiswal, A., Arun, C.J., Varma, A.: Rebooting employees: upskilling for artificial intelligence in multinational corporations. Int. J. Hum. Resour. Manage. **33**(6), 1179–1208 (2022)
20. Javaid, M., Haleem, A., Singh, R.P., Suman, R.: An integrated outlook of cyber–physical systems for industry 4.0: topical practices, architecture, and applications. Green Technol. Sustain. **1**(1), 100001 (2023)
21. Johnson, D.G., Wetmore, J.M.: Technology and society: building our sociotechnical future, MIT press (2021)
22. Kaasinen, E., Anttila, A.H., Heikkilä, P., Laarni, J., Koskinen, H., Väätänen, A.: Smooth and resilient human–machine teamwork as an industry 5.0 design challenge. Sustainability **14**(5), 2773 (2022)
23. Khan, M., Haleem, A., Javaid, M.: Changes and improvements in industry 5.0: a strategic approach to overcome the challenges of industry 4.0. Green Technol. Sustain. **1**(2), 100020 (2023)
24. Leenes, R., Palmerini, E., Koops, B.J., Bertolini, A., Salvini, P., Lucivero, F.: Regulatory challenges of robotics: some guidelines for addressing legal and ethical issues. Law Innov. Technol. **9**(1), 1–44 (2017)

25. Leon, R.D.: Employees' reskilling and upskilling for industry 5.0: selecting the best professional development programmes. Technol. Soc. **75**, 102393 (2023)
26. Li, N., Adepu, S., Kang, E., Garlan, D.: Explanations for human-on-the-loop: a probabilistic model checking approach. In: Proceedings of the IEEE/ACM 15th International Symposium on Software Engineering for Adaptive and Self-Managing Systems, pp. 181–187 (2020)
27. Longo, F., Padovano, A., Umbrello, S.: Value-oriented and ethical technology engineering in industry 5.0: a human-centric perspective for the design of the factory of the future. Appl. Sci. **10**(12), 4182 (2020)
28. Lu, Y., et al.: Outlook on human-centric manufacturing towards industry 5.0. J. Manuf. Syst. **62**, 612–627 (2022)
29. Luke, A.: Digital ethics now. Lang. Literacy **20**(3), 185–198 (2018)
30. Maddikunta, P.K.R., et al.: Industry 5.0: a survey on enabling technologies and potential applications. J. Ind. Inf. Integr. **26**, 100257 (2022)
31. Menezes, B., Yaqot, M., Hassaan, S., Franzoi, R., AlQashouti, N., Al-Banna, A.: Digital transformation in the era of industry 4.0 and society 5.0: a perspective. In: 2022 2nd International Conference on Emerging Smart Technologies and Applications (eSmarTA), pp. 1–6. IEEE (2022)
32. Mourtzis, D., Angelopoulos, J., Panopoulos, N.: A literature review of the challenges and opportunities of the transition from industry 4.0 to society 5.0. Energies **15**(17), 6276 (2022)
33. Murphy, C., Carew, P.J., Stapleton, L.: Ethical personalisation and control systems for smart human-centred industry 5.0 applications. IFAC-PapersOnLine **55**(39), 24–29 (2022)
34. Nayyar, A., Kumar, A.: A roadmap to industry 4.0: smart production, sharp business and sustainable development. Springer (2020). https://doi.org/10.1007/978-3-030-14544-6
35. Ordieres-Meré, J., Gutierrez, M., Villalba-Díez, J.: Toward the industry 5.0 paradigm: Increasing value creation through the robust integration of humans and machines. Comput. Ind. **150**, 103947 (2023)
36. Ozmen Garibay, O., et al.: Six human-centered artificial intelligence grand challenges. Int. J. Hum. Comput. Int. **39**(3), 391–437 (2023)
37. Panagou, S., Neumann, W.P., Fruggiero, F.: A scoping review of human robot interaction research towards industry 5.0 human-centric workplaces. Int. J. Prod. Res. **62**(3), 974–990 (2024)
38. Poláková, M., Suleimanová, J.H., Madzík, P., Copuš, L., Molnárová, I., Polednová, J.: Soft skills and their importance in the labour market under the conditions of industry 5.0. Heliyon **9**(8) (2023)
39. Popkova, E.G., Chatterji, M.: Technology, Society, and Conflict. Emerald Publishing Limited (2022)
40. Raja Santhi, A., Muthuswamy, P.: Industry 5.0 or industry 4.0 s? Introduction to industry 4.0 and a peek into the prospective industry 5.0 technologies. Int. J. Interact. Des. Manuf. (IJIDeM) **17**(2), 947–979 (2023)
41. Raso, F.A., Hilligoss, H., Krishnamurthy, V., Bavitz, C., Kim, L.: Artificial intelligence & human rights: opportunities & risks. Berkman Klein Cent. Res. Publ. (2018-6) (2018)
42. Rodgers, W., Murray, J.M., Stefanidis, A., Degbey, W.Y., Tarba, S.Y.: An artificial intelligence algorithmic approach to ethical decision-making in human resource management processes. Hum. Resour. Manag. Rev. **33**(1), 100925 (2023)

43. Saikia, B.: Industry 5.0–its role toward human society: obstacles, opportunities, and providing human-centered solutions. In: Fostering Sustainable Businesses in Emerging Economies: The Impact of Technology, pp. 109–126. Emerald Publishing Limited (2023)
44. Salvatore, M., et al.: Smart operators: how industry 4.0 is affecting the worker's performance in manufacturing contexts. Procedia Comput. Sci. **180**, 958–967 (2021)
45. Saniuk, S., Grabowska, S., Straka, M.: Identification of social and economic expectations: contextual reasons for the transformation process of industry 4.0 into the industry 5.0 concept. Sustainability **14**(3), 1391 (2022)
46. Santhoshkumar, S., Susithra, K., Prasath, T.K.: An overview of artificial intelligence ethics: issues and solution for challenges in different fields. J. Artif. Intell. Capsule Netw. **5**(1), 69–86 (2023)
47. Schuett, J., Reuel, A., Carlier, A.: How to design an AI ethics board. arXiv preprint arXiv:2304.07249 (2023)
48. Shan, S., Wen, X., Wei, Y., Wang, Z., Chen, Y.: Intelligent manufacturing in industry 4.0: a case study of Sany heavy industry. Syst. Res. Behav. Sci. **37**(4), 679–690 (2020)
49. Tóth, A., Nagy, L., Kennedy, R., Bohuš, B., Abonyi, J., Ruppert, T.: The human-centric industry 5.0 collaboration architecture. MethodsX **11**, 102260 (2023)
50. Uygun, Ö., Aydin, M.E.: Digital transformation: industry 4.0 for future minds and future society (2021)
51. Véliz, C.: Three things digital ethics can learn from medical ethics. Nat. Electron. **2**(8), 316–318 (2019)
52. Vrana, J., Singh, R.: Digitization, digitalization, and digital transformation. Handb. Nondestr. Eval. **4**, 1–17 (2021)
53. Wang, B., Tao, F., Fang, X., Liu, C., Liu, Y., Freiheit, T.: Smart manufacturing and intelligent manufacturing: a comparative review. Engineering **7**(6), 738–757 (2021)
54. Whittlestone, J., Nyrup, R., Alexandrova, A., Dihal, K., Cave, S.: Ethical and societal implications of algorithms, data, and artificial intelligence: a roadmap for research. Nuffield Found. London (2019)
55. Wulandari, E., Winarno, W., Triyanto, T.: Digital citizenship education: shaping digital ethics in society 5.0. Univ. J. Edu. Res. **9**(5), 948–956 (2021)
56. Yao, X., Zhou, J., Lin, Y., Li, Y., Yu, H., Liu, Y.: Smart manufacturing based on cyber-physical systems and beyond. J. Intell. Manuf. **30**, 2805–2817 (2019)
57. Yaqot, M., Menezes, B.: The good, the bad, and the ugly: review on the social impacts of unmanned aerial vehicles (UAVS). In: International Conference of Reliable Information and Communication Technology, pp. 413–422. Springer (2021). https://doi.org/10.1007/978-3-030-98741-1_34
58. Yaqot, M., Menezes, B.C.: Unmanned aerial vehicle (UAV) in precision agriculture: business information technology towards farming as a service. In: 2021 1st international conference on emerging smart technologies and applications (eSmarTA), pp. 1–7. IEEE (2021)
59. Yaqot, M., Menezes, B.C.: Mining process systems in the industry 4.0 mandate towards the plant of future. In: Proceedings of the 5th European International Conference on Industrial Engineering and Operations Management, pp. 1–8. IEOM Society International (2022)
60. Zhang, C., et al.: Towards new-generation human-centric smart manufacturing in industry 5.0: a systematic review. Adv. Eng. Inf. **57**, 102121 (2023)

61. Zhou, J., Zhou, Y., Wang, B., Zang, J.: Human-cyber-physical systems (HCPSS) in the context of new-generation intelligent manufacturing. Engineering **5**(4), 624–636 (2019)
62. Zizic, M.C., Mladineo, M., Gjeldum, N., Celent, L.: From industry 4.0 towards industry 5.0: a review and analysis of paradigm shift for the people, organization and technology. Energies **15**(14), 5221 (2022)
63. Zolotová, I., Papcun, P., Kajáti, E., Miškuf, M., Mocnej, J.: Smart and cognitive solutions for operator 4.0: laboratory h-CPPS case studies. Comput. Ind. Eng. **139**, 105471 (2020)

Impact of Collaborative Robots on Human Trust, Anxiety, and Workload: Experiment Findings

Elias Montini[1,2(✉)], Giovanni Ploner[2(✉)], Davide Matteri[1], Vincenzo Cutrona[1], Paolo Rocco[2], Andrea Bettoni[1], and Paolo Pedrazzoli[1]

[1] Department of Innovative Technologies, University of Applied Sciences and Arts of Southern Switzerland, Lugano, Switzerland
{elias.montini,davide.matteri,vincenzo.cutrona,andrea.bettoni, paolo.pedrazzoli}@supsi.ch

[2] Dipartimento di Elettronica, Informazione e Bioingegneria, Politecnico di Milano, Milano, Italy
{elias.montini,giovanni.ploner,paolo.rocco}@polimi.it

Abstract. This work proposes an experiment setup and its protocols to investigate the impact of cobot's size, speed and collaboration modes on different human factors including trust, propensity to trust, anxiety, and mental workload. The setup and the protocols supported the execution of different experiments where the 29 participants were asked to complete the Tower of Hanoi in collaboration with a cobot. The setup and the protocols provide a ready-to-use solution to expand experiments for further studies. Moreover, statistical analysis of the results shows higher cobot speeds increased trust propensity despite not significantly affecting overall trust or anxiety. Collaboration modes significantly influenced perceived workload and task performance, with the "Collaboration with Trigger" mode resulting in lower mental workload but longer task completion times. No significant differences were found in human factors concerning cobot size, indicating that variations in size do not significantly impact trust, propensity to trust, anxiety, or workload. Additionally, the collaboration mode with cobots notably affects workload perception and task performance, with specific modes reducing perceived effort but not necessarily improving task efficiency.

Keywords: Cobots · Human Robot Collaboration · Collaborative robots · Human factors · Ergonomics

Published by Springer Nature Switzerland AG 2024
M. Thürer et al. (Eds.): APMS 2024, IFIP AICT 729, pp. 401–415, 2024.
https://doi.org/10.1007/978-3-031-65894-5_28

1 Introduction

Industrial robots were initially developed to substitute human labour in tasks deemed hazardous or monotonous [13]. However, over the past decade, there has been a notable pivot towards fostering coexistence and collaboration between humans and robots with the advent of collaborative robots, also known as cobots. This shift aims to enhance operational efficiency and flexibility, concurrently augmenting employees' working conditions and well-being [33].

However, despite offering soft interaction modes supporting collaboration [20], a cobot remains fundamentally an automation device. Consequently, as with any automation device, the effective and safe integration of cobots into industrial environments requires an understanding of how workers react and operate with it and what Human Factors (HFs) play a role in the interaction.

HFs are studied within a scientific discipline that focuses on the interactions between humans and system elements, aiming to mitigate human error, which can lead to adverse outcomes if not addressed [14]. This field, known as HFs engineering, involves designing equipment and systems that align with operators' mental and physical characteristics, thereby optimising system performance and human well-being.

This work and its associated experiment concentrate on examining *mental workload* [25], *trust* [45], including a specific focus on the *propensity to trust* [15] and *anxiety* [10] within collaborative robotics and Human Robot Interaction (HRI), employing the Tower of Hanoi (TOH) as the experimental framework. The selection of these particular aspects follows from extensive studies in the fields of HFs in collaborative robotics and HRI, which encompass a broad range of dimensions such as, in addition to the four already mentioned, physical ergonomics, acceptance, usability, and many others [12,32]. The experiment results provide statistical evidence on the dynamics of the selected HFs in relation to the size, speed, and collaborative modes of cobots.

This paper is organised as follows. Section 2 investigates the role of HF in HRI evaluating previous works that propose TOH. Section 3 presents the setup of the experiments with all the different collaboration modes, while Sect. 4 outlines the results of the experiments, with valuable insights regarding HRI. Conclusions and future research directions are presented in Sect. 5.

2 Human Factors, Serious Games and Cobots

2.1 Human Factors

Collaborative robotics brought attention and interest to the centrality of people's feelings and well-being in industrial environments. This leads to more recent in-depth studies on HFs. This research builds on these studies by exploring various interaction modes using different robots and variables. It also establishes a flexible experimental framework, facilitating future investigations into additional aspects and factors.

Mental workload. It refers to the cognitive demands imposed on an individual during task performance [25]. *Mental workload* encompasses various cognitive processes such as attention, memory, decision-making, and problem-solving. A collaborative robot acting as a leader and follower has been used to study the impacts on stress and trust [30]. A few frameworks have been developed to assess the *mental workload* in working with a cobot [35,43]. Furthermore, research on HRI indicates that designing intuitive interfaces and providing adequate support can mitigate *mental workload* and enhance user performance [46].

Trust. It refers to an individual's belief or confidence in another entity or system's reliability, integrity, and competence [45]. *Trust* is crucial in HRI, influencing user acceptance, cooperation, and reliance on robotic systems. Properly calibrated *trust* in robotic systems is critical and can be enabled through robot explanations of their behaviour [11]. As robots perform increasingly complex tasks in environments shared with human teammates, the need for explainable robot behaviour will gain relevance. Previous work in HFs has explicitly explored the importance of user *trust* in automated systems [24]. The focus is not always on increasing user *trust*, but on properly calibrating *trust* to use robotic systems. For this purpose, many different questionnaires and tests exist to estimate the *trust* felt by a human being, including Interpersonal Trust Questionnaire [41]; International Personality Item Pool [9]; Negative Attitudes Towards Robots Scale [36]; Trust Perception Scale for Human-Robot Interaction (TPS-HRI) [42].

Propensity to trust. It refers to an individual's inherent inclination or predisposition to trust other entities or systems [15]. The *propensity of trust* varies among individuals and is influenced by factors such as personality traits, past experiences, cultural background, and situational contexts. Few studies have examined how the *propensity to trust* influences user reliance on automation. Recent works have adopted the Propensity to Trust Machines Scale (PTM) to measure *propensity to trust* [21,29], which typically consists of a series of statements or items that respondents rate based on their level of agreement or disagreement.

Anxiety. It refers to a state of uneasiness, apprehension, or worry experienced by individuals when faced with perceived threats or uncertainties [10]. *Anxiety* may arise due to factors such as unfamiliarity with robots, concerns about privacy or safety, or uncertainty about the robot's intentions or capabilities and it is probably one of the human emotions that is most present during interactions of every nature [31]. Studies on anxiety towards robots produced various results [34], showing how this factor is related to cultural differences [2], age [22] and anxiety attitudes [37]. In the industrial context, *anxiety* is mainly measured through the State-Trait Anxiety Index (STAI), a subjective test that discerns between "state anxiety", i.e., the anxiety that someone feels at the moment he/she is filling in the questionnaire, and "trait anxiety", which is, instead, the feeling of anxiety that is congenital and typical of somebody's character [23]. Furthermore, an anxiety scale was developed for robotic systems [38] and used to measure anxiety that prevents people from interacting with robots.

2.2 The Tower of Hanoi

To investigate HFs in HRI, it is beneficial to utilise a controlled setting where consistent rules, behaviours, and protocols can be applied, similar to those used in a serious game. Serious games are games designed for purposes beyond mere entertainment [26], including education, training, research, and in this case, the exploration of HRI dynamics. In the context of HRI, serious games usually involve the so-called *social robots*. Examples of these social robots are *iCat* [1] and *Leonardo* [8], applied using the Wizard of Oz method. This method involves a human operator (the "wizard"), who controls the robot responses, often behind the scenes. At the same time, experiment participants interact with the robot as if it is fully automated [7]. This method is widely used since it provides a flexible way to implement complex robot behaviour within a quick time-scale [48].

Recent advancements in robotics, such as the extreme ease of programming, make less relevant this control scheme, empowering the use of autonomous robot systems directly to engage in interactive in-game scenarios with humans [27]. Consequently, this research opts out of the Wizard of Oz methodology, favouring a fully automated approach. This enables the realistic simulation of an industrial context that performs predefined automation tasks without the flexibility typically introduced by human intervention, as in the Wizard of Oz method. This environment is set up through a simple yet meaningful game where an autonomous cobot and a human collaborate to execute a joint task: the TOH.

The TOH, traditionally recognised as a mathematical puzzle, acquires a broader function as a serious game when applied in studying HRI. The TOH uses three vertical pegs and a number d of disks with a central hole and increasing external diameter. Typically, the game begins with all disks stacked on one peg to form a complete tower, although there are no strict rules requiring this starting arrangement as long as it adheres to the game's rules. The final goal of the game is to build a full stack in one of the pegs making one move at a time, following three rules: i) Only one disk can be moved at a time; ii) Each move consists of taking the top disk from one of the pegs and placing it in another peg; iii) No disk may be placed on top of a smaller disk.

The TOH can be considered complex from a problem-solving perspective because it requires strategic planning and foresight. Even if the game has a recursive solution, the time framework and the pressure to solve the puzzle in a minimum number of moves make it mentally demanding [18].

Many studies using the TOH have been performed in the HRI field. Several works adopted this game as a task to be performed by a robot in collaboration with a child to assess i) children's learning process with different levels of robot vocalisation [49], ii) robot intervention effectiveness regarding a voluntary turn-taking interaction [5] and iii) completion rate changes with the introduction of a pre-task instruction session with the robot as the teacher [40]. A separate study, not focusing on children, illustrated that the physical involvement of a robot in task execution enhances both performance and social perception by humans [47]. Conversely, research into the impacts of anthropomorphism in robotics unexpectedly showed that the propensity to offer assistance and aid to a robot facing malfunctions diminishes when the robot exhibits more human-like characteristics, following a collaboration task on the TOH [39]. In other researches, the TOH has been used to measure stress and fatigue levels of operators, while doing a task in collaboration with a robotic arm, assessing the cognitive fatigue and workload [3,16]. Likewise, this work expects to extend the use of TOH to study HFs in collaborative robotics.

3 System and Experiment Setup

This work sheds light on the HRI focusing on collaborative robots. It proposes a robotic system that enables a set of experiments to investigate HRI through the TOH. The system is complemented by the formulation of various experiments alongside their execution protocols, which encompass questionnaires and analysis methods. This supported the experimental activities detailed in this study and opened pathways for further extension and diverse analyses in the future.

The system's setup includes two different types of cobots, an igus ReBeL and an ABB GoFa 15000, 3D-printed TOH's disks and pegs, a vision system, a Graphical User Interface (GUI) and ArUco markers, as depicted in Fig. 1. Using two different cobots enables the study of the influence of robot characteristics, such as size and speed, on HFs.

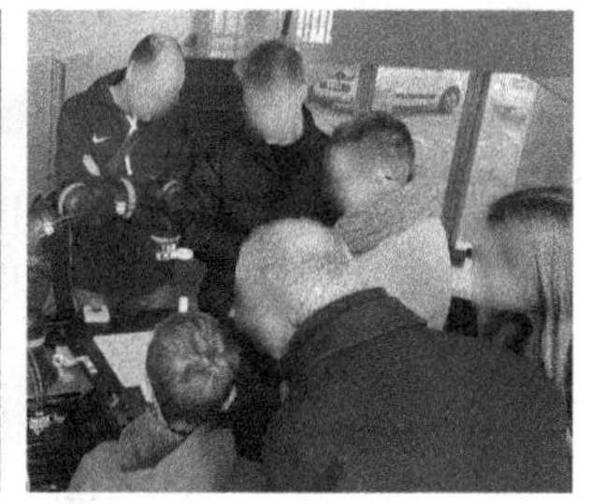

Fig. 1. ABB GoFa and Igus ReBeL setups; participants engaging with the cobot

3.1 Experiment Protocol

The experiment protocol is divided into three phases: i) pre-task phase, ii) task phase and iii) post-task phase.

Volunteers learn the TOH rules, research goals, and the cobot type in the pre-task phase. After consenting, they wear devices to record baseline physiological data for 3–4 minutes. A simplified TOH helps to acquaint themselves with the rules and the task. This phase ends with a questionnaire on prior robot interactions, initial trust, and anxiety levels.

During the task phase, participants enter the workspace with the robot to commence the task. They select collaboration modes (triggered, alternated with trigger, alternated with countdown or mixed, later detailed) from the GUI before initiating the game. The participants make the first move in any collaboration mode, while physiological data collection, game events and Key Performance Indicators (KPIs) are collected.

The study considers three variables: cobot model (i.e., igus ReBeL, ABB GoFa 15000) and hence its size (respectively small and big), cobot speed (fast (1500 mm/s) or slow (300 mm/s)), and collaboration mode.

Triggered Collaboration defines a collaboration mode where the robot intervenes only when the participant wants, using the GUI. *Alternated Collaboration with Trigger* is a turn-taking-based collaboration where the robot intervenes only after the participants' commands (the participants have the time they want to think and complete their move). *Alternated Collaboration with Countdown* is a turn-taking-based collaboration where the cobot moves every 5 s from when it has completed its previous move. *Mixed Collaboration* involves turn-taking between human and cobot after 7 consecutive moves. Experiments are categorised based on the combination of these variables, enabling statistical analysis to identify differences caused by varying experimental conditions.

Once the task phase has been completed, the participant moves toward the third and last phase of the experiment where the participant fills in the post-task questionnaires. The overall duration of the three steps is around 25–30 minutes.

Table 1 resumes all experiments with their main characteristics. The current study makes statistical analysis (Sect. 3.2) by comparing one variable at a time and detecting possible differences caused by the different experiment conditions.

Table 1. Experiments

ID	Participants	Size	Speed	Collaboration Mode
E1	4	Igus (small)	Slow (300 mm/s)	Triggered Collaboration
E2	6	Igus (small)	Slow (300 mm/s)	Alternated with Countdown
E3	3	Gofa (big)	Slow (300 mm/s)	Alternated with Countdown
E4	5	Gofa (big)	Fast (1500 mm/s)	Alternated with Countdown
E5	6	Gofa (big)	Fast (1500 mm/s)	Alternated with Trigger
E6	5	Gofa (big)	Fast (1500 mm/s)	Mixed Collaboration

3.2 Analysis Protocol

During the pre-task phase, participants complete several questionnaires: an initial demographic survey and three tests assessing HFs: the PTM [29], STAI [44], and the TPS-HRI [42]. The same questionnaires are administered in the post-task phase, except for the demographic survey. They are supplemented with the NASA Task Load Index (NASA-TLX) [17] and questions about the experiment and HRI quality. Table 2 provides an overview of questionnaires included in the protocol proposed by this work.

Table 2. Questionnaires

Questionnaire	Description (Phases)
Demographic questionnaire	A series of questions for the identification of the participant characteristics (e.g., age, sex) (Pre-task)
PTM	A series of statements regarding their *propensity to trust* the collaborative robot that respondents rate based on their level of agreement or disagreement. (Pre-task, post-task)
STAI (short)	A subjective test that discerns between "state anxiety" and "trait anxiety". The version used in this research, however, is a shortened one, that pivots only on "state anxiety", leaving the other trait apart. (Pre-task, post-task)
TPS-HRI (short)	A series of Likert-scale-based statements to measure the extent to which individuals *trust* a robot in various aspects of interaction, such as reliability, competence, predictability, and intentionality (Pre-task, post-task)
NASA-TLX (short)	An assessment on six dimensions that capture different aspects of perceived *mental workload.* (Post-task)
HRI quality questionnaire	Questions to investigate how the human has perceived the collaboration between the operator and the cobot. (Post-task)

In addition, a set of KPIs reported has been defined to assess the quality and efficiency of the task execution (Table 3).

Table 3. Key Performance Indicators

KPI	Description
Effectiveness	Whether the game has been solved or not
Time-to-task	Time to complete the game
Human Errors	Number of non-legit configurations made by volunteer
Robot Errors	Number of disks misplaced by the robot
COP[a]	Efficiency of the solution in terms of the number of moves
Cobot moves	Number of times the cobot makes a move

[a] Coefficient Of Performance: the ratio between optimal and total moves.

The analysis of data is performed by making use of various statistical tests: i) Student's t-test, ii) One-way ANOVA, iii) Mixed ANOVA. T-test is used to compare the means of two normally distributed samples; ANOVA and Mixed ANOVA detect differences between groups divided into two independent variables: the between-subject variable and the within-subject variable (*i.e.,* the repeated measures). These statistical tests were chosen because they are well-suited for analysing differences between groups (t-tests and ANOVA) and examining interactions between multiple factors (mixed ANOVA), which is essential for understanding collaborative robotics's complex and multifaceted impacts on human factors. Alternative tests might not provide the same level of detail or may be less appropriate for the experimental design used in this study.

The following statistical terms have been used for the results presented in Sect. 4: *F (F-statistic)*, *p (p-value)*, η_p^2 *(Partial Eta Squared)* and *Time*.

F: the F-statistic is a ratio used in the context of ANOVA to compare the variance among group means to the variance within the groups. A higher F-value indicates a more significant disparity between group means, suggesting that at least one group mean is significantly different. The F-statistic helps to determine if these differences are statistically significant. It is calculated based on the degrees of freedom between groups (numerator) and the variance within groups (denominator).

p: it represents the probability of obtaining the observed results, or more extreme, assuming that the null hypothesis is true. It is a key indicator of statistical significance. A low p-value (typically less than 0.05) suggests that the observed data are unlikely under the null hypothesis, leading to its rejection in favour of the alternative hypothesis. Conversely, a high p-value indicates that the observed data are consistent with the null hypothesis, suggesting no statistically significant effect.

η_p^2: it is a measure of effect size used in the context of ANOVA to quantify the proportion of the total variance in the dependent variable that is attributable to an independent variable, after accounting for other variables or factors in the model. It ranges from 0 to 1, with higher values indicating a larger effect size or greater explanatory power of the independent variable on the dependent variable. According to conventional benchmarks, a value of 0.01 is considered a small effect, 0.06 a medium effect, and 0.14 a significant effect.

Time: the within-variable is called *Time*, and its values are "Before" or "After" (*i.e.*, pre-task and post-task).

3.3 System Architecture

The architecture developed connects all the system components, as depicted in Fig. 2a. An orchestrator is in charge of the coordination of the task execution and the communication between the vision system, the cobot and the GUI, depicted in Fig. 2b.

The participants can interact directly with the TOH or with the GUI, sending activation signals to the cobot, which the orchestrator elaborates. The orchestrator receives the current TOH state thanks to the vision system, computes the optimal next move, and, if required, sends the instructions to the cobot. All the software components communicate by exchanging JSON messages over MQTT.

The communication method between the orchestrator and the robot depends on the collaborative robot model used in the experiment. The ABB GoFa 15000 establishes a direct connection with the orchestrator using Socket communication. Conversely, with the igus ReBeL, the orchestrator employs the MQTT protocol to communicate with an industrial Raspberry Pi board, triggering the cobot's physical digital inputs.

The system is also capable of collecting physiological data (e.g., using Polar Verity Sense armband, Polar H10 chest band) from participants to analyse heart rate variability and physiological data, relying on an IIoT platform to create human-digital twins [6].

4 Results and Discussion

The experiments involved a total of $V = 29$ volunteers ($M = 20$, $F = 9$). The demographic composition suggests a youthful participant pool, with almost 70% of people younger than 30. 62% of the participants reported that they had never worked with collaborative robots. Among those who had previous interactions with cobots, the experiences were largely positive, except for one participant who recounted an adverse incident in a laboratory setting.

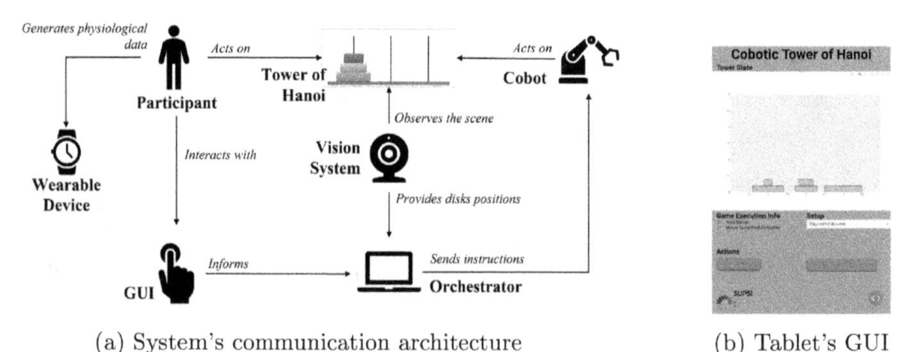

(a) System's communication architecture (b) Tablet's GUI

Fig. 2. Architecture and GUI

4.1 Speed Impact on Human Factors

Contrary to expectations, the tests revealed that higher cobot speed positively influences the *propensity to trust*, measuring increased levels post-task. Mixed-design ANOVA revealed a significant main effect ($F(1,6) = 21.440$, $p = .004$, $\eta_p^2 = 0.781$), which underlines that the *propensity to trust* changed from the pre-task phase to the post-task one. However, speed does not significantly impact *trust* or *anxiety*. These findings suggest that while cobot speed influences the *propensity to trust*, other factors such as predictability, safety and reliability may play a more substantial role in determining *trust*. Additionally, it is important to note that numerous participants expressed concerns about the cobot's pace in the post-task assessments of experiments with a slow-speed setting. Many stated, *"The cobot is too slow. I prefer to work alone since I am faster, even if I am not sure I am making the correct move."*

Further exploration of this finding might also consider findings from additional studies, suggesting that *trust* is influenced not merely by the speed but broadly by the robots' movement patterns and their predictability [4,28]. In the TOH, the predictability of each cobot's movement is notably high, as the participant can observe the cobot's movements before the commencement of the activity and is familiar with its designated operational zone.

4.2 Collaboration Mode Impact on Human Factors

The results show that *trust* ($F(1,13) = 1307.80$, $p < .001$, $\eta_p^2 = 0.629$) and *propensity to trust* ($F(1,13) = 272.305$, $p < .001$, $\eta_p^2 = 0.954$) increased after collaboration in all modes, with no significant differences between them ((Fig. 3a, Fig. 3b). The *mental workload* was lowest in the *Alternated Collaboration with Trigger* ($F(2,13) = 4.899$, $p < .05$, $\eta_p^2 = 0.430$).

The ANOVA revealed a significant effect of the collaboration mode on the perceived workload ($F(2,13) = 4.899$, $p < .05$, $\eta_p^2 = 0.430$). Figure 3c qualitatively shows the difference between the means of the three samples, underlying how the *Collaboration with Trigger* has a lower NASA-TLX score. Results suggest a significant effect of collaboration mode on the subjective assessment regarding the perceived workload, with post hoc analyses revealing a lower level of mental workload in volunteers that worked in *Collaboration with Trigger*.

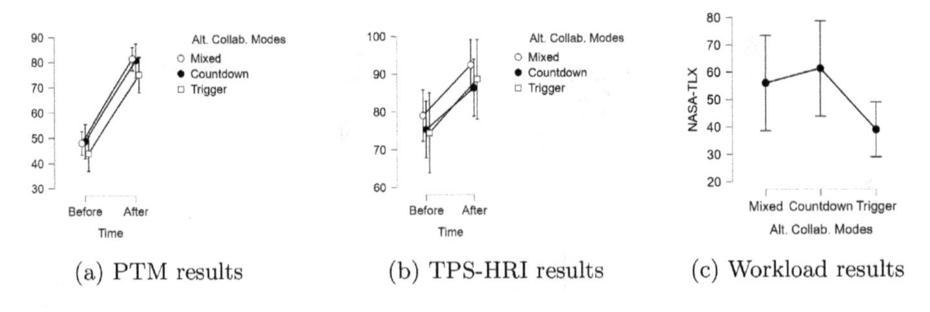

(a) PTM results (b) TPS-HRI results (c) Workload results

Fig. 3. Results

In terms of performance, the *Alternated Collaboration with Trigger* resulted in significantly longer task completion times ($F(2,13) = 4.624$, $p < .05$, $\eta_p^2 = 0.416$). Responses to the HRI questionnaire revealed differences in perceptions of task-solving abilities and willingness to use the system in everyday life across collaboration modes: participants in *Alternated Collaboration with Trigger* showed a more neutral opinion on the idea that working alone is better than working with the robot ($p < .05$), while the other volunteers disagreed with it; regarding usability, cobot in *Mixed Collaboration* resulted in the one that people would use more in their everyday life ($p < .05$). There was also a discrepancy between perceived understanding of the task algorithm and actual task performance: when asked to indicate the optimal successive move in a random TOH configuration with 6 disks, only 60% guessed the right choice, suggesting a potential false belief about comprehension induced by robot collaboration.

Alternated Collaboration with Trigger resulted in fewer cobot moves but more total moves to complete tasks than in other *Alternated Collaboration* modes ($U = 24$, $p < .05$,). In *Alternated Collaboration with Trigger*, participants preferred to solve the game alone as a personal challenge and were less motivated to activate the robot. This conclusion agrees with other research exploring the differences between similar collaboration modes (voluntary interaction and turn-taking) [5]. Nevertheless, there exists a discrepancy when looking at performance results: people who worked voluntary interaction also performed better, in terms of the number of moves, with respect to the turn-taking, while in our research, the analysis on COP showed the opposite outcomes, detecting better levels of performance in volunteers in alternated collaboration.

4.3 Size Impact on Human Factors

Analysis of different cobot sizes shows no significant differences in propensity to trust, trust, anxiety, or mental workload. Despite variations in cobot size, human perceptions remain consistent. Hence, it is possible to conclude that cobot size does not significantly affect human factors.

At first glance, these results seem to be at odds compared to existing works [19], which identified the robot's size as a factor that considerably impacts human anxiety. Nonetheless, on a deeper inspection, *anxiety* has been identified only with robot models with more than 1.8 m reach, while both Gofa and ReBeL are much smaller. Furthermore, according to this work, the most significant change in *anxiety* happened while the volunteers were sitting; the standing position returned lower changes in *anxiety* instead. The volunteers used a standing position in every experiment of the current study, which could be another explanation for these different results.

4.4 Summary of Results

Table 4 summarises the experimental results, providing a comprehensive overview of the key findings and observations from the study.

Table 4. Summary of Results

Variable	IDs	Main Results
Triggered Collaboration	E1	No effects on HFs; in Triggered Collaboration mode, humans tend not to activate the cobot
Size	E2, E3	Size does not affect any HF
Velocity	E3, E4	Velocity affects partially propensity to trust
Alternated and *Mixed Collaboration*	E4, E5, E6	*Alternated with Trigger* has the lowest mental workload, the highest time to complete the task and the most neutral opinion on the importance of the collaboration with the robot; *trust* and *propensity to trust* greatly increases after the task in all modes; mixed collaboration has the highest perceived usability

Despite the thoroughness of the experiments, limitations include a relatively small sample size and young participants. However, the setup and the protocols developed in this research hold relevance for future works as they provide a ready-to-use solution to expand experiments for further studies. Future research will involve a broader range of participants and incorporate tasks more realistic for an industrial environment.

5 Conclusions

This work investigates the impact of cobot size, speed, and collaboration modes on HFs like *trust*, *propensity to trust*, *anxiety*, and *mental workload*. *Trust* and *propensity to trust* cobots increased across all three *Alternated Collaboration* modes, while *mental workload* was lower in *Collaboration with Trigger*, albeit at the expense of time. *Anxiety* remained mostly unaffected. Experiments showed that the collaboration robots led to an increased perceived understanding of TOH solution algorithm, but further trials on increased difficulty tasks suggest that this sensation could be a false belief.

From the practical perspective, the study highlights that collaboration modes significantly affect perceived workload and task performance. Specifically, the *Collaboration with Trigger* mode, which resulted in lower mental workload but longer task completion times, suggests that different modes of interaction need to be carefully chosen based on the specific task requirements and worker preferences. This insight can help managers design work processes that balance efficiency with employee well-being.

Furthermore, understanding that cobot *size* does not significantly impact *trust*, *propensity to trust*, *anxiety*, or *workload* can simplify decision-making regarding deploying different models. Managers can focus on other factors, such as speed and collaboration mode, to improve human-robot collaboration without worrying about the physical size of the cobots affecting human factors adversely.

Acknowledgments. This work has been partly supported by the European Union's research and innovation programme under project XR5.0 (Grant n. 101135209).

References

1. Bartneck, C., Hoek, M., Mubin, O., Mahmud, A.: Daisy, daisy, give me your answer do!: switching off a robot. In: HRI 2007 - Proceedings of the 2007 ACM/IEEE Conference on Human-Robot Interaction - Robot as Team Member, pp. 217–222 (2007). https://doi.org/10.1145/1228716.1228746
2. Bartneck, C., Suzuki, T., Kanda, T., Nomura, T.: The influence of people's culture and prior experiences with AIBO on their attitude towards robots. AI and Soc. **21**, 217–230 (2007). https://doi.org/10.1007/s00146-006-0052-7
3. Chacón, A., Ponsa, P., Angulo, C.: Cognitive interaction analysis in human-robot collaboration using an assembly task. Electronics **10**(11) (2021). https://doi.org/10.3390/electronics10111317
4. Charalambous, G., Fletcher, S.: Trust in industrial human-robot collaboration. 21st Century Ind. Rob. Tools Become Collaborators. Intell. Syst. Control Auto. Sci. Eng. **81** (2022). https://doi.org/10.1007/978-3-030-78513-0_6

5. Charisi, V., Gomez, E., Mier, G., Merino, L., Gomez, R.: Child-robot collaborative problem-solving and the importance of child's voluntary interaction: a developmental perspective. Front. Rob. AI **7** (2020). https://doi.org/10.3389/frobt.2020.00015
6. Cutrona, V., Bonomi, N., Montini, E., Ruppert, T., Delinavelli, G., Pedrazzoli, P.: Extending factory digital twins through human characterisation in asset administration shell. Int. J. Comput. Integr. Manuf. 1–18 (2023). https://doi.org/10.1080/0951192X.2023.2278108
7. Dahlbäck, N., A, J., Ahrenberg, L.: Wizard of OZ studies: why and how. Knowl. Based Syst. **6**, 258–266 (1993)
8. Dang, T., Tapus, A.: Coping with stress using social robots as emotion-oriented tool: potential factors discovered from stress game experiment. Social Robotics. ICSR **8239** (2013). https://doi.org/10.1007/978-3-319-02675-6_16
9. Donnellan, M., Oswald, F., Baird, B., Lucas, R.: The mini-IPIP scales: tiny-yet-effective measures of the big five factors of personality. Psychol. Assess. **18**, 192–203 (2006). https://doi.org/10.1037/1040-3590.18.2.192
10. Duvall, A., Roddy, C.: What is anxiety? Managing Anxiety Sch. Settings (2020). https://doi.org/10.1037/e558982009-001
11. Dzindolet, M.T., Peterson, S.A., Pomranky, R.A., Pierce, L.G., Beck, H.P.: The role of trust in automation reliance. Int. J. Hum. Comput. Stud. **58**(6), 697–718 (2003). https://doi.org/10.1016/S1071-5819(03)00038-7
12. Faccio, M., et al.: Human factors in COBOT era: a review of modern production systems features. J. Intell. Manuf. **34**, 1–22 (2022). https://doi.org/10.1007/s10845-022-01953-w
13. Gasparetto, A., Scalera, L.: From the unimate to the delta robot: the early decades of industrial robotics. In: Explorations in the History and Heritage of Machines and Mechanisms: Proceedings of the 2018 HMM IFToMM Symposium on History of Machines and Mechanisms, pp. 284–295. Springer (2019). https://doi.org/10.1007/978-3-030-03538-9_23
14. Glavin, R., Maran, N.: Integrating human factors into the medical curriculum. Med. Edu. **37** (2003). https://doi.org/10.1046/j.1365-2923.37.s1.5.x
15. Hancock, P.A., Billings, D.R., Schaefer, K.E., Chen, J.Y., De Visser, E.J., Parasuraman, R.: A meta-analysis of factors affecting trust in human-robot interaction. Hum. Factors **53**(5), 517–527 (2011). https://doi.org/10.1177/0018720811417254
16. Hardy, D., Wright, M.: Assessing workload in neuropsychology: an illustration with the tower of Hanoi test. J. Clin. Exp. Neuropsychol. **40**(10), 1022–1029 (2018). https://doi.org/10.1080/13803395.2018.1473343
17. Hart, S.G., Staveland, L.E.: Development of NASA-TLX (task load index): results of empirical and theoretical research. In: Advances in psychology, vol. 52, pp. 139–183. Elsevier (1988). https://doi.org/10.1016/S0166-4115(08)62386-9
18. Hinz, A.M., Klavžar, S., Milutinović, U., Petr, C.: The tower of Hanoi-Myths and maths. Springer (2013). https://doi.org/10.1007/978-3-0348-0237-6
19. Hiroi, Y., Ito, A.: Are bigger robots scary? - The relationship between robot size and psychological threat -. IEEE/ASME International Conference on Advanced Intelligent Mechatronics, AIM (2008). https://doi.org/10.1109/AIM.2008.4601719
20. ISO/TC 299: Iso/ts 15066:2016(en), robots and robotic devices. Tech. rep., ISO (2016)
21. Jessup, S.A., Schneider, T.R., Alarcon, G.M., Ryan, T.J., Capiola, A.: The measurement of the propensity to trust automation. In: Virtual, Augmented and Mixed Reality. Applications and Case Studies: 11th International Conference,

VAMR 2019, pp. 476–489. Springer (2019). https://doi.org/10.1007/978-3-030-21565-1_32
22. Klamer, T.: Acceptance and use of a social robot by elderly users in a domestic environment. In: 2010 4th International Conference on Pervasive Computing Technologies for Healthcare, Pervasive Health 2010, pp. 1 – 8 (2010). https://doi.org/10.4108/ICST.PERVASIVEHEALTH2010.8892
23. Knowles, K.A., Olatunji, B.O.: Specificity of trait anxiety in anxiety and depression: meta-analysis of the state-trait anxiety inventory. Clin. Psychol. Rev. **82**, 101928 (2020). https://doi.org/10.1016/j.cpr.2020.101928
24. Lee, J.D., See, K.A.: Trust in automation: designing for appropriate reliance. Hum. Factors **46**(1), 50–80 (2004)
25. Longo, L., Wickens, C.D., Hancock, P.A., Hancock, G.M.: Human mental workload: a survey and a novel inclusive definition. Front. Psychol. **13** (2022). https://doi.org/10.3389/fpsyg.2022.883321
26. Mattei, G., Pedrazzoli, P., Landolfi, G., Daniele, F., Montini, E.: Introducing active learning and serious game in engineering education: experience from lean manufacturing course. In: IFIP International Conference on Advances in Production Management Systems. Springer (2023). https://doi.org/10.1007/978-3-031-43666-6_25
27. Matuszek, C., et al.: Gambit: an autonomous chess-playing robotic system. In: 2011 IEEE international conference on robotics and automation, pp. 4291–4297. IEEE (2011). https://doi.org/10.1109/ICRA.2011.5980528
28. Mayer, M.P., Kuz, S., Schlick, C.M.: Using anthropomorphism to improve the human-machine interaction in industrial environments (part ii). In: Digital Human Modeling and Applications in Health, Safety, Ergonomics, and Risk Management. Human Body Modeling and Ergonomics: 4th International Conference, DHM 2013. Springer (2013). https://doi.org/10.1007/978-3-642-39182-8_11
29. Merritt, S.: Affective processes in human-automation interactions. Hum. Factors **53**, 356–70 (2011). https://doi.org/10.1177/0018720811411912
30. Messeri, C., et al.: On the effects of leader-follower roles in dyadic human-robot synchronization. IEEE Trans. Cogn. Dev. Syst. **15**(2), 434–443 (2023). https://doi.org/10.1109/TCDS.2020.2991864
31. Miller, L., Kraus, J., Babel, F., Baumann, M.: More than a feeling-interrelation of trust layers in human-robot interaction and the role of user dispositions and state anxiety. Front. psychol. **12** (2021). https://doi.org/10.3389/fpsyg.2021.592711
32. Montini, E., Cutrona, V., Dell'Oca, S., Landolfi, G., Bettoni, A., Rocco, P., Carpanzano, E.: A framework for human-aware collaborative robotics systems development. Procedia CIRP **120**, 1083–1088 (2023). https://doi.org/10.1016/j.procir.2023.09.129, 56th CIRP International Conference on Manufacturing Systems 2023
33. Montini, E., Cutrona, V., Dell'Oca, S., Landolfi, G., Bettoni, A., Rocco, P., Carpanzano, E.: An industrial human-robot collaboration case study for workers' well-being. In: Procedia CIRP (2024), 57th CIRP International Conference on Manufacturing Systems (2024)
34. Naneva, S., Sarda Gou, M., Webb, T., Prescott, T.: A systematic review of attitudes, anxiety, acceptance, and trust towards social robots. Int. J. Soc. Rob. **12** (2020). https://doi.org/10.1007/s12369-020-00659-4
35. Nenna, F., Orso, V., Zanardi, D., Gamberini, L.: The virtualization of human-robot interactions: a user-centric workload assessment. Virtual Reality **27**(2), 553–571 (2023). https://doi.org/10.1007/s10055-022-00667-x

36. Nomura, T., Kanda, T., Suzuki, T.: Experimental investigation into influence of negative attitudes toward robots on human-robot interaction. AI Soc. **20**, 138–150 (2006). https://doi.org/10.1007/s00146-005-0012-7
37. Nomura, T., Kanda, T., Suzuki, T.: Do people with social anxiety feel anxious about interacting with a robot? AI and Soc. 381–390 (2020). https://doi.org/10.1007/s00146-019-00889-9
38. Nomura, T., Suzuki, T., Kanda, T., Kato, K.: Measurement of anxiety toward robots. In: Proceedings - IEEE International Workshop on Robot and Human Interactive Communication, pp. 372 – 377 (2006). https://doi.org/10.1109/ROMAN.2006.314462
39. Onnasch, L., Roesler, E.: Anthropomorphizing robots: The effect of framing in human-robot collaboration. In: Proceedings of the Human Factors and Ergonomics Society Annual Meeting. vol. 63, pp. 1311–1315 (2019). https://doi.org/10.1177/1071181319631209
40. Resing, W., V., B., Elliott, J.G.: Children's solving of 'tower of Hanoi' tasks: dynamic testing with the help of a robot. Edu. Psychol. **40**(9), 1136–1163 (2020). https://doi.org/10.1080/01443410.2019.1684450
41. Robinson, J.P., Shaver, P.R., Wrightsman, L.S.: Chapter 1 - criteria for scale selection and evaluation. In: Robinson, J.P., Shaver, P., Wrightsman, L. (eds.) Measures of Personality and Social Psychological Attitudes, pp. 1–16. Academic Press (1991). https://doi.org/10.1016/B978-0-12-590241-0.50005-8
42. Schaefer, K.: The perception and measurement of human-robot trust. Ph.D. thesis, University of Central Florida (2013)
43. Sotirios, P., Fabio, F., Francesco, M.: A methodological framework to assess mental fatigue in assembly lines with a collaborative robot. In: International Conference on Flexible Automation and Intelligent Manufacturing, pp. 297–306. Springer (2022). https://doi.org/10.1007/978-3-031-17629-6_31
44. Spielberg, C., Gorsuch, R., Lushene, R.: The state-trait anxiety inventory (test manual). Palo Alto, California: Consulting Psycholgists Press **53** (1970)
45. Hoff, K.A., Bashir, M.: Trust in automation: integrating empirical evidence on factors that influence Trust. Hum. Factors **57**(3), 407–434 (2015). https://doi.org/10.1177/0018720814547570
46. Thorpe, A., Nesbitt, K., Eidels, A.: A systematic review of empirical measures of workload capacity. ACM Trans. Appl. Percept. (TAP) **17**(3), 1–26 (2020). https://doi.org/10.1145/3422869
47. Wainer, J., Feil-Seifer, D., Shell, D., Mataric, M.: The role of physical embodiment in human-robot interaction. In: IEEE Proceedings of the International Workshop on Robot and Human Interactive Communication, pp. 117 – 122 (2006). https://doi.org/10.1109/ROMAN.2006.314404
48. Walters, M.L., Woods, S., Koay, K.L., Dautenhahn, K.: Practical and methodological challenges in designing and conducting human-robot interaction studies. In: Procs of the AISB 05 Symposium on Robot Companions. AISB (2005)
49. Wright, L.L., Kothiyal, A., Arras, K.O., Bruno, B.: How a social robot's vocalization affects children's speech, learning, and interaction. In: 2022 31st IEEE International Conference on Robot and Human Interactive Communication (RO-MAN), pp. 279–286. IEEE (2022). https://doi.org/10.1109/RO-MAN53752.2022.9900811

Integrating Industry 5.0 Competencies: A Learning Factory Based Framework

Lorenzo Agbomemewa(✉), Fabio Daniele, Michele Foletti, Matteo Confalonieri, and Paolo Pedrazzoli

Department of Innovative Technologies, University of Applied Sciences and Arts of Southern Switzerland, Lugano, Switzerland
lorenzo.agbomemewa@supsi.ch

Abstract. Industries across the world are witnessing profound technological paradigm changes, necessitating a new set of engineering skills and capabilities for the future workforce. This paper investigates the needed engineering competencies required to bridge the skill gap related to Industry 5.0. Our work analyses current educational frameworks on the topic and proposes a new concept based on the exploitation of learning factories. The paper showcases the conception, execution, and evaluation of a summer school program dedicated to such an objective. It discusses the structure, methodologies, and outcomes of this course, demonstrating the effectiveness of the proposed didactic framework. The findings offer valuable insights for educators and industry professionals alike in preparing the future workforce for the challenges and opportunities of Industry 5.0.

Keywords: Industry 5.0 Framework · Industry 4.0 skills · Educational Framework · Skill Gap · Upskilling · Experiential Learning · Learning Factory

1 Introduction

The advent of Industry 5.0 heralds a significant transformation in manufacturing, emphasizing sustainability, resilience, human-centricity [16], personalized production, and enhanced human-machine collaboration. This evolution builds upon the digital foundation laid by Industry 4.0, integrating advanced technologies with a human-centric approach [49]. However, the swift pace of change and the broadening scope of required competencies have revealed a significant gap in current engineering education and professional training.

This gap hinders the adoption of Industry 5.0 principles and limits innovation and efficiency improvements. Addressing this challenge necessitates the development of educational frameworks and training programs tailored to equip both the current and future workforce with the comprehensive skills and knowledge demanded by this new industrial paradigm [18].

This era of digital transformation is defined by a phenomenon of convergence, where various disciplines merge in a transdisciplinary manner, challenging traditional boundaries and necessitating a reevaluation of engineering and educational paradigms [9].

Published by Springer Nature Switzerland AG 2024
M. Thürer et al. (Eds.): APMS 2024, IFIP AICT 729, pp. 416–429, 2024.
https://doi.org/10.1007/978-3-031-65894-5_29

Over the past two decades, this convergence has catalyzed the emergence of new technologies and necessitated a shift in how engineering and sciences are perceived and taught. The erosion of clear distinctions between disciplines such as mechanical, electrical, and computer engineering, alongside the blending of engineering with basic and social sciences, underlines the complexity of defining modern engineering education. As highlighted by leading voices in education, without a significant reform in teaching methodologies, the future ability to address industry needs and global challenges may be compromised [48].

To meet this challenge, educational institutions and industries must embrace innovative educational models like learning factories [14]. These environments have proven effective in imparting both theoretical and practical knowledge within real production settings, offering a hands-on approach to education that aligns with the future trends reshaping the manufacturing industry. Learning factories serve as a critical bridge, enabling the validation of new technologies and the development of a workforce equipped with the necessary skills to navigate the complexities of the modern industry [1].

In light of the transformative shifts characterizing the transition to Industry 5.0, this paper aims to pave the way toward addressing the competency gap observed in engineering education and professional training in a learning factory environment. Recognizing the critical need for an educational framework that not only encompasses the requisite competencies but also embodies effective didactic paths to disseminate this knowledge, two fundamental questions drive the research:

RQ1. What competencies are required by engineers to work effectively within an Industry 5.0 environment, and under which category could they be framed?

RQ2. How can educators integrate Industry 4.0 and 5.0 topics within a didactic framework to successfully transfer them to future engineers?

Addressing the aforementioned research inquiries is crucial in formulating an educational structure aimed at effectively aiding the closing of the existing skill gap while aligning with the advancing requirements of Industry 5.0. This study establishes itself as a significant contributor to the continual dialogue concerning the reform of engineering education by offering practical perspectives and a scalable blueprint. Noteworthy is the strategic methodology that guarantees the adaptability of the framework across various scenarios, striving to meet the evolving necessities of both the workforce and the wider industry.

The paper continues with Sect. 2 examining existing scholarly works related to engineers' needed competencies and educational frameworks for teaching Industry 5.0. Section 3 describes the Didactic Framework proposed in this work while Sect. 4 demonstrates practical implementation. Section 5 summarizes key findings and future research.

2 Literature Review

This literature review was conducted using a systematic approach. Keywords such as 'Industry 4.0', 'Industry 5.0', 'digital transformation', 'engineering education', 'human-centric manufacturing', and 'advanced manufacturing technologies' were utilized to search databases including IEEE Xplore, ScienceDirect, Scopus and Google Scholar.

The inclusion criteria focused on peer-reviewed articles published mainly in the last decade, ensuring the relevance and currency of the sources. Quantitative and qualitative studies were considered to provide a comprehensive understanding of the subject matter.

The advent of Industry 4.0 has effectively broken down the silos that traditionally segmented departments within manufacturing organizations, fostering enhanced connectivity and coordination across personnel, systems, and operations [8]. This transformation is principally facilitated through the automation of manufacturing processes, leading to a comprehensive revolution in the supply chain dynamics [19]. The affordability and accessibility of cutting-edge technologies, such as cloud computing and big data analytics, have enabled their widespread adoption across the manufacturing sector [46]. This era is marked by its unique methodology of integrating technologies like the Internet of Things (IoT), big data, and Artificial Intelligence (AI), which harmonize their capabilities within both the physical and digital spheres, setting Industry 4.0 apart from its predecessors [4].

Transitioning into Industry 5.0, a new paradigm emerges, placing human interaction at the forefront of technological integration [53]. Unlike its predecessor, Industry 5.0 diverges by prioritizing efficient and user-friendly production solutions through the symbiosis of human expertise and intelligent devices [50]. This human-centric strategy paradigm underscores people-oriented, environment-oriented, and resilience-oriented concepts, fundamentally embedding core values of human centrality, resilience, and sustainability into the fabric of industrial processes [41]. It champions the deployment of advanced technologies such as digital twins, collaborative robots, mass customization, and hyper-personalization, marking a distinct shift from the emphasis on the interconnectivity of cyber-physical systems in Industry 4.0 [24]. Instead of focusing solely on the relationship between "man and machine," Industry 5.0 fosters a collaborative partnership between humans and machines, aiming for mutual advancement [28]. This represents a significant paradigm shift towards a more coordinated, human-centred approach, enhancing the principles established by Industry 4.0 to achieve greater sustainability and resilience [5]. This paradigm emphasizes the crucial role of engineers, who are directly engaged with and significantly affected by the introduction and application of these advanced digital technologies. Engineers are required to delve deep into the intricacies of these technologies, overseeing the creation and refinement of digital applications by leading a team [37]. This involves integrating new technologies into existing frameworks, analyzing their benefits, and understanding financial dimensions like ROI to ensure economic viability. Additionally, engineers must address the human aspect by assessing workforce responses, developing skills, and ensuring ethical considerations in technology adoption.

2.1 Engineers' Needed Competencies

In the transition to Industry 5.0, identifying specific competencies for engineers becomes essential. The skill gap in Industry 5.0 refers to the disparity between the skills required by employers and those possessed by employees [38]. This gap, driven by rapid technological advancements, necessitates a workforce proficient in technical and human-centric skills. Various methods, including employer surveys, workforce analytics, and competency assessments, measure this gap and identify areas for focused training [30].

Addressing this gap presents challenges like rapid skill obsolescence, the need for continuous learning, and interdisciplinary knowledge integration. Key stakeholders in bridging the skill gap include educational institutions, industry, government, and employees, each playing a crucial role in developing relevant curricula, providing industry insights, shaping policies, and engaging in lifelong learning. Universities play a critical role in this process by offering dedicated and specialized courses, utilizing effective methods to educate engineers who can leverage the opportunities arising from the integration of technologies into industrial processes [34].

The role of universities extends to ensuring that engineering students acquire these essential competencies through a curriculum that is both innovative and responsive to the demands of Industry 5.0 [22]. This involves not only the provision of knowledge on advanced digital tools but also the application of these technologies in ways that enhance human capabilities and foster innovation [23]. The shift to Industry 5.0 highlights the symbiosis between humans and technology, requiring engineers to master advanced digital tools to enhance human capabilities. [21].

Engineers lead Industry 5.0 by navigating digital technologies, heading development teams, and ensuring smooth integration with current systems. Their role spans technical expertise, evaluating new technologies' benefits and economic impact, and addressing the human side of tech adoption, including workforce responses, skill development, and ethical standards [32]. Universities, therefore, must adapt their engineering programs to instil these multi-dimensional competencies, ensuring that graduates are not only technically proficient but also adept at managing the human dimensions of technological integration [52].

Engineers in Industry 5.0 need managerial, technical, and cultural skills. Managerial skills help navigate complex systems and ensure ethical project execution. Technical skills are essential for integrating advanced digital technologies with human processes. Cultural skills foster an inclusive, collaborative, and flexible work environment, aligning with Industry 5.0's focus on human-centeredness and sustainability. From a managerial point of view, engineers must learn a variety of skills to navigate the complex landscape of digital transformation effectively. These include the ability to conduct economic evaluations [40] to ensure projects are financially viable and cost-effective, coupled with strong project management skills [47] to plan, execute, and finalize projects within strict timelines and budgets. They must also possess innovation management competencies [27] to oversee the entire innovation process from idea generation to product development and market introduction. Additionally, risk management [47] becomes crucial in identifying, assessing, and mitigating risks associated with digital system deployment. Engineers must also be adept at stakeholder management [40], efficiently managing expectations and communications with all project stakeholders, and resource allocation [12], ensuring time, budget, and manpower are utilized effectively to maximize project outcomes. Furthermore, ethical and social responsibility [27] becomes even more critical when deploying advanced digital systems like AI. The implications of using AI extend beyond technical proficiency, requiring engineers to navigate ethical considerations, such as data privacy, algorithmic bias, and the potential impact on employment.

Engineers must understand these implications to ensure that AI and other digital systems are developed and deployed in a way that aligns with societal values and ethical standards [33].

Another skill is proficiency in deploying advanced digital technologies. Advanced skills in data analysis are vital for deciphering intricate datasets and converting them into actionable insights [40]. IoT devices are pivotal in efficient data gathering, guaranteeing a continuous flow of real-time information that bolsters dynamic decision-making [22]. Cyber-physical systems (CPS) epitomize the fusion of physical processes and digital technologies, augmenting efficiency and fostering innovation [10]. Familiarity with digital twin technology enables the simulation and enhancement of processes in a virtual setting [22, 29]. In contrast, expertise in additive manufacturing unveils novel prospects for design and production, facilitating swift prototyping and the formation of intricate structures [51]. Machine learning (ML) programming is indispensable as it facilitates the establishment of systems capable of predictive analytics and autonomous operations [21].

In conjunction with managerial and technical skills, cultural competencies are crucial for effective functioning in Industry 4.0 and 5.0 contexts. Engineers need to possess a heightened cultural intelligence and competencies to successfully navigate diverse and globally interconnected work environments [6]. This involves understanding the significance of cultural competence in the industry, and acknowledging both its benefits and limitations [26]. Engineers also need to be knowledgeable about the various generational aspects of professional competencies, motivational drivers, and behavioural patterns within the workforce, as this knowledge informs effective team dynamics and leadership approaches [36]. Moreover, active cultural engagement is recognized as a key factor in the cultivation of general competencies, underscoring the importance of interaction with wider cultural and societal norms [44]. Additionally, considerations regarding individuals and culture in the industrial metamorphosis underscore the importance for engineers to adjust to evolving organizational cultures, particularly as industries incorporate new technologies and methodologies [42].

2.2 Educational Frameworks for 5.0

The educational frameworks for Industry 4.0 have been extensively examined, with research emphasizing innovative pedagogical approaches that focus on entrepreneurship and the widespread adoption of intelligent technologies in engineering programs [20]. However, as attention turns towards Industry 5.0, characterized by integrating advanced manufacturing methods with human-centred technology, there is a noticeable gap in comprehensive educational strategies tailored to this new paradigm. While there is recognition of the necessity to adjust educational systems to meet the evolving industrial demands, as evidenced in the analysis and comparison of educational needs between Industry 4.0 and 5.0, explicit and actionable frameworks are lacking [9].

Research addressing the upskilling of workers for Industry 5.0 is a step towards workforce readiness. Still, it falls short of providing a holistic educational model for students preparing to enter these upcoming industries [39]. This oversight underscores a crucial need for developing pedagogical strategies that impart technical knowledge and nurture soft skills and critical thinking essential for innovation in Industry 5.0.

Recent studies attempt to bridge this gap by exploring the correlation between Industry 5.0 and future skills, proposing a shift towards tailored, high-quality educational approaches that emphasize creativity and human-centred problem-solving [7]. Conversations surrounding the evolving role of online education suggest potential methods for delivering such content, albeit lacking specificity in the context of Industry 5.0 [43]. The proposed Industry 5.0 framework seeks to establish a more tangible link between higher education and industry demands, advocating for an integrated approach encompassing technological proficiency and developing soft skills [35]. Despite these advancements, educators' perceptions regarding the adoption of Education 5.0 technologies vary, indicating a need for consensus and collaboration in developing effective educational strategies [31]. Moreover, the necessity for a paradigm shift in higher education to align with the imperatives of Industry 4.0 and 5.0 is clear, highlighting the urgency for academic institutions to reassess and update curricula and teaching methods to equip students for the complexities of the future industrial landscape [45].

2.3 Positioning Within Existing Literature

This study extends the current body of knowledge on Industry 4.0 and Industry 5.0 competencies by addressing the integration of these competencies into workforce development. Previous research has focused on the technological advancements and educational frameworks required for Industry 4.0 [49]. However, there is a notable gap in literature addressing the human-centric and sustainable aspects of Industry 5.0 [53]. Our research diverges from previous studies by emphasizing the importance of learning factories as a practical method for implementing Industry 5.0 competencies [1]. This approach provides a handson, experiential learning environment that bridges the gap between academic knowledge and industrial application, thus preparing the future workforce for the challenges of Industry 5.0.

3 Didactic Framework

Engineering vocation is being transformed by new industrial policies and social megatrends. Traditionally, engineers have honed their expertise within distinct sectors, such as mechanics, energy, electronics, materials, and management. Yet, the advent of policies such as Industry 5.0 presents challenges for engineers, including introducing complex new paradigms and technologies and pressing ethical considerations that transcend conventional disciplinary boundaries. Additionally, future employees need skills and competencies covering all hierarchy levels [2]. In this regard, the new educational framework depicted in Fig. 1 has been proposed.

The framework was developed from researchers' previous experiences in teaching Industry 4.0 topics and the need to categorize the activities proposed during the summer school 4. The framework is composed of five different layers. On the external layer of the framework, Industry 4.0 and 5.0, are the two current paradigms revolutionizing the manufacturing sector. Then, the engineering workforce, through their education in university, is the key element to acquire and diffuse the knowledge required to fill the industrial knowledge gap. For Industry 5.0, this gap is related to the human-centric approach and the increased digital and sustainable production. The next layer represents the different learning approaches. The framework argues that learning factories are the ideal tool for transferring knowledge to students. Abele et al. [1] introduced learning factories, which utilize virtual value chains for remote, ICT-enabled training, effectively bridging the theory-practice gap. The framework proposes Learning Factories as the optimal approach to develop essential Industry 5.0 competencies due to their highly interactive nature and comprehensive coverage of real manufacturing environments. In these regards, learning factories offer students a more effective learning approach than traditional classroom settings or E-learning. The next layer consists of Industry 5.0 competencies, identified in the literature review and categorized under managerial, technical, and cultural aspects. These Industry 5.0 skills are merged with Industry 4.0 skills to create five sets of skills covering all hierarchy levels in today's manufacturing environment. In the framework, skills are defined as a set of domain-specific capabilities, whereas competencies are broader and cover a set of skills. Each skill is described in the next paragraphs.

Operational Technologies (OT) Skills. Operation Technology involves the utilization of hardware and software systems for real-time monitoring and control of industrial processes. This includes Control Systems like Programmable Logic Controllers (PLCs), Distributed Control Systems (DCS), and Supervisory Control and Data Acquisition (SCADA). While industry standards drive these systems, engineers must stay updated on emerging control paradigms, architectures, and standards proposed by practitioners to ensure efficient operation.

Information Technology (IT) Skills. Engineers increasingly rely on IT skills to analyze complex processes and integrate diverse systems, responding to emerging trends like the Industrial Internet of Things and advancements in artificial intelligence. The convergence of IT and OT is propelled by the imperative for seamless integration and optimization across various facets of modern industries, driven by the pursuit of enhanced efficiency, robust data integration, streamlined automation, continuous innovation, and fortified cybersecurity measures. The digitalization of industries further underscores the indispensability of IT skills in contemporary industrial landscapes.

Physical Layer-Related Skills. Competence in the physical layer encompasses a variety of devices like sensors, actuators, motors, and connectivity, operating based on fundamental physical laws. Despite technological advancements, the core principles remain unchanged. Industry 4.0 has notably improved the management of these components through innovations like digital twins, aiding in predictive maintenance and assisted commissioning. Students must develop proficiency in these physical elements for the effective implementation of industrial automation devices.

Conceptual Skills: Future engineers must acquire conceptual skills regarding Industry 4.0 and 5.0. A full understanding of these conceptual frameworks equips engineers to innovate and adeptly respond to emerging challenges and opportunities in the digital era. By mastering theoretical knowledge of complex paradigms, engineers can seamlessly transfer their expertise to industry, enabling the integration of advanced technologies, optimization of processes, and the advancement of sustainable and efficient solutions tailored to the evolving demands of tomorrow's industrial sectors.

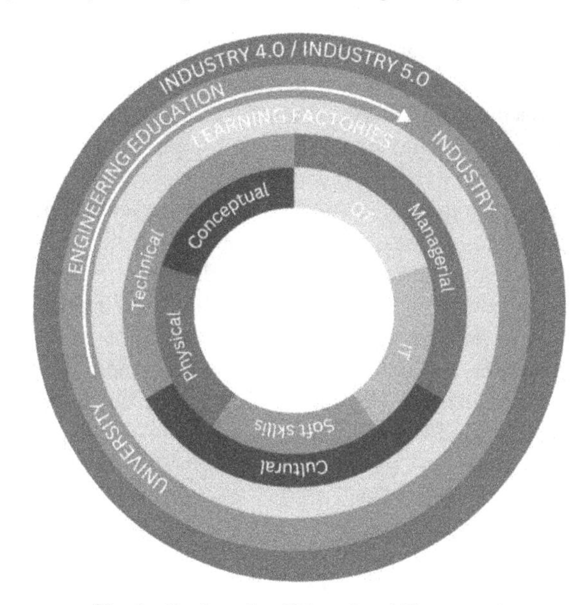

Fig. 1. Engineering Educational Framework

Soft Skills: Industries and employers increasingly require soft skills from graduates, for example, the National Association of Colleges and Employers [11] has consistently highlighted critical thinking and problem-solving abilities, teamwork and collaboration skills, and effective oral and written communication as key competencies for employees. Additionally, well-developed critical thinking skills are essential for addressing the multi-dimensional nature of engineering problems, as developing optimum solutions typically relies on structured and complex thought processes involving evaluation, interpretation, and opinion [3]. Soft skills can be developed in several ways, mostly by exercising and practice. For example, communication oral skills can be increased by presenting topics or critical thinking by exposing two conflicting arguments of the same concept.

In summary, this framework defines the necessary skills for Industry 4.0 and 5.0 across all levels, divided into Managerial, Technical, and Cultural categories. These are further subdivided into OT skills, IT skills, Physical layer-related skills, soft skills, and Conceptual skills. Learning Factories are recommended for their alignment with the manufacturing environment. The framework can be used to develop learning modules for Industry 4.0 and 5.0. The next section discusses its application during the Hands-on Industrial IoT Summer School in July 2023 at the SUPSI Mini-Factory.

4 Framework Application

Hosted at the SUPSI Mini-Factory laboratory by the Institute of System and Technology for Sustainable Production (ISTePS), the Summer School leveraged the Mini-Factory's proven track record in merging educational and research initiatives. As highlighted in

[15], the SUPSI Mini-Factory, part of the University of Applied Sciences and Arts of Southern Switzerland, functions as a live manufacturing environment equipped with operational production machines. It addresses the challenges faced in manufacturing through a hands-on approach, allowing students to apply theoretical knowledge in a real-world setting and researchers to test Industry 4.0 innovations. The Mini-Factory's dual role facilitates both the practical application of industrial engineering principles and the exploration of new solutions across the automation pyramid, promoting innovation transfer and serving as a pilot site for industrial applications, including public-private funded projects [13]. Its educational philosophy, combining traditional automation with the latest technologies, is detailed in [17], showcasing its effectiveness in bridging theoretical learning with practical application, thereby preparing students for the complexities of modern engineering challenges. This innovative educational model made the SUPSI Mini-Factory an ideal venue for the Summer School, which was structured into five modules reflecting the framework's comprehensive approach to engineering education in the era of Industry 4.0 towards Industry 5.0.

4.1 Methodology

To test the effectiveness of the Didactic Framework, a summer school for engineering students was organized in July 2023. Integrating this module into the Master in Engineering program underscores the framework's goal of fostering knowledge transfer to support the engineering education trajectory within an Industry 5.0 context. This week-long event attracted 25 engineering students from various undergraduate engineering disciplines. The structure of the summer school was based on two pillars: the five competence categories of the proposed Didactic Framework and five trending topics chosen by the research team of the SUPSI Mini-Factory.

Regarding data collection, different data were collected to validate the application of the Didactic Framework and to assess the value, relevance, and quality of the SUPSI Hands-on Industrial IoT Summer School.

The 25 participants were asked to complete a tailored evaluation form. The form included an assessment of the learning outcomes regarding the five trending topics. Each topic had some theoretical and practical exercises. Students were assessed with a score between 1 and 6 for each exercise. This result provides an overview of the degree to which competencies and knowledge have been transferred to the students. Additionally, the form included an open-ended feedback question to collect impartial answers regarding the quality of the summer school. To ensure the validity and reliability of the assessment, the exams were reviewed by SUPSI's subject matter experts. The next section presents summer school activities in detail.

4.2 Hands-On Industrial IoT Summer School

Conceptual. The summer school prioritized conceptual competencies, giving engineering students a foundation in Industry 4.0 and emerging Industry 5.0. Introductory sessions covered digital transformation and industry trends, preparing students for future engineering demands. Students visited the MADE Competence Center 4.0 in Milan [25], which offers advanced equipment and technology for research and prototyping. This experience enriched their understanding and provided practical insights into Industry 4.0 technologies, equipping them with a global perspective essential for success in the evolving industrial landscape.

Soft. During the summer school, crucial soft skills such as critical thinking and problem-solving were developed through targeted methodologies. Critical thinking was enhanced in in the Control Standards lecture, described in the next paragraph. Problem-solving skills were cultivated in the Industrial Internet of Things module, where students tackled a practical problem involving data collection and analysis from a conveyor in the Mini-Factory. This hands-on approach solidified technical competencies and instilled a proactive problem-solving mindset, giving students valuable experience in applying theoretical concepts to real-world industrial contexts.

OT. The Control Standards module emphasizes the shift to decentralized event-based control systems. It covers the evolution from programmable logic controllers (PLCs) to distributed control systems under IEC standards IEC 61131-3 and IEC 61499. Practical exercises at the SUPSI Mini-Factory enhance critical thinking and explore advantages and challenges. The module is divided into two sections. The first covers classic PLC programming using Sysmac Studio, focusing on Ladder Logic, Structured Text, and Function Block Diagrams. The second part explores modern PLC programming in 4DIAC, emphasizing extended Function Blocks, Event Execution Control (ECC), and distributed systems principles and communication protocols.

IT. The summer school curriculum focused on IoT and AI, teaching students to build data pipelines from machinery to AI-driven applications. The module emphasized monitoring production processes in the Mini-Factory, integrating with 3D printers and the injection moulding machine BOY E35. Students collected data from this equipment to analyze and identify production anomalies. This hands-on training explored IoT technologies in manufacturing, showing how data flows from industrial equipment to analytical tools. Practical exercises taught participants to connect devices, manage information flow, and apply AI to analyze data, addressing inefficiencies in the production process.

Physical. This module focused on one aspect of the physical layer. In a dedicated lecture, students delved into the intricacies of setting up a brushless motor and fine-tuning its control parameters. This task necessitated a solid foundation in control loop theory and the ability to apply this knowledge to practical exercises. Students were challenged to collaborate in groups of three, leveraging their collective expertise to tackle the task at hand. The exercise involved precisely tuning the motor's parameters to achieve an optimal motion sequence, a task requiring careful analysis and strategic adjustment. Through hands-on experimentation and collaboration, students not only honed their technical abilities but also developed essential teamwork and problem-solving skills crucial for success in the field of industrial control.

Table 1 shows each competency category of the framework is in conjunction with the content of the summer school program. The column score gives the overall average score of the students for each activity. The scores are the results of exercises shared in the evaluation form described in the methodology chapter. Overall, the results demonstrate that after the summer school, students have a good understanding of the five industry 5.0 topics. These results confirm the effectiveness of the didactic framework.

Table 1. Acquired skills per module

	Conceptual	Soft Skills	OT	IT	Physical	Score
Introductory Lecture	•					NA
Control Standards	•	•	•			5.23
IIoT	•	•		•		5.19
Brushless Motor	•				•	5.08
Visit at Made CC	•	•				NA

5 Conclusion

This study aims to facilitate the transition from Industry 4.0 to Industry 5.0 by equipping engineers with key competencies through an educational framework based on learning factories. It structures these competencies across technical, cultural, and managerial dimensions to enhance knowledge transfer from academia to industry. The framework details five essential skill areas for engineers in this evolving environment: understanding the physical layer, mastering OT, excelling in IT, grasping Industry 4.0 and 5.0 concepts, and developing soft skills such as critical thinking and communication. These skills ensure engineers are prepared to lead in technological integration, process improvement, and sustainable development.

Through the implementation of a detailed case study titled "Hands-on Industrial IoT: Mastering the Mini-Factory," this paper has demonstrated the efficacy of the proposed educational framework in meeting the evolving needs of the industry. The research has addressed competency gaps related to the transition to Industry 5.0 and presented effective methods for pertinent industrial education. The results of a summer school involving 25 participants have shown the framework's effectiveness in bridging these competency gaps across physical, IT, OT, and conceptual domains.

This study has made significant contributions to practitioners involved in knowledge transfer in the field of engineering, as well as to education providers who can use the didactic framework to design their programs. However, the study acknowledges certain limitations, such as the absence of direct feedback from the industry. Future research should aim to expand the application of this framework to various educational initiatives and assess its effectiveness in retraining programs for the current workforce.

References

1. Abele, E., et al.: Learning factories for research, education, and training. Procedia CiRp **32**, 1–6 (2015)
2. Adolph, S., Tisch, M., Metternich, J.: Challenges and approaches to competency development for future production. J. Int. Sci. Publ. Educ. Altern. **12**, 1001–1010 (2014)
3. Ahern, A., Dominguez, C., McNally, C., O'Sullivan, J.J., Pedrosa, D.: A literature review of critical thinking in engineering education. Stud. High. Educ. **44**(5), 816–828 (2019)
4. Alcácer, V., Cruz-Machado, V.: Scanning the industry 4.0: a literature review on technologies for manufacturing systems. Eng. Sci. Technol. Int. J. **22**(3), 899–919 (2019)

5. Alves, J., Lima, T.M., Gaspar, P.D.: Is industry 5.0 a human-centred approach? A systematic review. Processes **11**(1), 193 (2023)
6. Ang, S., Rockstuhl, T., Tan, M.L.: Cultural intelligence and competencies. In: International Encyclopedia of the Social & Behavioral Sciences, pp. 433–439. Elsevier (2015). https://doi.org/10.1016/B978-0-08-097086-8.25050-2
7. Bakkar, M., Kaul, A.: Education 5.0 serving future skills for industry 5.0 era, pp. 130–147 (2023). https://doi.org/10.4018/978-1-7998-8805-5.ch007
8. Bartodziej, C.J.: The Concept Industry 4.0. Springer, Wiesbaden (2017). https://doi.org/10.1007/978-3-658-16502-4
9. Broo, D.G., Kaynak, O., Sait, S.M.: Rethinking engineering education at the age of industry 5.0. J. Ind. Inf. Integr. **25**, 100311 (2022)
10. Caratozzolo, P., Smith, C.J., Munoz-Escalona, P., Menbrillo-Hernandez, J.: Exploring engineering skills development through a comparison of institutional practices in Mexico and Scotland. In: 50th Annual Conference of the European Society for Engineering Education, pp. 1878–1883. European Society for Engineering Education (SEFI) (2022)
11. Carnevale, A.P., Fasules, M.L., Peltier Campbell, K.: Workplace basic: the competencies employers want. Technical report. Georgetown University Center on Education and the Workforce (2020)
12. Cerezo-Narváez, A., Otero-Mateo, M., Pastor-Fernández, A.: Project management competences for next engineers in the industry 4.0 era. a case study. In: Kumar, V., Rezaei, J., Akberdina, V., Kuzmin, E. (eds.) Digital Transformation in Industry. NISO, vol. 44, pp. 203–216. Springer, Cham (2021). https://doi.org/10.1007/978-3-030-73261-5_19
13. Daniele, F., et al.: Collaborative robotics adoption: KITT4SME report 2022 in collaboration with trinity robotics. Technical report, University of Applied Science and Arts of Southern Switzerland (2022). https://doi.org/10.13140/RG.2.2.24439.50081
14. Daniele, F., et al.: The innovative educational approach in the SUPSI mini-factory. In: INTCESS 2022 - 9th International Conference on Education & Education of Social Sciences (2019)
15. Daniele, F., Confalonieri, M., Pedrazzoli, P., Graf, A., Ferrario, A., Foletti, M.: Implementation of a learning factory for research, education and training: the SUPSI mini-factory. In: Proceedings of the Conference on Learning Factories (CLF) 2021 (2021). https://doi.org/10.2139/ssrn.3858404
16. Esmaeilian, B., Behdad, S., Wang, B.: The evolution and future of manufacturing: a review. J. Manuf. Syst. **39**, 79–100 (2016)
17. Ferrario, A., Confalonieri, M., Barni, A., Izzo, G., Landolfi, G., Pedrazzoli, P.: A multipurpose small-scale smart factory for educational and research activities. Procedia Manuf. **38**, 663–670 (2019). https://doi.org/10.1016/j.promfg.2020.01.085
18. González-Pérez, L.I., Ramírez-Montoya, M.S.: Components of education 4.0 in 21st century skills frameworks: systematic review. Sustainability **14**(3), 1493 (2022)
19. Hofmann, E., Sternberg, H., Chen, H., Pflaum, A., Prockl, G.: Supply chain management and industry 4.0: conducting research in the digital age. Int. J. Phys. Distrib. Logist. Manag. **49**(10), 945–955 (2019)
20. on Industrie 4.0, S.G.C.W.G., 4.0, I.M.A.E.G.T.: Employee qualification as key success factor in digitalised factories. Technical report, Deutsche Gesellschaft für Internationale Zusammenarbeit (GIZ) GmbH (2024)
21. Jiménez López, E., et al.: Technical considerations for the conformation of specific competences in mechatronic engineers in the context of industry 4.0 and 5.0. Processes **10**(8), 1445 (2022)
22. Kaynak, O.: Engineering education at the age of industry 5.0. In: 2023 IEEE 21st World Symposium on Applied Machine Intelligence and Informatics (SAMI), pp. 000015–000016. IEEE (2023)

23. Lantada, A.D.: Engineering education 5.0: strategies for a successful transformative project-based learning. In: Insights Into Global Engineering Education After the Birth of Industry 5.0. IntechOpen (2022)
24. Maddikunta, P.K.R., et al.: Industry 5.0: a survey on enabling technologies and potential applications. J. Ind. Inf. Integr. **26** (2022)
25. MADE Competence Center 4.0. https://www.made-cc.eu/it/. Accessed 27 Mar 2024
26. Mal'shina, N.A.: Competence approach to culture industry: advantages and disadvantages (2017). https://doi.org/10.18384/2310-6646-2017-3-44-48
27. Suarez-Fernandez de Miranda, S., Aguayo-González, F., Ávila-Gutiérrez, M.J., Córdoba-Roldán, A.: Neuro-competence approach for sustainable engineering. Sustainability **13**(8), 4389 (2021)
28. Montini, E., et al.: A smart work cell to reduce adoption barriers of collaborative robotics. In: Alfnes, E., Romsdal, A., Strandhagen, J.O., von Cieminski, G., Romero, D. (eds.) APMS 2023. IFIPAICT, vol. 689, pp. 702–715. Springer, Cham (2023). https://doi.org/10.1007/978-3-031-43662-8_50
29. Montini, E., et al.: The human-digital twin in the manufacturing industry: current perspectives and a glimpse of future. Trusted artificial intelligence in manufacturing: a review of the emerging wave of ethical and human centric AI technologies for smart production, pp. 132–147 (2021)
30. Morrar, R., Arman, H., Mousa, S.: The fourth industrial revolution (industry 4.0): a social innovation perspective. Technol. Innov. Manag. Rev. **7**(11), 12–20 (2017)
31. Muzira, D.R., Bondai, B.: Perception of educators towards the adoption of education 5.0 technologies. Eur. J. Educ. Soc. Sci. **1**(2), 43–53 (2020). https://doi.org/10.46606/EAJESS2020V01I02.0020
32. Olsson, A.K., Eriksson, K.M., Carlsson, L.: Management toward industry 5.0: a co-workership approach on digital transformation for future innovative manufacturing. Eur. J. Innov. Manag. (2024)
33. Orchard, A., Radke, D.: An analysis of engineering students' responses to an AI ethics scenario. In: Proceedings of the AAAI Conference on Artificial Intelligence, vol. 37, pp. 15834–15842 (2023)
34. Pacher, C., Woschank, M., Zunk, B.M.: The role of competence profiles in industry 5.0-related vocational education and training: exemplary development of a competence profile for industrial logistics engineering education. Appl. Sci. **13**(5), 3280 (2023)
35. Patil, K.K., Szymański, J., Zurek-Napierała, M., Trivedi, B.: I5 framework: institutions-industries-interaction-innovation-integration for education 5.0. Int. J. Emerg. Technol. Learn. **18**(8), 148–163 (2023). https://doi.org/10.3991/ijet.v18i08.36647
36. Pistrui, D., Kleinke, D.K., Das, S., Bonnstetter, R.J., Gehrig, E.T.: The industry 4.0 talent pipeline: a generational overview of the professional competencies, motivational factors, and behavioral styles of the workforce (2020)
37. Rajumesh, S.: Promoting sustainable and human-centric industry 5.0: a thematic analysis of emerging research topics and opportunities. J. Bus. Soc. Econ. Dev. **4**(2), 111–126 (2024). https://doi.org/10.1108/JBSED-10-2022-0116
38. Rikala, P., Braun, G., Järvinen, M., Stahre, J., Hämäläinen, R.: Understanding and measuring skill gaps in industry 4.0 — a review. Technol. Forecast. Soc. Change **201**, 123206 (2024). https://doi.org/10.1016/j.techfore.2024.123206
39. Romero, D., Stahre, J.: Towards the resilient operator 5.0: the future of work in smart resilient manufacturing systems. Procedia CIRP **104**, 1089–1094 (2021). https://doi.org/10.1016/j.procir.2021.11.183
40. Saniuk, S., Grabowska, S.: Development of knowledge and skills of engineers and managers in the era of industry 5.0 in the light of expert research. Zeszyty Naukowe. Organizacja i Zarządzanie/Politechnika Śląska (2022)

41. Sarıoğlu, C.İ.: Industry 5.0, digital society, and consumer 5.0. In: Handbook of Research on Perspectives on Society and Technology Addiction, pp. 11–33. IGI Global (2023)
42. Sihombing, M., Riyanto, S.: Aspects of people and culture in industrial transformation 4.0 in banking industry. SIMAK (2022). https://doi.org/10.35129/simak.v20i02.366
43. Talanov, M.: Online Education in Industry 5.0, pp. 22–35. IGI Global (2022). https://doi.org/10.4018/978-1-7998-8077-6.ch002
44. Temmink, L.: Cultural participation and the development of generic competencies (2016)
45. Thiyagarajan, R., Harish, V.: The Paradigm Shift in Higher Education and Impact of Distance Learning in Era of Industry 4.0 and Society 5.0, pp. 135–154. Chapman and Hall/CRC (2023)
46. Wang, G., Gunasekaran, A., Ngai, E.W., Papadopoulos, T.: Big data analytics in logistics and supply chain management: certain investigations for research and applications. Int. J. Prod. Econ. **176**, 98–110 (2016)
47. Werner-Lewandowska, K., Radecki, A., Więcek-Janka, E.: Body of management competencies for engineer 4.0 (BoMC4E4. 0): a model proposal. Eur. Res. Stud. J. (2022)
48. World Economic Forum: Top quotes from Davos on the future of education (2018). https://www.weforum.org/agenda/2018/01/top-quotes-from-davos-on-the-future-of-education/. Accessed 21 Mar 2024
49. Xu, L.D., Xu, E.L., Li, L.: Industry 4.0: state of the art and future trends. Int. J. Prod. Res. **56**(8), 2941–2962 (2018)
50. Yadav, M., Vardhan, A., Chauhan, A.S., Saini, S.: A study on creation of industry 5.0: new innovations using big data through artificial intelligence, Internet of Things and next-origination technology policy. In: 2023 IEEE International Students' Conference on Electrical, Electronics and Computer Science (SCEECS), pp. 1–12 (2023)
51. Yanuardi, R., Widiaty, I., Komaro, M.: Vocational high school technical skills facing the industrial era 4.0. In: 4th International Conference on Innovation in Engineering and Vocational Education (ICIEVE 2021), pp. 25–29. Atlantis Press (2022)
52. Zhanna, M., Nataliia, V.: Development of engineering students competencies based on cognitive technologies in conditions of industry 4.0. Int. J. Cognit. Res. Sci. Eng. Educ. **8**(S), 93–101 (2020)
53. Zizic, M.C., Mladineo, M., Gjeldum, N., Celent, L.: From industry 4.0 towards industry 5.0: a review and analysis of paradigm shift for the people, organization and technology. Energies **15**(14), 5221 (2022)

Strategies for Managing the Ageing Workforce in Manufacturing: A Survey-Based Analysis

Andrea Rubini, Claudia Piffari, Alexandra Lagorio, and Chiara Cimini(✉)

Department of Management, Information and Production Engineering, University of Bergamo, Viale Marconi 5, Dalmine, BG, Italy
a.rubini1@studenti.unibg.it, {claudia.piffari,alexandra.lagorio, chiara.cimini}@unibg.it

Abstract. In recent years, the phenomenon of an ageing workforce has emerged as a critical area of concern, drawing attention from various stakeholders, including policymakers, researchers, and industrial practitioners. The rapid advancements in digitalisation within the manufacturing sector have compounded this issue, highlighting the need for a comprehensive understanding of its contextual underpinnings and implications. As a result, the responsibility of addressing these shifts in the human resource markets and, more importantly, their implication for working conditions falls on industrial engineering and production research communities as well as national and corporate policymakers. Addressing these problems is particularly important in manufacturing or assembly lines where workers must execute manual and cognitive tasks requiring both their full physical and mental faculties. A research approach based on a targeted questionnaire surveys was conducted within an Italian province. By engaging directly with industry stakeholders and workforce representatives, the research sought to extract insights regarding the awareness of the ageing workforce phenomenon and the practices employed to address its challenges. This study reveals that half of the surveyed manufacturing companies lack targeted strategies due to resource limitations, resistance to change, and competing priorities. However, companies implementing flexible working practices, training programs, job rotation, and enlargement report positive outcomes and emphasize the importance of preserving older workers' expertise.

Keywords: Workforce ageing · Manufacturing · Survey

1 Introduction

The global population is marked by a shift in the distribution towards older ages. This pattern began on a global scale around the middle of the twentieth century and is expected to intensify in the decades ahead. Between 2021 and 2050, the global share of the older population, defined as people aged 65 years or over, is projected to increase from less than 10 per cent to around 17 per cent. The number of older people is expected to more than double from 761 million to 1.6 billion during the same period [1]. Three main causes drive this demographic transition. First, fertility declines cause a decrease

Published by Springer Nature Switzerland AG 2024
M. Thürer et al. (Eds.): APMS 2024, IFIP AICT 729, pp. 430–443, 2024.
https://doi.org/10.1007/978-3-031-65894-5_30

in the proportion of younger individuals and an increase in the share of older people. This phenomenon is defined as ageing "from the bottom". The second cause is ageing "from the top"; as individuals live longer, more people are becoming older. Third, the ageing population will increase due to the "cohort effect," which occurs when the group of individuals approaching old age is more significant than that of their ancestors [2]. Many people who reach their 60s and beyond do so because of earlier drops in mortality at younger ages, but other causes include historically high or variable levels of fertility, migrant inflows, or a combination of variables [3]. The result is a highly asymmetrical workforce structure: while the 20–54-year-old cohort is progressively shrinking, the 55–69-year-old age group is becoming increasingly populous. Therefore, the increase in workers over 55 does not correspond to an equal entry of young workers into the job market. Consequently, people over 55 have become more relevant to the job market; they are no longer outgoing people but are an active part of the market, namely older workers.

The phenomenon is unfolding in the manufacturing sector. For example, by analysing the overall employment data of the European manufacturing sector, breaking them down into three age groups (15–39 years, 40–64 years and over 64), it is possible to observe a tendency towards a progressive increase in the weight of 40–64-year age group, which went from 50.4% in 2008 to 59% in 2022 [4]. Additionally, another trend affecting the manufacturing sector is its digitalisation. The digitalisation process in the manufacturing sector represents a fundamental transformation that redefines the traditional concept of industrial production. This evolution concerns adopting new technologies and implies a cultural and organisational change involving the entire production chain. From implementing advanced automation systems to connecting machinery and processes through the Internet of Things (IoT), the digitalisation of the manufacturing sector creates opportunities for optimisation, innovation, and customisation. Real-time data collection and analysis become essential pillars, allowing a more in-depth and predictive vision of production activities. In this context, adopting emerging technologies such as artificial intelligence and augmented reality further amplifies the potential for improvement. The goal is to create an interconnected digital ecosystem where manufacturing businesses can survive and thrive in the era of Industry 4.0. These advancements lead to three significant changes in the workplace: an increase in the cognitive complexity of human work, a reshaping of the notion of "work" around tasks rather than jobs, and a rise in the significance of technology-related and cross-functional skills. These changes will impact the ageing workforce.

The objective of this research is to analyse the phenomenon of workforce ageing in the manufacturing sector, highlighting the level of awareness of this phenomenon among manufacturing companies and consequently analysing how this portion of the workforce is managed in an evolving work context, specifically in relation to the advent of new technological innovations.

The paper is structured as follows. Section 2 reports the theoretical background on the ageing manufacturing workforce. The methodology adopted, based on a survey administered to 50 manufacturing companies in northern Italy, is explained in Sect. 3. Section 4 analyses the results of the survey. Discussion and future research directions are reported in Sect. 5. Finally, limitations and conclusions are addressed in the last section.

2 Background

In the literature regarding the ageing manufacturing workforce, it is possible to identify the recurrence of three macro-themes analysed in the following sections. The literature search was conducted via the Scopus platform, and the keywords used were "aging/ageing/elderly" combined with "workforce/worker" and "manufacturing/manufacture/industry" or "HR/Human-resources". The macro-themes were inductively extracted from the paper corpus with the aim of identifying first the relevant effects of aging on workers in order to provide a better understanding about how to monitor and further mitigate them. To this purpose, Sect. 2.1 provides an overview of the psycho-physical changes of elderly workers and their performance. Section 2.2 proposes monitoring systems regarding the critical aspects of elderly workers. Section 2.3 analyses the strategies for workforce ageing management. These themes are the same that guided the development of the questionnaire proposed to companies.

2.1 Physiological Factors and Performance

The concept of human factors refers to the interaction between people and the systems, products or environments they interact with. They encompass various aspects of human behaviour, capabilities, limitations, and preferences, aiming to optimise the design of systems and environments to enhance human performance, safety, and well-being [5]. One of the critical human factors that significantly affect human-centred assembly performance is the effect of ageing [6]. The International Labour Organization (ILO) defines ageing workers as those likely to face challenges in employment and occupation as they age [7]. Ageing can result in a persistent loss of biological components due to internal physiological degeneration. Individual performance deteriorates with age due to the natural reduction of physical and physiological functions such as visual capacity, musculoskeletal force, flexibility, motion capability, memory, concentration, and thermoregulation. Hearing and vision loss are two practically ubiquitous symptoms of the ageing process [8].

While older workers confront unique obstacles on the job due to deteriorating motor and psychomotor skills, the capacity to work successfully is determined by various factors which substantially impact the physiological reaction to manufacturing tasks. Older workers may suffer higher fatigue or stress levels than younger workers, requiring special attention to task design and workplace fatigue management measures [9].

Although alterations in the body and mind are unavoidable, they do not always result in decreased employee efficiency and output. Neural plasticity indicates that older adults can still learn; however, it takes longer and requires more sophisticated material. While acquired abilities, like vocabulary and language use, are retained until later in life, ageing impairs abilities that depend on quick processing, precise reasoning, and spatial awareness. Older workers experience no such evident loss of productivity in terms of work compared to younger colleagues. This result can be explained by the fact that the inevitable decline illustrated previously is mitigated by the know-how and experience acquired over the years of work. This aspect might be critical in modern distributed manufacturing networks, virtual manufacturing, and supply chain integration, for instance, in the importance of collaboration, cooperation and mutual understanding in collaborative engineering [10, 11].

While accident rates decline with age, the severity and frequency of injuries rise due to declines in functional capacity. This fact results in higher injury costs and longer recovery times. The impact of age on injury risk is influenced by job demands and workers' health [19]. As older workers may experience increased risks of injury due to age-related factors such as decreased muscle density, balance, and reaction times, it becomes imperative for companies to implement effective measures to mitigate these risks. Ergonomics, the science of designing workplaces to fit the capabilities and limitations of workers, emerges as a crucial solution to address these challenges. Elderly workers in the manufacturing sector may be susceptible to environmental factors, material handling, and working posture. Unhealthy environmental conditions and improper material handling can negatively impact their well-being and productivity, while poor working postures may lead to musculoskeletal pain or stress [19]. Therefore, work design should carefully consider these factors to provide a supportive work environment for elderly workers. Indeed, the extant research examined the impact of ergonomics on absenteeism and job performance, finding that a high physical workload increases absenteeism and higher error rates in assembly tasks due to discomfort and fatigue [20]. Implementing ergonomic interventions can help reduce absenteeism and improve job performance, especially among older workers. In this regard, a theoretical framework and a digital guidance tool for industrial designers were developed to create ergonomic work environments suitable for older and younger workers. This approach facilitates user-sensitive design, considering physical, cognitive, and emotional ergonomics, thereby improving workplace productivity and well-being [21]. Additionally, a mathematical framework for designing activity plans for elderly workers facing repetitive tasks was developed [22]. This model optimises job rotations to reduce the risk of musculoskeletal disorders caused by repetitive movements. Considering workers' competencies and physical characteristics, this approach ensures better person-job fit and enhances overall productivity while minimising ergonomic risks.

2.2 Monitoring Systems

In recent years, the industrial sector has been confronted with various issues related to workforce ageing. This phenomenon, stemming from changes in the demographic composition of productive organisations and the workforce, may lead to a widespread skills shortage, resulting in the loss of extensive knowledge and expertise. Several measures to manage the ageing workforce across all production sectors were proposed to address this situation. For instance, the Work Individual Performance (WIP) instrument was designed to evaluate individual characteristics and the working environment, offering practical solutions, instructions, and recommendations. Regardless of age, this tool assists employees, employers, health and safety professionals, and human resource (HR) managers in optimising well-being and performance while enhancing occupational safety and health performance [12]. Other scholars proposed an age management framework, involving a highly skilled and interdisciplinary team of professionals. This collaborative and multidisciplinary approach ensures a thorough understanding of the organisation's workforce ageing dynamics [13]. Finally, a conceptual decision support system (DSS) was proposed to aid decision-makers in improving operational performance and occupational safety and health for older workers [14]. This DSS considers the interdependencies between operational performance, occupational safety and health, and intervention key performance indicators (KPIs).

In summary, these studies offer valuable resources for businesses seeking to address workforce ageing proactively and comprehensively. Companies can create a dynamic and diverse workplace that promotes workers' performance and well-being throughout their careers by employing a cooperative strategy based on in-depth analysis and targeted actions.

2.3 Ageing Workforce Management

Another relevant topic in the literature concerns HR strategies to increase engagement, performance, resource optimisation and knowledge transfer in older workers. Flexibility is one of these strategies. It is defined as the employees' opportunities to choose when and how they work [15]. When employees have access to flexibility in their work, they obtain a higher correspondence between the job demands and their private lives. In this way, employees can decide how to allocate time, energy, and attention to their work, which enables them more control and autonomy, leading to higher work engagement. Finally, when people become older, they may also have obligations in other domains, such as eldercare, through which their preference for adjusted work schedules increases [16].

For older workers, research emphasises the importance of job reassignment in addition to flexibility. It refers to the practice of redistributing or reallocating the work tasks and responsibilities of older employees to accommodate their evolving needs and capabilities [17]. This practice includes a variety of strategies and interventions to ensure that older workers maintain a meaningful and productive role in the work environment (i.e., task repositioning, job enlargement, training and upskilling, ergonomic adjustments). Research has demonstrated that workplace wellness programs and promoting healthy lifestyles, including exercise and dietary habits, can help slow down physical function deterioration [18]. By doing this, the ageing workforce's functional capacity can be increased, making it easier for them to undertake physically demanding tasks and improve their work abilities.

Finally, the intersection of increased workplace automation and an ageing workforce raises important questions regarding the implications for older workers. While some argue that automation could be a solution for replacing retiring workers without displacing older employees, concerns persist regarding its impact on labour returns and opportunities for older individuals. Despite the potential benefits of automation in increasing economic productivity and creating new job types, several factors may limit its effectiveness for older workers. Three main challenges are highlighted in this regard. Firstly, a disparity exists between pay and productivity, with automation often leading to a shift in rewards from labour to capital. Secondly, there are skill deficits among elderly workers, particularly in fields like information and communication technology, necessitating increased investment in education and retraining. Thirdly, older workers face unique challenges in the job market, including higher rates of long-term unemployment and age discrimination, which automation-driven displacement may exacerbate

[23]. The importance of job redesign in accommodating new technologies like automation and artificial intelligence is underscored by the recognition that experienced workers contribute valuable skills and experiences to enhance service delivery. Technology serves a dual role. It can assist older workers in overcoming age-related physical limitations and enhancing their employability. However, challenges arise when interfaces are not user-friendly for older individuals, particularly in manufacturing sectors. Smaller enterprises struggle with training older workers on digital technologies, leading to disparities in device and application usage compared to younger workers.

Nonetheless, older workers with extensive experience and qualifications are willing to undergo training and can leverage digital technologies to contribute significantly to organisations [24]. Furthermore, the relationship between older workers and new technologies encompasses skill imbalances and obsolescence management. The growing demand for knowledge and skill-intensive labour underscores the importance of investing in education and training to enhance human capital and foster innovation. While younger workers may be perceived as more adaptable to learning, older workers possess valuable soft skills and experience that positively impact productivity. Therefore, a balanced approach recognising the strengths of both age groups is crucial for maximising organisational effectiveness in the digital age.

Lastly, augmented reality technologies improved productivity, safety, and quality of work for senior employees. Augmented reality's customised instructions, real-time alerts, and interactive characteristics help older workers learn more efficiently, avoid mistakes, and ensure workplace safety [25].

3 Methodology

This paper aims to contribute to the evolving discourse on ageing in the manufacturing workforce. This topic is emerging as it is a phenomenon that manufacturing companies are beginning to deal with. Consequently, the impacts of this change are still at their early stage. For this reason, the survey method is most indicated for studying this topic [26]. A strength of the survey methodology is that it is easy to replicate, allowing it to perform longitudinal analysis suitable for studying a developing process [27]. Therefore, exploratory research was limited to the Italian province of Bergamo's manufacturing companies. We opted for a structured survey asking Human Resource (HR) managers to select an answer from 21 multiple-choice questions. The questionnaire was administered by e-mail. In addition to a description of the composition of the labour force and the challenges organisations are experiencing, the questions followed the themes that emerged from the literature review. The questionnaire is available at this link.

When conducting surveys, it is important to consider that they may be subject to biases that could invalidate the results obtained. In particular, such biases are the non-response bias (i.e., when the response rate is too low), sampling bias (i.e., caused by errors in the selection of the sample of respondents), social desirability bias (i.e., respondents avoid too negative judgements about their company), and finally recall bias (caused by the respondents' lack of motivation, memory, ability to respond) [26, 28]. To ensure replicability and consistency in the application of the survey methodology, we have followed the guidelines for the implementation of survey research in operations management [28] reported in Fig. 1.

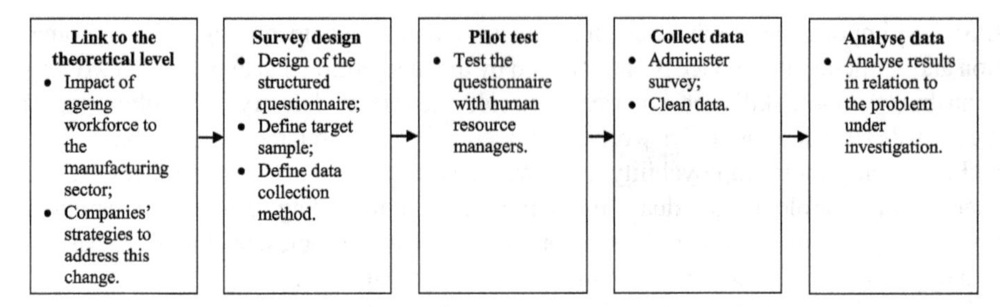

Fig. 1. Survey research flowchart.

To avoid sampling bias, the sample was restricted to companies working in the Italian Bergamo province, and the questionnaire was sent to the human resources managers of each selected company. The questionnaire did not include direct questions about company performance and employer satisfaction to avoid social desirability bias. Additionally, the questionnaire was pilot-tested with human resource managers from two companies in the province to avoid recall and non-response bias. Regarding the pilot testing phase, a semi-structured interview protocol was developed regarding various aspects of the workforce ageing phenomenon emerging from literature analysis (available upon request). The interviews were used to test the questionnaire by identifying areas that needed further investigation and improving unclear questions. The interview feedback was used to develop the final version of the structured questionnaire. Finally, the questionnaire was developed on the Google Forms platform and administered via e-mail to 50 manufacturing companies in the Bergamo area. After one month, the questionnaire was closed, followed by the analysis phase of the results obtained from the questionnaire completion.

4 Results Analysis

In this paragraph, the results of the questionnaires sent to manufacturing companies located in the province of Bergamo are reported and analysed. Out of the 50 companies to which the questionnaire was sent, 20 responded, equivalent to 40% of the total. Of these, 19 accepted the privacy terms and proceeded with the compilation, while one company did not accept the privacy terms. The 40% response rate indicates a widespread interest in providing feedback and sharing information, suggesting an active involvement from the responding companies. This result allows for a preliminary understanding of the dynamics and challenges manufacturing companies face in the local context regarding an ageing workforce.

The first part of the questionnaire aimed to investigate general information regarding the responding companies operating in the manufacturing sector, such as the total number of employees and the percentage of the workforce over 55 years old employed in the company. The results of this descriptive analysis are summarised in Fig. 2. The predominant group of interviewed companies belong to the metalworking sector, accounting for 32% of respondents. Regarding the number of employees of the companies involved in the questionnaire, it can be observed that the majority, namely 79%, have a number of employees ranging from 101 to 500. Finally, regarding the portion of the workforce over 55 years old, the majority of companies (58%) has a percentage ranging between 25% and 50%, followed by 26% of the companies with a percentage lower than 25% and 16% of companies with a percentage ranging between 50% and 75%. No company has a percentage of workers over 55 years old exceeding 75%.

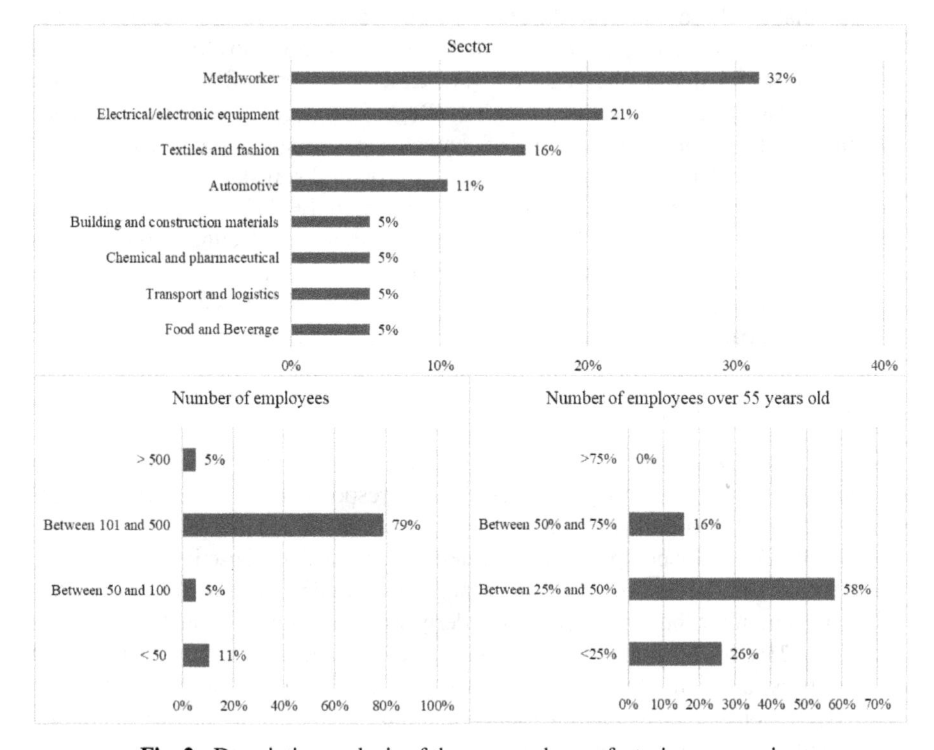

Fig. 2. Descriptive analysis of the surveyed manufacturing companies.

The subsequent part of the questionnaire, comprising 18 multiple-choice questions, aimed to investigate various aspects related to managing the elderly workforce in the manufacturing context. The responses to the questions, which mainly focus on the themes addressed in the theoretical background and interviews, support the development of a knowledge base on the performance of the ageing workforce (Sect. 4.1), how it is monitored (Sect. 4.2) and managed (Sect. 4.3). In order to increase the contribution of the research in guiding companies to face the ageing phenomenon, we explored the challenges related to ageing workforce that are expected to be faced in the long term (Sect. 4.4).

4.1 Physiological Factors and Performance

As expected, regarding the productivity of elderly workers and its possible decrease, the results reflect and align with what emerged from the analysis of scientific literature outlined in the background section. Indeed, 79% of the companies confirmed that "The elderly worker is considered a valuable resource for their experience and knowledge of the sector". Therefore, many companies (37%) offer financial incentives and benefits to encourage elderly workers not to leave. This result may be justified by the fact that the worker in question incorporates and holds indispensable technical knowledge, difficult to replace, or the company has not yet developed adequate strategies to safeguard the know-how that would be lost. Regarding the hiring process, 47% of companies responded that they rarely recruit elderly workers, while the remaining 53%, to attract experienced talents in the sector due to the years of work experience, offer financial incentives, bonuses, and career development opportunities. Regarding selection criteria, experience accumulated in the manufacturing sector of competence is primarily preferred, followed by the ability to work in a team, thus investigating workers' soft skills.

Regarding the issue of psycho-physical decline, only 38% of the companies expressed their opinion, of which 16% stated that although a decline is observed, it is mitigated. 16% did not notice any difference from other workers, which could mean they do not have any measure of this aspect or the work is not physically demanding. Finally, one company affirmed that psycho-physical decline affects production.

4.2 Monitoring Systems

Companies were asked if and how they monitor the performance of their employees, in particular older workers. Among the questionnaire respondents, it emerged that companies mainly prefer to evaluate their employees through ad-hoc assessments and continuous feedback (42%); companies may prefer this approach because it allows them to provide timely and specific feedback to employees regarding their performance. 26% of companies monitor the performance of elderly workers through annual competency assessments. 21% of companies prefer to analyse performance through productivity and quality metrics, and only 11% do not have structured processes for evaluating their employees. Differences in evaluation criteria could be due to various factors, including corporate culture, industry sector, specific job requirements, and managerial preferences. The choice of performance evaluation method often depends on the company's goals, available resources, and adopted personnel management practices.

4.3 Ageing Workforce Management

Regarding the policies and practices implemented by companies, the results show that 47% of companies does not have policies or practices aimed at managing the ageing workforce phenomenon, while the remaining 53% of companies mainly applies reassignment practices to different tasks. In literature, it emerged that work flexibility could be a winning strategy to engage and motivate elderly workers to remain active in the sector and, at the same time, be productive for the company. In this case, two opposing trends are observed, with a slight imbalance towards those who do not practice any strategy aimed at making work more flexible (63% of companies); the remaining 37%, on the other hand, implements practices focused mainly on job sharing or task rotation, and of these, 21% of companies offers elderly workers the opportunity to work part-time or with flexible hours. Job rotation and job enlargement are two other versatile processes for companies, as they allow for practical skill development and facilitate succession to new entrants. Job rotation involves rotating employees through different tasks or roles within the organisation. Job enlargement involves assigning additional or more complex tasks to an elderly employee, thus expanding their role and responsibilities. These approaches can help keep elderly employees engaged and motivated, stimulating them to learn new skills and adapt to new tasks.

Regarding the management of skills development programs for elderly workers, it emerges that in the responding companies, there are no programs specifically targeted at the elderly workforce; indeed, only one company provides this type of service. The remaining companies standardise training programs for all employees without any age distinction. In both processes, mentoring programs are implemented for training and succession of skills, and an experienced worker follows one in the insertion field for a while.

Moving into health and well-being, companies were asked what the main factor influencing elderly workers in the workplace was. Companies identified four main factors: work environment, interpersonal relationships, workload, and stress. However, when asked about the initiatives they proposed to their employees to address these factors, most companies, namely 68%, responded that they do not propose any initiatives. Instead, the remaining 32% of companies implements policies for stress management, offer counselling and psychological support, or provide fitness programs or company sports activities.

Regarding adopting innovative technologies, such as automation, augmented/virtual reality, and monitoring systems, to improve operations and support elderly workers, questionnaire respondents generally showed little inclination to invest part of their budget (63% of companies), while the remaining 37% of companies were more inclined to adopt them. Among these, most are directed towards restructuring the production process regarding automation and implementing health and safety monitoring systems. As for the adoption of augmented reality technologies, none of the respondent companies mentioned its application, while 32% of companies responded that none of the mentioned technologies is implemented in their company.

To conclude this section regarding age management practices, companies were asked if the implemented strategies positively impacted the workforce in question. As expected, 47% of companies that previously stated that they did not apply particular company practices confirmed the impossibility of evaluating these practices as they are not applied. On the other hand, the remaining 53% confirmed the success of the strategies for managing the elderly workforce.

4.4 Ageing Workforce Challenges

Afterwards, an investigation was conducted to identify the main challenges and issues related to adapting tasks and responsibilities for elderly workers. In this case, almost all respondents, except for 16% of companies without challenges, expressed concern about adapting tasks. In particular, 42% of companies perceived the risk of obsolescence of specific skills due to the rapid processes inherent in digitisation and technological changes. Similarly, 47% of companies encountered resistance to change and a need for continuous training among elderly workers when implementing innovative processes to increase the company's competitiveness. Finally, 37% of companies encountered critical issues in redistributing work responsibilities based on the new skills and knowledge that various digitalisation and innovation processes require.

The concluding part of the questionnaire aimed to investigate and seek further reflections regarding the consequences and challenges of workforce ageing. In this regard, the respondent companies have a heterogeneous view of the demographic composition of the future workforce. The different recruitment policies probably cause this because they have not initiated specific paths regarding this phenomenon. 26% of companies do not foresee a demographic change in their workforce, 42% of companies, instead, expect a significant increase in the number of elderly workers, and, on the contrary, 32% of companies anticipate a decrease.

5 Discussion

The ageing workforce poses significant challenges for the manufacturing sector, impacting companies, workers, and society. Using a methodological framework based on a structured survey, this study allows us to gain insights into how manufacturing companies perceive and manage the ageing workforce phenomenon. In particular, the research suggests that several strategic approaches could be used to face it and to mitigate the potential negative effects on the manufacturing performance. Companies can address ageing workforce by adopting solutions in the organisational domain, by reviewing job profiles, task, roles and responsibilities or can choose to design more inclusive workplaces by supporting ageing workforce through technical and technological means. Nevertheless, the best solutions could be envisioned in a mixed organizational-technical approach to the problem.

By comparing the obtained findings with existing literature, it is possible to identify promising areas for future research directions and suggestions for practitioners. Despite widespread acknowledgement of the ageing workforce, approximately half of the surveyed companies lacks targeted strategies to address this issue. Resource limitations, resistance to change, and competing priorities contribute to this scenario. However, companies that implement targeted strategies, such as flexible working practices and training programs, report positive outcomes and highlight the importance of preserving and leveraging the expertise of older workers. Strategies such as job rotation and job enlargement facilitate organisational skill development and knowledge transfer. Moreover, practices like job enlargement align with the need for organisational flexibility and continuous training.

It should be noted that the companies that responded to the questionnaire were mostly large companies. While larger companies tend to implement such strategies more readily, smaller companies may face more significant challenges due to resource constraints. Consequently, supporting even small and medium-sized enterprises in this transition is an interesting area for future research. Furthermore, succession planning and skills development programs are essential to mitigate the effects of the ageing workforce and ensure a smooth transfer of knowledge, and academic researchers should address them in future, delving deeper into potential solutions to address the manufacturing workforce ageing.

Considering solutions more related to the technical and technological domain, another promising future research direction is the evaluation of health and safety monitoring systems for solving ergonomic issues and optimising workplace adaptation, which are still underexplored in the industrial practice.

This research also underscores the importance of understanding the impact of technological innovation, such as Industry 4.0, on older workers. While the literature suggests positive outcomes, the survey results indicate limited adoption of assistive technologies, raising questions about missed opportunities and potential barriers to implementation. This is also an interesting avenue for future research, as exploring the reasons behind the limited adoption of assistive technologies among older workers could provide valuable insights for organizations looking to support their ageing workforce.

Classifying the different approaches used for the management of the ageing workforce could be another interesting avenue to pursue in the future. In this way, one could map the situation of organisations with respect to the ageing workforce and formalise the strategies to be implemented. From a managerial perspective, companies must invest in continuous training, promote inclusivity, and foster an organisational culture that values learning and adaptation also for older workers. By embracing innovation and prioritising human resource development, companies can navigate the complexities of demographic change and thrive in an evolving manufacturing landscape.

6 Conclusion

The present research has extensively explored the phenomenon of workforce ageing in the manufacturing sector. Through a literature review and contextual analysis of the problem, we have highlighted the growing importance of understanding and managing the implications of the ageing workforce. The structured survey provided a preliminary overview of the awareness of the problem and the effective practices used by manufacturing companies to tackle this problem. It has become clear that the challenges associated with the ageing workforce require strategic approaches and targeted policies to maximise the contribution of older workers and ensure the sustainability of business operations in the long term. Furthermore, data analysis has highlighted the importance of adopting job redesign strategies and continuous training programs to foster an inclusive and dynamic work environment. Such approaches can help mitigate disparities and fully harness the potential of diverse skills and experiences in the workplace. However, despite the awareness of the problem, from a practical point of view, practices to manage this transition are still limited.

This paper has some limitations. In particular, the survey was submitted to companies in the Italian province of Bergamo, which may impact the generalisability of the presented results. The extension to manufacturing companies in different regions and countries could improve the generalisability of the results and represent a future development of this research. Additionally, given the nature of the surveys conducted primarily towards human resources personnel, there may be gaps in the comprehensive representation of company policies and practices related to the ageing workforce. Conducting specific interviews or surveys targeted at workers could be helpful to fully understand their needs and assess the effectiveness of current workplace safety policies and procedures. Finally, workforce ageing remains a significant challenge for manufacturing companies. However, by adopting innovative practices and implementing targeted policies, organisations can transform this challenge into an opportunity to promote diversity, improve performance, and ensure long-term sustainability.

References

1. World Social Report 2023: Leaving No One Behind In An Ageing World. DESA Publications. https://desapublications.un.org/publications/world-social-report-2023-leaving-no-one-behind-ageing-world. Accessed 8 Apr 2024
2. Eurostat: Statistics Explained (2024). https://ec.europa.eu/eurostat/statistics-explained/index.php?title=Population_structure_and_ageing&oldid=584064#The_share_of_elderly_people_continues_to_increase
3. Sudharsanan, N., Bloom, D.E.: The demography of aging in low - and middle-income countries: chronological versus functional perspectives. In: Future Directions for the Demography of Aging: Proceedings of a Workshop, National Academies Press, US (2018)
4. Hurley, J., Boskovic, S., Bisello, M., Vacas Soriano, C., Fana, M., Macias, E.: European jobs monitor 2021: gender gaps and the employment structure employment and labour markets (2022). https://doi.org/10.2806/16416
5. Cimini, C., Lagorio, A., Piffari, C., Galimberti, M., Pinto, R.: Human Factors and Change Management: A Case Study in Logistics 4.0. In: Kim, D.Y., Von Cieminski, G., and Romero, D. (eds.) Advances in Production Management Systems. Smart Manufacturing and Logistics Systems: Turning Ideas into Action. pp. 461–468. Springer Nature Switzerland, Cham (2022). https://doi.org/10.1007/978-3-031-16411-8_53
6. Abubakar, M.I., Wang, Q.: Key human factors and their effects on human centered assembly performance. Int. J. Ind. Ergon. **69**, 48–57 (2019). https://doi.org/10.1016/j.ergon.2018.09.009
7. Samorodov, A.: Ageing and labour markets for older workers. Employment and Training Department, International Labour Office, Geneva (1999)
8. Strasser, H.: The art of aging from an ergonomics viewpoint – wisdoms on age. Occup. Ergon. **13**, 1–24 (2018). https://doi.org/10.3233/OER-170250
9. Nardolillo, A.M., Baghdadi, A., Cavuoto, L.A.: Heart rate variability during a simulated assembly task; influence of age and gender. Proc. Hum. Factors Ergon. Soc. Annu. Meet. **61**, 1853–1857 (2017). https://doi.org/10.1177/1541931213601943
10. Binoosh, S.A., Mohan, G., Bijulal, D.: Assessment and prediction of industrial workers' fatigue in an overhead assembly job. S. Afr. J. Ind. Eng. **28**, 164–175 (2017). https://doi.org/10.7166/28-1-1697
11. Xu, X., Qin, J., Zhang, T., Lin, J.H.: The effect of age on the hand movement time during machine paced assembly tasks for female workers. Int. J. Ind. Ergon. **44**, 148–152 (2014). https://doi.org/10.1016/j.ergon.2013.11.010
12. Varianou-Mikellidou, C., Boustras, G., Nicolaidou, O., Dimopoulos, C., Mikellides, N.: Measuring performance within the ageing workforce. Saf. Sci. **140** (2021). https://doi.org/10.1016/j.ssci.2021.105286
13. De Felice, F., Longo, F., Padovano, A., Falcone, D., Baffo, I.: Proposal of a multidimensional risk assessment methodology to assess ageing workforce in a manufacturing industry: a pilot case study. Saf. Sci. **149**, 105681 (2022). https://doi.org/10.1016/j.ssci.2022.105681
14. Peron, M., Arena, S., Micheli, G.J.L., Sgarbossa, F.: A decision support system for designing win–win interventions impacting occupational safety and operational performance in ageing workforce contexts. Saf. Sci. **147** (2022). https://doi.org/10.1016/j.ssci.2021.105598

15. Jeffrey Hill, E., et al.: Defining and conceptualizing workplace flexibility. Commun. Work Fam. **11**, 149–163 (2008). https://doi.org/10.1080/13668800802024678
16. Bal, P.M., De Lange, A.H.: From flexibility human resource management to employee engagement and perceived job performance across the lifespan: a multisample study. J. Occup. Organ. Psychol. **88**, 126–154 (2015). https://doi.org/10.1111/joop.12082
17. Nekola, M., Principi, A., Švarc, M., Nekolová, M., Smeaton, D.: Job change in later life: a process of marginalization? Educ. Gerontol. **44**, 403–415 (2018). https://doi.org/10.1080/03601277.2018.1490162
18. Rožman, M., Treven, S., Mulej, M., Čančer, V.: Creating a healthy working environment for older employees as part of social responsibility. Kybernetes **48**, 1045–1059 (2018). https://doi.org/10.1108/K-12-2017-0483
19. Bures, M., Simon, M.: Adaptation of production systems according to the conditions of ageing population. MM Sci. J. **2015**, 604–609 (2015). https://doi.org/10.17973/MMSJ.2015_06_201513
20. Fritzsche, L., Wegge, J., Schmauder, M., Kliegel, M., Schmidt, K.H.: Good ergonomics and team diversity reduce absenteeism and errors in car manufacturing. Ergonomics **57**, 148–161 (2014). https://doi.org/10.1080/00140139.2013.875597
21. Gonzalez, I., Morer, P.: Ergonomics for the inclusion of older workers in the knowledge workforce and a guidance tool for designers. Appl. Ergon. **53**, 131–142 (2016). https://doi.org/10.1016/j.apergo.2015.09.002
22. Botti, L., Mora, C., Calzavara, M.: Design of job rotation schedules managing the exposure to age-related risk factors. IFAC-PapersOnLine **50**(1), 13993–13997 (2017). https://doi.org/10.1016/j.ifacol.2017.08.2420
23. University of Oxford: Ageing populations and automation: how will these trends shape the future of work? https://www.ageing.ox.ac.uk/blog/ageing-populations-and-automation. Accessed 8 Apr 2024
24. Bianco, A.: Ageing workers and digital future. RTSA **3**(3), 1–22 (2021). https://doi.org/10.32049/RTSA.2021.3.04
25. Moghaddam, M., Wilson, N.C., Modestino, A.S., Jona, K., Marsella, S.C.: Exploring augmented reality for worker assistance versus training. Adv. Eng. Inf. **50**, 101410 (2021). https://doi.org/10.1016/j.aei.2021.101410
26. Malhotra, M.K., Grover, V.: An assessment of survey research in POM: from constructs to theory. J. Oper. Manag. **16**, 407–425 (1998). https://doi.org/10.1016/S0272-6963(98)00021-7
27. Dale, A.: Quality issues with survey research. Int. J. Soc. Res. Methodol. **9**, 143–158 (2006). https://doi.org/10.1080/13645570600595330
28. Forza, C.: Survey research in operations management: a process-based perspective. Int. J. Oper. Prod. Manag. **22**, 152–194 (2002). https://doi.org/10.1108/01443570210414310

Contemporary and Future Manufacturing – Unveiling the Skills Palette for Thriving in Industry 5.0

Marta Pinzone[1](✉), Greta Braun[2], and Johan Stahre[2]

[1] Department of Management Engineering, Politecnico di Milano, Via Lambruschini 4b, 20156 Milan, Italy
marta.pinzone@polimi.it

[2] Chalmers University of Technology, 412 96 Gothenburg, Sweden
{greta.braun,johan.stahre}@chalmers.se

Abstract. When industry and society face major challenges, a strong and competent workforce is increasingly crucial. Throughout past crises and industrial revolutions, people have been at the core of supporting "business as usual", while simultaneously driving change. European industry is repositioning contemporary work towards value-based sustainability, resilience, and human-centricity, radically changing the workforce's task and skill needs. Unfortunately, there is a lack of understanding of which skills employees need to acquire to contribute to achieving a shift to Industry 5.0. This paper addresses this knowledge gap by putting forth the concept of an "Industry 5.0 Skills Palette", offering an overall understanding of the skills needed by the future workforce. The Industry 5.0 Skills Palette was developed through a literature review and expert interviews. It is divided into four main dimensions, i.e. Resilient Manufacturing, Green Industrial Transformation, Digital Human Work, and Technological Systems, and eighteen skills areas. The paper is a unique effort to identify essential skills required for the transition and successful integration of Industry 5.0. The holistic Industry 5.0 Skills Palette goes beyond previous studies, characterized by a limited focus on technological skills i.e. one pillar of Industry 5.0.

Keywords: Industry 5.0 · Skill · Human-centric · Sustainability · Resilience

1 Introduction

In 2021, the European Commission initiated an industrial paradigm shift, from being technology-centered in Industry 4.0 to putting sustainability, resilience, and humans at the core of Industry 5.0 [1]. Considering the new and value-based Industry 5.0 concept, the workforce needs to acquire relevant skills to perform in their new jobs and tasks. Succeeding with the transition is hindered by a lack of available staff with the right skills, pointed out as a major obstacle by 62% of manufacturing firms [2]. When industry transforms, so does the work the people do. Operator 4.0 [3] is transforming to Operator

Published by Springer Nature Switzerland AG 2024
M. Thürer et al. (Eds.): APMS 2024, IFIP AICT 729, pp. 444–456, 2024.
https://doi.org/10.1007/978-3-031-65894-5_31

5.0 [4], meaning that additional capabilities become crucial for their work. While the Operator 4.0 approach focuses on a demand for a human-centric cyber-physical system, Operator 5.0 makes use of the technologies that are enhancing Operator 4.0 too but has capabilities to create resilience and sustainability in industry. The paradigm shift impacts engineers and other job roles as well, as seen in the Swedish upskilling programme Ingenjör4.0 [5, 6].

To emphasize the commitment to learning and supporting people in industry, the European Commission made skills a special priority in 2023, naming it the "European Year of Skills" [7]. Similar policies and initiatives to improve the identification, development and use of skills can be found in many other countries worldwide, such as the U.S., Canada, Japan, South Korea, Singapore, etc. [8].

However, there remains a challenge to fully grasp the new industry we are entering and the new work that will be done by people, i.e. going from an abstract description to concrete implementation. We need to fully understand how work is changing and what skills the humans in Industry 5.0 need to have.

To this end, the ManuSkills project was funded by the EU-programme EIT Manufacturing (EITM) and focused on defining relevant skills and skill areas needed for manufacturing employees in a rapidly changing manufacturing industry.

This paper aims to report on a set of results from ManuSkills, answering the following research question: *What are the most valuable skills for manufacturing employees in Industry 5.0 applications?*

Accordingly, the remainder of the paper is structured as follows: first, relevant concepts and related work on Industry 5.0, Operator 5.0 and skill gaps are outlined in Sect. 2; then, the research methods applied to conduct the study are described in Sect. 3; the proposed Industry 5.0 Skills Palette is described in Sect. 4; finally, the results and their implications are discussed as well as the study limitations and directions for future research.

2 Related Concepts

To understand current trend shifts in industry and accompanying needs for Industry 5.0 skills, a set of conceptual descriptions is required for the reader.

2.1 Industry 4.0 and Industry 5.0

Industry has gone through four major revolutions, characterized by general-purpose technologies and connected energy sources, such as steam power, electricity, computers, and cyber-physical systems. Historically, the introduction of general-purpose technologies is often accompanied by a slowdown in productivity during the first phase of adoption. But, as intangible assets such as human capital are in place, productivity increases [9]. In 2021, the European Commission described a new and value-based vision for future industry, branding it Industry 5.0. This could be seen as a reaction to pressures on society and geopolitics [1, 10] that had not been resolved by the high technology levels of Industry 4.0. In comparison to previous industrial revolutions, the vision for Industry

5.0 is not characterized by certain technologies, but by three core values: sustainability, resilience, and human-centricity. In a way, this marks a pivot in industry – having common values as a base, not a certain technology. Also, the interdisciplinary research needed, and measuring the value of social and environmental aspects are new challenges within Industry 5.0 [1]. [10] discuss the coexistence of Industry 4.0 and Industry 5.0 and highlight that even though Industry 4.0 had a clear technology focus, some publications highlight sustainability or human-centric aspects too. Challenges needing to be addressed in Industry 5.0 are age demographics and the heterogeneity of values in society [1]. In Europe, but also in China and the USA, fewer young people are born, eventually resulting in a shrinking workforce [1].

2.2 Operator 4.0 and Operator 5.0

In the transition from Industry 4.0 to Industry 5.0, the scope of manufacturing work roles is significantly evolving, reflecting shifting paradigms in industrial operations, technology, and workforce dynamics. This evolution can be characterized by a marked progression in digitalization, resilience, sustainability, and human-centricity, each shaping the skills essential for future manufacturing roles [11]. The evolution from manufacturing roles 4.0 to 5.0 reflects a move towards roles that not only manage technology but also drive innovation, resilience, and sustainability in a human-centered manner [3, 4]. In the context of the fourth industrial revolution, the design of manufacturing work roles is mainly focused on managing and interacting with digital technologies, with a strong emphasis on technical skills to develop, operate, and maintain sophisticated cyber-physical systems [12]. The Operator 4.0 taxonomy is seen as a synthesis of various roles-e.g. augmented operator, collaborative operator, and analytical operator - each leveraging digital tools like augmented reality, collaborative robots, and big data analytics, to optimize human performance and decision-making [3]. In the more value-targeting Industry 5.0, digitalization takes a further step by integrating advanced technology with sustainability goals and actions into every level of decision-making, advocating for work roles that champion sustainability principles, renewable energy, and waste minimization [3]. Resilience is dynamic, incorporating advanced predictive analytics and continuous learning systems, enabling manufacturing systems and their human operators to anticipate disruptions, in real time. This shift requires employee roles that are not only reactive but also proactive in managing changes [4]. Transition to Industry 5.0 increasingly highlights social sustainability, prioritizing employee well-being, creativity, and collaboration alongside technological proficiency. Such roles emphasize emotional intelligence, distributed leadership, and collaborative problem-solving skills, aligning closely with organizational goals that promote a learning and innovative work culture and social sustainability.

2.3 Skill Gaps

Historically, industrial revolutions have been accompanied by emerging skill gaps, i.e. the skills of the workforce do not meet the skill requirements posed by employers, motivated by radically shifting business environments [9, 13]. Addressing a skill gap situation is complex, so employers, employees, and policymakers need to act and collaborate [14].

Thus, one of the major contemporary obstacles for manufacturing firms is to identify, hire, and retain employees with the right skills to support the transition [15].

Drastic skill gaps are also rapidly emerging from a global age demographics perspective. In the European Union, as well as also the US, China, and Japan, fewer people are born, resulting in a declining young workforce. To stay competitive, industry needs to prepare its workforce to do their jobs with fewer people [13, 16]. This results in skill shortages, i.e. companies struggle with recruiting the right-skilled people.

To bridge skill gaps, the first step is to understand and measure the extent of the skill gaps, but Rikala et al. find a lack of methods to do so [13]. As proposed by Rikala et al. [13], one way of measuring skill gaps is to create a skills framework, in which the skills for a certain job role or domain are synthesized (e.g. by focus group workshops) and in the next step the skills framework is used to analyze the performance and importance of the people. The result of this performance-importance analysis can be interpreted as the skill gap of these people by highlighting the skills that have high importance for the person, but they perform low on it [13, 17].

Skill frameworks exist for specific areas, e.g. skills needed for circular manufacturing [18], sustainability [19], and Industry 4.0 [20]. In 2017, the European Commission launched the ESCO project [21], to classify skills, competencies, and occupations. An alternative taxonomy is O*Net, offered by the US Department of labour [13].

Challenges involve developing methods defining and measuring skill gaps for companies, but also for individuals. Research has been approaching solutions towards measuring skill gaps [13] and determining the suitable way of upskilling [22].

What is missing in the literature is a model covering the 3 areas of Industry 5.0 in a skill framework, i.e. synthesizing skills needed for creating a resilient, sustainable, and human-centric industry.

3 Method

This study was conducted within the ManuSkills project with the aim of identifying the skills needed for employees entering the Industry 5.0 paradigm. The starting point was the European Commission's policy brief "Industry 5.0 - Towards a Sustainable, Human-centric and Resilient European Industry" [1], presenting the Industry 5.0 concept and outlining new directions for European industry. This visionary document left room for interpretation and concrete actions. In the ManuSkills project, the initial vision was a foundation for creating the Skills Palette for Industry 5.0, achieved through the following steps:

1. **Analysis of the state of the art**: Given the novelty of Industry 5.0, an exploratory literature review approach was used. The process was organized in a flexible way and involved multiple rounds of searching, reading, and analyzing relevant documents. Data was gathered from academic articles retrieved from Scopus and "grey literature" to understand the evolving manufacturing landscape, job roles, and changing skills. Five trend radars were consulted, to identify the most relevant trends influencing the emergence of new competency requirements in manufacturing. Trends with a medium/high impact in the short/medium term were considered as the most relevant to focus on. In addition, twenty-five documents presenting new job roles and/or skill

frameworks were reviewed. A thematic analysis was carried out [23] to identify recurring themes, relevant skills, and patterns.

2. **Formulation of the Industry 5.0 Skills Palette**: Data collected during the analysis of the state of the art were combined, prioritized, and formulated into the Industry 5.0 Skills Palette, comprising a set of four dimensions and eighteen skill areas.
3. **Expert validation**: In the final phase, the Industry 5.0 Skills Palette was presented to and discussed with the ManuSkills project partners to gather feedback and suggestions for improvement. In addition, five interviews were conducted with experts from manufacturing companies, research organizations and training providers to validate the skill areas. Then, the final version of Industry 5.0 Skills Palette was released.

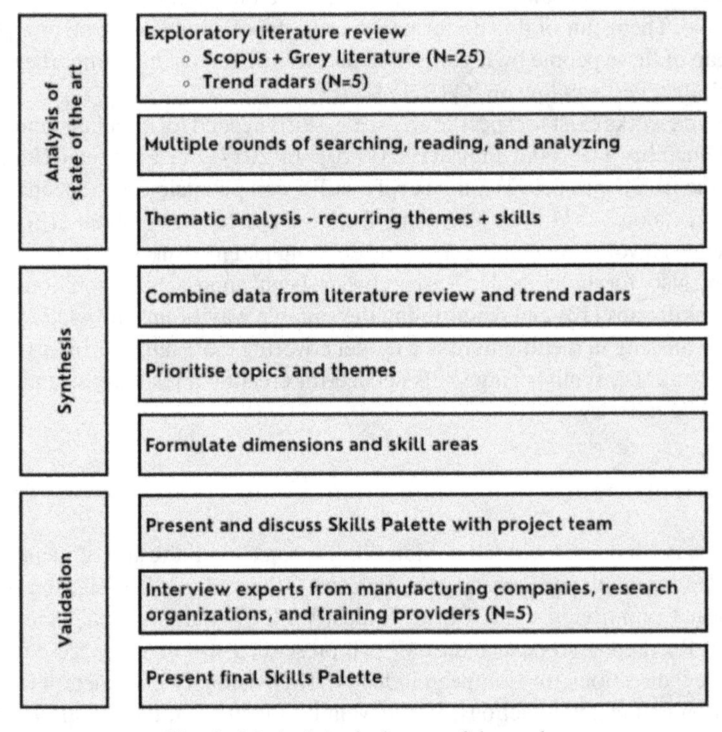

Fig. 1. Methodological steps of the study.

4 Results

In this chapter, the results for the initial question *"What are the skills that are most valuable for manufacturing employees in the Industry 5.0 paradigm?"* are presented.

After the research process explained above, the Industry 5.0 Skill Palette resulted in 18 skill areas, divided into four dimensions as presented in Fig. 1: (1) Digital and Resilient Manufacturing, (2) Green Industrial Transformation, (3) Digital Human Work Style, and [24] Technological Systems.

The blue dimension, "Digital and Resilient Manufacturing", contains four skill areas: 1) "Digital Strategy and Business model innovation"; 2) "Digital transformation of Manufacturing Processes & Systems"; 3) "Industrial Resilience"; and 4) "Entrepreneurship". Skills belonging to these four areas contribute to ensure the resilience pillar of Industry 5.0.

The green dimension of "Green Industrial Transformation" consists of three skill areas: "Circular Manufacturing"; "Energy and Resource efficiency"; and "Eco-innovation". These areas are identified as relevant for manufacturing employees to be knowledgeable about driving the green transformation in industry.

The yellow dimension of "Human Work Style" contains three skill areas: "Human Factor 4.0": "Work in the digital era"; and "Learn in the digital era". These skill areas ensure a human-centric industry, where lifelong learning and well-being of the human is in focus.

Lastly, the red dimension brings in technologies from Industry 4.0 needed to come up with smart solutions needed for Industry 5.0. Within the red area "Technological systems" eight skill areas arise: "Advanced HMI (AR/VR/wearables)", "New Production Technologies (e.g. Additive Manufacturing)", "Advanced automation and robotics", "Digital Twins and Simulation", "Artificial intelligence", "Cybersecurity", "Industrial Big data", and "IT/OT".

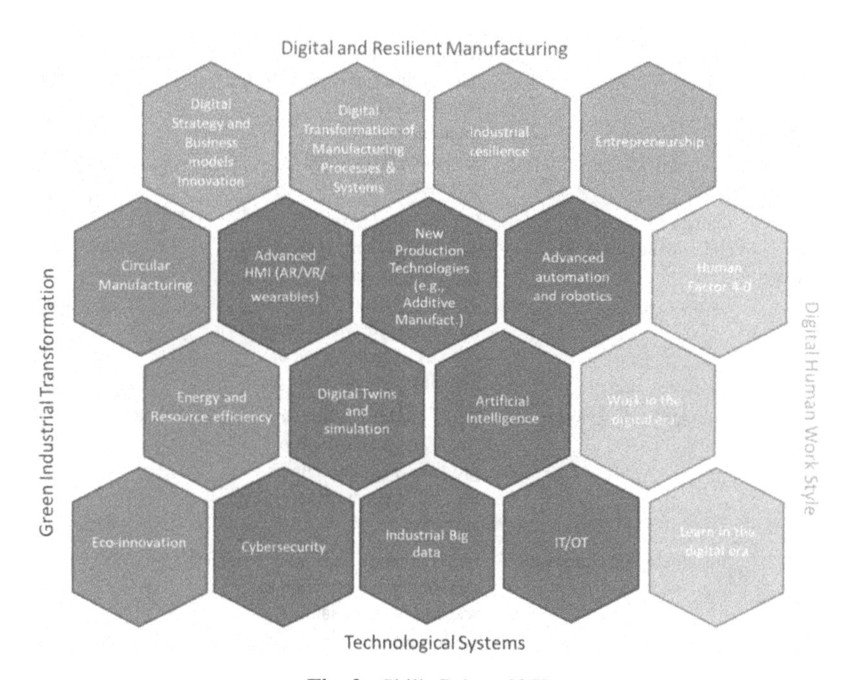

Fig. 2. Skills Palette [25]

Table 1 summarizes descriptions of each skill area and contains the references used to create the Skills Palette. Each skill area contains a list of skills, indicating more concrete what these skill areas mean.

Table 1. Description of skill areas and references

Dimension	Skill area	Description	References
Digital and resilient manufacturing	Digital Strategy and Business models Innovation	Skills related to strategy formulation and implementation, and new business models leveraging digital technologies	[26, 27]
	Digital Transformation of Manufacturing Processes & Systems	Skills linked to product-service design, engineering and lifecycle management, supply chain management, production, quality, maintenance, etc.	[20, 26, 28–31]
	Industrial resilience	Skills needed for the development of a higher degree of robustness in industrial production, by creating resilient strategic value chains, adaptable production capacity, flexible business processes, etc.	[4, 32]
	Entrepreneurship	Skills covering all phases from identification of opportunities to ideas development, to implementation and exploitation for value generation	[26, 30, 33, 34]
Green industrial transformation	Energy, resource efficiency and climate neutrality	Skills related to the definition and management of company's climate targets, through energy efficiency measures, renewable energy sources, compensation of unavoidable emissions, etc.	[35–37]
	Circular Manufacturing	Skills needed to eliminate waste and pollution, keep products and materials in use, and regenerate natural systems	[35, 37, 38]
	Sustainable/Eco-innovation	Skills for open and collaborative innovation to respond to sustainability challenges	[37]

(continued)

Table 1. (*continued*)

Dimension	Skill area	Description	References
Human Digital Work Style	Human Factor 4.0	Skills related to understanding of interactions among humans and other elements of a system, and the application of theory, principles, data and methods to design and optimize human well-being and overall system performance	[26, 30, 33, 39]
	Work in the digital era	Skills useful to conceive and exploit new ways of organizing and working such as remote work, industrial smart working, virtual teamwork, agile organization, etc.,	[26–28, 30, 33, 35, 40, 41]
	Learn in the digital era	Skills useful to conceive and exploit new ways of teaching and learning across the stages of life, especially those ones that are enabled by digital and other new technologies	[26–28, 30, 33, 35, 41, 42]
Industry 5.0-related technological system	Advanced Human Machine Interfaces (HMIs)	Skills related to Human-Machine-Interfaces, such as augmented, virtual, or mixed reality, and wearables	[26, 28, 30, 35]
	New Production Technologies	Skills related to the design and operation of new technologies used in production, e.g. additive manufacturing, bio-based manufacturing, etc.	[28, 30, 35, 43]
	Advanced automation and robotics	Skills related to advanced automation and robotics, which includes autonomous and collaborative robot systems	[28, 30, 35, 44, 45]
	IT/OT - IoT	Skills linked to IT-OT integration, which relates to the integration of a range of components, from IIoT devices, control systems, platforms, communications networks, business applications etc.	[20, 28, 30, 31, 35]

(*continued*)

Table 1. (*continued*)

Dimension	Skill area	Description	References
	Industrial Big Data	Skills for data management, data engineering, data analysis, data visualization and computer science	[20, 26–28, 35, 45–47]
	Digital Twins and simulation	Skills related to digital representations, models, shadows of physical system/process and the use of simulations that can predict how it will perform	[28–30, 35]
	Artificial Intelligence	Skills related to algorithms and systems capable of observing the environment, learning based on the knowledge and experience gained, taking intelligent action or proposing decisions	[28, 35, 45]
	Cybersecurity	Skills related to the body of technologies, processes, and practices to protect networks, devices, programs, and data from attack, damage, or unauthorized access	[26, 28, 30, 35, 48–50]

5 Discussion and Conclusions

5.1 Discussion of the Results and Method Used

This study aimed to map valuable skills for manufacturing employees in Industry 5.0. It represents an original effort to identify the essential skills needed for the transition and successful integration of Industry 5.0. By conceptualizing and developing the Industry 5.0 Skills Palette, the article addresses calls made by researchers [51], industrial stakeholders [52], and policymakers [1, 53] addressing skills for the future of manufacturing.

Moreover, this study extends the literature on Industry 5.0 by offering a holistic model and valuable insights that go beyond previous studies that often present a limited focus. While prior research has primarily concentrated on the technical skills associated with Industry 4.0, the Industry 5.0 Skills Palette broadens the scope to include new dimensions of skill requirements pertinent to the goals of Industry 5.0.

This work advances the state of the art by providing a comprehensive and forward-looking skill map that addresses the evolving needs of the industrial workforce. It highlights 18 key skill areas crucial for Industry 5.0. This granularity helps bridge the vision and abstract concepts of Industry 5.0 with more practical skill applications that can assist companies, workers, and other stakeholders in fostering more human-centric, sustainable, and resilient industrial systems enabled by digital technologies.

The method used for this study helped to answer the research question. Synthesizing and connecting literature and skill taxonomies built an understanding of the skill areas needed for Industry 5.0. There could have been different or additional ways of collecting data for creating the Skills Palette, i.e. collecting data from LinkedIn or other databases with job roles and skills such as ESCO (European Skills, Competences, Qualifications, and Occupations). In addition, it was important to validate the framework with a group of experts. This was done with five people, but the study could have benefitted from receiving feedback from more stakeholders.

5.2 Limitations and Future Research

In referring to Industry 5.0 Skills Palette it is necessary to consider the limitations of the research that has originated it. Researchers interested in the realm of Industry 5.0 are encouraged to address these limitations in future studies.

Future research can focus on establishing novel collaborative efforts between industry stakeholders, academic researchers, and training providers to delve deeper into the specific skills outlined in each area of the Industry 5.0 Skills Palette. This collaboration can involve conducting empirical studies to identify the most critical skills within each domain, as well as defining proficiency levels for each skill.

Researchers can develop more advanced models or frameworks to classify, organize, and prioritize the skills for Industry 5.0. Future models can offer more granularity in skill categorization, considering tasks and processes along an Industry 5.0 system lifecycle (e.g. beginning of life (design, development, etc.), middle of life (operation, management), end of life (change, decommissioning, etc.)), multiple levels of analysis (e.g., product-service, production/logistics system, value network), different units of analysis (individual, team, etc.), potential interdisciplinary connections, synergies or trade-offs within Industry 5.0.

Industry 5.0 Skills Palette gives an overview of the Industry 5.0 skill areas important for manufacturing employees in general. However, specific skills for a person vary depending on their job role and the industry they work in. Therefore, future studies could leverage this model to develop customized skill maps and frameworks to grasp the specificities of the different sectors and/or job roles.

5.3 Implications for Practice

The research holds significant practical implications for a range of stakeholders within the manufacturing industry, including managers, professionals, educators, policymakers, and individuals interested in or already working in manufacturing.

The identified skill areas pertinent to Industry 5.0 provide manufacturers with an initial roadmap to pinpoint the necessary skills within their organizations and formulate effective skills development strategies. These skill areas serve as a clear benchmark for identifying potential skill gaps and facilitating the creation of targeted learning and development initiatives. Furthermore, they can serve as guidelines for refining existing job descriptions or crafting new job profiles. Additionally, they aid in the recruitment process by simplifying the identification of suitable candidates for employment.

Similarly, employees can utilize the study findings to assess their skills and plan their professional development activities accordingly. Policymakers can leverage these insights to identify present and future skill requirements, enabling them to prioritize policies that prevent the emergence of skill gaps and mismatches in the labor market. Furthermore, education and training providers can use the research outcomes to update their curricula or develop new ones that address the demand for Industry 5.0 skills.

Finally, it's noteworthy that the above-mentioned findings have contributed to the establishment of the EITM Upskilling and Reskilling Competency Model [54] which aims to serve as a reference framework for the planning, development, the implementation of learning programs aligned with EIT Label standards, and the organization of learning nuggets in the EITM learning platform, skills.move [55].

5.4 Conclusions

Upskilling the workforce in the transformation towards Industry 5.0 will be key for successful change. The presented Industry 5.0 Skills Palette, including four dimensions and 18 skill areas related to Industry 5.0, is a compass for navigation toward correct learning paths. Figure 2 it serves not only as a guide for employers to upskill their employees but also for policymakers to introduce the right incentives and for education providers to adapt their course offerings and curricula. This set the stage for a joint effort between employers, employees, education providers, and policymakers that is needed to cultivate the relevant Industry 5.0 skills.

Acknowledgements. The authors would like to thank the European Institute of Innovation & Technology (EIT) Manufacturing and all partners of the ManuSkills project as well as all the participants in the different activities organized by the project for their valuable insights. This study has been co-funded by the European Commission and partially funded by the Horizon Europe project DaCapo - GA number: 101091780 – and Erasmus+ project GreenDiLT – GA number: 187 512 512. The study was supported by the HumanTech Project financed by the Italian Ministry of University and Research (MUR) 2023–2027 as part of the ministerial initiative "Departments of Excellence" (L. 232/2016).

References

1. Breque, M., de Nul, L., Petridis, A.: Industry 5.0 - towards a sustainable, human-centric and resilient European industry. Policy Brief European Commission (2021)
2. European Investment Bank: Resilience and renewal in Europe - Investment Report 2022/2023 (2023)
3. Romero, D., Stahre, J., Taisch, M.: The operator 4.0: towards socially sustainable factories of the future. Comput. Ind. Eng. **139** (2020)
4. Romero, D., Stahre, J.: Towards the resilient operator 5.0: the future of work in smart resilient manufacturing systems. Procedia CIRP **104**, 1089–1094 (2021)
5. European, C., European Innovation, C., Agency, S.M.E.: Pact for skills – Analysing of up- and reskilling policy initiatives and identifying best practices – Final report. Publications Office of the European Union (2024)

6. Braun, G., Stahre, J., Rosén, B.-G., Bokinge, M.: Ingenjör4.0 – a national upskilling programme to bridge industry's skill gap. Procedia CIRP **120**, 1286–1291 (2023)
7. https://ec.europa.eu/commission/presscorner/detail/en/speech_22_5493. Accessed 7 June 2024
8. https://www.oecd.org/skills/centre-for-skills/. Accessed 7 June 2024
9. Brynjolfsson, E., Rock, D., Syverson, C.: The productivity J-curve: how intangibles complement general purpose technologies. NBER Working Paper (2020)
10. Xu, X., Lu, Y., Vogel-Heuser, B., Wang, L.: Industry 4.0 and industry 5.0—inception, conception and perception. J. Manuf. Syst. **61**, 530–535 (2021)
11. Ciccarelli, M., Papetti, A., Germani, M.: Exploring how new industrial paradigms affect the workforce: a literature review of operator 4.0. J. Manuf. Syst. **70**, 464–483 (2023)
12. Pinzone, M., Fantini, P., Taisch, M.: Skills for industry 4.0: a structured repository grounded on a generalized enterprise reference architecture and methodology-based framework. Int. J. Comput. Integr. Manuf., 1–20 (2023)
13. Rikala, P., Braun, G., Järvinen, M., Stahre, J., Hämäläinen, R.: Understanding and measuring skill gaps in industry 4.0 — a review. Technol. Forecast. Soc. Change **201** (2024)
14. Braun, G., Rikala, P., Järvinen, M., Hämäläinen, R., Stahre, J.: Bridging skill gaps – a systematic literature review of strategies for industry. Proc. Swedish Production, pp. 687–696
15. World Economic Forum: Future of Jobs Report (2023)
16. European Commission, S.-G.: European Commission Report on the Impact of Demographic Change (2020)
17. Do, H.-D., Tsai, K.-T., Wen, J.-M., Huang, S.K.: Hard skill gap between university education and the robotic industry. J. Comput. Inf. Syst. **63**(1), 24–36 (2023)
18. Pinzone, M., Taisch, M.: Towards a circular manufacturing competency model: analysis of the state of the art and development of a model. In: Alfnes, E., Romsdal, A., Strandhagen, J.O., von Cieminski, G., Romero, D. (eds.) APMS 2023. IFIPAICT, vol. 692, pp. 189–199. Springer, Cham (2023). https://doi.org/10.1007/978-3-031-43688-8_14
19. Francesco Vona, G.M., Consoli, D., Popp, D.: Green skills. NBER (2015)
20. Pinzone, M., Fantini, P., Perini, S., Garavaglia, S., Taisch, M., Miragliotta, G.: Jobs and skills in industry 4.0: an exploratory research. In: Lödding, H., Riedel, R., Thoben, K.D., von Cieminski, G., Kiritsis, D. (eds.) APMS 2017. IFIPAICT, vol. 513, pp. 282–288. Springer, Cham (2017). https://doi.org/10.1007/978-3-319-66923-6_33
21. European Commission: ESCO - European Skills, Competences, and Occupations (2024)
22. Leon, R.D.: Employees' reskilling and upskilling for industry 5.0: selecting the best professional development programmes. Technol. Soc. **75** (2023)
23. Naeem, M., Ozuem, W., Howell, K., Ranfagni, S.: A step-by-step process of thematic analysis to develop a conceptual model in qualitative research. Int. J. Qual. Methods **22** (2023)
24. https://www.ingenjor40.se. Accessed 10 July 2023
25. https://www.eitmanufacturing.eu/news-events/activities/manufacturing-skills-observatory-and-competencies-framework/. Accessed 7 June 2024
26. Margherita, E.G., Braccini, A.M.: Managing the fourth industrial revolution: a competence framework for smart factory. In: Hamdan, A., Hassanien, A.E., Razzaque, A., Alareeni, B. (eds.) The Fourth Industrial Revolution: Implementation of Artificial Intelligence for Growing Business Success. SCI, vol. 935, pp. 389–402. Springer, Cham (2021). https://doi.org/10.1007/978-3-030-62796-6_23
27. https://sfia-online.org/en/sfia-7. Accessed 13 Apr 2024
28. Flores, E., Xu, X., Lu, Y.: Human capital 4.0: a workforce competence typology for industry 4.0. J. Manuf. Technol. Manag. (2020)
29. Baker, A., et al.: Preparing the acquisition workforce: a digital engineering competency framework. In: IEEE International Systems Conference, SysCon 2020, Montreal, QC, Canada (2020)

30. Blayone, T., van Oostveen, R.: Prepared for work in industry 4.0? Modelling the target activity system and five dimensions of worker readiness. Int. J. Comput. Integr. Manuf. (2020)
31. Antonucci, L., Fornasiero, M., Kowalski, R.: Partners in connection: MPG/UI LABS digital manufacturing and design job roles taxonomy and success profiles (2017)
32. Romero, D., Stahre, J., Larsson, L., Rönnbäck, A.Ö.: Building manufacturing resilience through production innovation. In: 2021 IEEE International Conference on Engineering, Technology and Innovation (ICE/ITMC), pp. 1–9 (2021)
33. Forum, W.M.: The 2019 world manufacturing forum report - skills for the future of manufacturing (2019)
34. Bacigalupo, M., et al.: EntreComp: The Entrepreneurship Competence Framework. Publication Office of the European Union, EUR 27939 EN, Luxembourg (2016)
35. Akyazi, T., et al.: Skills requirements for the european machine tool sector emerging from its digitalization. Metals **10**(12), 1665 (2020)
36. International Society of Sustainability Professionals (ISSP): Sustainability Practitioner Body of Knowledge (2013)
37. Cedefop: Skills for green jobs - European synthesis report (2010)
38. Sumter, D., De Koning, J., Bakker, C.A., Balkenende, R.: Key competencies for design in a circular economy: exploring gaps in design knowledge and skills for a circular economy. Sustainability **13** (2021)
39. International Labour Office, and International Ergonomics Association: Principles and guidelines for human factors/ergonomics (HFE) design and management of work systems (2021)
40. https://esco.ec.europa.eu/en. Accessed 16 Apr 2024
41. Sala, A., Punie, Y., Garkov, V., Cabrera Giraldez, M.: LifeComp: the European framework for the personal, social and learning to learn key competence (2020)
42. Redecker, C.: European framework for the digital competence of educators: DigCompEdu (2017)
43. CLLAIM: LOs' guideline for the AM qualifications (2019)
44. https://www.careeronestop.org/CompetencyModel/competency-models/advanced-manufacturing.aspx. Accessed 16 Apr 2024
45. https://www.project-drives.eu/en/aboutus. Accessed 16 Apr 2024
46. IBM Corporation: The data science skills competency model. In: The Data Science Skills Competency Model (2020)
47. DAMA International: DAMA-DMBOK Data Management body of knowledge. Technics Publications (2017)
48. https://niccs.cisa.gov/workforce-development/nice-framework. Accessed 16 Apr 2024
49. Martin, A., Rashid, A., Chivers, H., Danezis, G., Schneider, S., Lupu, E.: The cyber security body of knowledge v1.0, 2019. In: Book The Cyber Security Body of Knowledge v1.0, 2019 (2019)
50. SPARTA: D9.1 Cybersecurity skills framework: Book D9.1 Cybersecurity skills framework (2019)
51. Leng, J., et al.: Industry 5.0: prospect and retrospect. J. Manuf. Syst. **65**, 279–295 (2022)
52. World Economic Forum: Future of Jobs Report. World Economic Forum (2020)
53. European Commission: A new European innovation agenda. In: A New European Innovation Agenda. Publications Office of the European Union (2022)
54. EIT Manufacturing: Upskilling and reskilling quality system and competency model. In: Upskilling and Reskilling Quality System and Competency Model (2023)
55. https://www.skillsmove.eu/. Accessed 5 Mar 2024

Exploring the Cognitive Workload Assessment According to Human-Centric Principles in Industry 5.0

Ahmadreza Nadaffard[1], Ludovica Maria Oliveri[2], Diego D'Urso[2], Francesco Facchini[1], and Claudio Sassanelli[1,3](✉)

[1] Polytechnic University of Bari, Bari, Italy
a.nadaffard@phd.poliba.it, claudio.sassanelli@poliba.it
[2] University of Catania, Catania, Italy
[3] ReTech Center, École des Ponts Business School of École des Ponts ParisTech, Paris, France

Abstract. Industry 4.0 and 5.0 paradigms have been crucial for companies in employing digital technologies as an ally for men to free them from dangerous and routine tasks in favour of higher value tasks, putting humans at the centre of the organization as the decision maker. However, on the one hand, the new industrial systems shift to new tasks requiring more 'cognitive' than 'physical' efforts; on the other hand, the approaches to assess the cognitive workload and ensure the physical well-being of the operators are far to be considered easily applicable. For this reason, this research reveals current research trajectories and explores the cognitive workload using subjective and objective indicators. The discussion highlights cognitive ergonomics and advocates for a harmonious balance between human and machine capabilities. It identifies factors contributing to cognitive overload in manufacturing and maps their interconnections. The analysis of recent research trends reveals a growing adoption of new approaches requiring the adoption of physiological measurements (e.g., electrocardiogram (ECG), electroencephalography (EEG), Electromyography (EMG), etc.). Finally, this investigation offers insights into future research directions, urging a nuanced exploration of industrial activities and addressing cognitive workload across organisational layers in the context of Industry 5.0.

Keywords: Industry 5.0 · human-centric approach · human factors · cognitive workload assessment · smart manufacturing systems

1 Introduction

In 2021, the European Union introduced Industry 5.0, emphasising a shift towards human-centred production and aiming for a more sustainable and flexible European industry [1]. In Industry 5.0, the emphasis on improving collaboration between humans and advanced technologies is aligned with the key principles of Industry 4.0. This focus is evident in areas such as intelligent systems, augmented reality, system integration,

Published by Springer Nature Switzerland AG 2024
M. Thürer et al. (Eds.): APMS 2024, IFIP AICT 729, pp. 457–469, 2024.
https://doi.org/10.1007/978-3-031-65894-5_32

and human-robot interaction. The industrial landscape has evolved through five revolutions, recognising workers as an "investment" in development rather than just a "cost." This perspective underscores the utilisation of technology to serve industrial progress and community needs, moving beyond mere production [2]. The contemporary market's demand for personalised products [3, 4], each with a unique value, accentuates the necessity for human creativity in product development [5]. Therefore, recognising and addressing human states (categorised as physical, cognitive, and psychological) is pivotal.

Consistent with the new industrial era, the role of the operator has changed; the so-called 'Operator 4.0' is defined as a qualified operator who performs the work with the support of machines and who interacts with collaborative robots, advanced systems and sensors which use augmented and virtual reality and exploit the benefits of the enabling technologies to understand production through context-sensitive information [6]. In the Human-Cyber-Physical systems and adaptive automation concept, Operator 4.0 excels in utilising an automation-aided system toward a human-automation symbiosis system for a socially sustainable workforce in the human-centric smart factory [7]. According to Leng et al. (2022), the Fifth Industrial Revolution reintroduced human technicians to factory floors to increase process efficiency. The technology's rapid evolution empowers industrial workers to transition from mundane to engaging roles, integrating advanced technology with distinct human capabilities [8]. This integration will combine the creativity and brainpower of humans and machines.

It is possible to state that, on the one hand, the primary focus of Operator 3.0 was automation; on the other hand, Industry 4.0 and even more Industry 5.0 promote human and autonomous machine cooperation. The strong collaboration between humans and robots will lead to a highly effective manufacturing process with added value, thriving, trusted autonomy, and decreased waste and expenses [8]. The new approach of workers to the industrial system required research dedicated to sustaining industrial workers' physical and mental well-being by considering how this aspect assumed critical importance. Although various measures prioritise human well-being, including the Future Generations Well-Being Act, National Performance, and the OECD Better Life Index, studies concerning the methodologies and the strategies to assess and reduce the workers' physical and mental workload are in the experimental phase, their applications in industrial cases are not very frequent [9]. In most cases, the evaluations conducted were limited to a restricted sample of workers [10] or the behaviour of workers with different skills or ages was analysed in terms of human reliability [11].

Although diverse methods for measuring motor workload are widely available and frequently adopted in industrial environments [12], concerning the cognitive workload assessment, it is possible to observe that most methods are in an experimental phase. Indeed, in most cases, they are considered very hard to apply in the industrial work environment, and their validation is still one of the main challenges to be faced [13]. Therefore, on the one hand, the new industrial systems shift to new tasks requiring more 'cognitive' than 'physical' efforts; on the other hand, the approaches to assess the cognitive workload and ensure the physical well-being of the operators are far to be considered easily applicable.

The cognitive workload evaluation involves subjective indicators like self-reported questionnaires capturing individuals' perceptions of activity demands and psychological aspects and objective indicators such as physiological feedback, encompassing brain activity, pupillometer, body temperature, muscle activity, heart activity, respiratory activity, skin conductance, and positioning. Specific physiological signals, like skin conductance, heart rate variability (HRV), Oxygen Saturation (SPO2), and blood pressure, correlate with mental workload [2, 14]. In this context, the study wants to explore common causes underlying cognitive workload based on reviewed research and insights gathered from experienced technical personnel in the industry, with a focus on assembly, maintenance, manufacturing, and other relevant tasks. The study is structured around the following two research questions:

RQ1: What are the common factors contributing to cognitive workload in manufacturing?
RQ2: Which are the most frequent tools and techniques to assess the cognitive workload in current industrial scenarios?

The subsequent section of this paper presents the research method adopted to conduct the research. Then, the literature review results on cognitive workload are shown, including an exploration of the industrial context. The results are followed by a comprehensive discussion and conclusion highlighting potential avenues for future work.

2 Research Methodology

The literature review was conducted using the electronic database Scopus. Different keywords or search terms were adopted to identify works in the literature. These were combined with the Boolean operator "AND" to provide a greater spread. The implemented results of these combinations used in this research are: "Industry 4.0 and cognitive workload", "Industry 4.0 and mental workload", "Tools and cognitive workload", "tools and mental workload", "sensors and cognitive workload", and "sensors and mental workload".

The inclusion criteria adopted were:

- Peer-reviewed journal and conference articles;
- English language articles;
- documents published in the last five years (2019–2023).

The research led to the identification of 87 documents based on their relevance to the research objective. A manual selection based on the analysis of the title and abstract of each document was conducted. In this phase, 38 papers were excluded since they were considered not consistent with the core of the research in the contents shown in the title and the abstract, respectively. A total of 39 publications were considered eligible under the established criteria. No further publications identified through the snowballing technique from the reference lists of the selected literature were considered consistent with the selection criteria.

3 Results: Literature Review

3.1 Descriptive Analysis

The representation of the frequency of repeated keywords within the dataset selected was summarised in Fig. 1. Similarly, Fig. 2 provides insights into the 20 most frequently used words in the abstracts, outlining their prevalence in the selected studies.

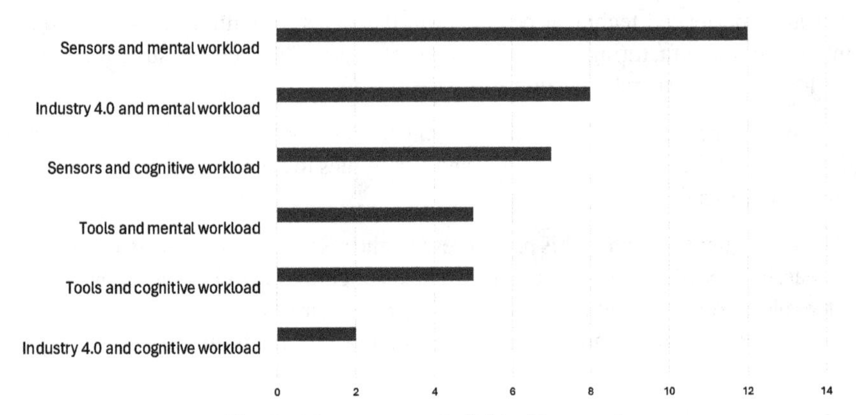

Fig. 1. Most common individual keywords

The analysis conducted proves that the EEG is one of the widest tests to assess how the perceived cognitive workload changes by varying the complexity of tasks assigned. EEG is used for real-time categorisation across varying cognitive workloads. Addressing the dynamic Industry 5.0 landscape, Chu et al. (2022) introduced multimodal techniques, combining EEG with fNIRS to enhance cognitive burden detection [15]. In assembly tasks, the cognitive burden is assessed using EEG and subjective measures, exploring human-automaton collaboration consequences, as emphasised by [16]. Although these studies offer a comprehensive view, integrating EEG and multimodal techniques and exploring human-automaton interaction implications, their applications are not tested in full industrial cases. The results were not validated to provide a reliable EEG-based Cognitive load assessment. EEG is also employed for task-switching workload measurement, utilising time-frequency analysis. These studies offer a comprehensive view, integrating EEG and multimodal techniques and exploring human-automaton interaction implications. Additionally, it contributes by focusing on wearable fNIRS wearability improvement, optimising sensor location for cognitive workload monitoring [16].

Stress analysis gaps were addressed in advanced manufacturing, induced by management practices, emphasising potential stressors in "lean" or "agile" systems; moreover, mental and physical workload among production line operators were evaluated, advocating for human-centred and socially sustainable production systems with flexible task scheduling [17]. The study aligns with International Labour Organization guidelines and categorises stress into physiological, physical, and psychological domains, using connected devices for data collection. This emphasis on stress factors complements the

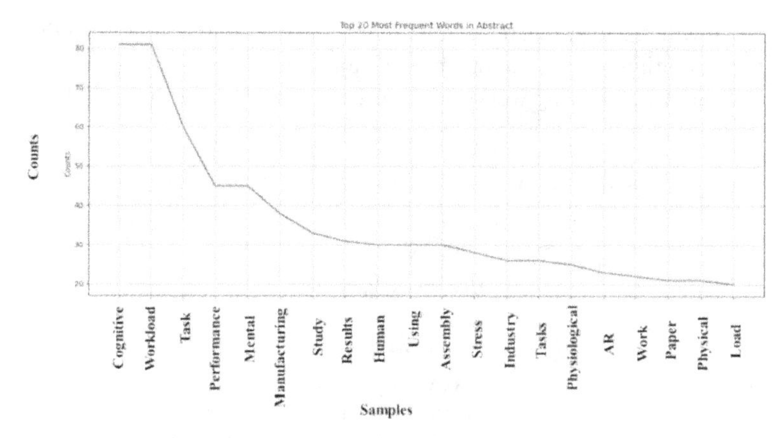

Fig. 2. Most frequent words in abstracts

theme of end-user involvement in the Norwegian oil industry, highlighting the importance of technology and human factors in implementing cognitive technology [18]. Perspectives from assembly operators, framed by the Job Demands-Resources model, underscore the collaborative approach needed for a balanced industrial assembly [19].

Addressing cognitive ergonomic requirements, criteria are categorised into design-oriented and user-oriented aspects, measured through efficient methods like eye-tracking and behavioural log data for performance monitoring [20]. In summary, these studies collectively illuminate the diverse effects of AR and VR technologies on cognitive load in industrial contexts, offering valuable insights for enhancement and optimisation. Stress analysis in manufacturing is stressed utilising wearables like eye-tracking and electrodermal monitoring for comprehensive physiological assessment [21]. Their focus underscores the importance of understanding stress factors in enhancing workplace well-being and performance.

3.2 Content Analysis

Cognitive ergonomics, an integral field in optimising work performance, delves into comprehending processes during work and addresses a spectrum of issues from attention distribution to the usability of human-computer systems. As underscored by [22] and [19], this discipline aims to establish a harmonious equilibrium between human cognitive abilities, machine characteristics, and task requirements to minimise errors and enhance performance in diverse work environments. Various approaches have been developed to address the importance of cognitive stress for workers' well-being, health, and safety. Many international standards (e.g., ISO-10075-3:2004) provide ergonomic design methods for assessing cognitive workload. These methods are categorised into four groups: physiology, subjective, performance, and job/task analysis.

Consistent with the identification of common factors contributing to cognitive workload in manufacturing (RQ1), the potential factors contributing to cognitive overload across common manufacturing workload areas have been summarised in Table 1.

Table 1. Cognitive factors in manufacturing environments

Factors	Sub-Factors
Time-Related	Deadlines, Pace Duration System Failures
Complexity	Task variety/Variance of product Problem-solving (e.g. Instruction) Precision (e.g. Instruction details) Recognition Requirements
Environmental	Noise, Lighting, Hazards
Organisational	Planning, Training Coordination/communication
Individual	Experience, Fatigue Age Gender Dexterity (Capability of worker)

Elements such as temporal pressures, intricacy, environmental stimuli, organisational deficiencies, and individual capacities collectively influence the degree of cognitive load experienced by workers. Each factor encompasses related sub-components that drive specific overload impacts. For instance, under the temporal factor, stringent deadlines, a rapid work pace, and production system failures, such as machine malfunction or downtime, correlate strongly with overload conditions during interruptions, training, manual labour, and order picking. Abrupt schedule changes from interruptions, combined with an already brisk pace, can overwhelm cognitive facilities. Similarly, attempting to learn intricate material under sharp time constraints hinders retention and comprehension. Picking at rapid rates often results in fixation errors and other manifestations of cognitive dysfunction.

Furthermore, within the intricacy sub-factor, problem-solving entails significant intrinsic load during maintenance, inspection, training, and custom manual work. Diagnosing equipment issues requires spatial reasoning, systemic thinking, and the identification of novel solutions. Detecting subtle product defects also burdens problem analysis more heavily than gross flaws. Crafting individualised formulations in assembly draws heavily on fluid reasoning and pattern recognition. This comprehensive understanding of cognitive ergonomics in the manufacturing context enhances our appreciation of the intricate interplay between cognitive factors and work demands.

The most frequent tools and techniques to assess the cognitive workload in current industrial scenarios (RQ2) have been provided in Table 2.

While the NASA-TLX remains the primary choice among researchers, there is a discernible upward trend in the adoption of electrocardiogram (ECG) and electroencephalography (EEG) approaches in recent years. However, their diffusion in natural work environments is strongly compromised by the complexity of using this kind of equipment for a long time, and in any case, the measurement can be affected by the rapid movements of the operators or by external factors.

Advancements influence this trajectory using wearable sensors and frugal technologies such as smartwatches and wristbands [23]. In the realm of continuous physiological measures, a chest band and wristband are employed to capture electrodermal activity (EDA) and interbeat interval (IBI), respectively, providing insights into autonomic arousal and heart rate variability (HRV) to gauge the operator's stress levels [16, 24]. Galvanic Skin Response (GSR), synonymous with EDA, is extensively studied for quantifying cognitive states, reflecting changes in skin electrical conductance due to variations in sweating activity [25]. The Multiple Attribute Task Battery (MATB) acts as a comprehensive tool, evaluating cognitive workload through multiple attributes and facilitating a nuanced understanding of mental demands [15, 18]. Eye tracking, integrated into a headset, records participants' eye movements and perspectives, enabling gaze-based interaction [26]. Analyses of ocular gaze behaviours and user perceptions capture factors influencing performance and safety, often disregarded in the evaluation of Augmented Reality (AR) usability in manufacturing and maintenance applications [24]. Ocular movement measures explore users' strategies in acquiring task-related information, with blinks and ocular movements correlated to cognitive aspects, where decreased blink rates indicate increased cognitive load [24, 27].

Table 2. The most frequent tools and techniques for cognitive workload evaluation

	Main Approach	Approach
1	EEG	EEG (based Bi-directional Gated Network (BDGN)), EEG, EEG-fNIRS
2	Electromyography (EMG)	EMG
3	fNIRS	EEG-fNIRS, fNIRS
4	NASA-TLX	NASA Task Load Index (NASA-TLX)
5	MATB	Multi-Attribute Task Battery
6	Questionnaire	SART (the situation awareness rating technique), PSSUQ (Post-Study System Usability Questionnaire), Questionnaire, PSS-10, SAM (Self-Assessment Manikin), INRS Method (Visual Fatigue), Questionnaire (Stanford), Negative Attitude toward Robots Scale (NARS)

(*continued*)

Table 2. (*continued*)

	Main Approach	Approach
8	Sensors	(Electrocardiogram (ECG), Skin Conductance), (wristband: IBI, EDA), ECG, ECG (Zephyr), Physiological biosensor (EDA, HRV), ECG, HRV, GSR, Sensors (wristband: SCR), Sensors(Chest Strap: HRV), ECG (Zephyr), PPG (Heart Rate Sensor), Sensor (Thermocouple), Sensors (Chest band, wristband), Physiological Biosensor (PPG, EDA), Cognex Scanner
9	Eye Tracker	Eye Tracker (Mixed Reality), Eye Tracker (AR), Eye Tracker (Tobii Eye Tracker), Eye Tracker (Glasses), Eye Tracker
10	HTA	Hierarchical Task Analysis (HTA)
11	Experimental Approach	Experimental Approach
12	Modelling	Simulation Modeling (IMPRINT), Neuro-Musculoskeletal Analysis,
13	MRT	Multiple Resource Theory (MRT)
14	HCCO	Human cognitive capacity occupancy
15	HCCR	Human cognitive capacity reserve
16	Performance Indexes	Performance Indexes
17	Interview	Interview
18	Literature Review	Literature Review
19	Camera	Camera (Thermal), Camera (Video), Camera (Motion Capture), Camera (Real Time Camera), 3D Camera
20	PVT	Psychomotor Vigilance Task (PVT)
21	CPT	Continuous performance test (CPT)
22	Conceptual Model	Conceptual Model, Linear Regression
23	Digital Twin	Asset Administration Shell (AAS)

4 Discussion

A novel online framework emerges, monitoring cognitive burden in industrial tasks through patterns of motion, offering promise for applications in manufacturing [25]. The impact of burden dimensions on indicators of cognitive performance is scrutinised, revealing intricate relationships between increased work demand, stress, and cognitive fatigue [28].

In the context of the exploration conducted, recent studies were dedicated to probing cognitive workload aspects in industrial activities, primarily emphasising assembly, maintenance, and manufacturing. These considerations lead to identifying the contributors to developing more effective policies to measure and mitigate issues arising from cognitive overload [29]. For instance, dynamically adjusting pick batch sizes or providing augmented equipment schematics could alleviate relevant contributors within order fulfilment and maintenance [30]. Furthermore, training programs could be adapted to sequence content, activities, and assessments to optimise working memory engagement without overtaxing mental resources [29]. The relationships characterised in this framework serve as an initial mapping of load contributors, with the understanding that further research should aim to elaborate on these findings through controlled studies and field observations. Another organisational approach to reducing the cognitive overload of the single operator is the job rotation application. This technique was widely used to reduce physical overload and was rearranged to avoid cognitive overload [31]. However, in many cases, the optimisation models reveal that the job rotation schedules and the human cognitive metrics influence the performance of the production lines. The emotional state of the workers can strongly compromise the repeatability of this approach. To this concern, there are stress factors whose impact is very hard to predict, allowing an increase in the perceived cognitive workload, with the consequence of drastically changing the performance of the same operator that accomplishes the same task. This effect is considerably reduced in experimental case studies, where the worker's behaviour is assessed in the same conditions and in short time windows.

This basic framework provides a valuable lens to recognise strain amplifiers on finite cognitive faculties within roles fundamental to manufacturing operations [32]. The careful management of these factors directly impacts productivity, safety, quality, and profitability. Considering the factors, such as organisational considerations and the nature of tasks, the decision-makers deliberate on the allocation of tasks between humans and machines [18]. Situational or contextual changes may influence this decision, and developers must prioritise understanding how users perceive new technologies in their work context. It is crucial to base technological proposals on user perceptions and consider how both technical and operational personnel process changes introduced by technology, as highlighted by [18].

This aspect is one of the more crucial since there is a strong relation between the technological complexity related to the use of a device and the corresponding cognitive workload. For instance, Augmented Reality (AR) and Artificial intelligence (AI) are leading technologies in maintenance activities, and this evidence is confirmed by several research projects and research dealing with these technologies [33]. Therefore, if, on the one hand, there are emerging technologies (e.g., AR, AI, etc.) to support the operator in "new" tasks, on the other hand, the complexity of tasks and the increasing use of innovative technologies could overload the operator, subjecting him to numerous options and efforts in a limited time, requiring decisions that lead to an excessive cognitive workload. This represents a real risk for the success of the most common tasks with high cognitive demand.

Recognising and addressing challenges arising from the interplay of humans, technology, and organisational dynamics is essential to fully harness the potential of advanced and complex [18]. To enhance technology utilisation and mitigate cognitive errors, it is recommended that end users be provided with training using simulators. Additionally, fostering active user participation in the learning and continuous improvement processes and encouraging feedback following technology implementation is vital [34]. The optimisation of manufacturing environments requires the provision of accurate and timely information, accompanied by effective communication channels, as pivotal measures for stress reduction. Conversely, instances of deficient communication, limited involvement in decision-making processes, or perceptions of inequitable treatment may contribute to heightened stress levels among operators [21].

Moreover, the assessment of one's dexterity, independent of experiential encounters with assembly tasks, assumes a crucial role. It is essential to ensure that the cognitive workload imposed on users aligns judiciously with their cognitive capacities, preventing the demands from exceeding their inherent limits. Within cognitive tasks, cognitive load is recognised as the capacity to deploy mental effort effectively [35]. This holistic approach fosters a conducive environment for optimal performance and well-being within diverse industrial contexts.

5 Conclusion

In conclusion, this study thoroughly examines cognitive workload within the framework of Industry 5.0, adopting a human-centric perspective. The literature review reveals prevalent keywords and abstract terms, providing valuable insights into current research trajectories. The analysis of factors contributing to cognitive workload underscores the significance of addressing time pressures, complexity, and individual capacities. Recommendations are put forth to advocate dynamic task allocation and user-centric technology implementation to optimise cognitive performance. The exploration of cognitive workload evaluation methods highlights a sustained interest in sensor applications, supplemented by a growing curiosity in AR, Ocular Tracker, and camera approaches. A more nuanced exploration of industrial activities, incorporating categorisation based on industry types, is essential for future research endeavours. Additionally, there is an urgent need to shift focus towards organisational layers, comprehending and alleviating cognitive workload from tactical to operational levels. This approach aims to cultivate a comprehensive understanding of cognitive workload within diverse industrial contexts, providing a foundation for informed interventions and optimisations in the evolving landscape of Industry 5.0.

Acknowledgements. This research work supported by the DESDEMONA project (DEcision Support system for the Diagnosis and Evaluation of the Maintenance OperatioNs Activities) of the Italian Ministry for Universities and Research MUR (Project PRIN – CUP D53D2301836 0001) funded by the European Union – NextGenerationEU, component M4C2, investment 1.1.

Disclosure of Interests. A. Nadaffard: Writing – original draft, Validation, Software, Methodology, Investigation, Formal analysis, Data curation, Conceptualization. L.M. Oliveri: Validation,

Conceptualization. D. D'Urso: Visualization, Supervision. F. Facchini: Writing – review & editing, Visualization, Supervision, Resources, Investigation, Conceptualization. C. Sassanelli: Validation, Supervision, Conceptualization.

References

1. Lu, Y., Zheng, H., Chand, S., et al.: Outlook on human-centric manufacturing towards Industry 5.0. J. Manuf. Syst. **62** (2022). https://doi.org/10.1016/j.jmsy.2022.02.001
2. Lin, C.J., Lukodono, R.P.: Classification of mental workload in human-robot collaboration using machine learning based on physiological feedback. J. Manuf. Syst. **65** (2022). https://doi.org/10.1016/j.jmsy.2022.10.017
3. Kotha, S., Pine, B.J.: Mass customization: the new frontier in business competition. Acad. Manag. Rev. **19** (1994). https://doi.org/10.2307/258941
4. Piller, F., Kumar, A.: For each, their own - the strategic imperative of mass customization. Ind. Eng. **38**, 40 (2006)
5. Simmons, C.H., Maguire, D.E., Phelps, N.: Product development and computer aided design. In: Manual of Engineering Drawing (2009)
6. Valentina, D.P., Valentina, D.S., Salvatore, M., Stefano, R.: Smart operators: how Industry 4.0 is affecting the worker's performance in manufacturing contexts. Procedia Comput. Sci. (2021)
7. Longo, F., Nicoletti, L., Padovano, A.: Smart operators in industry 4.0: a human-centered approach to enhance operators' capabilities and competencies within the new smart factory context. Comput. Ind. Eng. (2017). https://doi.org/10.1016/j.cie.2017.09.016
8. Leng, J., Sha, W., Wang, B., et al.: Industry 5.0: prospect and retrospect. J. Manuf. Syst. **65** (2022). https://doi.org/10.1016/j.jmsy.2022.09.017
9. Brunzini, A., Peruzzini, M., Grandi, F., et al.: A preliminary experimental study on the workers' workload assessment to design industrial products and processes. Appl. Sci. **11** (2021). https://doi.org/10.3390/app112412066
10. Digiesi, S., Lucchese, A., Mummolo, C.: A 'speed—difficulty—accuracy' model following a general trajectory motor task with spatial constraints: an information-based model. Appl. Sci. **10** (2020). https://doi.org/10.3390/app10217516
11. Digiesi, S., Cavallo, D., Lucchese, A., Mummolo, C.: Human cognitive and motor abilities in the aging workforce: an information-based model. Appl. Sci. **10** (2020). https://doi.org/10.3390/app10175958
12. Coronado, E., Kiyokawa, T., Ricardez, G.A.G., et al.: Evaluating quality in human-robot interaction: a systematic search and classification of performance and human-centered factors, measures and metrics towards an industry 5.0. J. Manuf. Syst. **63**, 392–410 (2022). https://doi.org/10.1016/j.jmsy.2022.04.007
13. Digiesi, S., Facchini, F., Mossa, G., Vitti, M.: A model to evaluate the human error probability in inspection tasks of a production system. Procedia Comput. Sci. (2022)
14. Argyle, E.M., Marinescu, A., Wilson, M.L., et al.: Physiological indicators of task demand, fatigue, and cognition in future digital manufacturing environments. Int. J. Hum. Comput. Stud. **145** (2021). https://doi.org/10.1016/j.ijhcs.2020.102522
15. Chu, H., Cao, Y., Jiang, J., et al.: Optimized electroencephalogram and functional near-infrared spectroscopy-based mental workload detection method for practical applications. Biomed. Eng. Online **21** (2022). https://doi.org/10.1186/s12938-022-00980-1
16. Gervasi, R., Capponi, M., Mastrogiacomo, L., Franceschini, F.: Manual assembly and human-robot collaboration in repetitive assembly processes: a structured comparison based on human-centered performances. Int. J. Adv. Manuf. Technol. **126** (2023). https://doi.org/10.1007/s00170-023-11197-4

17. Blandino, G., Montagna, F., Cantamessa, M.: Workload and stress evaluation in advanced manufacturing systems. Mater. Res. Proc. (2023)
18. Sætren, G.B., Ernstsen, J., Phillips, R., et al.: Cognitive technology development and end-user involvement in the Norwegian petroleum industry – Human factors missing or not? Saf. Sci. **170** (2024). https://doi.org/10.1016/j.ssci.2023.106337
19. Wollter Bergman, M., Berlin, C., Chafi, M.B., et al.: Cognitive ergonomics of assembly work from a job demands–resources perspective: three qualitative case studies. Int. J. Environ. Res. Public Health **18** (2021). https://doi.org/10.3390/ijerph182312282
20. Zhang, Y., Sun, J., Jiang, T., Yang, Z.: Cognitive ergonomic evaluation metrics and methodology for interactive information system. In: Advances in Intelligent Systems and Computing (2020)
21. Apraiz, A., Lasa, G., Montagna, F., et al.: An experimental protocol for human stress investigation in manufacturing contexts: its application in the NO-STRESS project. Systems **11** (2023). https://doi.org/10.3390/systems11090448
22. Carvalho, A.V., Chouchene, A., Lima, T.M., Charrua-Santos, F.: Cognitive manufacturing in industry 4.0 toward cognitive load reduction: a conceptual framework. Appl. Syst. Innov. **3** (2020). https://doi.org/10.3390/asi3040055
23. Agrawal, S., Chong, J., Yacoub, A.A., et al.: Physiological data measurement in digital manufacturing. In: 2021 24th International Conference on Mechatronics Technology, ICMT 2021 (2021)
24. Ariansyah, D., Erkoyuncu, J.A., Eimontaite, I., et al.: A head mounted augmented reality design practice for maintenance assembly: toward meeting perceptual and cognitive needs of AR users. Appl. Ergon. (2022)
25. Lagomarsino, M., Lorenzini, M., De Momi, E., Ajoudani, A.: An online framework for cognitive load assessment in industrial tasks. Robot. Comput. Integr. Manuf. **78** (2022). https://doi.org/10.1016/j.rcim.2022.102380
26. Yan, Z., Shan, Y., Li, Y., et al.: Gender differences of cognitive loads in augmented reality-based warehouse. In: Proceedings - 2021 IEEE Conference on Virtual Reality and 3D User Interfaces Abstracts and Workshops, VRW 2021 (2021)
27. Brunzini, A., Grandi, F., Peruzzini, M., Pellicciari, M.: An integrated methodology for the assessment of stress and mental workload applied on virtual training. Int. J. Comput. Integr. Manuf. (2023). https://doi.org/10.1080/0951192X.2023.2189311
28. Ghalenoei, M., Mortazavi, S.B., Mazloumi, A., Pakpour, A.H.: Impact of workload on cognitive performance of control room operators. Cognit. Technol. Work **24** (2022). https://doi.org/10.1007/s10111-021-00679-8
29. Kalatzis, A., Rahman, S., Girishan Prabhu, V., et al.: A Multimodal approach to investigate the role of cognitive workload and user interfaces in human-robot collaboration. In: ACM International Conference Proceeding Series (2023)
30. Alves, J.B., Marques, B., Ferreira, C., et al.: Comparing augmented reality visualization methods for assembly procedures. Virtual Real. **26** (2022). https://doi.org/10.1007/s10055-021-00557-8
31. Ayough, A., Farhadi, F., Zandieh, M.: The job rotation scheduling problem considering human cognitive effects: an integrated approach. Assem. Autom. **41** (2021). https://doi.org/10.1108/AA-05-2020-0061
32. Bommer, S.C., Fendley, M.: A theoretical framework for evaluating mental workload resources in human systems design for manufacturing operations. Int. J. Ind. Ergon. (2018). https://doi.org/10.1016/j.ergon.2016.10.007
33. Silvestri, L., Forcina, A., Introna, V., et al.: Maintenance transformation through Industry 4.0 technologies: a systematic literature review. Comput. Ind. **123** (2020). https://doi.org/10.1016/j.compind.2020.103335

34. Eversberg, L., Lambrecht, J.: Evaluating digital work instructions with augmented reality versus paper-based documents for manual, object-specific repair tasks in a case study with experienced workers. Int. J. Adv. Manuf. Technol. **127** (2023). https://doi.org/10.1007/s00170-023-11313-4
35. Van Acker, B.B., Parmentier, D.D., Conradie, P.D., et al.: Development and validation of a behavioural video coding scheme for detecting mental workload in manual assembly. Ergonomics **64** (2021). https://doi.org/10.1080/00140139.2020.1811400

Experiential Learning in Engineering Education

Understanding the Drivers of Lean Learning in Industrial Environments

Bruno Pereira, Luís Miguel D. F. Ferreira, and Cristóvão Silva(✉)

CEMMPRE, Department of Mechanical Engineering, University of Coimbra, Coimbra, Portugal
{luis.ferreira,cristovao.silva}@dem.uc.pt

Abstract. Organizations are increasingly recognizing the benefits of adopting lean tools. However, a successful lean transformation depends on the active involvement of all organizational members. Comprehensive employee training is, therefore, fundamental to the effective implementation of lean manufacturing practices. Although collaborative and dynamic practical sessions have been implemented in certain university settings for teaching lean principles, there is limited research exploring how learning outcomes are influenced by participants' characteristics. Examining these relationships is even scarcer in the literature concerning industrial training activities. This study examines the impact of lean training on employee learning within an industrial organization, looking at 177 participants and analyzing the relationship between learning outcomes and variables such as self-efficacy beliefs, prior knowledge, motivation, and enjoyment. The results show that the training significantly improved the participants' lean knowledge level. A significant relationship was found between self-efficacy beliefs and employees' motivation, which in turn had a positive impact on their learning. Based on these findings, companies should consider lean training that engages employees' curiosity using non-traditional teaching methods.

Keywords: lean training · learning · self-efficacy beliefs · motivation · enjoyment · prior knowledge

1 Introduction

Implementing lean principles improves production and reduces waste, cost, and delivery times. However, implementing lean principles faces some obstacles, such as the lack of appropriate training, resistance to change from the organization and employees, insufficient financial resources, and cultural barriers [1]. Training is acknowledged as a pivotal factor in the successful implementation of lean principles. Possessing both an understanding of lean principles and the skills required to execute necessary tasks is a key determinant of success in lean implementation [2]. Shrimali and Soni found that inadequate training emerged as an important constraint to successful lean implementation in SMEs [3]. Another study suggests that training and education are often cited as the most important success factors in the literature on improvement programs after management

Published by Springer Nature Switzerland AG 2024
M. Thürer et al. (Eds.): APMS 2024, IFIP AICT 729, pp. 473–487, 2024.
https://doi.org/10.1007/978-3-031-65894-5_33

commitment and involvement [4]. A survey presented by [5] shows that while UK companies demonstrate an interest in lean management, they have difficulties adapting to it due to a lack of workforce training. Thus, it is no surprise that lean teaching/training has become a subject of growing interest.

The literature on lean teaching has been prevalent in presenting a hands-on approach, showcasing innovative teaching methods like simulation or game-based techniques. Dinis-Carvalho presents an innovative framework for lean training, which has been tested in a higher education setting [6]. However, the impact of the proposed learning approach on learning outcomes is not presented, nor is the influence of learner characteristics considered. Choomlucksana and Doolen also explore innovative lean teaching techniques in engineering education [7]. They test two approaches: simulation and collaborative activities. They conclude that the best approach depends on the concepts being taught, but the collaborative approach shows statistically improved learning for both lean concepts considered – Jidoka and Pull. An interesting aspect of this study is that it examines the relationship between learner characteristics, like prior knowledge and self-efficacy beliefs, and learning outcomes. De Vin et al. present some game-based lean training, ranging from desktop games to full-scale simulators and actual manufacturing machinery, and test them with university students and industrial employees [8]. They conclude that desktop games are suitable for training people who already have a fair understanding of lean principles and that training outcomes for students and employees tend to be higher when using full-scale simulators. Tan et al. present a five-day lean innovation training developed for a public sector service company [9], concluding that the intervention positively impacted managers' transformational leadership, which in turn positively impacted employees' innovative work behavior.

Several approaches to teaching lean concepts have been developed and presented in literature, both in academic and industrial settings. Nevertheless, the impact of the proposed approaches on learning outcomes is often not presented, and the relationship between training, lean learning outcomes and learner characteristics is rarely considered. The relationship between training approaches, lean learning outcomes and learner characteristics provides a compelling rationale for investing in innovative teaching methods. Furthermore, understanding these relationships can help trainers decide which training practices to use in their organization.

This paper intends to contribute to a better understanding of innovative lean training techniques usefulness for industrial training. It investigates the impact of non-traditional lean training, considering learners' characteristics on the learning outcomes of industrial operators.

The article starts with a revision about lean training and the characteristics of employees that may influence their learning. Specifically, four key learner characteristics identified in the literature as potentially significant for the learning process are considered: self-efficacy beliefs [7, 18–20], motivation [22, 23], enjoyment [26, 27], and prior knowledge [7, 28]. A detailed description of the methodology used to investigate the relationships between the previously described variables is then presented. The article concludes with a discussion of the findings and conclusions, highlighting the main contributions and limitations of the present study.

2 Literature Review

The lean transition is a complex process that uses different methodologies and tools and requires adequately prepared employees. The lack of such preparation is one of the main obstacles to the lean transformation of organizations [5]. Adequate training makes employees understand how their work affects the entire process [6]. In addition, after training, employees can acquire the necessary skills to solve problems, individually or as a team. Furthermore, good training transmits to employees all the knowledge required to promote a correct lean philosophy within the corporation through individual commitment and improvements in their performance.

For effective lean training, it is crucial to use training tools that maintain employee interest and motivation [10]. In a traditional teaching system, a speaker delivers a lecture using a board or presentation, and students' success depends on their ability to absorb the information transmitted to them. According to Armstrong, this is often called conventional teaching [11]. In contrast, non-traditional teaching methods, also known as innovative methods, involve various techniques to stimulate students' interest in learning. These methods go beyond the exposition of ideas and may include problem-solving and group discussions to actively engage students in the learning process. One effective way of teaching is through dynamics and simulations, which provide practical experience for students to apply their knowledge. Although traditional teaching methods are well-organized and familiar to most students, previous studies have identified the benefits of using innovative teaching methods [12, 13].

The use of lean dynamics can simplify organizational processes [8]. Learning objectives can be effectively transferred by providing training that aligns with the skills and competencies required for each job. A personal reflection on concepts and discussion with peers can help ensure that the learning outcomes are in line with the objectives of the organization. The effectiveness of lean training depends on the suitability of the dynamics chosen, which determines how well the learning outcomes correspond to the required objectives [12]. However, the suitability of a lean dynamic depends not only on the aim of the training but also on the group of participants since distinct groups that conduct the same dynamic tend to exhibit different knowledge transfers [8].

Learning can be defined as a set of claims that the student is expected to be able to understand and perform after a training action [14]. The emotional state of individuals during training can affect their self-esteem and, consequently, their ability to learn [15]. The complexity of the activities conducted to transmit lean concepts may cause anxiety in employees, making their learning difficult [16]. In 1984, Kolb announced the existence of the so-called "flow channel" in learning using small dynamics [17]. He concludes that the participant will become frustrated or anxious if the task is too difficult, but he can be bored if the task is too trivial. Learning occurs between these two zones.

Lean learning outcomes might be influenced by the learner's characteristics, such as self-efficacy belief, motivation, enjoyment or prior knowledge. Self-efficacy belief can be defined as the belief that an individual has in being able to perform a particular task [16]. Self-efficacy beliefs influence the actions people are determined to take and the goals, commitment, and effort they invest in their activities, and it tends to increase throughout life through experience. People with high levels of self-efficacy believe they can complete a task successfully and demonstrate outstanding commitment and persistence in carrying it out. On the other hand, people with low levels of self-efficacy believe that they cannot complete a task and, as a result, try to avoid it [18]. Thus, self-efficacy beliefs impact the effort and time needed to perform a given operation. Self-efficacy beliefs are essential to improving learning environments and employee results [19]. This improvement is related to high levels of motivation and enjoyment resulting from high levels of self-efficacy. A high level of self-efficacy has a positive impact on trainees' motivation, enjoyment, and learning [20]. Moreover, individuals with a high level of self-efficacy tend to retain more knowledge during training sessions.

Motivation is a central workplace pillar. Various studies have been devoted to the link between employee motivation and performance in different organizations [21]. Motivation is the driving force within an individual who seeks to achieve a goal to satisfy a need. Three types of motivation can be defined: intrinsic motivation, extrinsic motivation, and task value. Intrinsic motivation refers to the degree to which a person perceives that they are engaged in a task because it is perceived as challenging, which arouses their curiosity. Extrinsic motivation is the degree to which someone perceives that they are involved in a task because the task itself is linked to an external motivating factor, such as a good grade in an assessment or a financial reward. Task value refers to the degree to which a person perceives that they are involved in a task because it is perceived as necessary [7]. A significant and positive improvement in motivation can lead to higher performance and increased learning outcomes [22]; on the other hand, students who show negative motivation throughout a course unit tend to receive lower scores in the final exam [23].

Enjoyment is a reliable indicator of employee retention. When managers encourage good working relationships, employee enjoyment improves because they believe that the organization uses and values their skills. In turn, greater enjoyment in the workplace results in employee retention [24]. In previous studies, a high level of enjoyment was also associated with high levels of learning [25]. However, no relationship between employee enjoyment and learning was found in other studies [26].

Prior knowledge was initially defined as "what is already known about a given subject" [27]. It represents all the knowledge that employees demonstrate when entering a learning environment, which is potentially relevant to acquiring new knowledge. According to the literature, prior knowledge is a factor that influences learning, as well as employees' self-efficacy, motivation, and enjoyment. It has been found that individuals with varying levels of prior knowledge display significant differences in performance and task completion [7]. People with high levels of prior knowledge before undergoing training tend to respond quickly and exhibit high confidence in learning new techniques, resulting in greater motivation and enjoyment [28].

3 Methodology

This study examines how nontraditional training, self-efficacy beliefs, prior knowledge, motivation, and enjoyment affect employee learning in an industrial setting. Thus, the research question is: What is the impact of nontraditional training, self-efficacy beliefs, prior knowledge, motivation, and enjoyment on the lean learning outcomes of employees in an organizational environment?

The study was conducted in a company in the aeronautical sector that specialized in maintenance and repair operations services, having approximately 1700 employees. The Operational Excellence System implemented at the company continually seeks business excellence through the practical application of the lean philosophy. To do so, it has implemented a Lean Academy aiming to streamline the cultural orientation and journey towards a lean attitude. The training sessions included in the training plan for the new company employees called the Dojo Experience, were used to carry out this research. The Dojo Experience involves theoretical exposition by the trainer and putting the acquired knowledge into practice through small simulation dynamics. Usually, there are ten participants in each group. The simulation dynamics are desktop Lego games: (1) a construction site layout is used to exemplify the concept of waste, (2) the assembly process of airplane helices to teach 5S, (3) the assembly process of a hypothetic product to transmit the concepts of takt time, cycle time and lead time and (4) a hangar layout with three airplanes to exemplify safety issues.

Data were collected through two questionnaires, one conducted before the training and one after. An exploratory data analysis was performed, since it is the best approach to identify patterns or relationships in the data without prior assumptions. Typical methods used in exploratory analysis, like descriptive statistics, data visualization or factor analysis, were applied.

3.1 Data Collection Method

Two questionnaires were developed for data collection: the first one was to measure participants' prior knowledge before the training session, and the second one was used to evaluate the learning outcomes of the lean training program. Additionally, the second questionnaire assessed the employees' motivation, self-efficacy beliefs, and enjoyment. Taking the case study into account, the questionnaires were applied to the employees who attended the Dojo Experience, a training session used by the company which includes some of the main lean tools. Figure 1 gives a better view of the content and timing of the two questionnaires. A section of the questionnaire was developed to gather demographic information about the participants, such as gender, age, level of education, position, and seniority in the company.

The first questionnaire consists of ten items referring to the theoretical contents covered in lean training. The second questionnaire contains ten theoretical-practical items to cover all the lean subjects considered in the training and to understand whether the knowledge was transmitted correctly to all trainees. Additionally, this questionnaire involves six items to assess self-efficacy beliefs, eleven to assess motivation, and four to evaluate enjoyment.

Two sets of multiple-choice questions were developed about the main concepts covered throughout the training to measure learning outcomes. The questions asked after training (learning) are applied to the organizational context to understand whether employees can transpose the lean concepts learned into their tasks. In contrast, those asked before (prior knowledge) have a more theoretical character. Table 1 provides some examples of the questions included to assess these constructs.

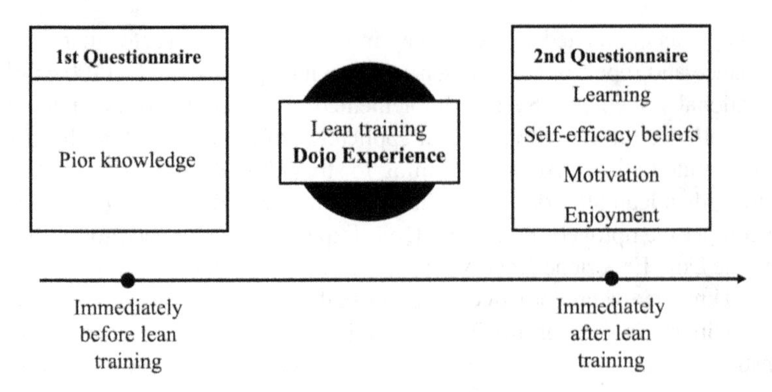

Fig. 1. Questionnaire application sequence

Table 1. Examples of questions used to assess prior knowledge and learning.

Inquiry	Question	Options
Before training (prior knowledge)	1. Which of the following terms is associated with overload on people or equipment?	(a) Mura (b) Muri (c) Muda (d) None of the above hypotheses
After training (Learning)	2. Each tool must be identified and have a defined location. Which step of the 5S tool does this action represent?	(a) Utilization (b) Ordering (c) Cleaning (d) Standardization

Although only two questions are presented in Table 1, 20 multiple-choice questions were created, which provide a complete coverage of the content of the training. Regarding the items assessing self-efficacy beliefs, these were developed based on previous studies [29]. To evaluate this construct, a 5-point Likert scale was used (1 = Strongly disagree; 2 = Disagree; 3 = Indifferent; 4 = Agree; 5 = Strongly agree). The items used in this investigation to assess employees' self-efficacy beliefs are presented in Table 2. The attitudes demonstrated by employees towards training actions help understand how individuals feel, think, and react to a given session. Two different attitudes were evaluated: employee motivation and enjoyment, which were adapted from [7]. The same 5-point Likert scale defined previously was used to assess the constructs. Table 3 presents the items used to assess motivation: items 1 to 4 related to intrinsic motivation, items 5 to 7 related to extrinsic motivation, and items 8 to 11 related to task value. Table 4 presents the items used to assess enjoyment.

3.2 Validation and Reliability of the Method

The participants were employees who directly experienced the Dojo Experience training. Self-efficacy, motivation and enjoyment were validated before data collection began.

Table 2. Items to evaluate the self-efficacy beliefs of the participants.

ID	Item
Selfef _1	As a result of the lean training, I could correctly answer the final quiz questions. *
Selfef _2	Lean training increased my confidence in understanding lean philosophy
Selfef _3	Lean training increased my confidence in myself regarding understanding the principles related to the lean philosophy
Selfef _4	I have no doubts about my ability to execute the concepts covered in my workplace correctly. *
Selfef _5	After this training, I can explain what I learned about lean thinking to my colleagues
Selfef _6	I am sure I have mastered all the concepts covered in lean training. *

Table 3. Items to evaluate the motivation of the participants.

ID	Item
Mot_1	I prefer training courses that have challenging dynamics so that I can learn new things
Mot_2	I prefer training that arouses my curiosity, even if it is difficult
Mot_3	I prefer training from which I will learn something, even if it requires more work
Mot_4	I prefer training courses where I can learn something, even if it doesn't guarantee a good grade on the final questionnaire
Mot_5	Learning from innovative dynamics helped me prepare for the final quiz
Mot_6	Learning from innovative dynamics helped me get a good grade on the final questionnaire
Mot_7	I participated in lean training because I had to. *
Mot_8	As a result of this lean training, I can apply what I learned in other training. *
Mot_9	I needed to learn what was transmitted throughout the lean training
Mot_10	What I learned in training is helpful for me and my training as a professional and worker
Mot_11	I can apply what I learned in lean training to problems in my workplace

Seven company members carried out this validation, along with two researchers with experience in developing questionnaires and teaching lean topics. The choice, clarity and understanding of each item were also assessed. The raters were asked to rate each item according to a 5-point Likert scale in the parameters. These items were validated using the content validation index (CVR), which can vary between -1 and $+1$. All the results presented satisfactory values, between 0.56 and 1, making all the content used to evaluate the participants' lean knowledge valid.

Table 4. Items to evaluate the enjoyment of the participants.

ID	Item
Enj_1	I enjoyed participating in lean training. *
Enj_2	I felt that "time flew by" throughout the lean training. *
Enj_3	After completing this training, I look forward to being invited to participate in an upcoming lean training
Enj_4	I would like to have the opportunity to participate in more lean training

Reliability refers to the consistency or stability of the measurement. Cronbach's alpha was used to evaluate the internal reliability of the survey items used for the individual constructs of each variable. The Cronbach's alpha value can vary between 0 and 1. A Cronbach's alpha value close to 0.7 or higher is considered satisfactory. In this way, each set of survey items was evaluated using the data collected for the present investigation. After using exploratory factor analysis, Cronbach's alpha was calculated for each construct (self-efficacy beliefs, motivation, and enjoyment).

4 Results

The sample has 177 respondents. About gender, the distribution of responses was not equitable, with only 6% being female and 94% being male. Most respondents only have an education equal to or less than the 12th year of schooling (83%), which allows us to assume that a large group of elements will have a low level of prior knowledge about lean.

4.1 Exploratory Factor Analysis

After verifying and removing outliers from the sample, several factor analyses were conducted to find the best statistical result. The proposed solution relies on five factors that characterize the sample, using the varimax rotation method, excluding coefficients with a value lower than 0.50. Next, the Kaiser-Meyer-Olkin (KMO) and Bartlett test was performed to verify whether the sample was suitable for the study to be carried out. As a result of this test, the sample adequacy measure is 0.839 and sigma 0.00, which indicates the rejection of the null hypothesis; that is, the factor analysis is appropriate for the sample.

After a preliminary sample analysis, several exploratory factor analyses were undertaken to eliminate items with little relevance, aiming to obtain results aligned with the objective of the present study. After this iterative process, seven items were removed from a total of twenty-one due to them having loading coefficients lower than 0.50 or appearing isolated in independent factors. The removed items are identified with a "*" in Tables 2, 3 and 4. A new exploratory factor analysis was conducted after excluding these items and Cronbach's alpha was calculated to understand the reliability of each construct.

After calculating Cronbach's alpha for each of the constructs, the range of values, except for extrinsic motivation, is between 0.74 and 0.91, which indicates that they all presented good levels of reliability. Extrinsic motivation presents a moderate internal consistency (0.65), which can be acceptable in exploratory research (Table 5).

Table 5. Results of exploratory factor analysis

Construct	Item	Load Coefficient	Cronbach's Alpha
Self-efficacy	Selfef_2	0,751	0,74
	Selfef_3	0,749	
	Selfef_5	0,692	
Intrinsic Motivation	Mot_1	0,760	0,82
	Mot_2	0,857	
	Mot_3	0,769	
	Mot_4	0,722	
Extrinsic Motivation	Mot_5	0,565	0,65
	Mot_6	0,665	
Task Value	Mot_9	0,661	0,80
	Mot_10	0,779	
	Mot_11	0,722	
Enjoyment	Enj_3	0,869	0,91
	Enj_4	0,871	

Table 6 shows each factor's mean and standard deviation. This table also includes prior knowledge and learning, which are vital variables for the present study.

Table 6. Mean/Standard Deviation of Factors Related to Dojo Experience

Variable	Average	Standard Deviation
Prior Knowledge	3,62	1,70
Learning	7,74	1,43
Self-efficacy	4,04	0,40
Intrinsic Motivation	4,15	0,57
Extrinsic Motivation	4,13	0,50
Task Value	4,20	0,46
Enjoyment	4,06	0,75

4.2 Correlations Analysis

The two questionnaires that were given to the participants were analyzed to evaluate the impact of lean training on their understanding of lean concepts. T-student tests were carried out to determine whether there was a significant difference between the knowledge demonstrated before and after the training. Results are presented in Table 7 and Fig. 2.

Table 7. Summary of t-student tests to assess learning outcomes.

Inquiry	Answers	Average	t-value	p-value
Before training	177	3,62	– 4,12	<0,001
After training	177	7,74		

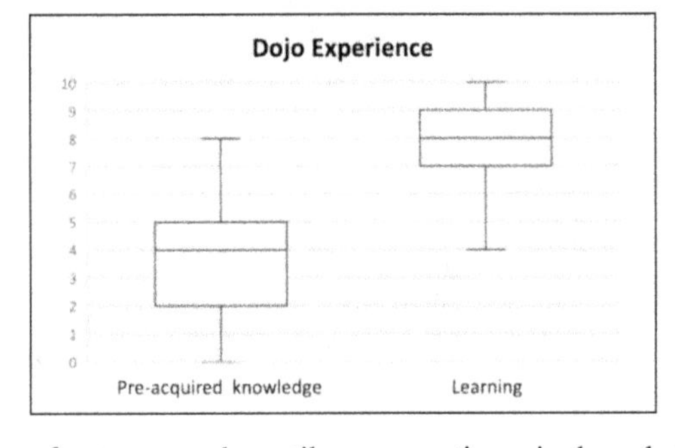

Fig. 2. Diagrams of extremes and quartiles representing prior knowledge and learning.

Employees showed improved understanding of lean concepts after training, with the average score rising from 3.62 to 7.74. Additionally, by analyzing the extreme and quartile diagrams, a reduction in dispersion can be seen in the lean knowledge demonstrated by the participants after they have gone through the training action. This indicates that the training serves to homogenize and level the lean knowledge of all participants.

Next, the impact of the different variables (self-efficacy, intrinsic motivation, extrinsic motivation, task value and enjoyment) on learning and the relationship between them is presented. Pearson's correlation coefficient was used to explore the relationship between the variables measured in this study (Table 8).

A positive and significant correlation was found between self-efficacy and the various types of motivation that were studied. The analysis showed a weak positive correlation between self-efficacy and intrinsic motivation ($r = 0.290$). Additionally, self-efficacy and extrinsic motivation had a strong positive correlation ($r = 0.601$). This suggests that employees who have high self-efficacy beliefs tend to carry out their tasks not only for their personal satisfaction but also because they are motivated by external factors.

It was also found that there is a significant correlation between an individual's self-efficacy beliefs and the value they place on a particular task ($r = 0.546$). This indicates

Table 8. Summary of t-student tests to assess learning.

		Self-efficacy	Intrinsic Motivation	Extrinsic Motivation	Task Value	Enjoyment	Learning
Self-efficacy	r	1	**0,290****	**0,601****	**0,546****	**0,362****	0,106
	p		**<0,001**	**<0,001**	**<0,001**	**<0,001**	0,155
	n	177	**177**	**177**	**177**	**177**	177
Intrinsic Motivation	r	**0,290****	1	**0,348****	**0,344***	**0,407****	0,178*
	p	**<0,001**		**<0,001**	**<0,001**	**<0,001**	0,017
	n	**177**	177	**177**	**177**	**177**	177
Extrinsic Motivation	r	**0,601****	**0,348****	1	**0,555****	**0,285****	0,081
	p	**<0,001**	**<0,001**		**<0,001**	**<0,001**	0,387
	n	**177**	**177**	177	**177**	**177**	177
Task Value	r	**0,546****	**0,344***	**0,555****	1	**0,420****	**0,150***
	p	**<0,001**	**<0,001**	**<0,001**		**<0,001**	**0,045**
	n	**177**	**177**	**177**	177	**177**	**177**
Enjoyment	r	**0,362****	**0,407****	**0,285****	**0,420****	1	−0,002
	p	**<0,001**	**<0,001**	**<0,001**	**<0,001**		0,983
	n	**177**	**177**	**177**	**177**	177	177
Learning	r	0,106	0,178*	0,081	**0,150***	−0,002	1
	p	0,155	0,017	0,387	**0,045**	0,983	
	n	177	177	177	**177**	177	177

**Correlation is significant at the 0.01 level.
*Correlation is significant at the 0.05 level.

that employees who have high self-efficacy beliefs are more likely to engage in tasks that they find important. Finally, self-efficacy and enjoyment also correlate (r = 0.362). This proves a moderate positive relationship between an individual's ability to perform a task and personal and professional satisfaction.

There is no correlation between self-efficacy and learning. This means that, for the sample under study, there is no direct relationship between the employee's ability to be successful in a task and the learning resulting from the training action. Analyzing the relationships between motivation and learning, and between enjoyment and learning is essential. In the case of motivation, this correlation is partially verified, given that there is a significant, albeit weak, correlation between intrinsic motivation and learning (r = 0.178) and between the value of the task and learning (r = 0.150). However, no correlation was found between extrinsic motivation and learning. Therefore, the fact that respondents enjoy their work tasks without considering any reward allows them to obtain greater learning and see training as an opportunity for personal and professional improvement.

Finally, regarding employee enjoyment of the training action, although this variable presents a relatively high value (4.06), indicating that employees enjoyed the training, there is no significant correlation with the resulting learning. This means that even if the respondents enjoy the training, they may not gain the expected knowledge. The study examined whether prior knowledge affects the ability to learn. The difference in test scores was analyzed by dividing participants into two groups based on their average prior knowledge score - low prior knowledge (<= 3.62) and high prior knowledge (>3.62). The results showed that there was no significant difference in learning between the two groups. This means that the training served to level the playing field, regardless of the participants' prior knowledge. The study also found that the participants' prior knowledge did not affect their self-efficacy, motivation, or enjoyment.

5 Discussion of Results

The objective of this study was to collect and analyze data that addresses the importance of lean training in an industrial organization and identify factors that can influence learning (self-efficacy beliefs, motivation, prior knowledge, and enjoyment). 177 responses obtained from new employees attending a lean training program were analyzed. Results indicate that this group of employees may never have had contact with lean concepts. By undergoing the training, they are now able to identify waste and voluntarily and spontaneously seek continuous improvement opportunities.

Regarding self-efficacy beliefs, 95% of respondents agreed or strongly agreed that lean training increased their confidence in understanding the concepts covered. This high confidence is portrayed in another item, where 89% of respondents agreed or strongly agreed that they can explain to their colleagues what they have learned about lean. This value reinforces the idea that employees can transmit what they have learned to their colleagues, who have not yet undergone training, to disseminate the lean culture throughout the organization.

Regarding motivation, after going through the training, 87% of respondents agreed or strongly agreed that the use of challenging dynamics is an added value for their learning, and 89% of respondents agreed or strongly agreed that they prefer a training action that arouses their curiosity, even if they are difficult.

Regarding the applicability of the concepts transmitted during the training, 95% of respondents agreed or strongly agreed that what they learned is helpful for their role in the organization. Furthermore, 96% of respondents agreed or strongly agreed that they can apply what they learned to problems in their workplace, which supports the importance of lean training delivered to all employees in the organization.

Finally, about the enjoyment shown by employees regarding the training they received, 98% of respondents agreed or strongly agreed that they enjoyed participating in lean training. They tend to believe that the organization is using and valuing their skills. Additionally, 84% of respondents agreed or strongly agreed that they would like to participate in more lean training.

The first objective of this research was to understand whether training using innovative dynamics influences employee learning. The participants' average scores showed statistically significant differences before and after participating in the lean training activities. These results validate the value of such training, which is supported by previous research conducted by [7]. One reason why training actions can lead to high learning outcomes is that they involve various educational approaches, such as brainstorming, group discussions, and small team dynamics. These findings imply that innovative training methods can serve as a complement to traditional teaching techniques and can help support lean training for an organization's workforce.

A direct correlation between self-efficacy and learning was not verified; however, the results indicate that this relationship may be mediated by motivation (in particular, intrinsic motivation and the value of the task) since self-efficacy presents a positive correlation with these two types of motivation [weak positive correlation ($r = 0.290$) and moderate positive correlation ($r = 0.546$), respectively], which in turn have a positive impact on employee learning [weak positive correlation ($r = 0.178$) and weak positive correlation ($r = 0.150$), respectively]. About enjoyment, it presents a moderate positive correlation with self-efficacy ($r = 0.362$). However, there is no relationship between this construct and employee learning.

The study's results show that the level of prior knowledge did not significantly impact learning, nor did the other variables that were studied. Therefore, these findings contradict previous research that suggested that different levels of prior knowledge played a crucial role in learning [29]. This suggests that the effect of prior knowledge may vary depending on the specific context of the case study.

6 Conclusion

This study surveyed 177 employees from a company who have undergone a lean training program. The analysis demonstrates a significant increase in employees' knowledge of lean concepts following training. Moreover, the reduction in dispersion among participants' lean knowledge indicates that training serves to standardize and enhance overall comprehension of lean principles within the workforce.

The study reveals a positive and significant correlation between self-efficacy and various types of motivation, including intrinsic and extrinsic motivation, as well as task value. Employees with higher self-efficacy beliefs tend to be driven both intrinsically and extrinsically and are more likely to engage in tasks they perceive as important. However, contrary to expectations, there is no direct correlation between self-efficacy and learning outcomes. Instead, intrinsic motivation and the perceived value of tasks correlate weakly but significantly with learning, suggesting that employees driven by personal satisfaction and task importance tend to exhibit greater learning outcomes.

Although employees report relatively high levels of enjoyment during training, this does not significantly correlate with learning outcomes. This suggests that while enjoyable training experiences contribute to overall employee satisfaction, they may not necessarily translate into improved knowledge acquisition.

The study finds no significant difference in learning outcomes between participants with low and high levels of prior knowledge. This indicates that training effectively levels the playing field, regardless of employees' pre-existing knowledge levels. Furthermore, prior knowledge does not influence self-efficacy, motivation, or enjoyment, indicating that training has a consistent impact on these variables across all participants.

This study was based on a single case study and should be replicated in other industrial settings, including a more diverse demographic to generalize the findings. Furthermore, the study should be replicated with different lean concepts taught during training programs to ascertain whether the results are dependent on the specific lean concept being taught. Given that previous studies suggest higher training outcomes when utilizing full-scale simulators rather than desktop games, it would be beneficial to replicate the study while considering both approaches for comparison. We intend to replicate the second questionnaire a few months after the training to evaluate its long-term impact. Finally, in this paper, an exploratory analysis was conducted to establish relationships between the collected data, and a confirmatory analysis will be put in place to further validate the findings.

Acknowledgments. This research is sponsored by national funds through FCT –Fundação para a Ciência e a Tecnologia, under the project UIDB/00285/2020 and LA/P/0112/2020.

Disclosure of Interests. The authors have no competing interests to declare that are relevant to the content of this article.

References

1. Lodgaard, E., Ingvaldsen, J., Gamme, I., Aschehoug, S.: Barriers to lean implementation: perceptions of top managers, middle managers and workers. Procedia CIRP **57**, 595–600 (2016)
2. Achanga, P., Shehab, E., Roy, R., Nelder, G.: Critical success factors for lean implementation within SMEs. J. Manuf. Technol. Manag. **17**(4), 460–471 (2006)
3. Shrimali, A., Soni, V.: A review on issues of lean manufacturing implementation by small and medium enterprises. Int. J. Mech. Prod. Eng. Res. Dev. **7**(3), 283–300 (2017)
4. Netland, T.: Critical success factors for implementing lean production: the effect of contingencies. Int. J. Prod. Res. **54**(8), 2433–2448 (2016)
5. Ichimura, M., Arunachalam, S., Page, T.: An emerging training model for successful lean manufacturing - an empirical study. i-manag. J. Manag. **2**(4), 29–40 (2008)
6. Dinis-Carvalho, J.: The role of lean training in lean implementation. Prod. Plann. Control **32**(6), 441–442 (2021)
7. Choomlucksana, J., Doolen, T.: An exploratory investigation of teaching innovations and learning factors in a lean manufacturing system engineering course. Eur. J. Eng. Educ. **42**(6), 829–843 (2016)
8. De Vin, L., Jacobsson, L., Odhe, J.: Game-based lean production training of university students and industrial employees. Procedia Manuf. **25**, 578–585 (2018)
9. Tan, A., van Dun, D., Wilderom, C.: Lean innovation training and transformational leadership for employee creative role identity and innovative work behavior in a public service organization. Int. J. Lean Six Sigma **15**(8), 1–31 (2024)
10. Maware, C., Parsley, D.: The challenges of lean transformation and implementation in the manufacturing sector. Sustainability **14**(10), 6287 (2022)
11. Armstrong, K.: Applications of role-playing in tourism management teaching: an evaluation of a learning method. J. Hosp. Leis. Sport Tour. **2**(1), 5–16 (2003)
12. Martínez, P., Moyano, J.: Learning to teach lean management through games: systematic literature review. In: Working Papers on Operations Management, vol. 8, p. 164 (2017)

13. Badurdeen, F., Marksberry, P., Hall, A., Gregory, B.: Teaching lean manufacturing with simulations and games: a survey and future directions. Simul. Gaming **41**(4), 465–486 (2010)
14. Donnelly, R., Fitzmaurice, M.: designing modules for learning. In: O'Neill, G., Moore, S., McMullin, B. (eds.) Emerging Issues in the Practice of University Learning and Teaching, All Ireland Society for Higher Education, Dublin (2005)
15. Bandura, A.: Self-efficacy: toward a unifying theory of behavioral change. Psychol. Rev. **84**(2), 191–215 (1977)
16. Vaz, C., et al.: Lean learning academy: an innovative framework for lean manufacturing training. In: Proceedings of the 1st International Conference of the Portuguese Society for Engineering Education, Porto, Portugal, pp. 1–5 (2013)
17. Kolb, D.A.: Experiential Learning: Experience as the Source of Learning and Development. Prentice Hall, Englewood Cliffs (1984)
18. Hmieleski, K., Corbett, A.: The contrasting interaction effects of improvisational behavior with entrepreneurial self-efficacy on new venture performance and entrepreneur work satisfaction. J. Bus. Ventur. **23**, 482–496 (2008)
19. Lorsbach, A., Jinks, J.: Self-efficacy theory and learning environment research. Learn. Environ. Res. **2**, 157–167 (1999)
20. Lunenburg, F.C.: Expectancy theory of motivation motivating by altering expectations. Int. J. Manag. Bus. Adm. **15**, 1–9 (2011)
21. Vo, T., Tuliao, K., Chen, C.: Work motivation: the roles of individual needs and social conditions. Behav. Sci. **12**(2), 49 (2022)
22. Luckie, D., Aubry, J., Marengo, B., Rivkin, A., Foos, L., Maleszewski, J.: Less teaching, more learning: 10-yr study supports increasing student learning through less coverage and more inquiry. Adv. Physiol. Educ. **36**(4), 325–335 (2012)
23. Depaolo, C., Mclaren, C.: The relationship between attitudes and performance in business calculus. INFORMS Trans. Educ. **6**(2), 8–22 (2006)
24. Malik, P., Garg, P.: Learning organization and work engagement: the mediating role of employee resilience. Int. J. Hum. Resour. Manag. **31**(8), 1–24 (2019)
25. Hernik, J., Jaworska, E.: The effect of enjoyment on learning. In: Proceedings of the 12th International Technology, Education and Development Conference, Valencia, Spain, pp. 508–514 (2018)
26. Rieber, L., Noah, D.: Games, simulations, and visual metaphors in education: antagonism between enjoyment and learning. Educ. Media Int. **45**(2), 77–92 (2008)
27. Stevens, K.: The effect of background knowledge on the reading comprehension of ninth graders. J. Read. Behav. **12**(2), 151–154 (1980)
28. Yates, G., Chandler, M.: Prior knowledge and how it influences classroom learning. What does the research tell us? Res. Inf. Teach. **2**, 1–8 (1994)
29. Pintrich, P., Smith, D., Garcia, T., McKeachie, W.: Reliability and predictive validity of the motivated strategies for learning questionnaire (MSLQ). Educ. Psychol. Mea. **53**, 801–813 (1993)

Designing an Online Workshop for Creativity and Value Co-creation: Three Case Studies in Gastronomic Sciences on Viewpoint Setting and Sustainability Education

Tomomi Nonaka[1(✉)], Masayoshi Ishida[2], Seiko Shirasaka[3], Tomomi Honda[4], Masami Oginuma[5], Kan Yoshitake[6], and Kazuki Taniguchi[6]

[1] Department of Industrial and Management Systems Engineering, School of Creative Science and Engineering, Faculty of Science and Engineering, Waseda University, 3-4-1, Okubo, Shinjuku-ku, Tokyo 169-8555, Japan
nonaka@waseda.jp
[2] College of Gastronomy Management, Ritsumeikan University, Kyoto, Japan
[3] Graduate School of System Design and Management, Keio University, Minato, Japan
[4] Department of Innovative Food Sciences School of Food Sciences and Nutrition, Mukogawa Women's University, Nishinomiya, Japan
[5] The Asahi Shimbun Company, Osaka, Japan
[6] COMARS Co., Ltd., Chiyoda, Japan

Abstract. To co-create value in sustainable communities, it is essential to understand the nature of social issues from multiple perspectives and to communicate and dialogue with people with different values and expertise from a comprehensive and integrated perspective. This study poses the following two research questions 1) What are the key components of an effective online sustainability education workshop that encourages active participation and dialogue? and 2) How can the integration of food and local resources in online workshops enhance participants' awareness and understanding of sustainability? In this research, we develop an online workshop to connect the world on the theme of food as knowledge acquisition and exchange (in food contexts). We focus on communication and interaction between different actors in the creation of local industries using food and local resources. The aim of this research project, called "GAstroEdu", is to develop an online workshop for creativity and value co-creation that supports value co-creation among different stakeholders in the region and encourages participants to set their own points of view and learn from them. As an implementation, we present three GAstroEdu projects. Tomato Adventure for elementary school students, Lemon Adventure2 held in Setoda, Onomichi, Hiroshima, and Potato Adventure Workshop held in Kutchan-cho, Hokkaido.

Keywords: Creativity · Co-creation · Online workshop · Sustainability · Education · Food

M. Thürer et al. (Eds.): APMS 2024, IFIP AICT 729, pp. 488–500, 2024.
https://doi.org/10.1007/978-3-031-65894-5_34

1 Introduction

To achieve a sustainable society, sustainable & responsible management is required in the area of education and co-creation of value in local communities, and it is essential to grasp the essence of social issues from multiple perspectives. To balance environmental, economic and social aspects and to discuss sustainability from a comprehensive perspective with short, medium and long term time horizons, it is necessary to engage in dialogue with people who embrace diversity and have a variety of values. However, sustainability solutions are not clearly defined as in the natural sciences but are often evaluated differently from different perspectives. Furthermore, what kind of metric to use to assess sustainability is itself a research question, and there is not necessarily one right answer that can be said to be right in the future.

Although the importance of sustainability education for elementary and junior high school students has been pointed out, there has been little research on teaching methods and educational content [1]. It is believed that it is valuable for upper elementary school students to learn through experience that social problems can be solved with their own hands. In Japan, the Basic Law on Nutrition Education was enacted in 2005, and nutrition improvement activities have been carried out for many years based on terms such as nutrition counseling, nutrition education, and dietary education. Since the enactment of the Basic Law on Nutrition Education, nutrition education has been promoted in a way that promotes the field of nutrition [2]. Some studies [3] interpret sustainability education as part of environmental education, while others [4] argue that sustainability education has evolved from environmental education. UNESCO states that education for sustainable development is about empowering people to address global challenges constructively and imaginatively, now and in the future, and to create more sustainable and resilient societies [5]. However, in many cases only some of the knowledge, values and theories related to sustainable development are addressed, and there are few examples of comprehensive education [1].

Next, we consider value co-creation with multiple actors. Innovative breakthroughs require co-creation by diverse entities [6], and support for co-creation and dialogue using collective knowledge is needed [7]. For creative value creation, the importance of integrative thinking has long been highlighted in design thinking [8], and Paulus showed that interaction in groups and teams can be an important source of creative ideas and innovations through their review and analysis. A theoretical model of creativity in idea-generating groups is developed and implications for teamwork and future research are presented [9]. In addition, Florini and Pauli examined how and why these cross-sector collaborations are developing and what steps can or should be taken to ensure that partnerships create public and private value [10].

To co-create value in local communities for the realization of a sustainable society, it is essential to grasp the essence of social issues from multiple perspectives, and it is necessary to engage in communication and dialogue with people with different values and expertise, embracing diversity from a comprehensive and integrated thinking perspective. Therefore, this study poses the following two research questions 1) What are the key components of an effective online sustainability education workshop that encourages active participation and dialogue? and 2) How can the integration of food and local resources in online workshops enhance participants' awareness and understanding of sustainability? The workshop developed by the GAstroEdu project then takes food as a starting point for knowledge acquisition and sharing (in food contexts) and uses food as a subject to explore social issues from different perspectives to design programmes from different angles to promote broad learning, to interact with people using online technology to connect with people around the world, and to provide a wide range of learning opportunities. By using online technology and connecting with people around the world, we can interact with people and experience the reality of local and on-the-ground situations together. The aim is to foster creativity through cross-cultural understanding and the cultivation of intrinsic motivation. This paper focuses on communication between different actors in the creation of local industries based on food and local resources. We develop an online creativity and value co-creation workshop that supports the co-creation of value in the community for the realization of a sustainable society and encourages participants to set their own points of view and learn from them. As an implementation, we present three GAstroEdu projects. Tomato Adventure for elementary school students, Lemon Adventure2 held in Setoda, Onomichi, Hiroshima, and Potato Adventure Workshop held in Kutchan-cho, Hokkaido.

2 Workshop Design

This section describes the workshop design for the GAstroEdu project. In this design, the workshop design considers the process of knowledge acquisition and knowledge sharing through online communication.

2.1 GAstroEdu Project

GAstroEdu is an acronym of Gastronomic Sciences, Astronomy and Education and was chosen by the authors as the name of the project. In this project, facing food is facing society and community, and the aim is to confront social issues more deeply and to deepen learning by using food as a starting point. Creativity education is to cultivate the ability to create the future, to understand society deeply and to think about the future together with children, junior and senior high school students, university students and adults in a multi-generational way, about what kind of society and how they want to live. In a large, complex and rapidly changing society, social issues are seen differently depending on one's position and time, and solving sustainability and social issues requires not only science and technology, but also the ability to discern the essence of things from multiple perspectives, including culture and history. This project aims to foster creativity through the following studies, using food as a starting point [11].

- Inspire your imagination.

  Connect to the world's most advanced knowledge online.
  Courage to face complexity with a rich imagination.

- See the Essence

  Understand sustainability.
  See things from multiple perspectives and understand them deeply.

- Cultivate the seeds of curiosity that becomes intrinsic motivation for the future.

  Deep online connections with people from all over the world.
  Diversity and cross-cultural understanding.
  Cultivate a desire to know.

2.2 Workshop Design Considering Knowledge Acquisition and Knowledge Sharing Processes

In the creation of local industries based on food and local resources, business and product development is promoted by taking advantage of the local climate, historical characteristics and strengths. At present, knowledge acquisition and value creation beyond expert bias is required through interactions with various entities in the local community, including government and the private sector.

Expert knowledge representation has been studied for many years. Chan-drasekaran proposed generic tasks from a problem solving perspective [12]. Taki et al. proposed an expert model as a knowledge representation for knowledge acquisition and showed the knowledge acquisition process [13]. Referring to this knowledge acquisition process, we add (8) and (10) to define the knowledge acquisition process as follows. Communication and dialogue using online technology can extend (8) temporal and spatial expansion of knowledge sources and (10) knowledge exchange in the knowledge acquisition process. These two extensions are considered to be the characteristics and strengths of this workshop design. While the scope of spatial and temporal exchange was defined by business and environmental constraints, online exchange allows this extension to be realised with less cost and effort.

(1) problem formulation, (2) evaluation of existing technologies, (3) identification of knowledge sources, (4) identification of expert models, (5) identification of user models, (6) selection of knowledge representation, (7) knowledge extraction, (8) temporal and spatial extension of knowledge sources, (9) knowledge transformation, (10) knowledge sharing, (11) knowledge base management, and (12) performance evaluation. To facilitate these activities, a workshop is designed (Fig. 1).

This figure was created by the authors by adding the workshop design considering the knowledge acquisition and knowledge exchange process to the system conceptual diagram [14]. To realize the temporal and spatial expansion of knowledge sources, online technology is used to transfer information between multiple locations and to facilitate interaction among participants.

A workshop is designed by setting goals after analyzing the requirements for what is to be achieved in the workshop using a systems engineering approach. We design a workshop to achieve interaction between stakeholders in local society as knowledge acquisition and exchange (in food contexts). Through the online workshop to be designed, we aim to implement a place for clarification and realization of how knowledge is shared and how it can be better shared by different stakeholders in the local community. The online workshops are not limited to online communication and exchange, but are also designed to encourage interactions with the real physical space, such as hybrid interactions and later visits to the actual site, etc. The design of the site is required to allow for interaction between cyberspace and physical space.

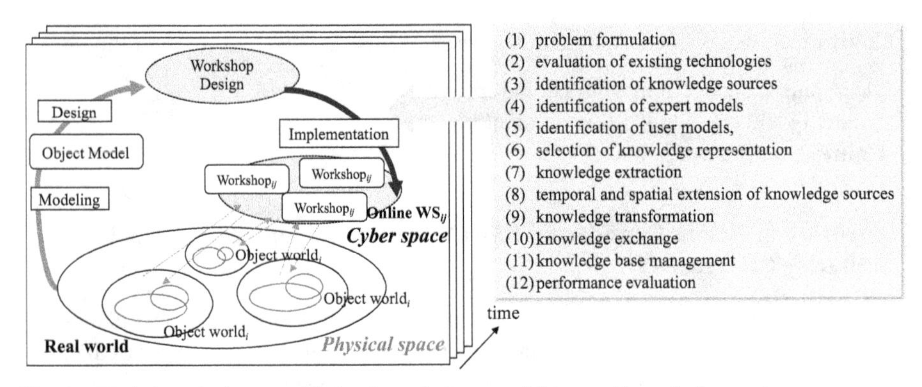

Fig. 1. Workshop design considering knowledge acquisition and knowledge exchange processes.

3 Workshop Implementation (1) Tomato Adventure

Tomato Adventure was conducted entirely online, from the planning and design meetings to the preparation and operation on the day of the workshop [11]. The workshop was held on three days, Saturday, August 1, Sunday, August 2, and Saturday, August 8, 2020, for fifth- and sixth-grade students of upper elementary schools. Seven students from Ritsumeikan Elementary School participated. In addition to Ritsumeikan University and its affiliated elementary school in Japan, a domestic company, the True Neapolitan Pizza Association (Associazione Verace Pizza napoletana, AVPN), and the True Neapolitan Pizza Association Japan Chapter also participated in the program. AVPN is a non-profit organization founded in Naples in June 1984. The workshop was conducted in a completely online format using Zoom, except that the elementary school students participated online from their homes, and the teachers also participated from their workplaces or homes, except for some administrative offices set up at the university.

This workshop is designed to develop innovative human resources who can grasp the essence of problems not only from the viewpoints of science and technology, but also from those of culture and history, through the experience of creating a solution, however small, to social problems facing the world with their own hands through familiar foods, by conducting a workshop with hands-on experience. The program is designed to cultivate innovative human resources who can grasp the essence of problems not only from the viewpoints of science and technology, but also from those of culture and history.

The requirements for sustainable development and education [5] were set as the functional requirements for the workshop, and the functional and physical design were conducted in the architectural process of systems engineering. In the physical design, we used tomatoes, a global food familiar to upper elementary school students and popular around the world as tomato sauce, tomato ketchup, and raw tomatoes. To address social issues with familiar ingredients and wisdom, many hands-on sessions such as cooking and science experiments were included in the program. In addition, to connect with the real world and realize the function of coming into contact with the reality of the real world, we set up a program that connects producers and consumers through education by broadcasting the sites of food producers in Japan and overseas online and allowing them to come into contact with the real thing.

3.1 Program Overview of Tomato Adventure Workshop

Table 1 and Fig. 2 provide an overview of the program and its progress. At the beginning of Day 1, the first day of the workshop, the target picture and problems are shared to gain a deeper understanding of the social problems targeted by the workshop. In this workshop, we focus on the problem of food loss among the social issues related to food and learn to understand the problem to be solved and its relationship to food through the use of tomatoes. In particular, we incorporated many programs for upper elementary students to learn through hands-on activities. At the beginning of the workshop, the participants were asked to cook a familiar tomato dish to stimulate their interest and attachment to tomatoes and to make them reevaluate and become aware of familiar foods that they usually eat without thinking about them. In Science of Tomatoes, students learn the characteristics of tomato cooking from a scientific perspective. Specifically, through experiments in which tomatoes are dipped in sugar water of different concentrations and observed to rise to the surface, the sugar content of tomatoes is determined from the relationship between the amount of nutrients contained in the tomatoes and their weight. We also learn about the relationship between cooking with heat and oil and nutrients, and think about the secret of the deliciousness of tomatoes from a scientific point of view.

Table 1. Overview of the program of Tomato Adventure workshop [11].

Day1	Share goal images and challenges. A deeper understanding of the relationship between food and the world's challenges to be saved - Let's make a familiar tomato dish <Practicum> - The science of tomatoes learn about the characteristics of tomato cooking - Pizza making: pizza dough (live online with Italy)
Day2	Where does the deliciousness of tomatoes come from? Live lessons with Italy, the home of tomatoes - Learn about the realities of food loss and waste around the world and the cases that are being solved. - What is the appeal of Italian tomatoes from the home of tomato farming? (live online with Italy) - Try making Neapolitan pizza together with an authentic Neapolitan pizza chef using tomatoes from Italy (live online with Italy) - Creative cooking plan: How to make a menu to solve food loss and waste
Day3	Can you save the world with tomatoes? Propose original tomato dishes and original pizza to Italy - Propose Operation Change the World (live online with Italy) - Feedback from the President of Associazione Verace Pizza Napoletana (live online with Italy) - Overall summary

At the end of Day 1, students learn how to make pizza dough from an authentic Neapolitan Pizza Association pizza maker via an online connection to Naples, Italy. Students make the pizza dough at home while learning from an authentic local instructor through consecutive interpretation in Italian. The ingredients used to make the pizza are the same as those used in Naples, creating a tangible but realistic experience that connects participants to the local area. Italian flour, salt, yeast, and canned tomatoes made from tomatoes harvested in the tomato fields that are visited on day 2 are sent to the participants' homes in advance, and these ingredients are used in the cooking exercise. In addition, although the exact same ingredients are used, the amount of water used to make the dough needs to be fine-tuned according to the local climate and environment. For this reason, a Japanese university student acts as a mentor, using the Zoom Breakout function to assist in the dough-making process as the cooking progresses.

On Day 2, we conduct an online classroom-style training with the goal of gaining a deeper understanding of the social issues of food loss. Participants learn about the current state of food loss in the world with data and receive presentations on the latest examples of food loss solutions and upcycling efforts around the world. At the end of the session, participants are invited to propose their own ideas for upcycling, making the session more interactive than a one-way classroom lecture. Next, as in Day 1, we connect to Naples, Italy, via an online webcast and have a grower speak from the tomato fields of San Marzano Tomatoes. The growers speak directly about the environmental issues, such as soil and climate, that produce the tomatoes used for the world-famous canned tomatoes that are hard to find in Japan, and their ingenuity. The simultaneous live broadcast with Italy is provided with consecutive interpretation to realize a live online practical training. Next, using the pizza dough made the day before, the dough is rolled out and the pizza is baked. As on the previous day, a lecture is given by a pizza chef from Naples and mentoring support is provided by Japanese university students.

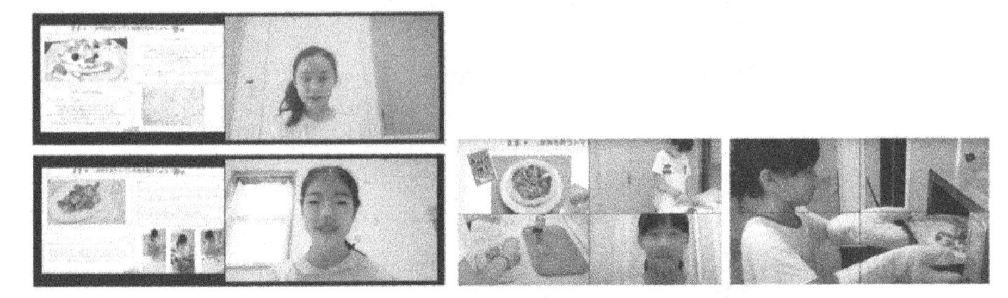

Fig. 2. Photos of the Tomato Adventure workshop [11].

Based on what they have learned over the past two days, the participants propose a new dish to Italy that solves the social problem of food waste. The participants are divided into two groups and the proposal is either an original tomato dish (creative cooking) or a proposal for a new pizza topping. On the final day, Day 3, participants present their creative cooking proposals, which they have been working on as homework since Day 2, as a "Proposal for a Strategy to Change the World". The presentations are critiqued by the instructors and Mr. Pace, President of the True Neapolitan Pizza Association. The best proposal are offered for a limited time on the menu of the Association's historic restaurant in Naples, and the profits are used to purchase compost, a direct implementation of an SDG solution.

3.2 Design of Online Communication Initiatives for Workshops

In addition to the main facilitation by the instructors, the university students are provided with detailed follow-up and encouragement during the group exercises using Zoom's breakout function, so that the participants, who are elementary school students, can have a stronger sense of connection to the real world, as they are participating from their own homes. In addition, local ingredients are delivered to the participants in advance, creating a situation in which they have the exact same ingredients that they see as scenery from overseas, connected via a live broadcast. The experience of working with chefs in Naples, far from home, is created with an online communication tool that transcends time and space and uses food that has a tangible presence.

3.3 Results of Participant Survey

After the workshop, a questionnaire was answered by the participants. The results showed that the satisfaction of the participating children and their parents was generally high. In the children's free-written responses, the presentations at the end of the workshop made them realize that they had such an idea. It made me want to go to Italy. I felt that Italy was close to me. I want to serve okonomiyaki to people from Italy. I want to do the experiment of becoming sweet again. I want to do the sweetening experiment again. Parents commented that it was very good for them to be able to cook while connecting with their teachers and friends, even though it was difficult for them to go out due to the situation of COVID-19. Parents said that they were able to connect with their friends from elementary school, teachers from the university, and people from Italy from the comfort of their own homes, and that it was satisfying to be able to communicate face to face even though it was online. These are some of the comments we heard from the participants.

In the questionnaire, the following results were obtained: 1) 72% strongly agree, 14% agree, and 14% neither agree nor disagree that the workshop was exciting and fun, 2) 43% strongly agree, 43% agree, and 14% disagree that the tomatoes helped them understand different issues in the world, and 3) 43% strongly agree and 57% agree that the tomatoes made them want to take on challenges in the world. 3) Did it make you want to tackle global issues? 43% strongly agree and 57% agree. The results indicate that the workshop design, which aimed to develop an online workshop to confront social issues with familiar ingredients and wisdom and to foster creativity, had some positive effects in fostering a sense of familiarity with social issues and a sense of being connected to the world and the real world. Further design and verification of the effectiveness of the workshop is an issue to be addressed in the future as the number of participants and applicable technologies are increased and the workshop is conducted more frequently.

4 Workshop Implementation (2) Lemon Adventure2

Setoda, Onomichi City, Hiroshima Prefecture, Japan, is known as the birthplace of the domestic lemon, and its mild climate and low rainfall have long been favorable for citrus production. In this workshop [15], lemon growers, chefs, processors, and government officials from Setoda and Amalfi, Italy, are invited to Setoda, Onomichi City, Hiroshima Prefecture, Japan, on December 18, 2021, in person or virtually. The aim is to discuss local value creation and industry through dialogue and discussion among various stakeholders. The project was held in Setoda, Onomichi City, Hiroshima Prefecture, on December 18, 2021, in a hybrid online (Zoom) and face-to-face format (Fig. 3). The goal was to engage participants in thinking about the richness of food and the region in the future, while experiencing the beautiful scenery of lemons and the sea through the live broadcast between Amalfi and Setoda. Lemon Adventure 2 was designed and implemented in response to the research question 2) How can the integration of food and local resources in online workshops increase participants' awareness and understanding of sustainability?

In the Lemon Adventure 2 workshop, we implemented the workshop design described in Sect. 2.2 drawn in Fig. 1 by (8) extending knowledge sources temporally and spatially, (9) knowledge exchange, and (10) knowledge exchange through online interaction and communication with the real world, including the region, industry, and sector to which the participants themselves belong. (9) knowledge exchange, and (10) knowledge sharing. Amalfi is an area that Onomichi city officials had visited five years ago, and we tried to re-establish online connections between the areas that had been exchanged at that time and to develop new exchanges after a lapse of five years. The result was a case of knowledge acquisition and exchange about the culture and technical characteristics of Amalfi and Italian confectionery production (space) and food history (time and space), which led to the development and sale of new products through collaboration between confectioners and researchers in Setoda, triggered by the exchange at the workshop. After Lemon Adventure 2, the following year the workshop participants went to Italy, where they had an online exchange to develop a new recipe with a local gelateria. In addition to the lemon production areas visited in Lemon Adventure 2, workshop participants visited lemon production areas throughout Italy to expand the local exchange of lemon products. The knowledge was brought back to Onomichi City, Hiroshima Prefecture, where it was transformed and developed through local knowledge exchange with local stakeholders.

Fig. 3. Photos of the Lemon Adventure2 workshop.

5 Workshop Implementation (3) Potato Adventure

The Potato Adventure workshop was held in November 2021 in Ecuador and Argentina and in Kutchan-cho, Hokkaido, Japan [15]. Potato growers in Ecuador, Argentina and Kutchan-cho had an online dialogue with the aim of considering the future of rich food and agriculture based on the diversity of potatoes. Participants were elementary school students from Hokuyo Elementary School in Kutchan Town, Hokkaido. More than 400 varieties of potatoes are still grown in Ecuador. Most of the native varieties are grown at altitudes of more than 3,000 m above sea level, and are nurtured in the strong sunlight and organic Andean soil that give potato production a special natural quality. In this workshop, we connected far-flung places in the country and abroad, and while witnessing agricultural practices with very different attitudes toward mechanization and breeding, we considered the richness of the food and agriculture of the future in terms of the diversity of potatoes and production. Potato Adventure tries to encourage participants to have a holistic view using systems engineering approach with online workshop technologies such as drone and camera works. The workshop programs were designed and implemented in response to the research question "1) What are the key components of an effective online sustainability education workshop that encourages active participation and dialogue?

Fig. 4. Setting up and managing a comprehensive view of the region using drone movies [15].

The workshop was held at Hokuyo Elementary School in Kutchan City, where the participating elementary school students attended in person. In addition, students from Ritsumeikan Keisho Junior and Senior High School and students from Ritsumeikan University participated as mentors. At the venue, an online filming team participated in the workshop, filming and projecting the participants' discussions and work in real time. The video streaming was conducted using Zoom, which combined individual reports from each participant, a bird's-eye view of the entire venue, and a cameraman's viewpoint while interacting with the participants, and projected them on a display in the venue. The participating elementary school students were influenced by the camera crew's intervention as an opportunity to reflect on why they were being observed and focused on by the cameraman, and were observed to have a bird's eye view or an objective view while looking at images of themselves, other participants, and the venue scenery.

At the beginning of the workshop, drone footage of Kutchan town, the surrounding area, and the potato production area was shown (Fig. 4). The video was composed of a bird's-eye view of the vast nature of the entire region, then the potato production area, and gradually approaching the elementary school where the workshop was held, so that the participants could realize that potatoes, a local specialty, are produced right on their land.

Then, after the drone image over the elementary school, we switched to an image from the on-site camera, which was recording in real time (Fig. 5). At the beginning of the workshop, we produced a series of images to recapture the place where we are now from a bird's eye view of the region and to recognize the characteristics of the region through a series of scenes. In addition, to create a sense of anticipation for the start of the workshop, we combined drone footage with live images from a camera in the venue. In addition, the opening scene was co-produced by the participants and the management after switching to the camera at the venue. Before the workshop started, we explained in advance that we would create the opening video together with the participants, and

Fig. 5. Setting up and managing a comprehensive viewpoint of participants [15].

elementary school students who attended the workshop joined the facilitators and the management in creating the video, which was broadcast live to the online participants of the workshop. An attempt was made to create an interactive co-creation between facilitators and participants from the very beginning of the workshop (Fig. 5).

6 Discussion

This section describes the design and implementation of three workshops conducted in the GAstroEdu project as knowledge acquisition and exchange (in food contexts): Tomato Adventure in Sect. 3, Lemon Adventure2 in Sect. 4, and Potato Adventure in Sect. 5. Tomato Adventure and Potato Adventure are online workshops aimed at creativity education for elementary school students and were designed and implemented in response to the research question "1) What are the key components of an effective online workshop for sustainability education that encourages active participation and dialog? Using the system architecture approach in systems engineering, the requirements for sustainability education were defined based on previous research [5], and functional design was performed as functional requirements. Then, a set of the designed functions was implemented as a concrete program in the workshop through physical design. Lemon Adventure 2 was designed and implemented in response to the research questions 2) How can the integration of food and local resources in online workshops enhance participants' awareness and understanding of sustainability? We extended the expert knowledge acquisition model and applied it to workshop design.

Because of COVID-19's situation, recent years have seen remarkable developments in video and audio technology, as well as machine and automatic translation technology. The development of online technologies make stunningly immersive and realistic experiences increasingly possible in the coming years. The Tomato Adventure described in Sect. 3 was conducted immediately after COVID-19, when face-to-face communication was almost impossible. Since then, online communication has become more common in education and business activities, and we are entering an era of finding ways to collaborate while taking advantage of the advantages and disadvantages of hybrid and face-to-face formats. The GAstroEdu project connects producers and consumers around the world who have had difficulty connecting directly around food. By connecting people and regions, we aim to build a platform that generates new co-creation through collaboration among diverse entities, and we continue to take on the challenge of social implementation while connecting with people around the world.

7 Conclusions

This research focused on communication and interaction among different actors in the creation of local industries using food and local resources. We designed an online workshop for creativity and value co-creation that supports value co-creation among diverse stakeholders in the region and induces participants to set their viewpoints and learn from them as knowledge acquisition and exchange (in food contexts). We designed an online workshop on creativity and value co-creation to support value co-creation in the community using food and local resources and to induce participants to set their viewpoints and learn from them. Three GAstroEdu projects were presented as their implementations. Three case studies were presented: Tomato Adventure for elementary school students, Lemon Adventure2 in Setoda, Onomichi, Hiroshima, and Potato Adventure Workshop in Kutchan-cho, Hok-kaido. The future challenge is to evaluate the effects of perspective setting, intervention, knowledge acquisition, and knowledge sharing, and to generalize the results quantitatively and qualitatively.

Acknowledgments. The GAstroEdu project was made possible by the cooperation of many people in Japan and abroad. We would like to express our sincere gratitude to them. This work was partially supported by JSPS KAKENHI Grant Number JP 23K22158.

References

1. Jeronen, E., Palmberg, I., Yli-Panula, E.: Teaching methods in biology education and sustainability education including outdoor education for promoting sustainability a literature review. Educ. Sci. **7**(1) (2017). https://doi.org/10.3390/educsci7010001
2. Miyuki, A., Kumi, E.: Expectations for "Shokuiku" (Food and nutrition education and promotion). Jpn. J. Nutr. Diet. **63**(4), 201–212 (2005). (in Japanese)
3. Wesselink, R., Wals, A.E.J.: Developing competence profiles for educators in environmental education organisations in the Netherlands. Environ. Educ. Res. **17**, 69–90 (2011)
4. Eilam, E., Trop, T.: ESD pedagogy: a guide for the perplexed. JEE **42**, 43–64 (2010)
5. UNESCO, Education for Sustainable Development: An Expert Review of Processes and Learning (2011). http://unesdoc.unesco.org/images/0019/001914/191442e.pdf. Accessed 1 Apr 2024
6. Fleming, L.: Perfecting cross-pollination. Harv. Bus. Rev., 22–24 (2004)
7. Woolley, A.W., Charbris, C.F., Pentland, A., Hashimi, N., Malone, T.W.: Evidence for a collective intelligence factor in the performance of human groups. Science **330**(6004), 686–688 (2010)
8. Graduate School of System Design and Management, Keio University, Innovation Dialogue Guidebook. Commissioned by Ministry of Education, Culture, Sports, Science and Technology, Japan (2023). (in Japanese)
9. Brown, T.: Change by design: how design thinking transforms organizations and inspires innovation. HarperBusines (2009)
10. Paulus, P.: Groups, teams, and creativity: the creative potential of idea-generating groups. Appl. Psychol. **49**(2), 237–262 (2000)
11. Florini, A., Pauli, M.: Collaborative governance for the sustainable development goals. Asia Pac. Policy Stud. **5**(3), 583–598 (2018)
12. Nonaka, T., Fukuda, K., Ishida, M., Koiwai, Y., Honda, T.: The future of online education by connecting advanced knowledge and practices across the world: GastroEdu project. Lect. Transcr. Ritsumeikan Gastron. Sci. Study **5**, 185–194. (in Japanese)
13. Chandrasekaran, B.: Generic tasks in knowledge-based reasoning: high-level building blocks for expert system design. IEEE Expert **1**(3), 23–30 (1986)
14. Taki, H., Tsubaki, K.: Expert model: a knowledge representation for knowledge acquisition **5**(2), 203–212 (1990). (in Japanese)
15. Sakakibara, K., Fujii, N., Tamaki, H., Kuroe, Y.: Thinking about systems in the age of systems. In: 2022 Annual Conference on Electronics, Information and Systems, Institute of Electrical Engineers of Japan(IEEJ), OS6-1, August–September (2022). (in Japanese)
16. Nonaka, T., et al.: Creativity and value co-creation online workshop to trigger participants' perspective setting and learning for food and local resource utilization. In: The 44th Annual Conference of the Japan Society Creativity Society (2022). (in Japanese)

Author Index

Published by Springer Nature Switzerland AG 2024
M. Thürer et al. (Eds.): APMS 2024, IFIP AICT 729, pp. 501–502, 2024.
https://doi.org/10.1007/978-3-031-65894-5

Printed in the USA
CPSIA information can be obtained
at www.ICGtesting.com
CBHW070138190924
14624CB00003B/183